INTERNATIONAL UNION OF PURE AND APPLIED CHEMISTRY

ANALYTICAL CHEMISTRY DIVISION
COMMISSION ON SOLUBILITY DATA

SOLUBILITY DATA SERIES

Volume 36

4-AMINOBENZENESULFONAMIDES

Part III

6-Membered Heterocyclic Substituents and Miscellaneous Systems

SOLUBILITY DATA SERIES

INTERNATIONAL UNION OF PURE AND APPLIED CHEMISTRY
IUPAC Secretariat: Bank Court Chambers, 2-3 Pound Way,
Cowley Centre, Oxford OX4 3YF, UK

A complete list of volumes published in the Solubility Data Series will be found on p. 523.

SOLUBILITY DATA SERIES

Editor-in-Chief
A.S. KERTES

Volume 36

4-AMINOBENZENESULFONAMIDES

Part III

6-Membered Heterocyclic Substituents and Miscellaneous Systems

Volume Editors

ANTHONY N. PARUTA
University of Rhode Island
Kingston, RI, USA

RYSZARD PIEKOS
Medical Academy
Gdansk, Poland

PERGAMON PRESS
OXFORD · NEW YORK · BEIJING · FRANKFURT
SÃO PAULO · SYDNEY · TOKYO · TORONTO

U.K. — Pergamon Press plc, Headington Hill Hall, Oxford OX3 0BW, England

U.S.A. — Pergamon Press, Inc., Maxwell House, Fairview Park, Elmsford, New York 10523, U.S.A.

PEOPLE'S REPUBLIC OF CHINA — Pergamon Press, Room 4037, Qianmen Hotel, Beijing, People's Republic of China

FEDERAL REPUBLIC OF GERMANY — Pergamon Press GmbH, Hammerweg 6, D-6242 Kronberg, Federal Republic of Germany

BRAZIL — Pergamon Editora Ltda, Rua Eça de Queiros, 346, CEP 04011, Paraiso, São Paulo, Brazil

AUSTRALIA — Pergamon Press Australia Pty Ltd., P.O. Box 544, Potts Point, N.S.W. 2011, Australia

JAPAN — Pergamon Press, 5th Floor, Matsuoka Central Building, 1-7-1 Nishishinjuku, Shinjuku-ku, Tokyo 160, Japan

CANADA — Pergamon Press Canada Ltd., Suite No. 271, 253 College Street, Toronto, Ontario, Canada M5T 1R5

First edition 1989

The Library of Congress has catalogued this serial title as follows:

Solubility data series.—Vol. 1—Oxford; New York;
Pergamon, c 1979-
v.; 28 cm.
Separately catalogued and classified in LC before no. 18.
ISSN 0191-5622 = Solubility data series.
1. Solubility—Tables—Collected works.
QD543.S6629 541.3'42'05-dc19 85-641351
AACR 2 MARC-S

British Library Cataloguing in Publication Data

4-Aminobenzenesulfonamides.
Pt. 3: 6-membered heterocyclic substituents and miscellaneous systems 1. Amino acids—Solubility
I. Paruta, Anthony N. II. Piekos, Ryszard III. Series
547.7'5045'47
ISBN 0-08-034710-X

Printed in Great Britain by A. Wheaton & Co. Ltd., Exeter

CONTENTS

FOREWORD

If the knowledge is
undigested or simply wrong,
more is not better

How to communicate and disseminate numerical data effectively in chemical science and technology has been a problem of serious and growing concern to IUPAC, the International Union of Pure and Applied Chemistry, for the last two decades. The steadily expanding volume of numerical information, the formulation of new interdisciplinary areas in which chemistry is a partner, and the links between these and existing traditional subdisciplines in chemistry, along with an increasing number of users, have been considered as urgent aspects of the information problem in general, and of the numerical data problem in particular.

Among the several numerical data projects initiated and operated by various IUPAC commissions, the *Solubility Data Project* is probably one of the most ambitious ones. It is concerned with preparing a comprehensive critical compilation of data on solubilities in all physical systems, of gases, liquids and solids. Both the basic and applied branches of almost all scientific disciplines require a knowledge of solubilities as a function of solvent, temperature and pressure. Solubility data are basic to the fundamental understanding of processes relevant to agronomy, biology, chemistry, geology and oceanography, medicine and pharmacology, and metallurgy and materials science. Knowledge of solubility is very frequently of great importance to such diverse practical applications as drug dosage and drug solubility in biological fluids, anesthesiology, corrosion by dissolution of metals, properties of glasses, ceramics, concretes and coatings, phase relations in the formation of minerals and alloys, the deposits of minerals and radioactive fission products from ocean waters, the composition of ground waters, and the requirements of oxygen and other gases in life support systems.

The widespread relevance of solubility data to many branches and disciplines of science, medicine, technology and engineering, and the difficulty of recovering solubility data from the literature, lead to the proliferation of published data in an ever increasing number of scientific and technical primary sources. The sheer volume of data has overcome the capacity of the classical secondary and tertiary services to respond effectively.

While the proportion of secondary services of the review article type is generally increasing due to the rapid growth of all forms of primary literature, the review articles become more limited in scope, more specialized. The disturbing phenomenon is that in some disciplines, certainly in chemistry, authors are reluctant to treat even those limited-in-scope reviews exhaustively. There is a trend to preselect the literature, sometimes under the pretext of reducing it to manageable size. The crucial problem with such preselection - as far as numerical data are concerned - is that there is no indication as to whether the material was excluded by design or by a less than thorough literature search. We are equally concerned that most current secondary sources, critical in character as they may be, give scant attention to numerical data.

On the other hand, tertiary sources - handbooks, reference books and other tabulated and graphical compilations - as they exist today are comprehensive but, as a rule, uncritical. They usually attempt to cover whole disciplines, and thus obviously are superficial in treatment. Since they command a wide market, we believe that their service to the advancement of science is at least questionable. Additionally, the change which is taking place in the generation of new and diversified numerical data, and the rate at which this is done, is not reflected in an increased third-level service. The emergence of new tertiary literature sources does not parallel the shift that has occurred in the primary literature.

With the status of current secondary and tertiary services being as briefly stated above, the innovative approach of the *Solubility Data Project* is that its compilation and critical evaluation work involve consolidation and reprocessing services when both activities are based on intellectual and scholarly reworking of information from primary sources. It comprises compact compilation, rationalization and simplification, and the fitting of isolated numerical data into a critically evaluated general framework.

The *Solubility Data Project* has developed a mechanism which involves a number of innovations in exploiting the literature fully, and which contains new elements of a more imaginative approach for transfer of reliable information from primary to secondary/tertiary sources. *The fundamental trend of the Solubility Data Project is toward integration of secondary and tertiary services with the objective of producing in-depth critical analysis and evaluation which are characteristic to secondary services, in a scope as broad as conventional tertiary services.*

Fundamental to the philosophy of the project is the recognition that the basic element of strength is the active participation of career scientists in it. Consolidating primary data, producing a truly critically-evaluated set of numerical data, and synthesizing data in a meaningful relationship are demands considered worthy of the efforts of top scientists. Career scientists, who themselves contribute to science by their involvement in active scientific research, are the backbone of the project. The scholarly work is commissioned to recognized authorities, involving a process of careful selection in the best tradition of IUPAC. This selection in turn is the key to the quality of the output. These top experts are expected to view their specific topics dispassionately, paying equal attention to their own contributions and to those of their peers. They digest literature data into a coherent story by weeding out what is wrong from what is believed to be right. To fulfill this task, the evaluator must cover *all* relevant open literature. No reference is excluded by design and every effort is made to detect every bit of relevant primary source. Poor quality or wrong data are mentioned and explicitly disqualified as such. In fact, it is only when the reliable data are presented alongside the unreliable data that proper justice can be done. The user is bound to have incomparably more confidence in a succinct evaluative commentary and a comprehensive review with a complete bibliography to both good and poor data.

It is the standard practice that the treatment of any given solute-solvent system consists of two essential parts: I. Critical Evaluation and Recommended Values, and II. Compiled Data Sheets.

The Critical Evaluation part gives the following information:

(i) a verbal text of evaluation which discusses the numerical solubility information appearing in the primary sources located in the literature. The evaluation text concerns primarily the quality of data after consideration of the purity of the materials and their characterization, the experimental method employed and the uncertainties in control of physical parameters, the reproducibility of the data, the agreement of the worker's results on accepted test systems with standard values, and finally, the fitting of data, with suitable statistical tests, to mathematical functions;

(ii) a set of recommended numerical data. Whenever possible, the set of recommended data includes weighted average and standard deviations, and a set of smoothing equations derived from the experimental data endorsed by the evaluator;

(iii) a graphical plot of recommended data.

The Compilation part consists of data sheets of the best experimental data in the primary literature. Generally speaking, such independent data sheets are given only to the best and endorsed data covering the known range of experimental parameters. Data sheets based on primary sources where the data are of a lower precision are given only when no better data are available. Experimental data with a precision poorer than considered acceptable are reproduced in the form of data sheets when they are the only known data for a particular system. Such data are considered to be still suitable for some applications, and their presence in the compilation should alert researchers to areas that need more work.

The typical data sheet carries the following information:

(i) components - definition of the system - their names, formulas and Chemical Abstracts registry numbers;
(ii) reference to the primary source where the numerical information is reported. In cases when the primary source is a less common periodical or a report document, published though of limited availability, abstract references are also given;
(iii) experimental variables;
(iv) identification of the compiler;
(v) experimental values as they appear in the primary source. Whenever available, the data may be given both in tabular and graphical form. If auxiliary information is available, the experimental data are converted also to SI units by the compiler.

Under the general heading of Auxiliary Information, the essential experimental details are summarized:

(vi) experimental method used for the generation of data;
(vii) type of apparatus and procedure employed;
(viii) source and purity of materials;
(ix) estimated error;
(x) references relevant to the generation of experimental data as cited in the primary source.

This new approach to numerical data presentation, formulated at the initiation of the project and perfected as experience has accumulated, has been strongly influenced by the diversity of background of those whom we are supposed to serve. We thus deemed it right to preface the evaluation/compilation sheets in each volume with a detailed discussion of the principles of the accurate determination of relevant solubility data and related thermodynamic information.

Finally, the role of education is more than corollary to the efforts we are seeking. The scientific standards advocated here are necessary to strengthen science and technology, and should be regarded as a major effort in the training and formation of the next generation of scientists and engineers. Specifically, we believe that there is going to be an impact of our project on scientific-communication practices. The quality of consolidation adopted by this program offers down-to-earth guidelines, concrete examples which are bound to make primary publication services more responsive than ever before to the needs of users. The self-regulatory message to scientists of the early 1970s to refrain from unnecessary publication has not achieved much. A good fraction of the literature is still cluttered with poor-quality articles. The Weinberg report (in 'Reader in Science Information', ed. J. Sherrod and A. Hodina, Microcard Editions Books, Indian Head, Inc., 1973, p. 292) states that 'admonition to authors to restrain themselves from premature, unnecessary publication can have little effect unless the climate of the entire technical and scholarly community encourages restraint...' We think that projects of this kind translate the climate into operational terms by exerting pressure on authors to avoid submitting low-grade material. The type of our output, we hope, will encourage attention to quality as authors will increasingly realize that their work will not be suited for permanent retrievability unless it meets the standards adopted in this project. It should help to dispel confusion in the minds of many authors of what represents a permanently useful bit of information of an archival value, and what does not.

If we succeed in that aim, even partially, we have then done our share in protecting the scientific community from unwanted and irrelevant, wrong numerical information.

A. S. Kertes

PREFACE

With few exceptions, these volumes of the solubility data series deal with solubilities of the derivatives of 4-aminobenzenesulfonamide, usually referred to as "sulfanilamide" (sulfanilic acid amide), a name coined in 1937 (1). The history of sulfanilamide begins in 1906, when Schroeter (2) synthesized the molecule containing a 4-acetylaminosulfanilamide portion.

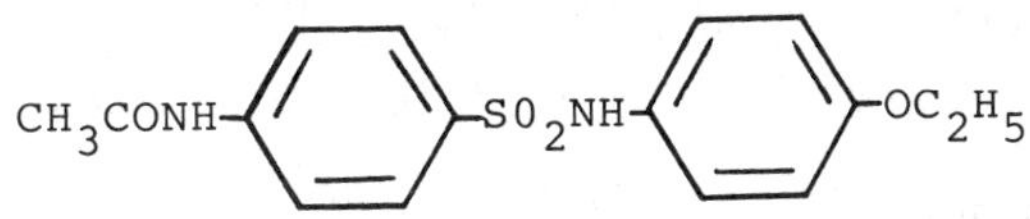

In 1908, Gelmo (3) described sulfanilamide and 13 of its derivatives and gave solubility values for these compounds. In 1935, Domagk (4) detected antibacterial activity of a synthetic azo dye, prontosil, with the structure.

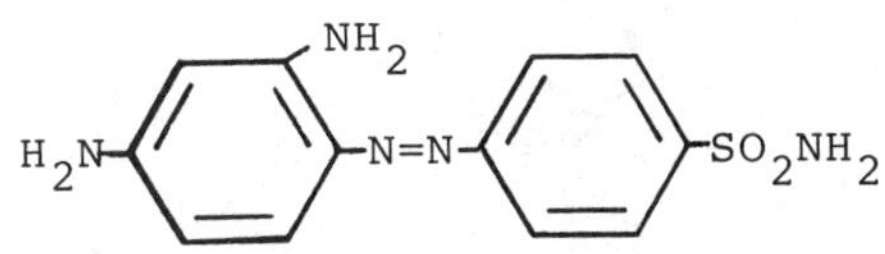

This compound had been tested for antibacterial activity (5), the "sulfanilamide" portion being responsible for its activity. This was confirmed (6) by isolation of sulfanilamide in the urine of patients. Fildes (7) and Wood (8), in 1940, demonstrated that the derivatives of sulfanilamide were antimetabolites of p-aminobenzoic acid (PABA) which is a step in the folic acid synthesis of bacteria. Thus, the structural similarity of PABA and sulfonamides caused interference by competitive antagonism and resulted in a bacteriostatic effect. The discoveries of antibacterial activity led to an exciting flood of research, and thousands of sulfanilamide derivatives have been synthesized. As early as 1948, the number of sulfonamide derivatives (9) was estimated to be several thousand. In the two decades after that, the number of synthesized sulfonamides have gone past 10,000(10)

Clinical trials of these sulfonamides and derivatives have been associated with low solubilities and some renal crystalluria. The low solubility, and its sensitivity to pH, could cause crystalline precipitation in the renal tubules in the filtration of blood into acidic urine. Some of the problems of limited solubility were overcome by complexation or salt formation, and solid state manipulations which in turn have stimulated investigations into solubility of the drugs in water, buffers and some binary solvent system. Analytical methodologies span a wide spectrum of techniques and the relevant references are in pharmaceutical, medical and chemical literature.

In all volumes the chemical structures, registry number and the molecular weight of the compounds considered are collected in the front of each volume. The compounds as they occur on the data sheets are given successively in each volume. In the first volume of this series there are 35 compounds. The second and third volumes have 58 compounds and 108 compounds, respectively.

NOMENCLATURE:

The nomenclature of sulfanilamide derivatives has conventionally been based on the following numbering system: substituents at the nitrogen atom of the amide group ($-SO_2NH_2$) are called $\underline{N}^1$-substituents, whereas substitutents at the 4-amino nitrogen ($4-H_2N-$) are called $\underline{N}^4$-substituents. Substitution in either or both of the two positions lead to compounds referred to as "sulfonamides" (sometimes "sulfanilamides" or even "sulfamides"). Here are illustrative examples of this nomenclature.

$\underline{N}^1$-substituted sulfonamides

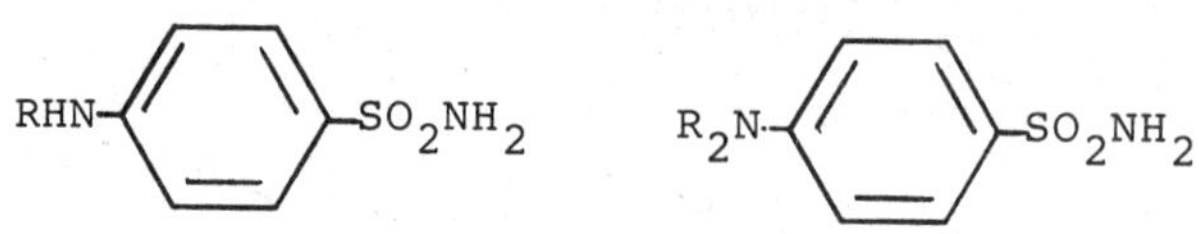

$\underline{N}^4$-substituted sulfonamides

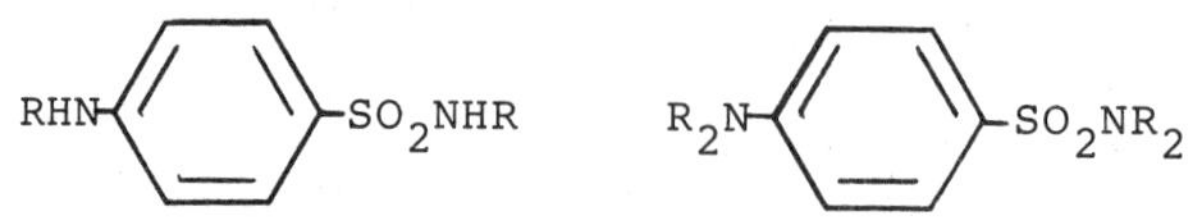

$\underline{N}^1,\underline{N}^4$-substituted sulfonamides

The 4-amino group can be diazotized to give derivatives of the formula

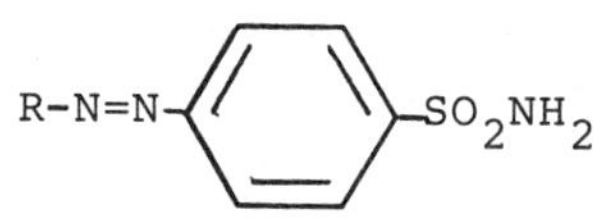

As the sulfonamide molecule carries a basic 4-NH_2 group and an acid $-SO_2NH_2$ one, it is capable of formation the respective salts or complexes, e.g.

where HX stands for an acid and M is a univalent metal atom.

In common use by health practitioners are nonproprietary names of sulfonamides which are brief and reflect the chemical nature of their molecules. Examples are: sulfacetamide, sulfapyridine, sulfathiazole, sulfadiazine, sulfaguanidine, etc. There are numerous trivial names; for example, sulfanilamide has as many as 140 synonyms, and sulfathiazole has 113. Negwer (11) has compiled an excellent guide to this nomenclature. In chemical literature, systematic names in line either with IUPAC (12) or Chemical Abstract rules (13) are used. The latter has been adopted in these volumes and the systematic name is, where appropriate, followed by the nonproprietary or trivial name.

ORGANIZATION OF THE VOLUMES:

The numerical data on the solubility of 2-aminobenzenesulfonamide, 3-aminobenzenesulfonamide, and 4-aminobenzenesulfonamide and its $\underline{N}^1$ and $\underline{N}^4$- derivatives, salts and complexes, compiled up to 1985 inclusive, have been divided into three volumes on the basis of chemical structure of the compounds.

The first volume includes the solubility of 2-aminobenzenesulfonamide, 3-aminobenzenesulfonamide, 4-aminobenzenesulfonamide and the derivatives of the last-named compound substituted at either of the nitrogen atoms, or both, with non-cyclic substituents (see System Index at the end of the first volume). The aroyl substituents, -C(:O)aryl, have also been included here. The second volume includes sulfanilamide derivatives substituted with 5-membered heterocyclic rings at either of the nitrogen atoms, and their derivatives. The third volume covers the solubilities of the derivatives substituted with 6-membered rings, mixtures of sulfonamides, and miscellanea. The compilations do not include compounds devoid of the $-NH_2$, -NHR or $-NR_2$ group in the benzene ring.

The solvent systems include all solvents with the exception of body fluids. The order of solvents for a particular solute are as follows: water; water-mineral acid; water-mineral base; water-mineral salt; water-miscellaneous mineral components; water-mineral and organic compounds; water-organic components; organic solvents; carboxylic acid and their salts; aliphatic acids; aromatic acids; other acids; alcohols, phenols (mono-, di-, polyhydric); amides; amines; aliphatic amines (primary, secondary, tertiary); aromatic amines (primary, secondary, tertiary); other amines; aminoalcohols; carboxylic acid esters; ethers (excluding tensides); hydrocarbons; aliphatic hydrocarbons; aromatic hydrocarbons; miscellaneous hydrocarbons; halogenated hydrocarbons (flouro-, chloro-, bromo-, iodo-); aliphatic halogenated hydrocarbons; aromatic halogenated hydrocarbons; ketones; tensides (surface-active agents); miscellaneous organic solvents.

SIGNIFICANT FIGURES AND GRAPHICAL DATA:

In most cases, solubility values given in the primary source by various workers are overstated with respect to significant figures. Since the author(s) original values are given on the data sheets, it is difficult to consider significant figures and analytical limitations in a completely consistant fashion. Therefore, the reader should be aware that in most cases the number of significant figures used for calculations was not that given by the original author(s). This was done to maintain coherence and consistency as data were given to varying significant figures. In many cases graphic data of sufficient size and clarity are reproduced. The data can be regarded of sufficient accuracy to serve as a starting point for more precise determinations. In many instances, the effect of additive concentration, pH, temperature, etc. can be depicted.

POLYMORPHISM:

Many sulfonamides exhibit several cyrstalline forms or polymorphs. There are several studies referenced in these volumes that specifically deal with the solubility difference between polymorphic modifications of the same compound. The solubility differences between polymorphs have been found to vary over a large range of values.

AMPHOLYTES:

Solubility of ampholytic sulfonamides as a function of pH varies enormously, sometimes by several orders of magnitude. Unless the pH is known experimentally, the solubility value may be suspect especially at low (1-3) and high pH (10-12) values. In these cases, the solubility is a rapidly changing value, frequently with small incremental changes of pH. The abrupt change of solubility with pH is usually associated with the formation of water soluble anionic and cationic species. Buffers, especially at higher concentrations may alter solubility by salting effect and the pH is also affected by ionic strength.

EQUILIBRATION TIME:

In general, it appears that many of these determined solubilities may not have been under equilibrium conditions. Unfortunately, in too many instances the equilibration time appears too low. Typically, solutes possess low aqueous solubilities and require long dissolution time to reach saturation. Saturation time should be experimentally determined in each case and for each compound. In many cases up to 24 hours may be required.

The editors consider the vast majority of the solubility values given in these volumes as tentative. It should be stressed, however, that they represent a useful starting point for more accurate determinations of a vast array of substituted 4-aminobenzene-sulfonamides, with many structurally and chemically related compound of various types. They amply illustrate the many factors and parameters affecting solubility and the direction and magnitude of these effects.

This compilation and evaluation is not only the result of the joint efforts of the compiler and evaluator, but also of all those who read the manuscripts, expressed their criticism, who procured copies of hard-to-get journals, who translated texts from Japanese as well as of those who in any other way assisted in the compilation and evaluation. We would like to express our gratitude in particular to the following colleagues: Prof. S. Kertes, Dr. M. Salomon, Prof. S. Yalkowsky, Prof. H. Akaiwa, Prof. C. Kalidas, Prof. W. Riess, Prof. A. Guerrero-Laverat, Prof. P. Rohdewald, Prof. J. Pütter, Dr.K. L. Loening, Dr. A. Brodin, Dr. D. Zimma, Mr. K. Hazelton, Dr. R. Fernandez-Prini, and Mr. E. MacMullan.

REFERENCES TO THE PREFACE:

1. Council on Pharmacy and Chemistry, J. Am. Med. Assoc. 1937, *108*, 1888.
2. Schroeter, G. Ber. Dtsch. Chem. Ges. 1906, *39*, 1559.
3. Gelmo, P. J. Prakt, Chem. *[2]*, 1908, *77*, 369.
4. Domagk, G. Dtsch. Med. Wschr. 1935, *61*, 250 and 829.
5. Tréfouël, T. J.; Nitti, F.; Bovet, D. Compt. Rend. Soc. Biol. 1935, *120*, 756.
6. Fuller, A. T. Lancet, 1937, 194.
7. Fildes, P. Lancet, 1940, 955.
8. Woods, D. D. J. Expt. Path. 1940, *21*, 74.
9. Langecker, H. Arch. Exptl. Pat. Pharmakol. 1948, *205*, 291.
10. Rolski, S. Chemia Srodkow Leczniczych (Chemistry of Medicinal Agents), 3rd ed., PZWL, Warsaw, 1968.
11. Negwer, M. Organisch-chemische Arzneimittel und ihre Synonyma (Organic-chemical Drugs and their Synonyms), Akademie-Verlag, Berlin, 1978.
12. Nomenclature of Organic Chemistry, Definitive Rules for Section C. Characteristic Groups Containing Carbon, Hydrogen, Oxygen, Nitrogen, Halogen, Sulfur, Selenium, and/or Tellurium IUPAC Commission on the Nomenclature of Organic Chemistry, London, Butterworths, 1971, rule 641.8.
13. J. Chem. Documentation 1974, *14*, 3.

INTRODUCTION TO THE SERIES ON SOLUBILITY OF SOLIDS IN LIQUIDS: SUBSERIES ON PHARMACEUTICALS

Nature of the Project

The Solubility Data Project (SDP) has as its aim a comprehensive search of the literature for solubilities of gases, liquids, and solids in liquids or solids. Data of suitable precision are compiled on data sheets in a uniform format. The data for each system are evaluated, and where data from different sources agree sufficiently, recommended values are proposed. The evaluation sheets, recommended values, and compiled data sheets are published on consecutive pages.

For *phamaceuticals*, the definitions, thermodynamics and methods of analysis are the same as those for the study of solubility of solids in liquids in general. For this subseries, special sections deal with matters of interest for pharmaceuticals, including discussions of polymorphism, factors influencing the rate of dissolution of drugs, and methods used to inhibit or enhance the rate of dissolution.

Definitions

A *mixture* (1, 2) describes a gaseous, liquid, or solid phase containing more than one substance, when the substances are all treated in the same way.

A *solution* (1, 2) describes a liquid or solid phase containing more than one substance, when for convenience one of the substances, which is called the *solvent*, and may itself be a mixture, is treated differently than the other substances, which are called *solutes*. If the sum of the mole fractions of the solutes is small compared to unity, the solution is called a *dilute solution*.

The *solubility* of a substance *B* is the relative proportion of *B* (or a substance related chemically to *B*) in a mixture which is saturated with respect to solid *B* at a specified temperature and pressure. *Saturated* implies the existence of equilibrium with respect to the processes of dissolution and precipitation; the equilibrium may be stable or metastable. The solubility of a substance in metastable equilibrium is usually greater than that of the corresponding substance in stable equilibrium. (Strictly speaking, it is the activity of the substance in metastable equilibrium that is greater.) Care must be taken to distinguish true metastability from supersaturation, where equilibrium does not exist.

Either point of view, mixture or solution, may be taken in describing solubility. The two points of view find their expression in the quantities used as measures of solubility and in the reference states used for definition of activities, activity coefficients and osmotic coefficients.

The qualifying phrase "substance related chemically to *B*" requires comment. The composition of the saturated mixture (or solution) can be described in terms of any suitable set of thermodynamic components. Thus, the solubility of a salt hydrate in water is usually given as the relative proportion of anhydrous salt in solution, rather than the relative proportions of hydrated salt and water.

For *pharmaceuticals*, the solubility of a drug substance in a given medium is of special importance in designing a suitable dosage form for a drug or in determination of a regimen for its administration. The solubility and rate of dissolution will determine the rate of appearance of the drug in various body fluids and at various sites of action. Therefore, the bioavailability of a drug is often determined by its solubility and rate of dissolution.

The solubility is a constant for a given substance in a given medium at constant temperature and pressure. Frequently it is possible to alter the solubility and rate of dissolution dramatically through changes in structure, degree of crystallinity or morphology, or by the addition of a solubilizing agent (cosolvent) to the dissolution medium. The appearance of a drug in adequate concentration at its site of action is a requirement for testing clinical efficiency; thus, enhancement of solubility may be required to render a substance clinically useful.

For reviews of recent literature on solubility and solubilization of drug substances, see (3, 4).

Quantities Used as Measures of Solubility

1. *Mole fraction* of substance *B*, x_B:

$$x_B = n_B / \sum_{s=1}^{c} n_s \qquad [1]$$

where n_s is the amount of substance of *s*, and *c* is the number of distinct substances present (often the number of thermodynamic components in the system). *Mole per cent* of *B* is 100 x_B.

2. *Mass fraction* of substance *B*, w_B:

$$w_B = m_B' / \sum_{s=1}^{c} m_s' \qquad [2]$$

where m_s' is the mass of substance *s*. *Mass per cent* is 100 w_B. The equivalent terms *weight fraction* and *weight per cent* are not used.

3. *Solute mole (mass) fraction* of solute *B* (5, 6):

$$x_{s,B} = n_B / \sum_{s=1}^{c'} n_s = x_B / \sum_{s=1}^{c'} x_s \qquad [3]$$

$$w_{s,B} = m_B' / \sum_{s=1}^{c'} m_s' = w_B / \sum_{s=1}^{c'} w_s \qquad [3a]$$

where the summation is over the solutes only. For the solvent *A*, $x_{S,A} = x_A/(1 - x_A)$, $w_{S,A} = w_A/(1 - w_A)$. These quantities are called *Jänecke mole (mass) fractions* in many papers.

4. *Molality* of solute *B* (1, 2) in a solvent *A*:

$$m_B = n_B / n_A M_A \qquad \text{SI base units: mol kg}^{-1} \qquad [4]$$

where M_A is the molar mass of the solvent.

5. *Concentration* of solute *B* (1, 2) in a solution of volume *V*:

$$c_B = [B] = n_B / V \qquad \text{SI base units: mol m}^{-3} \qquad [5]$$

The symbol c_B is preferred to [*B*], but both are used. The terms *molarity* and *molar* are not used.

Mole and mass fractions are appropriate to either the mixture or the solution point of view. The other quantities are appropriate to the solution point of view only. Conversions among these quantities can be carried out using the equations given in Table 1-1 following this Introduction. Other useful quantities will be defined in the prefaces to individual volumes or on specific data sheets.

In addition to the quantities defined above, the following are useful in conversions between concentrations and other quantities.

6. *Density*: $\rho = m/V \qquad$ SI base units: kg m^{-3} $\qquad$ [6]

7. *Relative density*: *d*; the ratio of the density of a mixture to the density of a reference substance under conditions which must be specified for both (1). The symbol $d_{t'}^{t}$ will be used for the density of a mixture at *t*°C, 1 bar divided by the density of water at *t'*°C, 1 bar. (In some cases, 1 atm = 101.325 kPa is used instead of 1 bar = 100 kPa.)

8. *A note on nomenclature*. The above definitions use the nomenclature of the IUPAC *Green Book* (1), in which a solute is called *B* and a solvent *A* In compilations and evaluations, the first-named component (component 1) is the solute, and the second (component 2 for a two-component system) is the solvent. The reader should bear these distinctions in nomenclature in mind when comparing nomenclature and theoretical equations given in this Introduction with equations and nomenclature used on the evaluation and compilation sheets.

Thermodynamics of Solubility

The principal aims of the Solubility Data Project are the tabulation and evaluation of: (a) solubilities as defined above; (b) the nature of the saturating phase. Thermodynamic analysis of solubility phenomena has two aims: (a) to provide a rational basis for the construction of functions to represent solubility data; (b) to enable thermodynamic quantities to be extracted from solubility data. Both these are difficult to achieve in many cases because of a lack of experimental or theoretical information concerning activity coefficients. Where thermodynamic quantities can be found, they are not evaluated critically, since this task would involve critical evaluation of a large body of data that is not directly relevant to solubility. The following is an outline of the principal thermodynamic relations encountered in discussions of solubility. For more extensive discussions and references, see books on thermodynamics, e.g., (7-14).

Activity Coefficients (1)

(a) *Mixtures.* The activity coefficient f_B of a substance B is given by

$$RT \ln (f_B x_B) = \mu_B - \mu_B{}^* \qquad [7]$$

where $\mu_B{}^*$ is the chemical potential of pure B at the same temperature and pressure. For any substance B in the mixture,

$$\lim_{x_B \to 1} f_B = 1 \qquad [8]$$

(b) *Solutions.*

(i) *Solute B.* The molal activity coefficient γ_B is given by

$$RT \ln(\gamma_B m_B) = \mu_B - (\mu_B - RT \ln m_B)^\infty \qquad [9]$$

where the superscript $^\infty$ indicates an infinitely dilute solution. For any solute B,

$$\gamma_B{}^\infty = 1 \qquad [10]$$

Activity coefficients y_B connected with concentrations c_B, and $f_{x,B}$ (called the *rational activity coefficient*) connected with mole fractions x_B, are defined in analogous ways. The relations among them (1, 9) are, where ρ^* is the density of the pure solvent:

$$f_B = (1 + M_A \sum_s m_s)\gamma_B = [\rho + \sum_s (M_A - M_s)c_s] y_B/\rho^* \qquad [11]$$

$$\gamma_B = (1 - \sum_s x_s) f_{x,B} = (\rho - \sum_s M_s c_s) y_B/\rho^* \qquad [12]$$

$$y_B = \rho^* f_{x,B}[1 + \sum_s (M_s/M_A - 1)x_B]/\rho = \rho^*(1 + \sum_s M_s m_s)\gamma_B/\rho \qquad [13]$$

For an electrolyte solute $B \equiv C_{\nu+}A_{\nu-}$, the activity on the molality scale is replaced by (11):

$$\gamma_B m_B = \gamma_\pm{}^\nu m_B{}^\nu Q^\nu \qquad [14]$$

where $\nu = \nu_+ + \nu_-$, $Q = (\nu_+{}^{\nu_+}\nu_-{}^{\nu_-})^{1/\nu}$, and $\gamma_\pm$ is the mean ionic activity coefficient on the molality scale. A similar relation holds for the concentration activity, $y_B c_B$. For the mole fractional activity,

$$f_{x,B} x_B = Q^\nu f_\pm{}^\nu x_\pm{}^\nu \qquad [15]$$

where $x_\pm = (x_+ x_-)^{1/\nu}$. The quantities x_+ and x_- are the ionic mole fractions (11), which are:

$$x_+ = \nu_+ x_B/[1 + \sum_s (\nu_s - 1)x_s]; \quad x_- = \nu_- x_B[1 + \sum_s (\nu_s - 1)x_s] \qquad [16]$$

where ν_s is the sum of the stoichiometric coefficients for the ions in a salt with mole fraction x_s. Note that the mole fraction of solvent is now

$$x_A{}' = (1 - \sum_s \nu_s x_s)/[1 + \sum_s (\nu_s - 1)x_s] \qquad [17]$$

so that

$$x_A{}' + \sum_s \nu_s x_s = 1 \qquad [18]$$

The relations among the various mean ionic activity coefficients are:

$$f_\pm = (1 + M_A \sum_s \nu_s ms)\gamma_\pm = [\rho + \sum_s (\nu_s M_A - M_s)c_s] y_\pm/\rho^* \qquad [19]$$

$$\gamma_\pm = \frac{(1 - \sum_s x_s) f_\pm}{1 + \sum_s (\nu_s - 1)x_s} = (\rho - \sum_s M_s c_s) y_\pm/\rho^* \qquad [20]$$

$$y_\pm = \frac{\rho^*[1 + \sum_s (M_s/M_A - 1)xs] f_\pm}{\rho[1 + \sum_s (\nu_s - 1)x_s]} = \rho^*(1 + \sum_s M_s m_s)\gamma_\pm/\rho \qquad [21]$$

(ii) *Solvent, A:*

The *osmotic coefficient*, ϕ , of a solvent *A* is defined as (1):

$$\phi = (\mu_A^* - \mu_A)/RT\, M_A \sum_s m_s \qquad [22]$$

where μ_A^* is the chemical potential of the pure solvent.

The *rational osmotic coefficient*, ϕ_x, is defined as (1):

$$\phi_x = (\mu_A - \mu_A^*)/RT\ln x_A = \phi M_A \sum_s m_s/\ln(1 + M_A \sum_s m_s) \qquad [23]$$

The activity, a_A, or the activity coefficient, f_A, is sometimes used for the solvent rather than the osmotic coefficient. The activity coefficient is defined relative to pure *A*, just as for a mixture.

For a mixed solvent, the molar mass in the above equations is replaced by the average molar mass; i.e., for a two-component solvent with components *J*, *K*, M_A becomes

$$M_A = M_J + (M_K - M_J)x_{v,K} \qquad [24]$$

where $x_{v,K}$ is the *solvent* mole fraction of component *K*.

The osmotic coefficient is related directly to the vapor pressure, *p*, of a solution in equilibrium with vapor containing *A* only by (14, p.306):

$$\phi M_A \sum_s \nu_s m_s = -\ln(p/p_A^*) + (V_{m,A}^* - B_{AA})(p - p_A^*)/RT \qquad [25]$$

where p_A^* is the vapor pressure of pure solvent *A*, $V_{m,A}^*$ is the molar volume of pure *A* in the liquid phase, and B_{AA} is the second virial coefficient of the vapor.

The Liquid Phase

A general thermodynamic differential equation which gives solubility as a function of temperature, pressure and composition can be derived. The approach is similar to that of Kirkwood and Oppenheim (9); see also (13, 14). Consider a solid mixture containing *c* thermodynamic components *i*. The Gibbs-Duhem equation for this mixture is:

$$\sum_{i=1}^{c} x_i'(S_i'dT - V_i'dp + d\mu_i') = 0 \qquad [26]$$

A liquid mixture in equilibrium with this solid phase contains c' thermodynamic components *i*, where $c' \geqslant c$. The Gibbs-Duhem equation for the liquid mixture is:

$$\sum_{i=1}^{c} x_i(S_i dT - V_i dp + d\mu_i') + \sum_{i=c+1}^{c'} x_i(S_i dT - V_i dp + d\mu_i) = 0 \qquad [27]$$

Subtract [26] from [27] and use the equation

$$d\mu_i = (d\mu_i)_{T,p} - S_i dT + V_i dp \qquad [28]$$

and the Gibbs-Duhem equation at constant temperature and pressure:

$$\sum_{i=1}^{c} x_i(d\mu_i')_{T,p} + \sum_{i=c+1}^{c'} x_i(d\mu_i)_{T,p} = 0 \qquad [29]$$

The resulting equation is:

$$RT\sum_{i=1}^{c} x_i'(d\ln a_i)_{T,p} = \sum_{i=1}^{c} x_i'(H_i - H_i')dT/T - \sum_{i=1}^{c} x_i'(V_i - V_i')dp \qquad [30]$$

where

$$H_i - H_i' = T(S_i - S_i') \qquad [31]$$

is the enthalpy of transfer of component *i* from the solid to the liquid phase at a given temperature, pressure and composition, with H_i and S_i the partial molar enthalpy and entropy of component *i*.

Use of the equations

$$H_i - H_i^o = -RT^2(\partial\ln a_i/\partial T)_{x,p} \qquad [32]$$

and

$$V_i - V_i^o = RT(\partial\ln a_i/\partial p)_{x,T} \qquad [33]$$

where superscript o indicates an arbitrary reference state gives:

$$RT\sum_{i=1}^{c} x_i' d\ln a_i = \sum_{i=1}^{c} x_i'(H_i^o - H_i')dT/T - \sum_{i=1}^{c} x_i'(V_i^o - V_i')dp \qquad [34]$$

where

$$d\ln a_i = (d\ln a_i)_{T,p} + (\partial \ln a_i/\partial T)_{x,p} + (\partial \ln a_i/\partial p)_{x,T} \quad [35]$$

The terms involving enthalpies and volumes in the solid phase can be written as:

$$\sum_{i=1}^{c} x_i' H_i' = H_s^* \qquad \sum_{i=1}^{c} x_i' V_i' = V_s^* \quad [36]$$

With eqn [36], the final general solubility equation may then be written:

$$R\sum_{i=1}^{c} x_i' d\ln a_i = (H_s^* - \sum_{i=1}^{c} x_i' H_i^0)d(1/T) - (V_s^* - \sum_{i=1}^{c} x_i' V_i^0)dp/T \quad [37]$$

Note that those components which are not present in both phases do not appear in the solubility equation. However, they do affect the solubility through their effect on the activities of the solutes.

Several applications of eqn [37] (all with pressure held constant) will be discussed below. Other cases will be discussed in individual evaluations.

(a) *Solubility as a function of temperature.*

Consider a binary solid compound A_nB in a single solvent A. There is no fundamental thermodynamic distinction between a binary compound of A and B which dissociates completely or partially on melting and a solid mixture of A and B; the binary compound can be regarded as a solid mixture of constant composition. Thus, with $c = 2$, $x_A' = n/(n + 1)$, $x_B' = 1/(n + 1)$, eqn [37] becomes:

$$d\ln(a_A^n a_B) = -\Delta H_{AB}^0 d(1/RT) \quad [38]$$

where

$$\Delta H_{AB}^0 = nH_A + H_B - (n + 1)H_s^* \quad [39]$$

is the molar enthalpy of melting and dissociation of pure solid A_nB to form A and B in their reference states. Integration between T and T_0, the melting point of the pure binary compound A_nB, gives:

$$\ln(a_A^n a_B) = \ln(a_A^n a_B)_{T=T_0} - \int_{T_0}^{T} \Delta H_{AB}^0 d(1/RT) \quad [40]$$

(i) Non-electrolytes

In eqn [32], introduce the pure liquids as reference states. Then, using a simple first-order dependence of ΔH_{AB}^* on temperature, and assuming that the activitity coefficients conform to those for a simple mixture (8):

$$RT\ \ln f_A = wx_B^2 \qquad RT\ \ln f_B = wx_A^2 \quad [41]$$

then, if w is independent of temperature, eqn [32] and [33] give:

$$\ln\{x_B(1-x_B)^n\} + \ln\left\{\frac{n^n}{(1 + n)^{n+1}}\right\} = G(T) \quad [42]$$

where

$$G(T) = -\left\{\frac{\Delta H_{AB}^* - T^*\Delta C_p^*}{R}\right\}\left\{\frac{1}{T} - \frac{1}{T^*}\right\} + \frac{\Delta C_p^*}{R}\ln(T/T^*) - \frac{w}{R}\left\{\frac{x_A^2 + nx_B^2}{T} - \frac{n}{(n + 1)T^*}\right\} \quad [43]$$

where ΔC_p^* is the change in molar heat capacity accompanying fusion plus decomposition of the pure compound to pure liquid A and B at temperature T^*, (assumed here to be independent of temperature and composition), and ΔH_{AB}^* is the corresponding change in enthalpy at $T = T^*$. Equation [42] has the general form:

$$\ln\{x_B(1-x_B)^n\} = A_1 + A_2/(T/K) + A_3\ln(T/K) + A_4(x_A^2 + nx_B^2)/(T/K) \quad [44]$$

If the solid contains only component B, then $n = 0$ in eqn [42] to [44].

If the infinite dilution reference state is used, then:

$$RT \ln f_{x,B} = w(x_A^2 - 1) \qquad [45]$$

and [39] becomes

$$\Delta H_{AB}^{\infty} = nH_A^* + H_B^{\infty} - (n+1)H_s^* \qquad [46]$$

where ΔH_{AB}^{∞} is the enthalpy of melting and dissociation of solid compound A_nB to the infinitely dilute reference state of solute B in solvent A; H_A^* and H_B^{∞} are the partial molar enthalpies of the solute and solvent at infinite dilution. Clearly, the integral of eqn [32] will have the same form as eqn [35], with ΔH_{AB}^{∞} replacing ΔH_{AB}^*, ΔC_p^{∞} replacing ΔCp^*, and $x_A^2 - 1$ replacing x_A^2 in the last term.

See (7) and (13) for applications of these equations to experimental data.

(ii) Electrolytes

(a) Mole fraction scale

If the liquid phase is an aqueous electrolyte solution, and the solid is a salt hydrate, the above treatment needs slight modification. Using rational mean activity coefficients, eqn [34] becomes:

$$\ln\left\{\frac{x_B^{\nu}(1-x_B)^n}{[1+(\nu-1)x_B]^{n+\nu}}\right\} - \ln\left\{\frac{n^n}{(n+\nu)^{n+\nu}}\right\} + \ln\left\{\left[\frac{f_B}{f_B^*}\right]^{\nu}\left[\frac{f_A}{f_A^*}\right]^{n}\right\}$$
$$= -\left\{\frac{\Delta H_{AB}^* - T^*\Delta C_p^*}{R}\right\}\left\{\frac{1}{T} - \frac{1}{T^*}\right\} + \frac{\Delta C_p^*}{R}\ln(T/T^*) \qquad [47]$$

where superscript * indicates the pure salt hydrate. If it is assumed that the activity coefficients follow the same temperature dependence as the right-hand side of eqn [47] (15-17), the thermochemical quantities on the right-hand side of eqn [47] are not rigorous thermodynamic enthalpies and heat capacities, but are apparent quantities only. Data on activity coefficients (11) in concentrated solutions indicate that the terms involving these quantities are not negligible, and their dependence on temperature and composition along the solubility-temperature curve is a subject of current research.

A similar equation (with $\nu = 2$ and without the heat capacity terms) or activity coefficients) has been used to fit solubility data for some $MOH-H_2O$ systems, where M is an alkali metal (15); enthalpy values obtained agreed well with known values. The full equation has been deduced by another method in (16) and applied to MCl_2-H_2O systems in (16) and (17). For a summary of the use of equation [47] and similar equations, see (18).

(2) Molality scale

Substitution of the mean activities on the molality scale in eqn [40] gives:

$$\nu\ln\left\{\frac{\gamma_{\pm}m_B}{\gamma_{\pm}^* m_B^*}\right\} - \nu(m_B/m_B^* - 1) - \nu\{m_B(\phi - 1)/m_B^* - \phi^* + 1)$$
$$= G(T) \qquad [48]$$

where $G(T)$ is the same as in eqn [47], $m_B^* = 1/nM_A$ is the molality of the anhydrous salt in the pure salt hydrate and $\gamma_{\pm}$ and ϕ are the mean activity coefficient and the osmotic coefficient, respectively. Use of the osmotic coefficient for the activity of the solvent leads, therefore, to an equation that has a different appearance to [47]; the content is identical. However, while eqn [47] can be used over the whole range of composition ($0 \leqslant x_B \leqslant 1$), the molality in eqn [48] becomes infinite at $x_B = 1$; use of eqn [48] is therefore confined to solutions sufficiently dilute that the molality is a useful measure of composition. The essentials of eqn [48] were deduced by Williamson (19); however, the form used here appears first in the *Solubility Data Series*. For typical applications (where activity and osmotic coefficients are not considered explicitly, so that the enthalpies and heat capacities are apparent values, as explained above), see (20).

The above analysis shows clearly that a rational thermodynamic basis exists for functional representation of solubility-temperature curves in two-component systems, but may be difficult to apply because of lack of experimental or theoretical knowledge of activity coefficients and partial molar enthalpies. Other phenomena which are related ultimately to the stoichiometric activity coefficients and which complicate interpretation include ion pairing, formation of complex ions, and hydrolysis. Similar considerations hold for the variation of solubility with pressure, except that the effects are relatively smaller at the pressures used in many investigations of solubility (7).

(b) *Solubility as a function of composition.*

At constant temperature and pressure, the chemical potential of a saturating solid phase is constant:

$$\mu_{A_nB}^* = \mu_{A_nB}(\text{sln}) = n\mu_A + \mu_B \qquad [49]$$

$$= (n\mu_A^* + \nu_+\mu_+^\infty + \nu_-\mu_-^\infty) + nRT\ \ln f_A x_A + \nu RT\ \ln(\gamma_\pm m_\pm Q)$$

for a salt hydrate A_nB which dissociates to water (A), and a salt (B), one mole of which ionizes to give ν_+ cations and ν_- anions in a solution in which other substances (ionized or not) may be present. If the saturated solution is sufficiently dilute, $f_A = x_A = 1$, and the quantity K_s in

$$\Delta G^\infty = (\nu_+\mu_+^\infty + \nu_-\mu_-^\infty + n\mu_A^* - \mu_{AB}^*)$$

$$= -RT\ \ln K_s$$

$$= -\nu RT\ \ln(Q\gamma_\pm m_B) \qquad [50]$$

is called the *solubility product* of the salt. (It should be noted that it is not customary to extend this definition to hydrated salts, but there is no reason why they should be excluded.) Values of the solubility product are often given on mole fraction or concentration scales. In dilute solutions, the theoretical behavior of the activity coefficients as a function of ionic strength is often sufficiently well known that reliable extrapolations to infinite dilution can be made, and values of K_s can be determined. In more concentrated solutions, the same problems with activity coefficients that were outlined in the section on variation of solubility with temperature still occur. If these complications do not arise, the solubility of a hydrate salt $C_\nu A_\nu \cdot nH_2O$ in the presence of other solutes is given by eqn [50] as

$$\nu\ \ln\{m_B/m_B(0)\} = -\nu \ln\{\gamma_\pm/\gamma_\pm(0)\} - n\ \ln\{a_A/a_A(0)\} \qquad [51]$$

where a_A is the activity of water in the saturated solution, m_B is the molality of the salt in the saturated solution, and (0) indicates absence of other solutes. Similar considerations hold for non-electrolytes.

Consideration of *complex mixed ligand equilibria* in the solution phase is also frequently of importance in the interpretation of solubility equilibria. For nomenclature connected with these equilibria (and solubility equilibria as well) see (21, 22).

(c) *Alteration of the dissolution medium for pharmaceuticals*

Many substances which are only slightly soluble in water may be made more soluble by the addition of a cosolvent, surface-active agents, or complexing agents.

(i) Addition of a cosolvent. It is frequently necessary to dissolve a quantity of drug in a small volume of liquid so that it may be administered parenterally by injection. If the drug is not sufficiently soluble in water because of its hydrophobicity, the addition of a quantity of water-miscible, but less polar solvent may render the drug soluble in a small quantity of the mixed solvent. Solvents used for this purpose have included propylene glycol, glycerol, ethanol, polyethylene glycol and glycofural. Solubilities of many drug substances in water-organic solvent mixtures have been tabulated by Yalkowsky and Roseman (23).

(ii) Surface-active agents. Another approach to increasing the solubility and rate of dissolution of drug substances is to add a surface-active agent. There is an extensive literature on the application of surfactants and micellar dissolution, which has been summarized recently by Florence (24). Cationic, anionic or neutral surfactants are available. In choosing a surfactant, the possibility of charge-charge interactions between the drug and the surfactant must be considered, as well as the degree of ionization of each species as a function of pH. Micellar dissolution of drugs or additives may protect the dissolved species from hydrolytic degradation by the aqueous solvent. The stability of drugs may therefore be enhanced considerably by the addition of a surfactant. Surfactants may also facilitate the transport of drugs across biological membranes. Examples of substantially improved bioavailability of drugs under the influence of micellar dissolution have been reported (24).

(iii) Other modifications of the dissolution medium. The solubility of weak acid and weak base drugs will usually depend on the pH of the medium. Within reasonable limits for pharmaceutical preparations, pH may be adjusted to obtain the drug in the charged (and usually more soluble) form. The addition of complexing agents such as chelating agents, organic salts, cyclodextrins, or ion-pairing agents may be used to enhance solubility and rate of dissolution. Examples are given in the chapter by A.J. Repta in (3).

The Solid Phase

The definition of solubility permits the occurrence of a single solid phase which may be a pure anhydrous compound, a salt hydrate, a non-stoichiometric compound, or a solid mixture (or solid solution, or "mixed crystals"), and may be stable or metastable. As well, any number of solid phases consistent with the requirements of the phase rule may be present. Metastable solid phases are of widespread occurrence, and may appear as polymorphic (or allotropic) forms or crystal solvates whose rate of transition to more stable forms is very slow. Surface heterogeneity may also give rise to metastability, either when one solid precipitates on the surface of another, or if the size of the solid particles is sufficiently small that surface effects become important. In either case, the solid is not in stable equilibrium with the solution. See (25) for the modern formulation of the effect of particle size on solubility. The stability of a solid may also be affected by the atmosphere in which the system is equilibrated.

Many of these phenomena require very careful, and often prolonged, equilibration for their investigation and elimination. A very general analytical method, the "wet residues" method of Schreinemakers (26), is often used to investigate the composition of solid phases in equilibrium with salt solutions. This method has been reviewed in (27), where [see also (28)] least-squares methods for evaluating the composition of the solid phase from wet residue data (or initial composition data) and solubilities are described. In principle, the same method can be used with systems of other types. Many other techniques for examination of solids, in particular X-ray, optical, and thermal analysis methods, are used in conjunction with chemical analyses (including the wet residues method).

Solid State Manipulation in Pharmaceuticals

(i) Polymorphism. Many drug substances may crystallize in more than one form, a phenomenon called *polymorphism*. The different modifications (polymorphs) arise because of the relative positions and bonding of the molecules in their crystal lattices; true polymorphs do not differ in chemical composition. Polymorphs of the same substance frequently have different physical properties such as solubility and rate of dissolution. Ultimately, the solubility of all forms will revert to that of the form with the lowest Gibbs energy; the solubility of a less-stable form will thus be an initial solubility. The rate of reversion to the most stable form is often very slow, and a form with higher Gibbs energy may exhibit its higher solubility for hours. This phenomenon may be used to advantage by choosing the polymorph with the desired solubility or rate of dissolution. Examples of polymorphism and methods of characterization have been reviewed by Haleblian (29) and Burger (30).

(ii) Crystallinity. In many cases, drug substances may occur in the solid state as amorphous or partly crystalline forms. This is a special case of polymorphism, and may result from rapid precipitation or from freeze-drying. These amorphous or partly crystalline materials are unstable relative to the crystalline form. However, reversion to the crystalline form may be slow, and the less stable forms may be used to enhance solubility and rate of dissolution (31).

(iii) Choice of salt form. Many drug substances are organic salts. In most cases the drug moiety is the organic cation or anion, such as a quaternary ammonium cation or a carboxylate or sulfonate anion. The counterion is frequently an inorganic ion such as sodium or chloride. It is possible to obtain large variations in initial solubility depending on the choice of the salt form of the drug.

COMPILATIONS AND EVALUATIONS

The formats for the compilations and critical evaluations have been standardized for all volumes. A brief description of the data sheets has been given in the FOREWORD; additional explanation is given below.

Guide to the Compilations

The format used for the compilations is, for the most part, self-explanatory. The details presented below are those which are not found in the FOREWORD or which are not self-evident.

Components. Each component is listed according to IUPAC or Chemical Abstracts (CA) name and CA Registry Number. The formula is given either in terms of the IUPAC or Hill (32) system and the choice of formula is governed by what is usual for most current users: i.e., IUPAC for inorganic compounds, and Hill system for organic compounds. Components are ordered according to:

(a) saturating components;
(b) non-saturating components in alphanumerical order;
(c) solvents in alphanumerical order.

The saturating components are arranged in order according to a 18-column periodic table with two additional rows:

Columns 1 and 2: H, alkali elements, ammonium, alkaline earth elements
3 to 12: transition elements
13 to 17: boron, carbon, nitrogen groups; chalcogenides, halogens
18: noble gases
Row 1: Ce to Lu
Row 2: Th to the end of the known elements, in order of atomic number.

Salt hydrates are generally not considered to be saturating components since most solubilities are expressed in terms of the anhydrous salt. The existence of hydrates or solvates is carefully noted in the text, and CA Registry Numbers are given where available, usually in the critical evaluation. Mineralogical names are also quoted, along with their CA Registry Numbers, again usually in the critical evaluation.

Original Measurements. References are abbreviated in the forms given by *Chemical Abstracts Service Source Index* (CASSI). Names originally in other than Roman alphabets are given as transliterated by *Chemical Abstracts*.

Experimental Values. Data are reported in the units used in the original publication, with the exception that modern *names* for units and quantities are used; e.g., mass per cent for weight per cent; mol dm^{-3} for molar; etc. Both mass and molar values are given. Usually, only one type of value (e.g., mass per cent) is found in the original paper, and the compiler has added the other type of value (e.g., mole per cent) from computer calculations based on 1983 atomic weights (33).

Errors in calculations and fitting equations in original papers have been noted and corrected, by computer calculations where necessary.

Method. Source and Purity of Materials. Abbreviations used in *Chemical Abstracts* are often used here to save space.

Estimated Error. If these data were omitted by the original authors, and if relevant information is available, the compilers have attempted to estimate errors from the internal consistency of data and type of apparatus used. Methods used by the compilers for estimating and and reporting errors are based on the papers by Ku and Eisenhart (34).

Comments and/or Additional Data. Many compilations include this section which provides short comments relevant to the general nature of the work or additional experimental and thermodynamic data which are judged by the compiler to be of value to the reader.

References. See the above description for Original Measurements.

Guide to the Evaluations

The evaluator's task is to check whether the compiled data are correct, to assess the reliability and quality of the data, to estimate errors where necessary, and to recommend "best" values. The evaluation takes the form of a summary in which all the data supplied by the compiler have been critically reviewed. A brief description of the evaluation sheets is given below.

Components. See the description for the Compilations.

Evaluator. Name and date up to which the literature was checked.

Critical Evaluation

(a) Critical text. The evaluator produces text evaluating *all* the published data for each given system. Thus, in this section the evaluator reviews the merits or shortcomings of the various data. Only published data are considered; even published data can be considered only if the experimental data permit an assessment of reliability.

(b) Fitting equations. If the use of a smoothing equation is justifiable the evaluator may provide an equation representing the solubility as a function of the variables reported on all the compilation sheets.

(c) Graphical summary. In addition to (b) above, graphical summaries are often given.

(d) Recommended values. Data are *recommended* if the results of at least two independent groups are available and they are in good agreement, and if the evaluator has no doubt as to the adequacy and reliability of the applied experimental and computational procedures. Data are considered as *tentative* if only one set of measurements is available, or if the evaluator considers some aspect of the computational or experimental method as mildly undesirable but estimates that it should cause only minor errors. Data are considered as *doubtful* if the evaluator considers some aspect of the computational or experimental method as undesirable but still considers the data to have some value in those instances where the order of magnitude of the solubility is needed. Data determined by an inadequate method or under ill-defined conditions are *rejected*. However references to these data are included in the evaluation together with a comment by the evaluator as to the reason for their rejection.

(e) References. All pertinent references are given here. References to those data which, by virtue of their poor precision, have been rejected and not compiled are also listed in this section.

(f) Units. While the original data may be reported in the units used by the investigators, the final recommended values are reported in S.I. units (1, 35) when the data can be accurately converted.

References

1. Whiffen, D.H., ed., *Manual of Symbols and Terminology for Physicochemical Quantities and Units*. *Pure Applied Chem.* 1979, 51, No. 1.
2. McGlashan, M.L. *Physicochemical Quantities and Units*. 2nd ed. Royal Institute of Chemistry. London. 1971.
3. Yalkowsky, S.H., ed. *Techniques of Solubilization of Drugs*. Marcel Dekker. New York. 1981.
4. Byron, S.R. *Solid State Chemistry of Drugs*. Academic Press. New York. 1982.
5. Jänecke, E. *Z. Anorg. Chem.* 1906, *51*, *132*.
6. Friedman, H.L. *J. Chem. Phys.* 1960, 32, 1351.
7. Prigogine, I.; Defay, R. *Chemical Thermodynamics*. D.H. Everett, transl. Longmans, Green. London, New York, Toronto. 1954.
8. Guggenheim, E.A. *Thermodynamics*. North-Holland. Amsterdam. 1959. 4th ed.
9. Kirkwood, J.G.; Oppenheim, I. *Chemical Thermodynamics*. McGraw-Hill. New York, Toronto, London. 1961.
10. Lewis, G.N.; Randall, M. (rev. Pitzer, K.S.; Brewer, L.). *Thermodynamics*. McGraw Hill. New York, Toronto, London. 1961. 2nd. ed.
11. Robinson, R.A.; Stokes, R.H. *Electrolyte Solutions*. Butterworths. London. 1959. 2nd ed.
12. Harned, H.S.; Owen, B.B. *The Physical Chemistry of Electrolytic Solutions*. Reinhold. New York. 1958. 3rd ed.
13. Haase, R.; Schönert, H. *Solid-Liquid Equilibrium*. E.S. Halberstadt, trans. Pergamon Press, London, 1969.
14. McGlashan, M.L. *Chemical Thermodynamics*. Academic Press. London. 1979.
15. Cohen-Adad, R.; Saugier, M.T.; Said, J. *Rev. Chim. Miner.* 1973, *10*, 631.
16. Counioux, J.-J.; Tenu, R. *J. Chim. Phys.* 1981, *78*, 815.
17. Tenu, R.; Counioux, J.-J. *J. Chim. Phys.* 1981, *78*, 823.
18. Cohen-Adad, R. *Pure Appl. Chem.* 1985, *57*, 255.
19. Williamson, A.T. *Faraday Soc. Trans.* 1944, *40*, 421.
20. Siekierski, S.; Mioduski, T.; Salomon, M. *Solubility Data Series*. Vol. 13. *Scandium, Yttrium, Lanthanum and Lanthanide Nitrates*. Pergamon Press. 1983.
21. Marcus, Y., ed. *Pure Appl. Chem.* 1969, *18*, 459.
22. IUPAC Analytical Division. *Proposed Symbols for Metal Complex Mixed Ligand Equilibria (Provisional)*. *IUPAC Inf. Bull.* 1978, No. *3*, 229.

23. Yalkowsky, S.H.; Roseman, T.J. *in* S.H. Yalkowsky, ed. *Techniques of Solubilization of Drugs*. Marcel Dekker. New York. 1981.
24. Florence, A.T. *in* S.H. Yalkowsky, ed. *Techniques of Solubilization of Drugs*. Marcel Dekker. New York. 1981.
25. Enüstün, B.V.; Turkevich, J. *J. Am. Chem. Soc.* 1960, *82*, 4502.
26. Schreinemakers. F.A.H. *Z. Phys. Chem., Stoechiom. Verwandschaftsl.* 1893, 11, 75.
27. Lorimer, J.W. *Can. J. Chem.* 1981, 59, 3076.
28. Lorimer, J.W. *Can. J. Chem.* 1982, *60*, 1978.
29. Haleblian, J.K. *J. Pharm. Sci.* 1975, 64, 1269.
30. Burger, A. *Pharm. Int.* 1982, 3, 158.
31. Shefter, E. in S.H. Yalkowsky, ed. *Techniques of Solubilization of Drugs*. Marcel Dekker. New York. 1981.
32. Hill, E.A. *J. Am. Chem. Soc.* 1900, *22*, 478.
33. IUPAC Commission on Atomic Weights. *Pure Appl. Chem.* 1984, 56, 653.
34. Ku, H.H., p. 73; Eisenhart, C., p. 69; in Ku, H.H., ed. *Precision Measurement and Calibration*. NBS Special Publication 300. Vol. 1. Washington. 1969.
35. *The International System of Units*. Engl. transl. approved by the BIPM of *Le Système International d'Unités*. H.M.S.O. London. 1970.

September, 1986

R. Cohen-Adad,
Villeurbanne, France

S. Lindenbaum,
Lawrence, Kansas, U.S.A.

J.W. Lorimer,
London, Ontario, Canada

A.N. Paruta,
Kingston, R.I., U.S.A.

R. Piekos,
Gdansk, Poland

M. Salomon,
Fair Haven, New Jersey, U.S.A.

Table I-1

Quantities Used as Measures of Solubility
Conversion Table for 2-Component Systems
Containing Solvent A and Solute B

	mole fraction x_B =	mass fraction w_B =	molality m_B =	concentration c_B =
x_B	x_B	$\dfrac{1}{1 - M_A(1 - 1/x_B)/M_B}$	$\dfrac{1}{M_A(1/x_B - 1)}$	$\dfrac{\rho}{M_B + M_A(1/x_B - 1)}$
w_B	$\dfrac{1}{1 + M_B(1/w_B - 1)}$	w_B	$\dfrac{1}{M_B(1/w_B - 1)}$	$\rho w_B/M_B$
m_B	$\dfrac{1}{1 + 1/m_B M_A}$	$\dfrac{1}{1 + 1/M_B m_B}$	m_B	$\dfrac{\rho}{M_B + 1/m_B}$
c_B	$\dfrac{1}{1 + (\rho/c_B - M_B)/M_A}$	$M_B c_B/\rho$	$\dfrac{1}{\rho/c_B - M_B}$	c_B

ρ = density of solution
M_A, M_B = molar masses of solvent, solute
Formulas are given in forms suitable for rapid computation; all calculations should be made using SI base units.

STRUCTURES

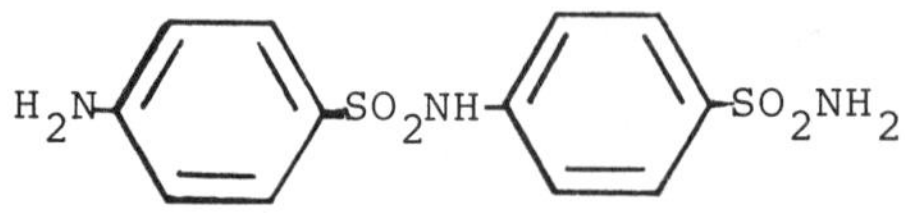 [547-52-4] M.W.=327.38	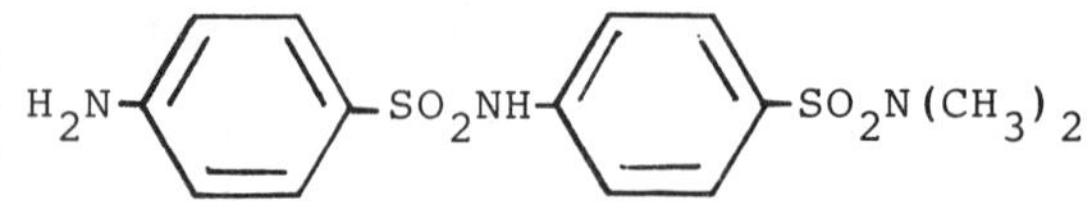 [515-67-3] M.W.=355.44
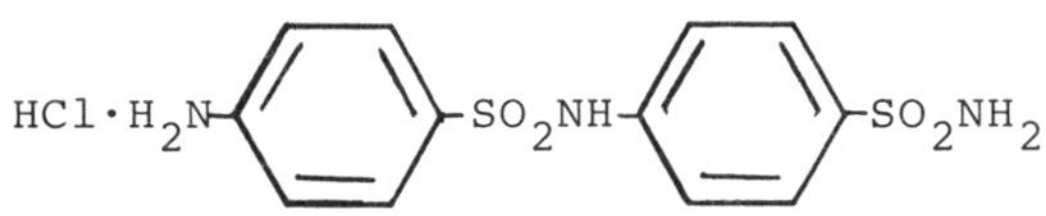 [77400-69-2] M.W.=363.84	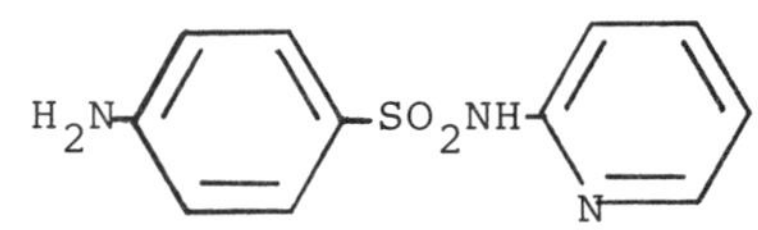 [144-83-2] M.W.=249.29
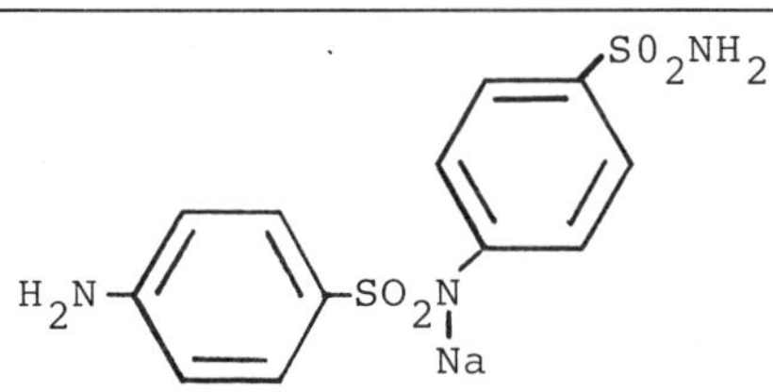 [77400-68-1] M.W.=349.36	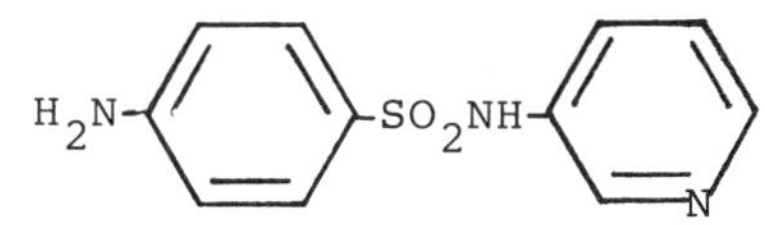 [599-81-5] M.W.=249.29
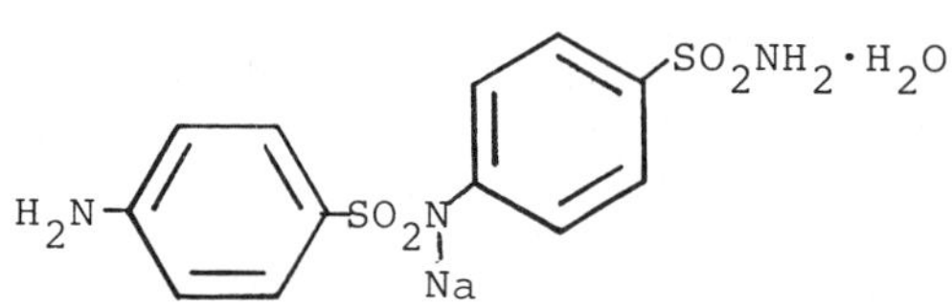 [81815-35-2] M.W.=367.38	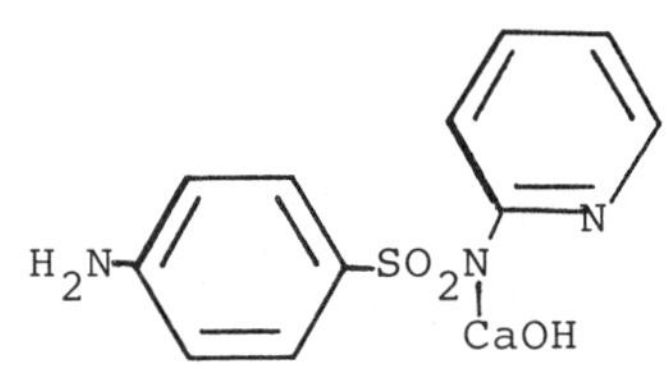 [77400-70-5] M.W.=305.37
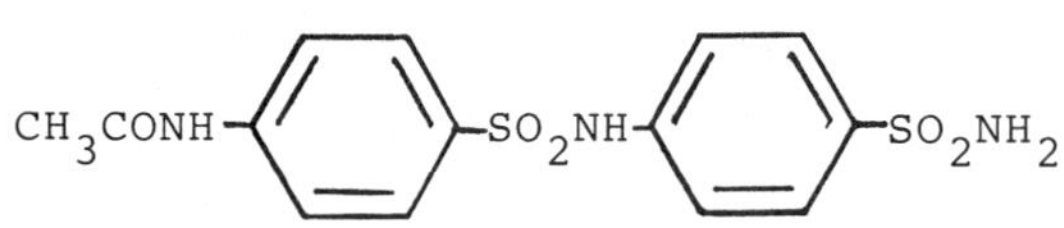 [56444-82-7] M.W.=369.42	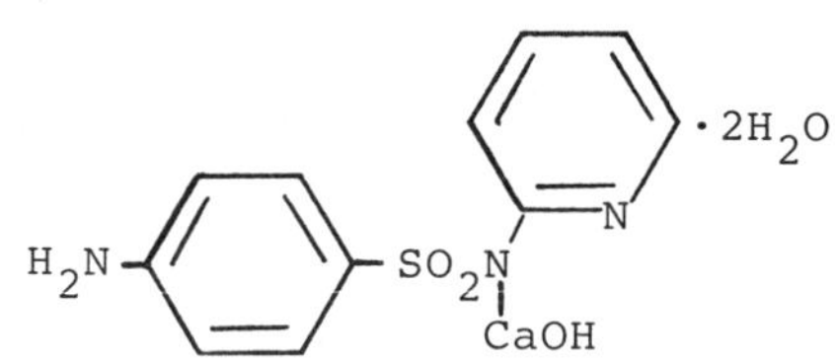 [77400-71-6] M.W.=341.40
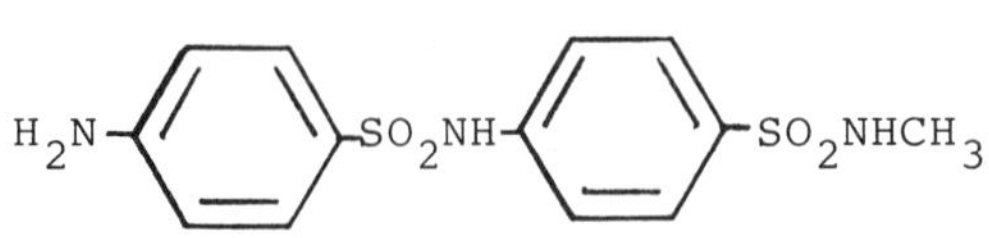 [547-53-5] M.W.=341.41	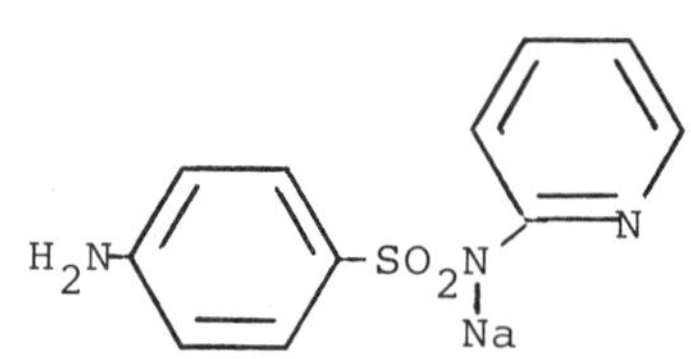 [127-57-1] M.W.=271.27
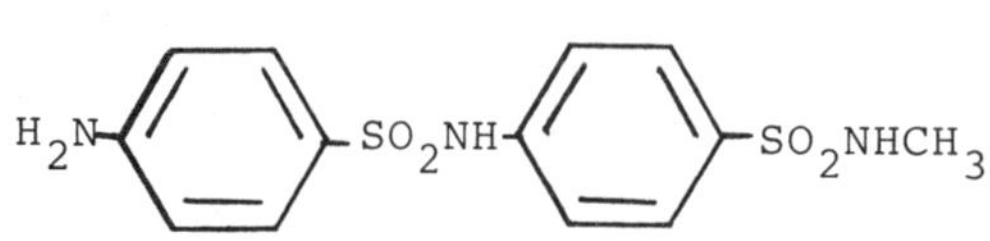 [71119-14-7] M.W.=383.44	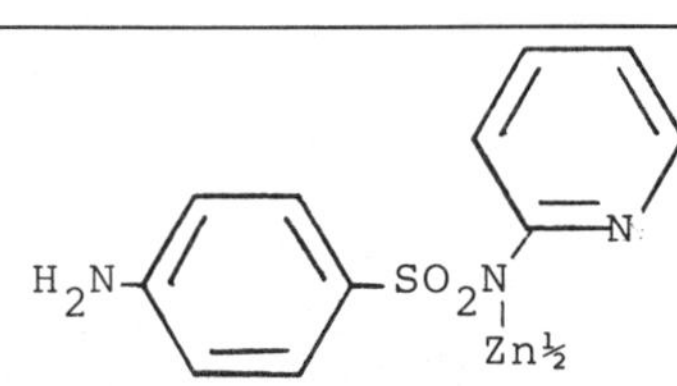 [71261-89-7] M.W.=561.94

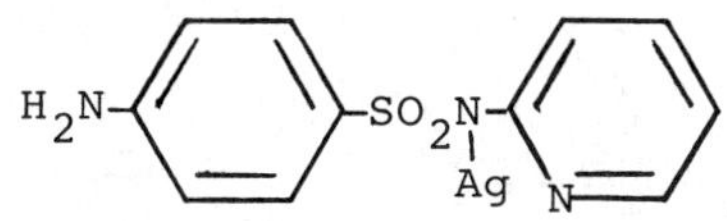

[24342-38-9] M.W.= 356.16

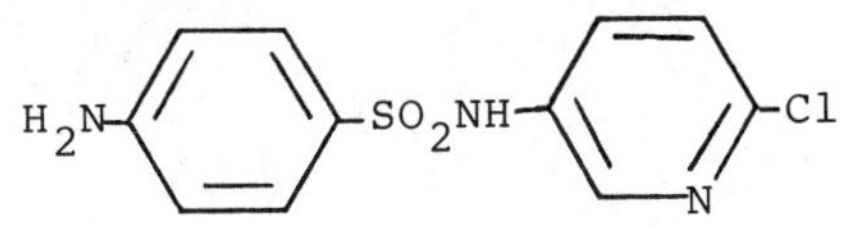

[34392-82-0] M.W.=283.73

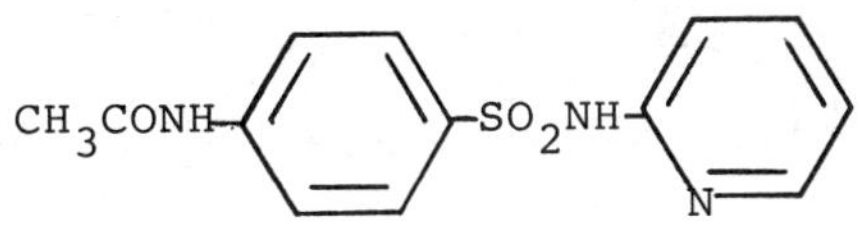

[19077-98-6] M.W.=291.33

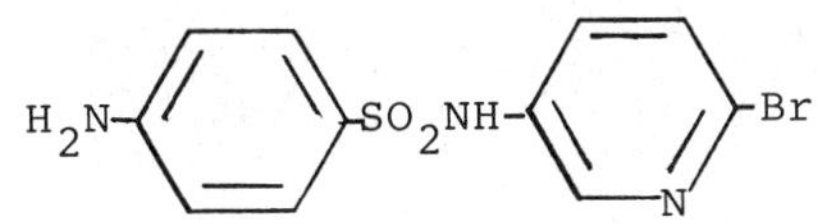

[17103-43-4] M.W.=328.19

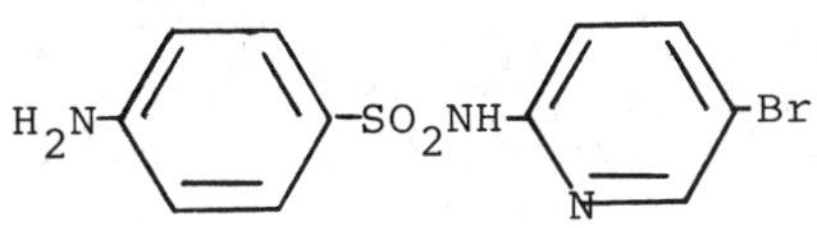

[16805-99-5] M.W.=328.19

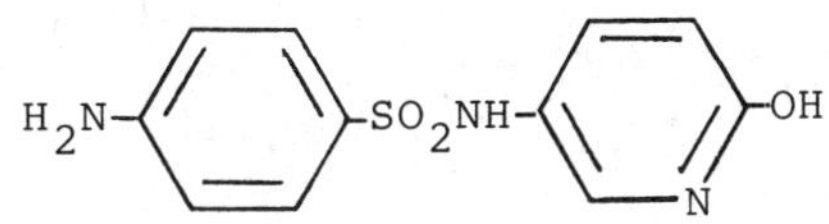

[71119-20-5] M.W.=265.29

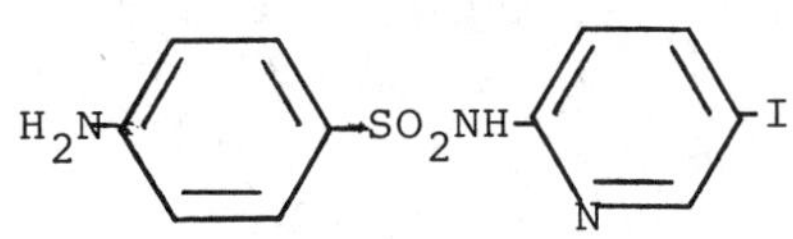

[71119-21-6] M.W.=375.19

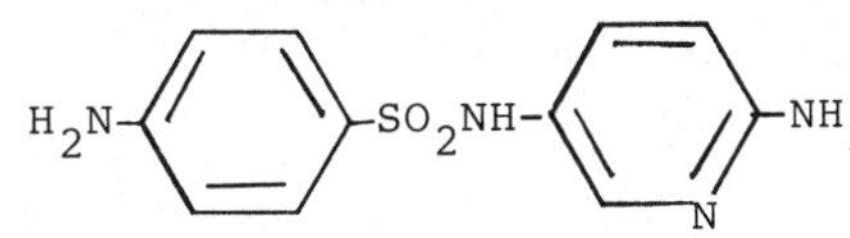

[17103-45-6] M.W.=264.30

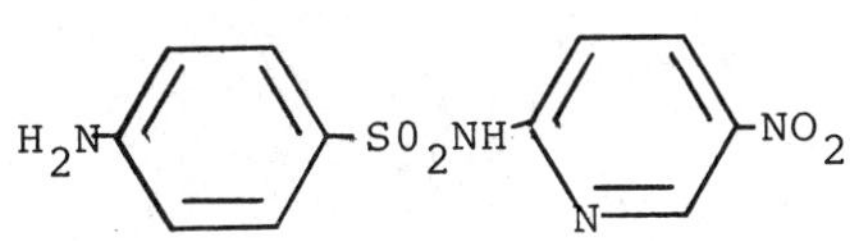

[39588-36-8] M.W.=294.29

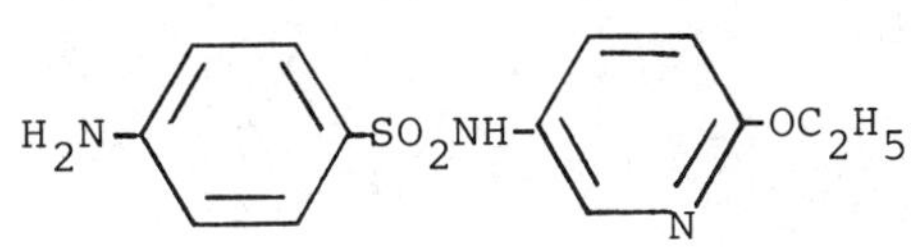

[71720-65-5] M.W.=293.34

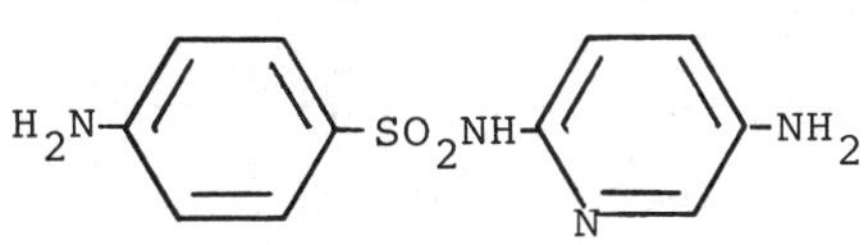

[16840-28-1] M.W.=264.30

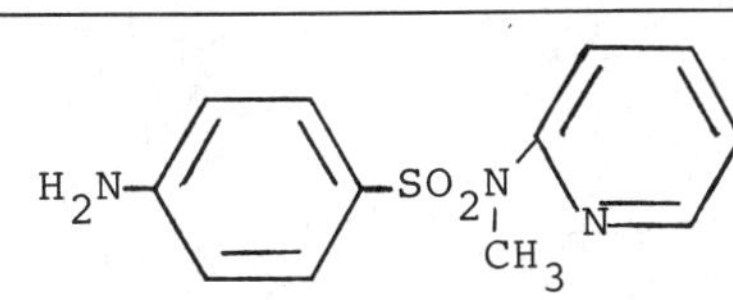

[51543-29-4] M.W.=263.32

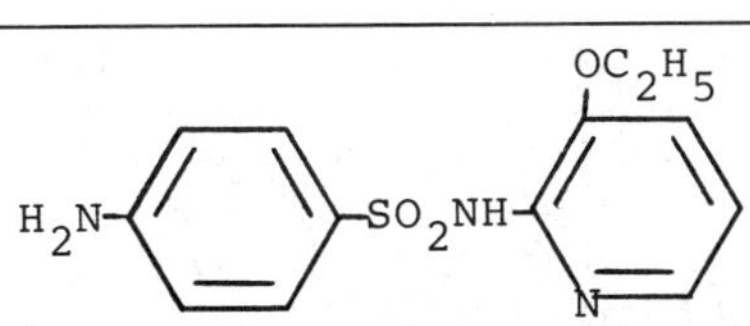

[71119-19-2] M.W.=293.34

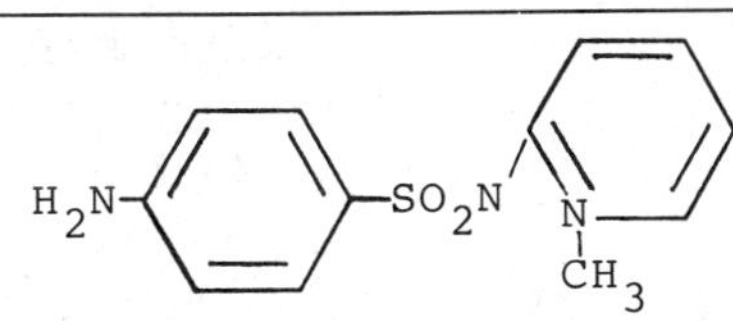

[51543-30-7] M.W.=265.33

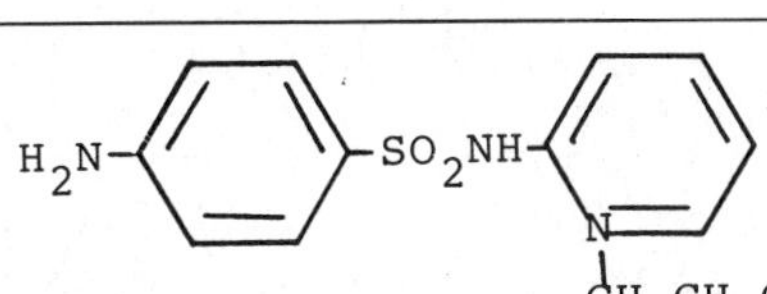

[71119-27-2] M.W.=294.35

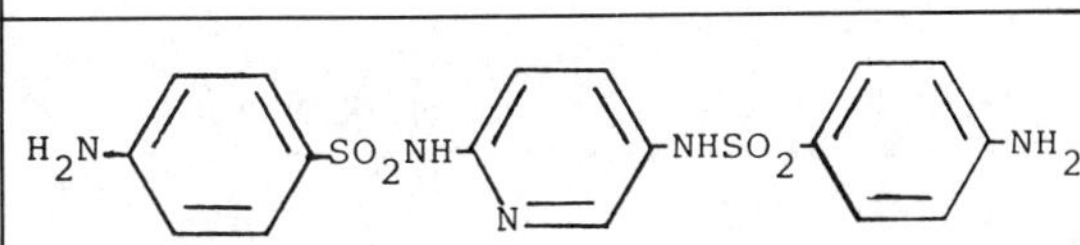

[71119-18-1] M.W.=419.48

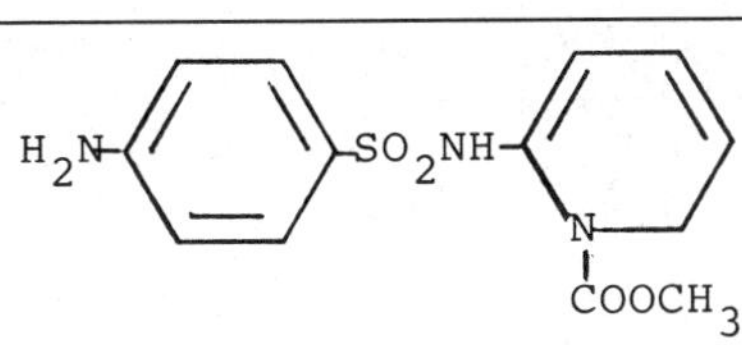

[71119-28-3] M.W.=297.33

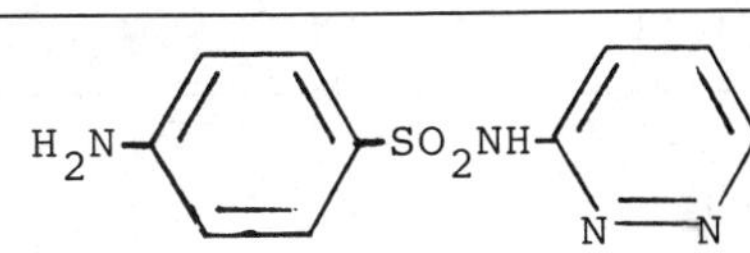

[515-62-8] M.W.=250.28

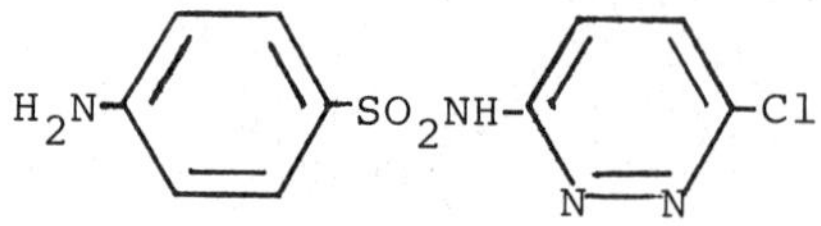

[80-32-0] M.W.=284.72

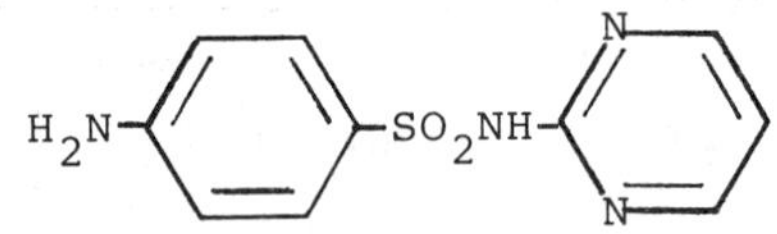

[68-35-9] M.W.=250.28

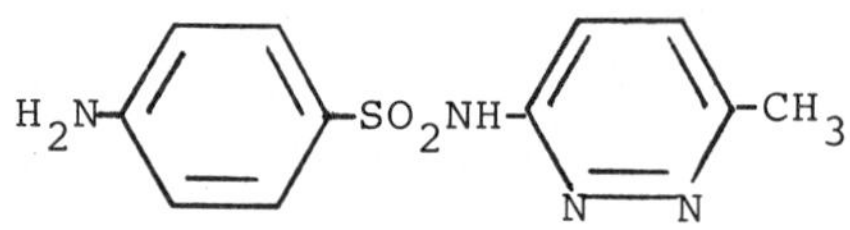

[5433-63-6] M.W.=264.31

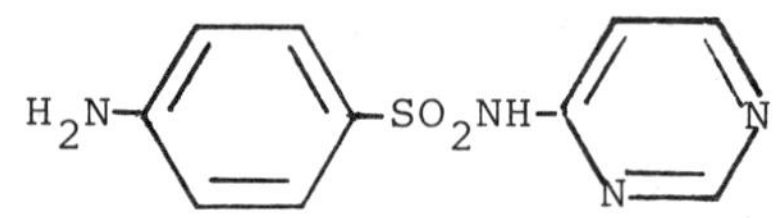

[599-82-6] M.W.=250.28

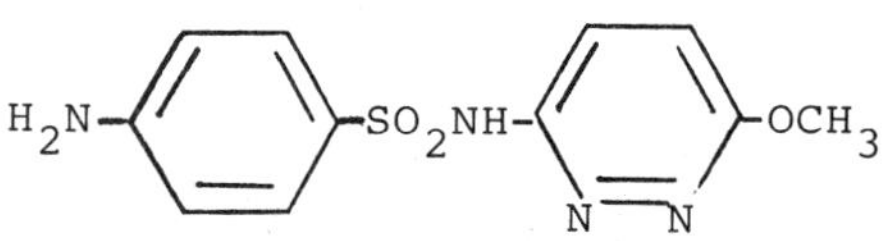

[80-35-3] M.W.=280.31

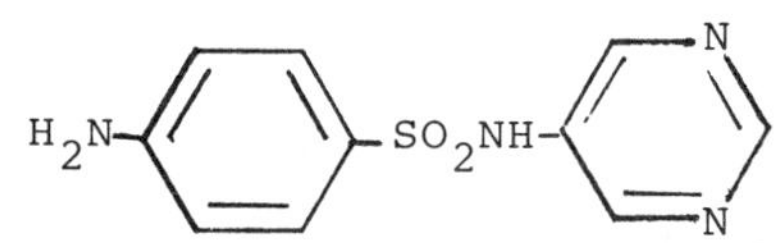

[17103-48-9] M.W.=250.28

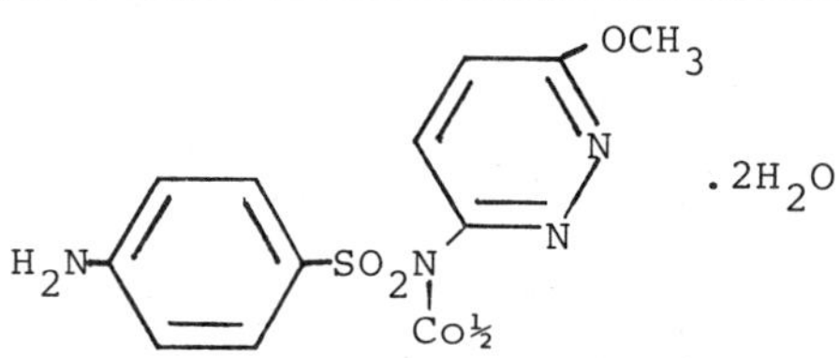

[72384-11-3] M.W.=653.57

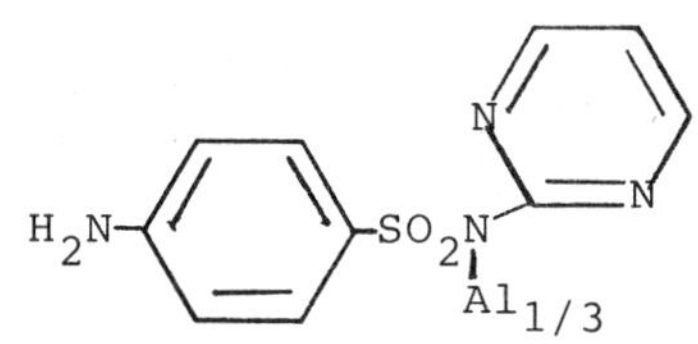

[71280-76-7] M.W.=774.80

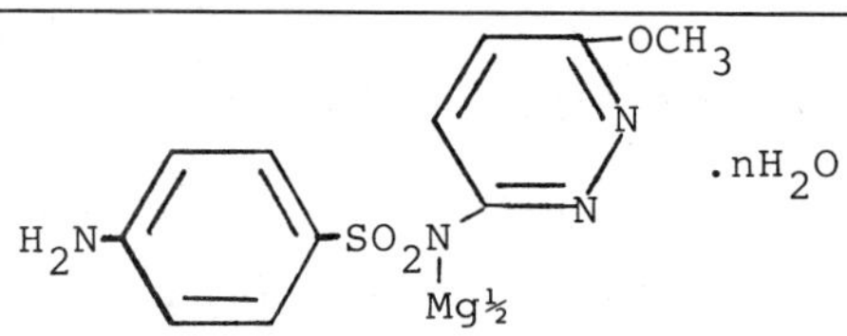

[84812-83-9] M.W.=UNSPECIFIED

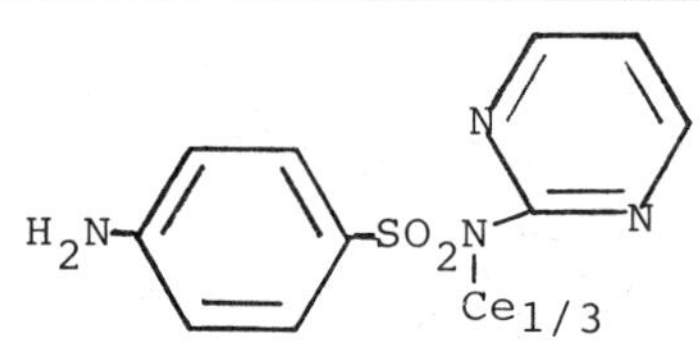

[66269-03-2] M.W.=887.94

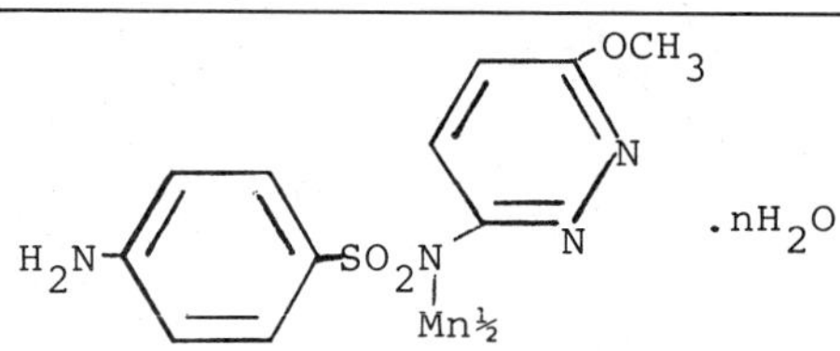

[84812-82-8] M.W.=UNSPECIFIED

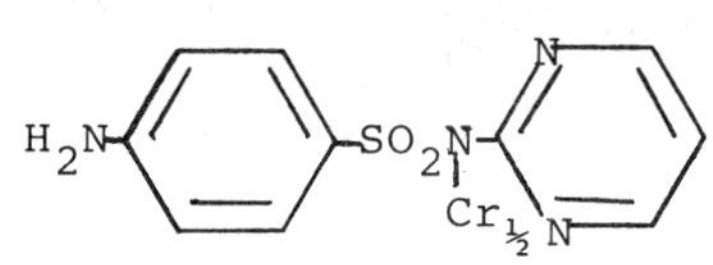

[71262-84-2] M.W.=550.54

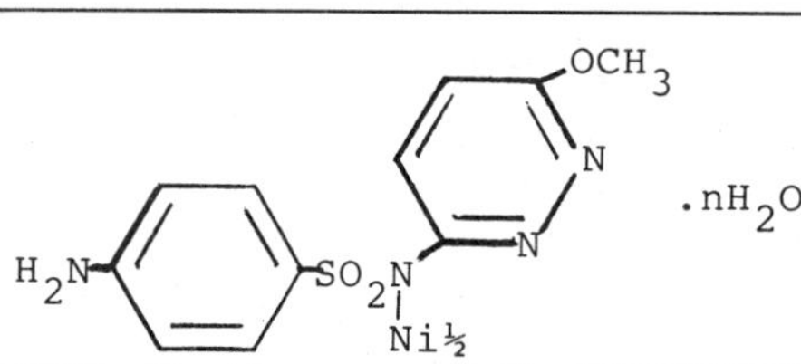

[84825-01-4] M.W.=UNSPECIFIED

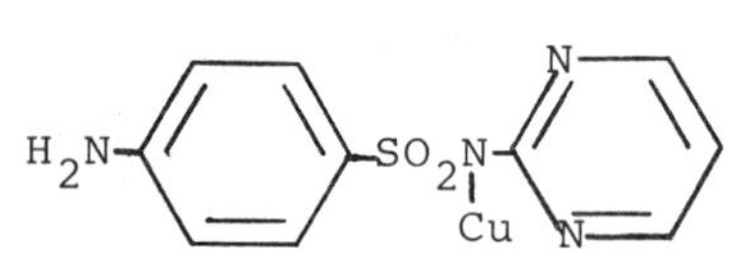

[71261-85-3] M.W.=312.82

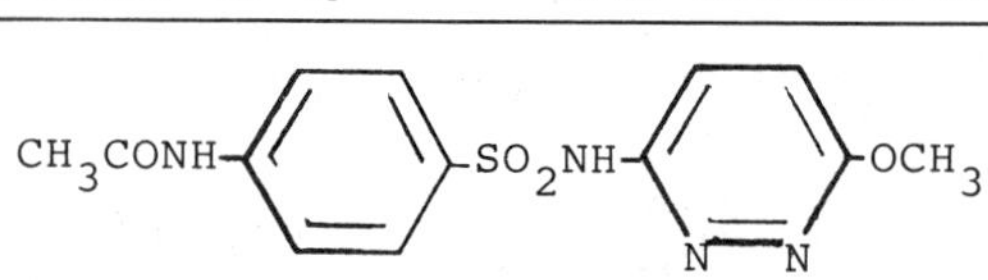

[127-75-3] M.W.=322.34

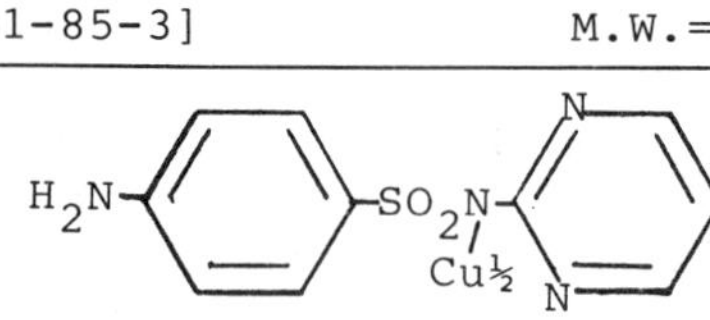

[12171-53-8] M.W.=562.09

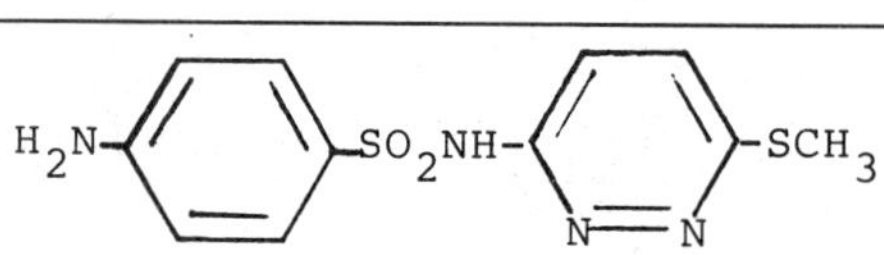

[7758-81-8] M.W.=296.37

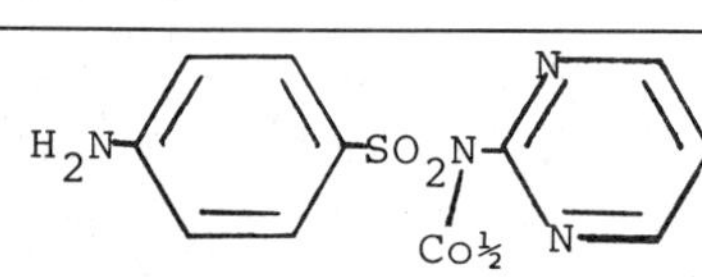

[71280-79-0] M.W.=557.48

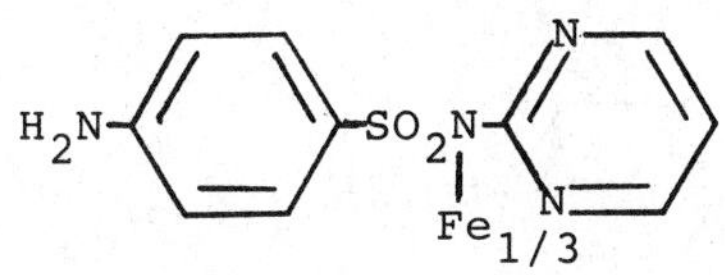

[71261-86-4] M.W.=803.66

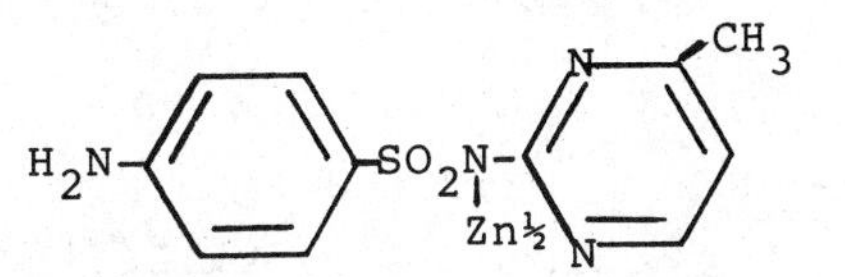

[71496-63-4] M.W.=591.98

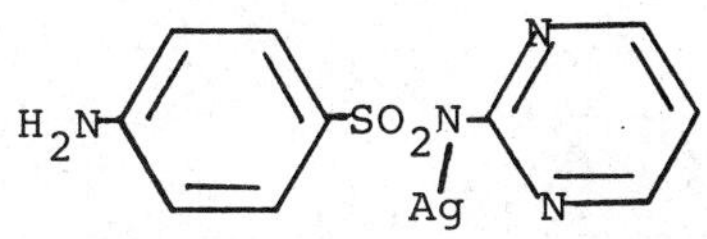

[22199-08-2] M.W.=357.14

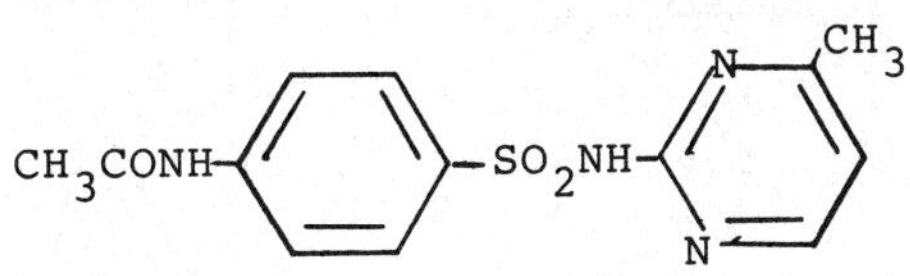

[127-73-1] M.W.=306.34

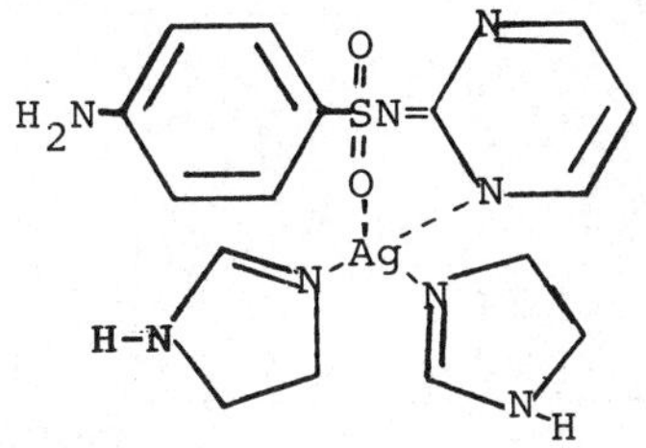

[76634-39-4] M.W.=493.29

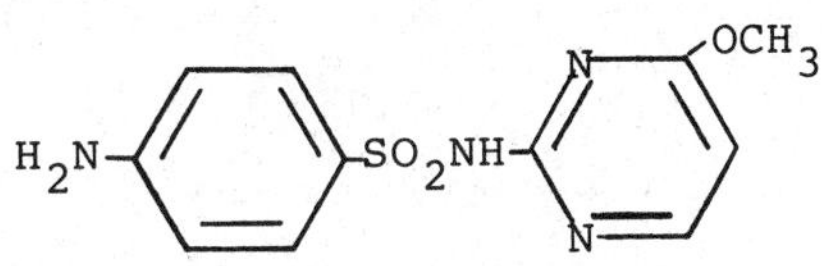

[3213-22-7] M.W.=280.31

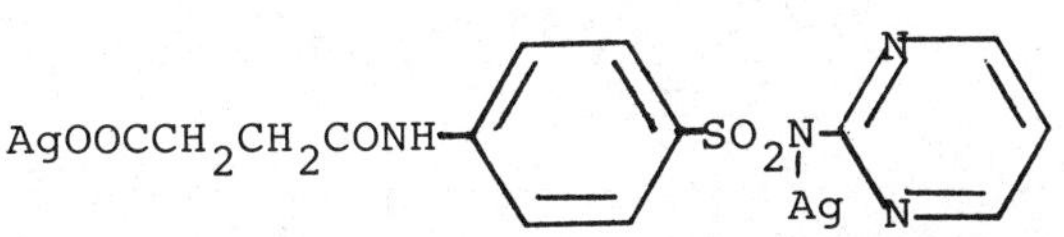

[76619-75-5] M.W.=564.07

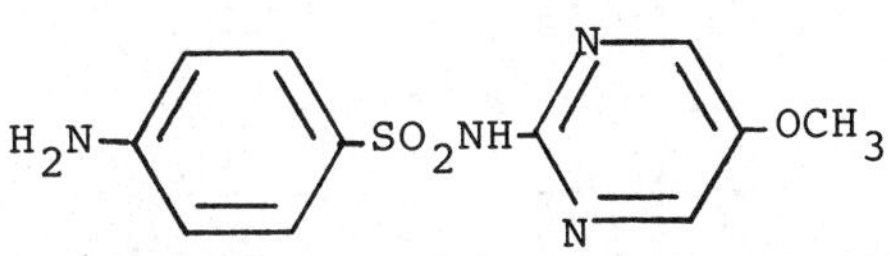

[651-06-9] M.W.=280.31

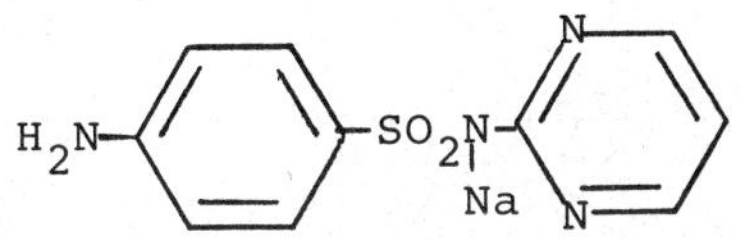

[547-32-0] M.W.=272.26

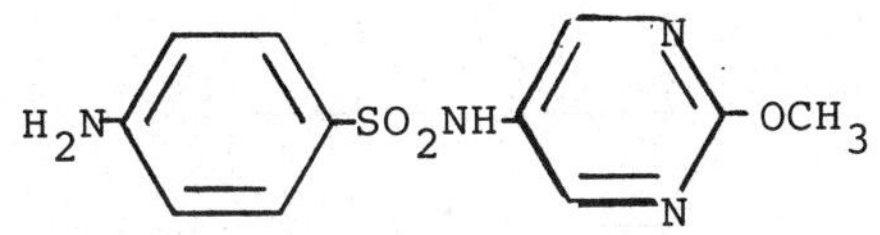

[71119-37-4] M.W.=280.31

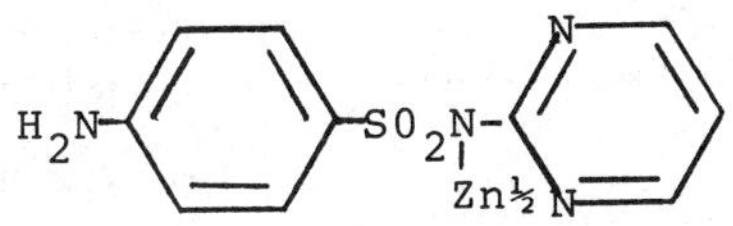

[66219-86-1] M.W.=563.92

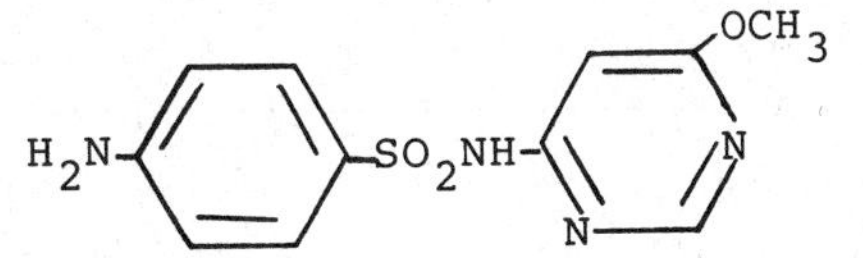

[1220-83-3] M.W.=280.31

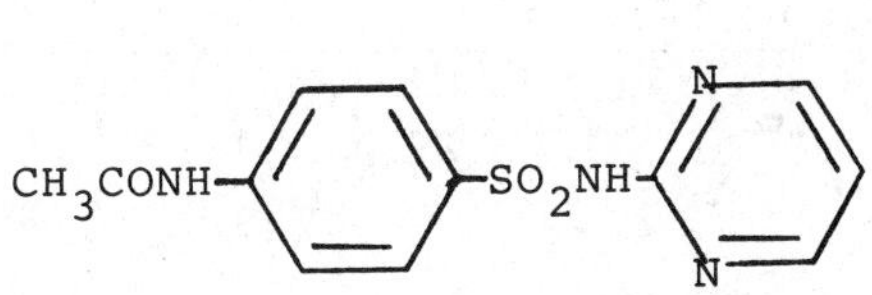

[127-74-2] M.W.=292.32

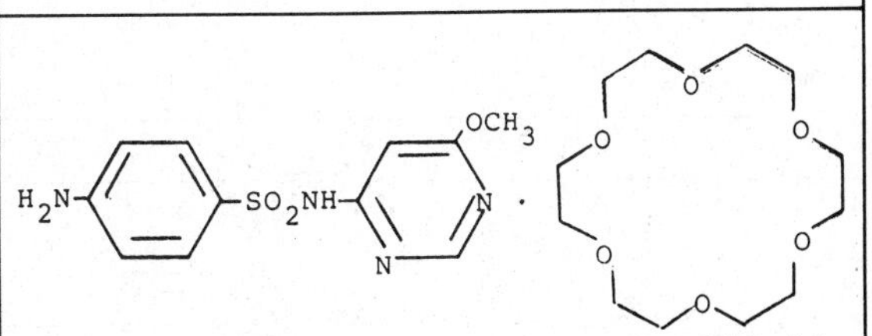

[65177-18-6] M.W.=544.62

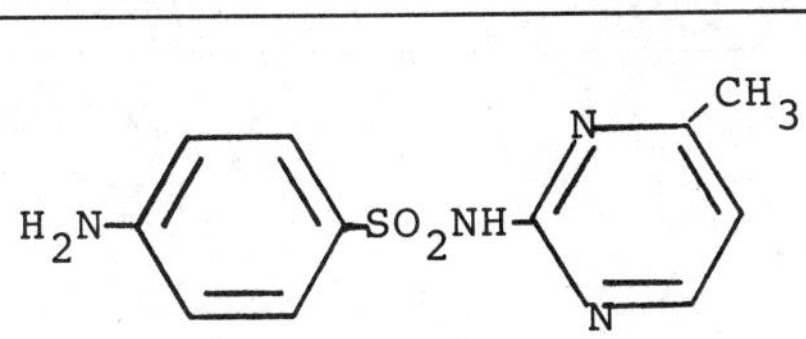

[127-79-7] M.W.=264.31

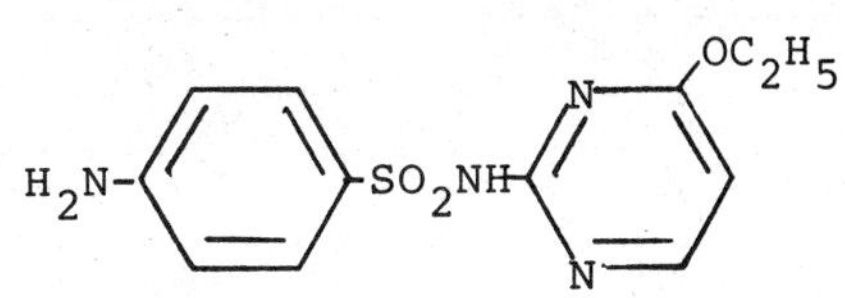

[71138-72-7] M.W.=294.34

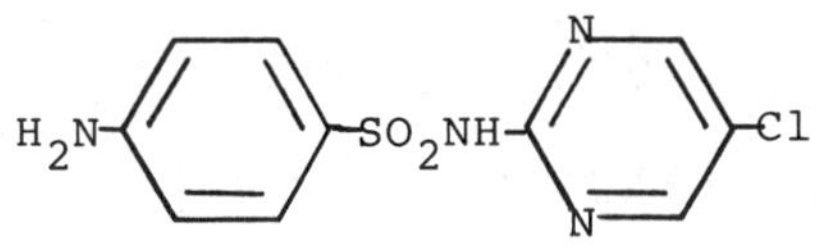

[4482-46-6] M.W.=284.72

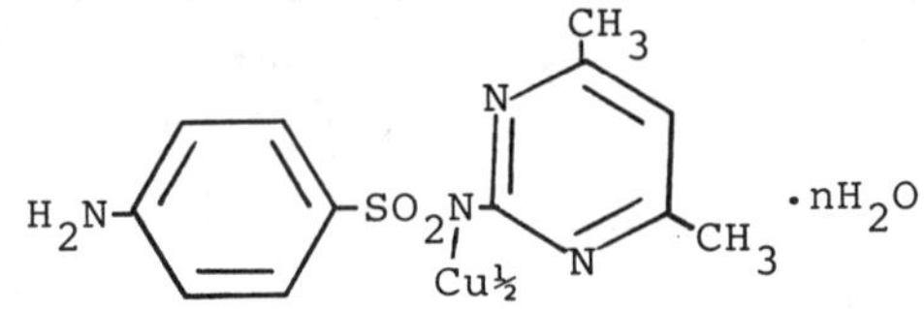

[86729-23-9] M.W.=UNSPECIFIED

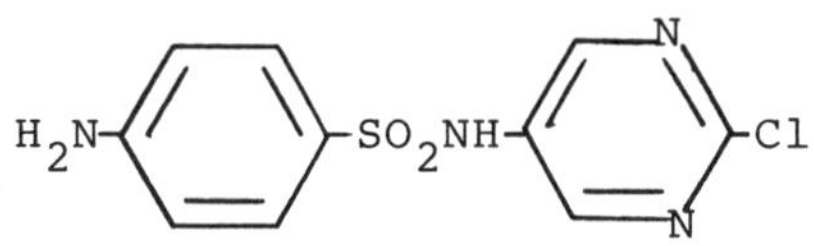

[17103-49-0] M.W.=284.72

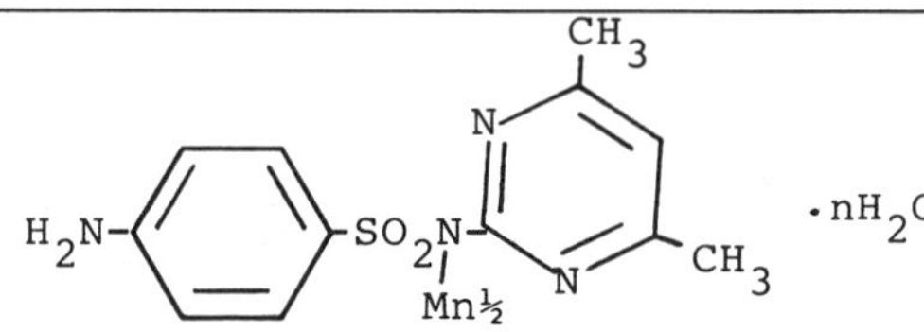

[84812-75-9] M.W.-UNSPECIFIED

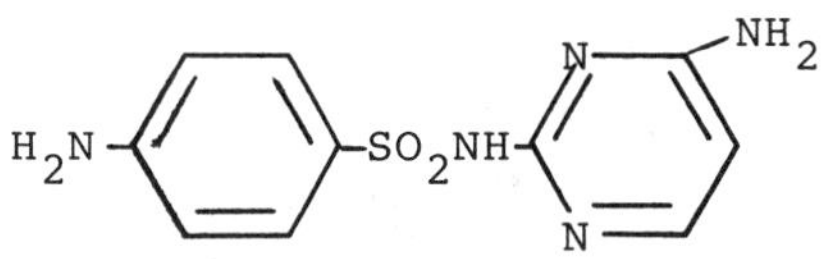

[16806-00-1] M.W.=265.29

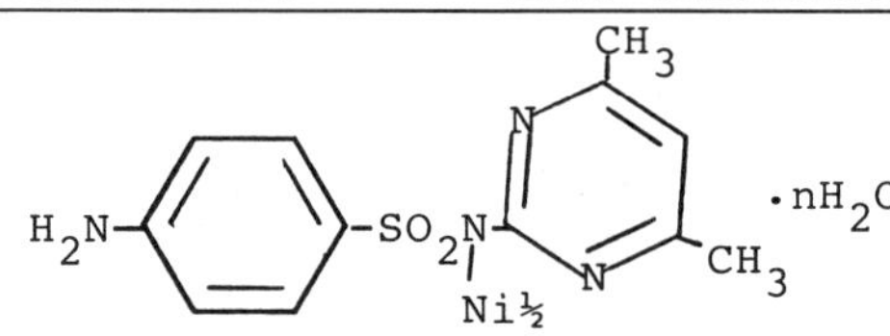

[84812-74-8] M.W.=UNSPECIFIED

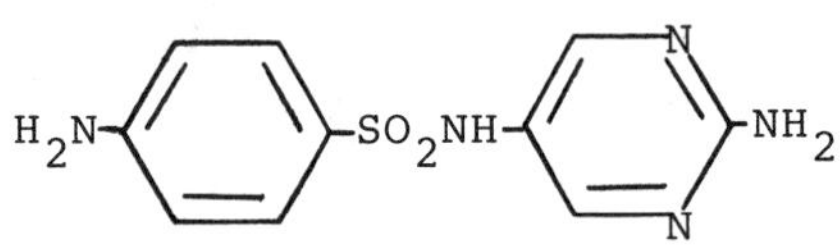

[71119-38-5] M.W.=265.29

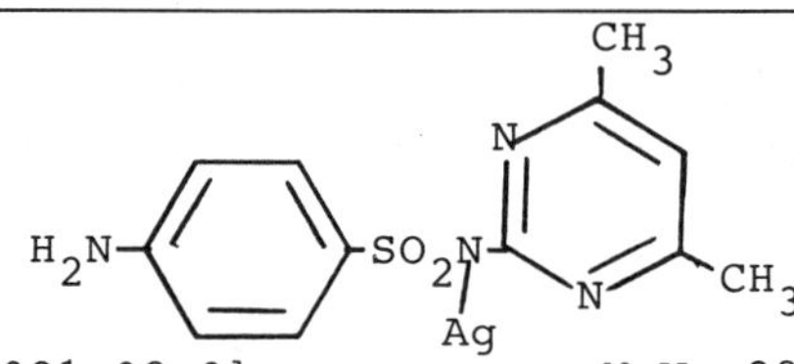

[53081-02-0] M.W.=385.19

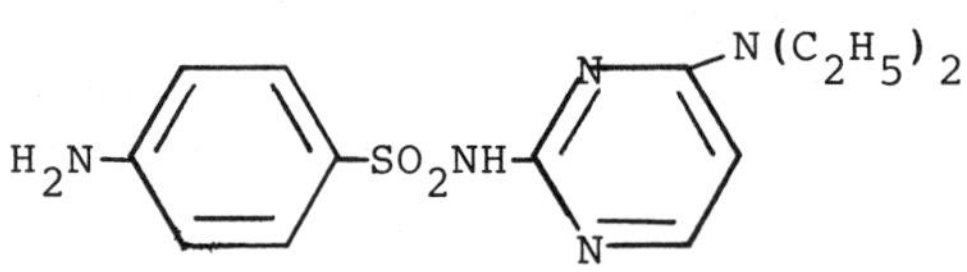

[71119-24-9] M.W.=321.40

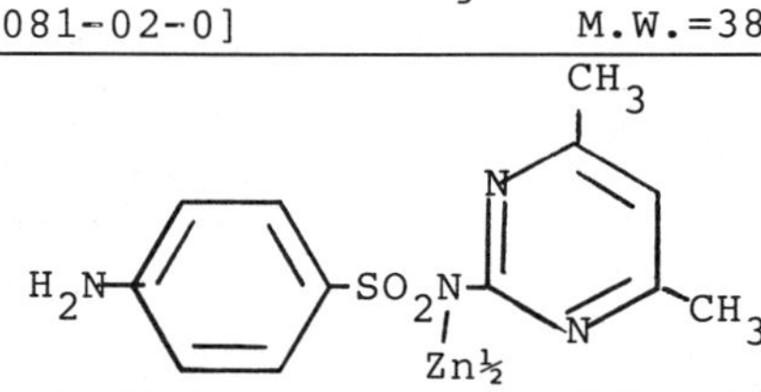

[71261-83-1] M.W.=620.02

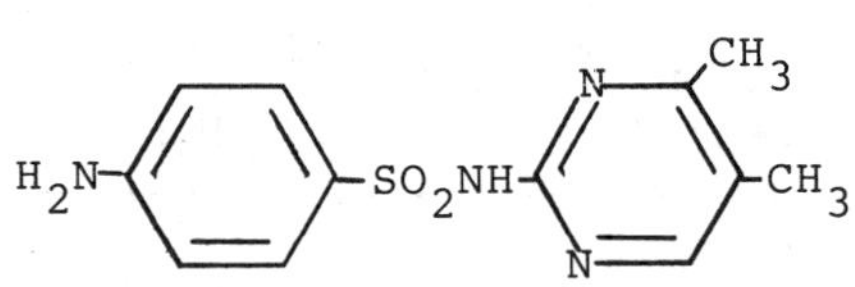

[4462-43-5] M.W.=278.33

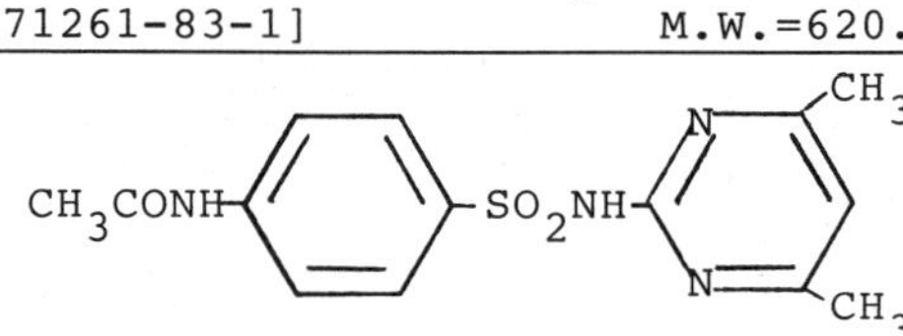

[100-90-3] M.W.=320.37

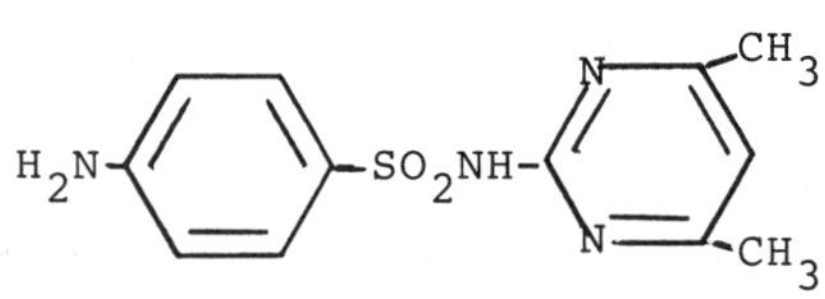

[57-68-1] M.W.=278.33

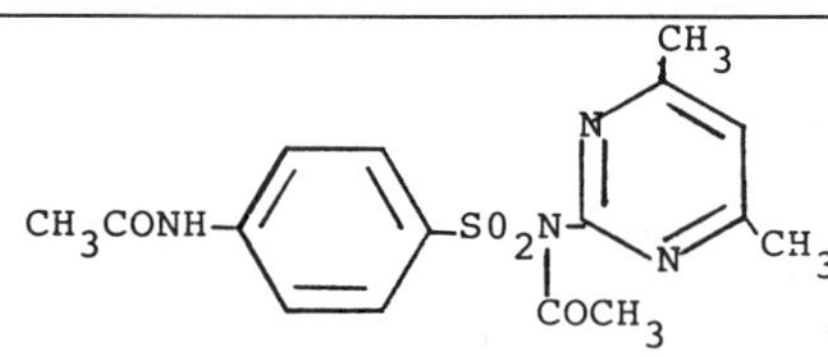

[59224-69-0] M.W.=362.41

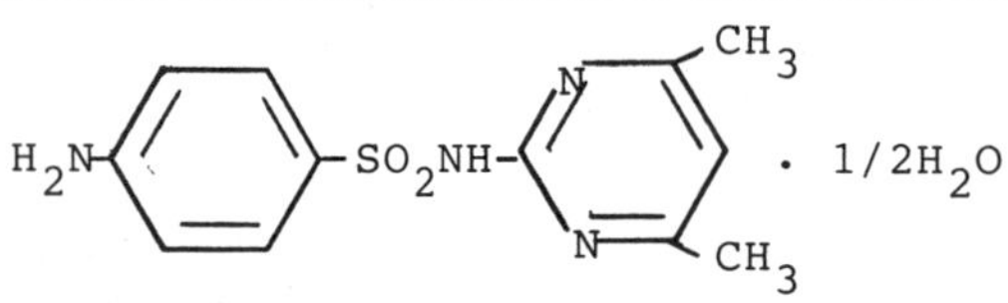

[82537-68-6] M.W.=287.34

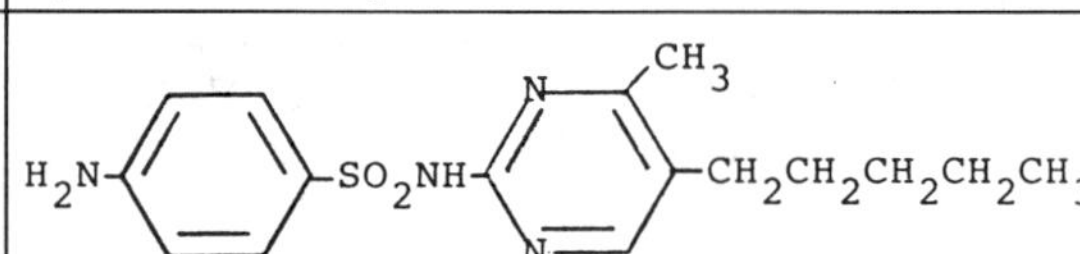

[71119-35-2] M.W.=334.44

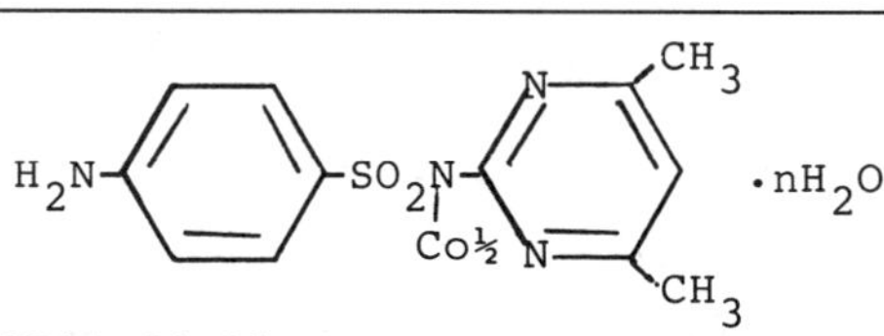

[86729-24-0] M.W.=UNSPECIFIED

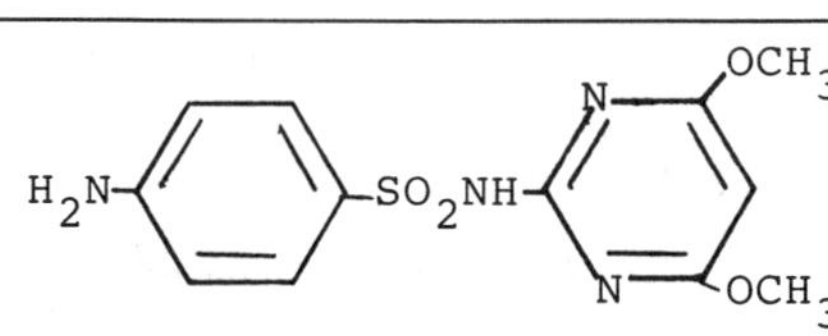

[155-91-9] M.W.=310.34

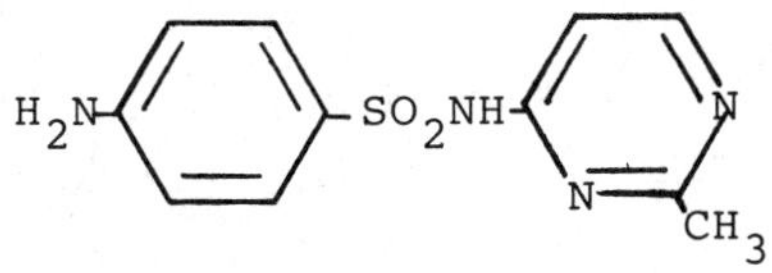 [599-84-8] M.W.=264.31	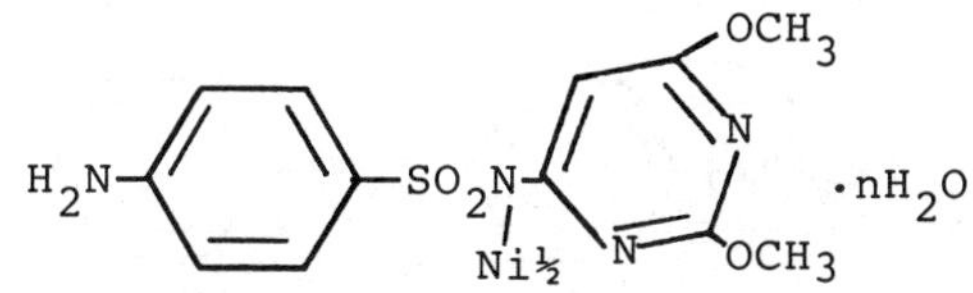 [84812-79-3] M.W.=UNSPECIFIED
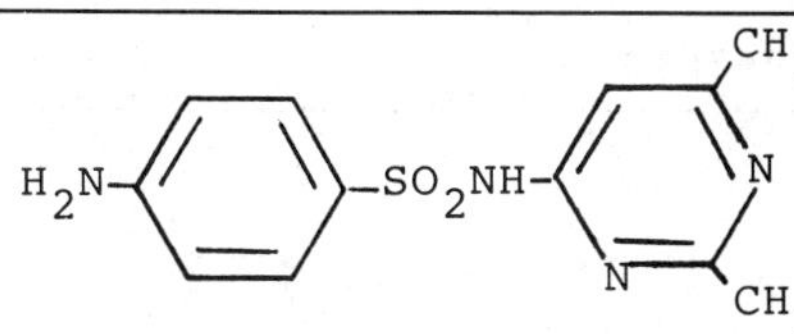 [515-64-0] M.W.=278.33	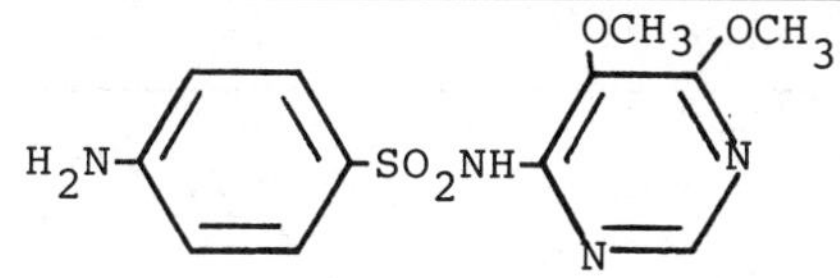 [2447-57-6] M.W.=310.34
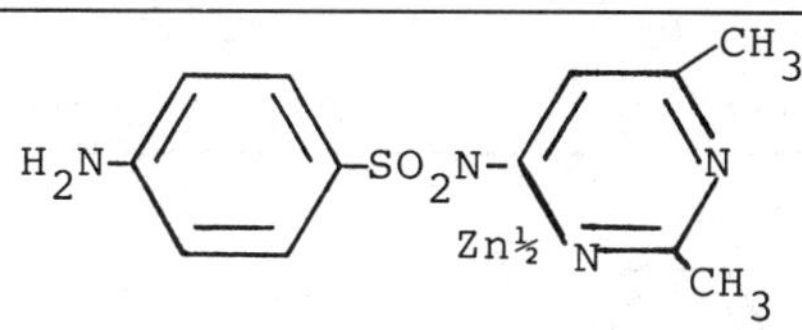 [71261-88-6] M.W.=620.02	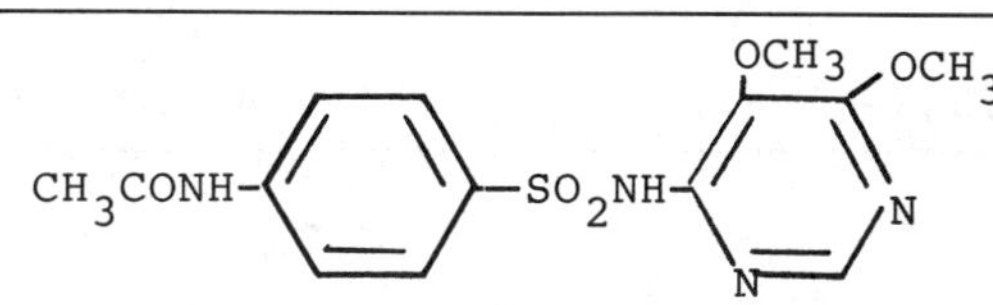 [5018-54-2] M.W.=352.38
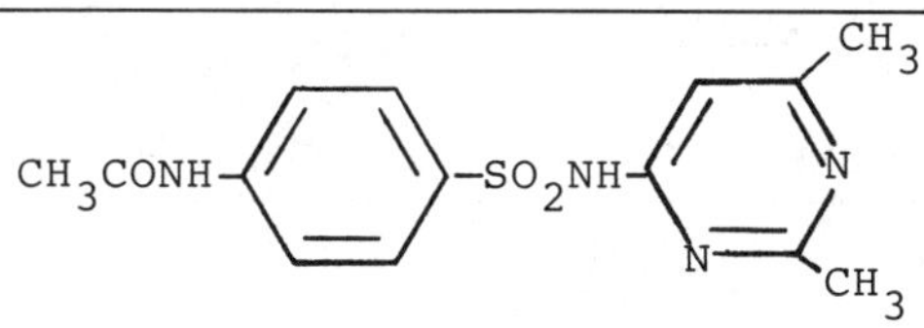 [3136-31-3] M.W.=320.37	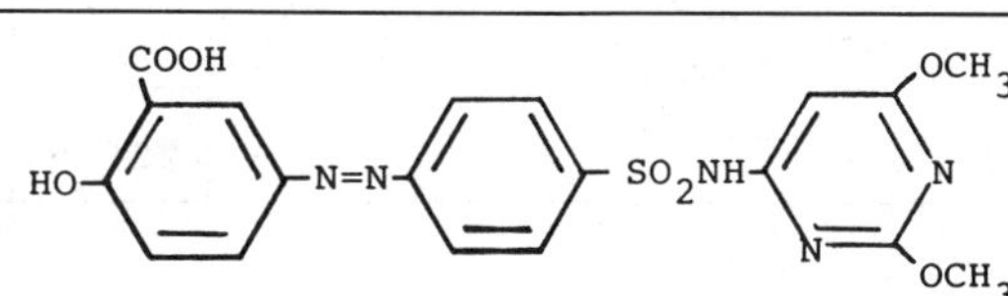 [40016-88-4] M.W.=459.44
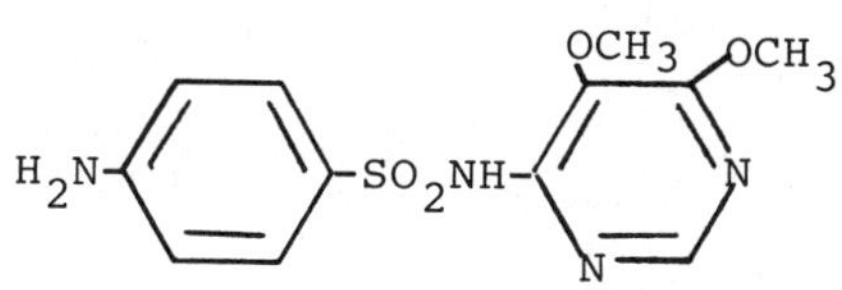 [122-11-2] M.W.=310.34	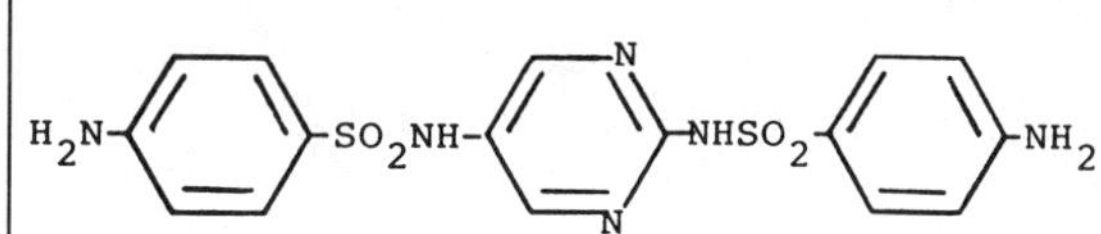 [71119-39-6] M.W.=420.47
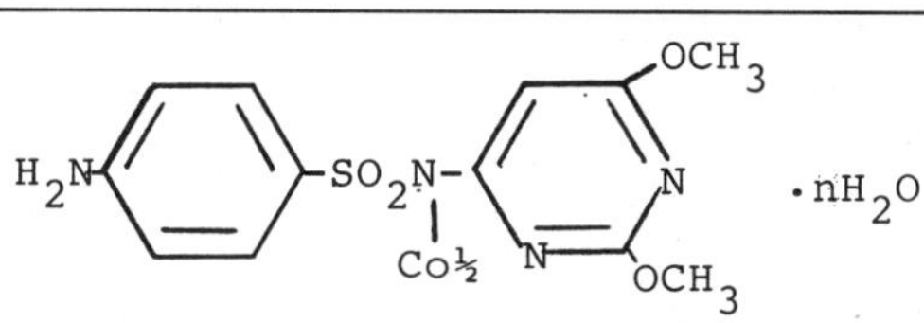 [86729-20-6] M.W.=UNSPECIFIED	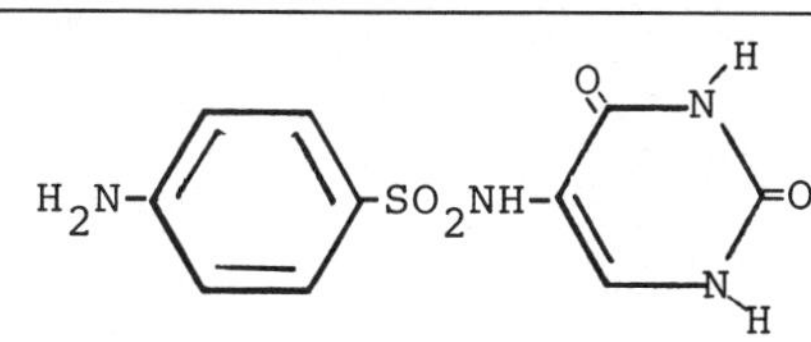 [6912-98-7] M.W.=282.28
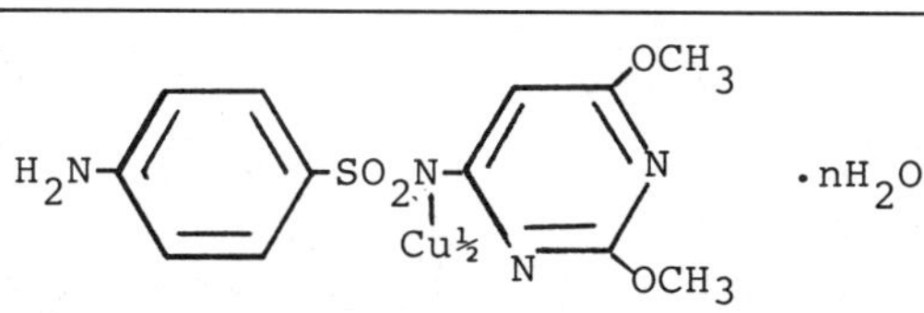 [86729-19-3] M.W.=UNSPECIFIED	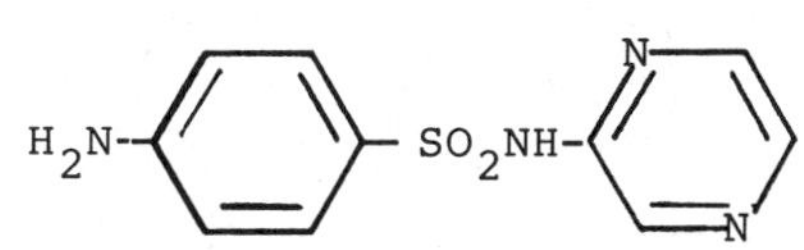 [116-44-9] M.W.=250.28
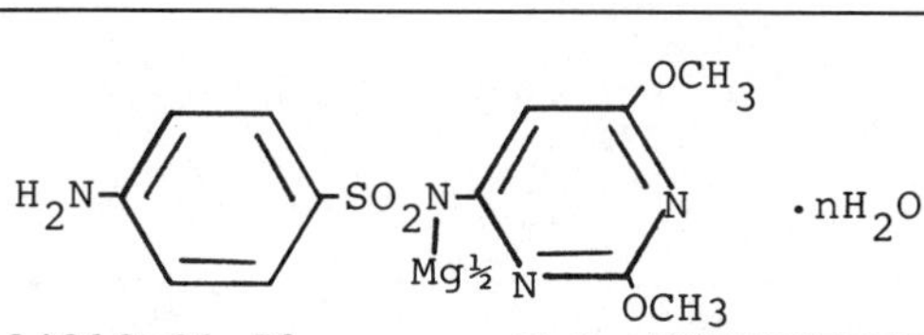 [84812-81-7] M.W.=UNSPECIFIED	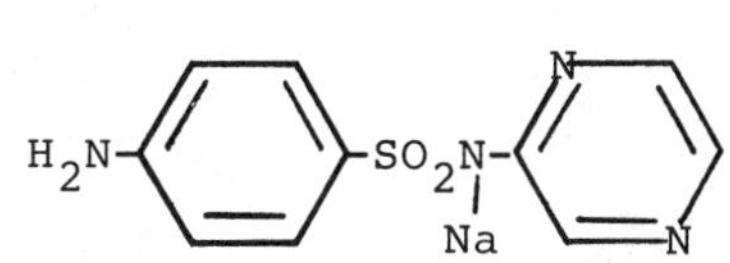 [547-31-9] M.W.=272.26
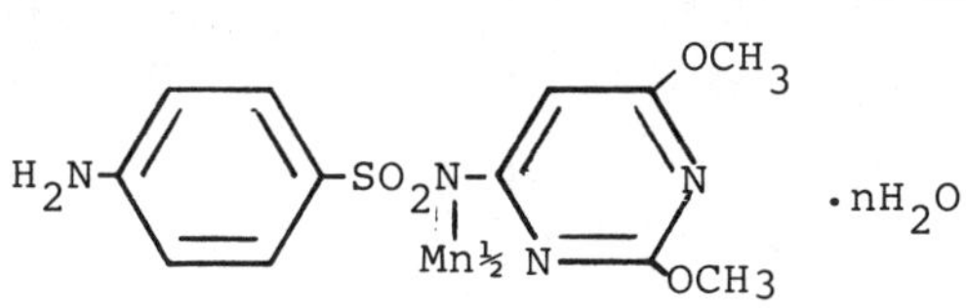 [84812-80-6] M.W.=UNSPECIFIED	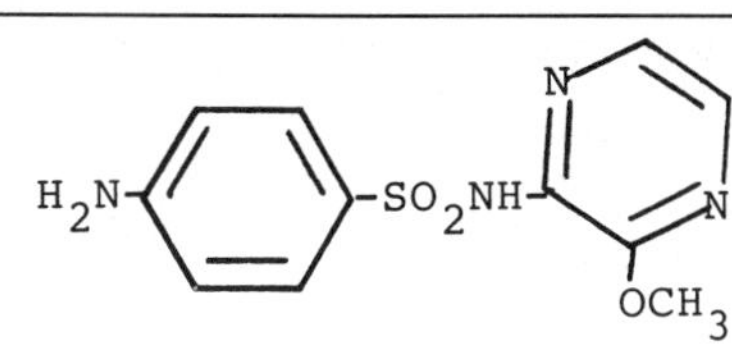 [152-47-6] M.W.=280.31

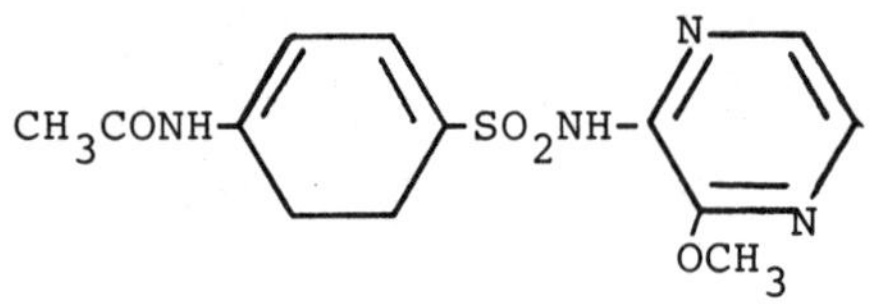

[655-78-7] M.W.=322.35

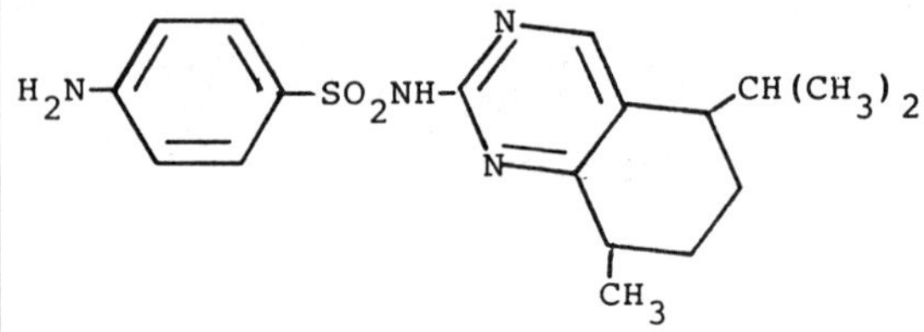

[71119-36-3] M.W.=360.47

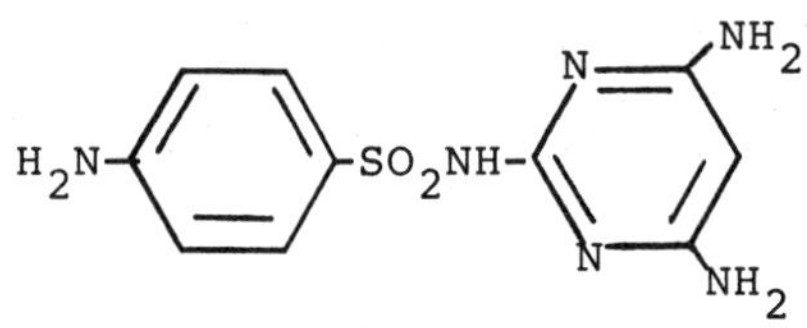

[51249-11-7] M.W.=281.29

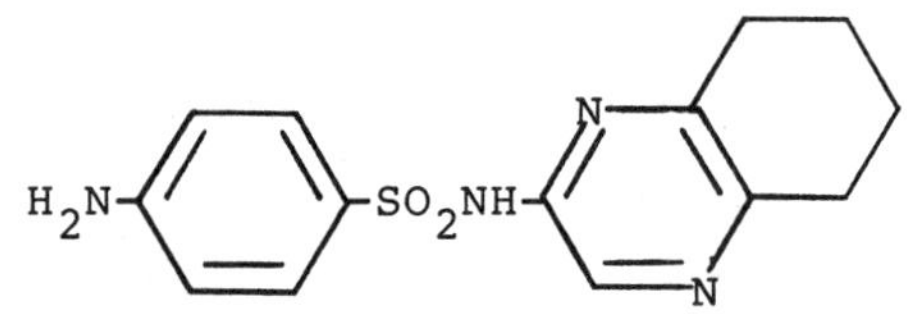

[71119-34-1] M.W.=304.37

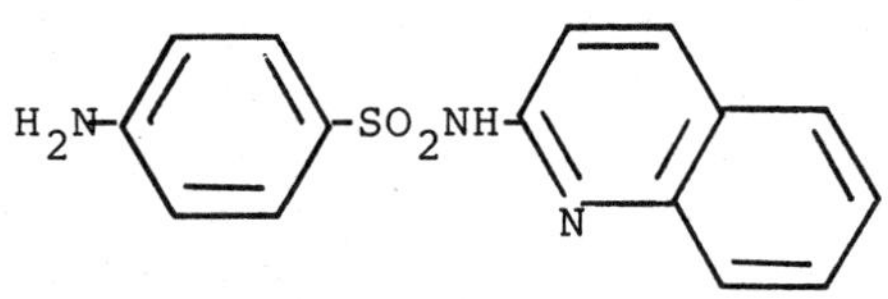

[59-40-5] M.W.=286.32

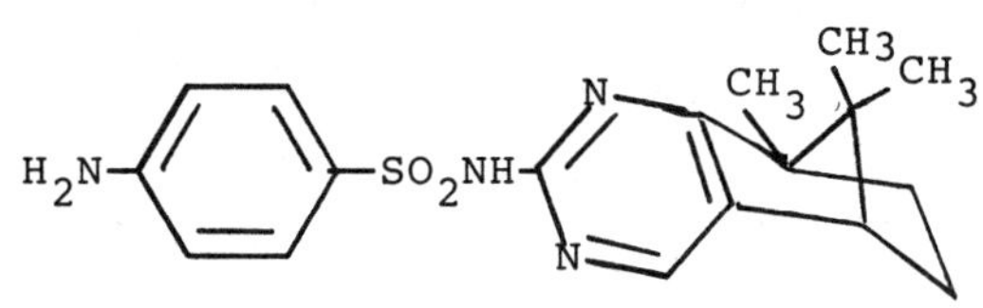

[71720-66-6] M.W.=358.46

COMPONENTS:

(1) Benzenesulfonamide, 4-amino-N-4-[(aminosulfonyl)phenyl]-; (disulfan) $C_{12}H_{13}N_3O_4S_2$; [547-52-4]

(2) Water; H_2O; [7732-18-5]

ORIGINAL MEASUREMENTS:

Becher, R. ; Leya, S. *Experientia* 1946, *2*, 459-60.

VARIABLES:

One temperature: 18-19°C

PREPARED BY:

R. Piekos

EXPERIMENTAL VALUES:

Solubility of 4-amino-N-4-[(aminosulfonyl)phenyl]benzenesulfonamide in water at room temperature (18-19°C) is 30 mg% (9.2×10^{-4} mol dm^{-3}, compiler).

AUXILIARY INFORMATION

METHOD/APPARATUS/PROCEDURE:

After standing for more than two days the soln of the sulfonamide in water was filtered and the sulfonamide was assayed in the filtrate colorimetrically by the method of Druey and Oesterheld (1).

SOURCE AND PURITY OF MATERIALS:

Nothing specified.

ESTIMATED ERROR:

Nothing specified.

REFERENCES:

1. Druey, J.; Oesterheld, G. *Helv. Chim. Acta.* 1942, *25*, 753.

COMPONENTS:

(1) Benzenesulfonamide, 4-amino-N-4-[(aminosulfonyl)phenyl]-;(disulfan) $C_{12}H_{13}N_3O_4S_2$; [547-52-4]

(2) Sodium chloride; NaCl; [7647-14-5]

(3) Water; H_2O; [7732-18-5]

ORIGINAL MEASUREMENTS:

Becher, R.; Leya, S. *Experientia* 1946, *2*, 459-60.

VARIABLES:

One temperature: 18-19°C

PREPARED BY:

R. Piekos

EXPERIMENTAL VALUES:

Solubility of 4-amino-N-4-[(aminosulfonyl)phenyl]benzenesulfonamide in a 5% NaCl solution at room temperature (18-19°C) is 28 mg% (8.6×10^{-4} mol dm^{-3}, compiler).

AUXILIARY INFORMATION

METHOD/APPARATUS/PROCEDURE:

After standing for more than two days the soln of the sulfonamide was filtered and the sulfonamide was assayed in the filtrate colorimetrically by the method of Druey and Oesterheld (1).

SOURCE AND PURITY OF MATERIALS:

Nothing specified.

ESTIMATED ERROR:

Nothing specified.

REFERENCES:

1. Druey, J.; Oesterheld, G. *Helv. Chim. Acta* 1942, *25*, 753.

COMPONENTS:

(1) Benzenesulfonamide, 4-amino-N-4-[(aminosulfonyl)phenyl]- (disulfan) ; $C_{12}H_{13}N_3O_4S_2$; [547-52-4]

(2) Pectin; $(C_{13}H_{18}O_{12})_n$; [9000-69-5]

(3) Water; H_2O; [7732-18-5]

ORIGINAL MEASUREMENTS:

Becher, R.; Leya, S. *Experientia* 1946, *2*, 459-60.

VARIABLES:

One temperature: 18-19°C

PREPARED BY:

R. Piekos

EXPERIMENTAL VALUES:

Solubility of disulfan in a 2.5% pectin solutiion ([pectin] = 6.8 x 10^{-2} mol kg^{-1}, compiler), of pH about 2.6 at room temperature (18-19°C) is 41 mg% (1.3 x 10^{-3} mol dm^{-3}, compiler).

AUXILIARY INFORMATION

METHOD/APPARATUS/PROCEDURE:

The soln was allowed to stand at room temp for more than 2 days. The soln was then filtered, and disulfan assayed in the filtrate colorimetrically by the method of Druey and Oesterheld (1).

SOURCE AND PURITY OF MATERIALS:

A high quality apple pectin was used: the rel viscosity of a 0.5% soln was 6.2, and for neutralization of 1 g of the pectin, 1.67 cm^3 of a 1 mol dm^{-3} NaOH soln was used. The source and purity of disulfan and water were not specified.

ESTIMATED ERROR:

Nothing specified.

REFERENCES:

1. Druey, J.; Oesterheld, G. *Helv. Chim. Acta* 1942, *25*, 753.

COMPONENTS:	ORIGINAL MEASUREMENTS:
(1) Benzenesulfonamide, 4-amino-N-4-[(aminosulfonyl)phenyl]-, (disulfan); $C_{12}H_{13}N_3O_4S_2$; [547-52-4] (2) Pectinic acid, sodium salt; $(C_{13}H_{17}NaO_{12})_n$; [9049-37-0] (3) Water; H_2O; [7732-18-5]	Becher, R.; Leya, S. *Experientia* <u>1946</u>, *2*, 459-60.
VARIABLES:	**PREPARED BY:**
One temperature: 18-19°C	R. Piekos

EXPERIMENTAL VALUES:

Solubility of disulfan in a 2.6% neutral sodium pectinate solution ([sodium pectinate] = 6.7×10^{-2} mol kg^{-1} (n = 1) - compiler) at room temperature (18-19°C) is 41 mg% (1.3×10^{-3} mol dm^{-3} - compiler).

AUXILIARY INFORMATION

METHOD/APPARATUS/PROCEDURE:

The soln was allowed to stand for more than 2 days at room temp. The soln was then filtered, and sulfonamide assayed in the filtrate colorimetrically by the method of Druey and Oesterheld (1).

SOURCE AND PURITY OF MATERIALS:

Nothing specified.

ESTIMATED ERROR:

Nothing specified.

REFERENCES:

1. Druey, J.; Oesterheld, G. *Helv. Chim. Acta* <u>1942</u>, *25*, 753.

COMPONENTS:

(1) Benzenesulfonamide, 4-amino-N-4-[(aminosulfonyl)phenyl]-; (disulfan) $C_{12}H_{13}N_3O_4S_2$; [547-52-4]

(2) D-Glucose; $C_6H_{12}O_6$; [50-99-7]

(3) Water; H_2O; [7732-18-5]

ORIGINAL MEASUREMENTS:

Becher, R.; Leya, S. *Experientia* 1946, *2*, 459-60.

VARIABLES:

One temperature: 18-19°C

PREPARED BY:

R. Piekos

EXPERIMENTAL VALUES:

Solubility of 4-amino-N-4-[(aminosulfonyl)phenyl]benzenesulfonamide in a 10% D-glucose solution at room temperature (18-19°C) is 30 mg% (9.2 x 10^{-4} mol dm^{-3}, compiler).

AUXILIARY INFORMATION

METHOD/APPARATUS/PROCEDURE:

After standing for more than two days the soln of 4-amino-N-4-[(aminosulfonyl)phenyl]-benzenesulfonamide was filtered and the sulfonamide was assayed in the filtrate colorimetrically by the method of Druey and Oesterheld (1).

SOURCE AND PURITY OF MATERIALS:

Nothing specified.

ESTIMATED ERROR:

Nothing specified.

REFERENCES:

1. Druey, J.; Oesterheld, G. *Helv. Chim. Acta* 1942, *25*, 753.

COMPONENTS:	ORIGINAL MEASUREMENTS:
(1) Benzenesulfonamide, 4-amino-N-4-[(aminosulfonyl)phenyl]- (disulfan); $C_{12}H_{13}N_3O_4S_2$; [547-52-4] (2) 2-Propanone (acetone); C_3H_6O; [67-64-1]	Gutierrez, F. H. *Anales fis. quim. (Madrid)* 1945, *41*, 537-60.
VARIABLES:	PREPARED BY:
Temperature	R. Piekos

EXPERIMENTAL VALUES:

t/°C	G[a]	E[b]	X_g/l[c]	mol/l[d] acetone	mmol/mol acetone	1:X_g^e	1+X_{cc}^f
0	49.983	33.325	407.161	1243	88.7	2.00	2.46
5	50.500	33.554	408.444	1248	89.6	1.98	2.45
10	50.998	33.772	409.514	1251	90.5	1.96	2.44
15	52.271	34.327	416.652	1272	92.7	1.91	2.40
20	53.312	34.735	421.805	1288	94.6	1.88	2.37
25	54.024	35.076	424.196	1295	95.8	1.85	2.36
30	55.934	35.742	435.894	1331	99.2	1.81	2.29
35	57.800	36.629	446.967	1365	102.5	1.73	2.24
40	60.198	37.577	461.959	1411	106.8	1.66	2.16
45	64.817	39.326	493.517	1507	114.9	1.54	2.03
50	69.244	40.914	523.139	1597	122.8	1.47	1.91

[a] $G = \frac{p\ 100}{P - p}$, where p and P are the weights of solute and solution, resp.

[b] $E = \frac{G\ 100}{G + 100}$; [c] g/l acetone; [d] should be mmol/l acetone (compiler);

[e] g of acetone required to dissolve 1 g of solute; [f] volume (cm^3) of acetone required to dissolve 1 g of solute.

AUXILIARY INFORMATION

METHOD/APPARATUS/PROCEDURE:

A special all-glass app was contructed enabling the prepn of satd solns, agitation by bubbling a stream of acetone-satd N, filtration, and distn off the solvent without contact with air. Two exchangeable dissoln vessels of 15 and 8 cm^3 working capacity were used depending on the soly of solute. The app was immersed in a thermostat. The vols of acetone used were 15 or 5 cm^3, and the equilibration time was 2-2.5 h. The satd solns were filtered, weighed and the solvent was distd off, the residues were dried at 105 °C, weighed, examd for the presence of solvated acetone.

SOURCE AND PURITY OF MATERIALS:

The source of the materials was not specified. Pure, anhyd acetone was used. The absence of impurities and water was confirmed by procedures of the German Pharmacopeia VI and Spanish Pharmacopeia VIII.
The purity of disulfan was not specified.

ESTIMATED ERROR:

Soly: measurements were repeated until 2 values not differing in the second decimal were obtained (authors).

Temp: ±0.1°C (authors).

REFERENCES:

COMPONENTS:

(1) Benzenesulfonamide, 4-amino-N-4-[(aminosulfonyl)phenyl]-, monohydrochloride (disulfan-HCl); $C_{12}H_{13}N_3O_4S_2 \cdot HCl$; [77400-69-2]

(2) 2-Propanone (acetone); C_3H_6O; [67-64-1]

ORIGINAL MEASUREMENTS:

Gutierrez, F. H.
Anales fis. quim. (Madrid) 1945, *41*, 537-60.

VARIABLES:

Temperature

PREPARED BY:

R. Piekos

EXPERIMENTAL VALUES:

t/°C	G^a	E^b	$X_g/1^c$	mol/1[d] acetone	mmol/mol acetone	$1{:}X_g^e$	$1 + X_{cc}^f$
15	0.406	0.404	3.236	8.89	0.65	246.31	309.02
20	0.420	0.418	3.323	9.13	0.67	238.09	300.93
25	0.433	0.431	3.400	9.34	0.69	230.95	294.12

[a] $G = \frac{p\ 100}{P - p}$, where p and P are the weights of solute and solution, resp.

[b] $E = \frac{G\ 100}{G + 100}$; [c] g/1 acetone; [d] should be mmol/1 acetone (compiler);

[e] g of acetone required to dissolve 1 g of solute; [f] volume (cm^3) of acetone required to dissolve 1 g of solute.

AUXILIARY INFORMATION

METHOD/APPARATUS/PROCEDURE:

A special all-glass app was constructed enabling the prepn of satd solns, agitation by bubbling a stream of acetone-satd N, filtration, and distn off the solvent without contact with air. Two exchangeable dissoln vessels of 15 and 8 cm^3 working capacity were used depending on the soly of solute. The app was immersed in a thermostat. The vols of acetone used were 15 or 5 cm^3, and the equilibration time was 2-2.5 h. the satd solns were filtered, weighed, the solvent was distd off, the residues were dried at 105°C, weighed, and examd for the presence of solvated acetone.

SOURCE AND PURITY OF MATERIALS:

The source of the materials was not specified. Pure, anhyd acetone was used. The absence of impurities and water in it was confirmed by procedures of the German Pharmacopeia VI and Spanish Pharmacopeia VIII.
The purity of disulfan-HCl was not specified.

ESTIMATED ERROR:

Soly: measurements were repeated until 2 values not differing in the second decimal were obtained (author).
Temp: ±0.1°C (author).

REFERENCES:

COMPONENTS:	ORIGINAL MEASUREMENTS:
(1) Benzenesulfonamide, 4-amino-N-4-[(aminosulfonyl)phenyl]-, monosodium salt (Na disulfan); $C_{12}H_{12}N_3NaO_4S_2$; [77400-68-1] (2) 2-Propanone (acetone); C_3H_6O; [67-64-1]	Gutierrez, F. H. *Anales fis. quim. (Madrid)* 1945, *41*, 537-60.
VARIABLES:	PREPARED BY:
Temperature	R. Piekos

EXPERIMENTAL VALUES:

$t/^{\circ}C$	G[a]	E[b]	X_g/l[c]	mol/l[d] acetone	mmol/mol acetone	$1:X_g$[e]	$1 + X_{cc}$[f]
0	0.144	0.144	1.173	3.3	0.24	694.44	852.52
10	0.161	0.161	1.293	3.7	0.26	621.12	773.39
20	0.174	0.174	1.377	3.9	0.29	574.71	726.21
30	0.191	0.190	1.488	4.2	0.32	523.56	672.04
40	0.206	0.205	1.581	4.5	0.34	485.43	632.51
50	0.220	0.219	1.655	4.7	0.36	454.54	604.24

[a] $G = \frac{p\ 100}{P - p}$, where p and P are the weights of solute and solution, resp.

[b] $E = \frac{G\ 100}{G + 100}$; [c] g/l acetone; [d] should be mmol/l acetone (compiler);

[e] g of acetone required to dissolve 1 g of solute; [f] volume (cm^3) of acetone required to dissolve 1 g of solute.

AUXILIARY INFORMATION

METHOD/APPARATUS/PROCEDURE:

A special all-glass app was contructed enabling the prepn of satd solns, agitation by bubbling a stream of acetone-satd N, filtration, and distn off the solvent without contact with air. Two exchangeable dissoln vessels of 15 and 8 cm^3 working capacity were used depending on the soly of solute. The app was immersed in a thermostat. The vols of acetone used were 15 or 5 cm^3, and the equilibration time was 2-2.5 h. The satd solns were filtered, weighed, the solvent was distd off, the residues were dried at 105°C, weighed, and examd for the presence of solvated acetone.

SOURCE AND PURITY OF MATERIALS:

The source of the materials was not specified. Pure, anhyd acetone was used. The absence of impurities and water was confirmed by procedures of the German Pharmacopeia VI and Spanish Pharmacopeia VIII. The purity of Na disulfan was not specified.

ESTIMATED ERROR:

Soly: measurements were repeated until 2 values not differing in the second decimal were obtained (authors).
Temp: ±0.1°C (authors).

REFERENCES:

COMPONENTS:

(1) Benzenesulfonamide, 4-amino-N-[4-(aminosulfonyl)phenyl]-,monosodium salt, monohydrate (Na disulfan monohydrate); $C_{12}H_{12}N_3NaO_4S_2 \cdot H_2O$; [81815-35-2]

(2) Water; H_2O; [7732-18-5]

ORIGINAL MEASUREMENTS:

Crossley, M. L.; Northey, E. H.; Hultquist, M. E.

J. Am. Chem. Soc. 1938, *60*, 2222-4.

VARIABLES:

Temperature

PREPARED BY:

R. Piekos

EXPERIMENTAL VALUES:

t/°C	Solubility g/100 cm³	Solubility mol dm⁻³ [a]
10	9.6	0.26
37	20.0	0.54

[a] Calculated by compiler

AUXILIARY INFORMATION

METHOD/APPARATUS/PROCEDURE:

Nothing specified.

SOURCE AND PURITY OF MATERIALS:

The Na disulfan monohydrate was prepd by the authors and purified by recrystn from concd aq soln with use of activated charcoal. Analysis - calcd 6.28% Na, found 6.35% Na. Assay by nitrite 100.2% Purity of the water was not specified.

ESTIMATED ERROR:

Nothing specified.

REFERENCES:

COMPONENTS:	ORIGINAL MEASUREMENTS:
(1) Acetamide, [4-[[[4-(aminosulfonyl)phenyl]-amino]sulfonyl]phenyl]- (acetyl disulfanilamide); $C_{14}H_{14}N_3O_5S_2$; [56444-82-7] (2) 2-Propanone (acetone); C_3H_6O; [67-64-1]	Gutierrez, F. H. *Anales fis. quim. (Madrid)* 1945, *41*, 537-60.
VARIABLES:	PREPARED BY:
Temperature	R. Piekos

EXPERIMENTAL VALUES:

t/°C	G[a]	E[b]	X_g/1[c]	mol/1[d] acetone	mmol/mol acetone	1:X_g[e]	1 + X_{cc}[f]
0	1.271	1.255	10.354	28.0	2.0	78.52	96.58
5	1.402	1.382	11.339	30.7	2.2	71.32	88.19
10	1.608	1.563	12.912	34.9	2.5	62.19	77.45
15	1.687	1.659	13.457	36.2	2.7	59.28	74.32
20	1.749	1.719	13.838	37.5	2.8	55.75	72.26
25	1.984	1.945	15.578	42.2	3.1	50.40	64.19
30	2.234	2.185	17.410	47.1	3.5	44.76	57.44
35	2.515	2.453	19.448	52.6	3.9	39.76	51.36
40	2.854	2.775	21.902	59.3	4.5	35.03	45.65
45	3.250	3.138	24.746	66.9	5.1	30.77	40.41
50	3.679	3.548	27.795	75.2	5.8	25.55	39.13

[a] $G = \frac{p\ 100}{P - p}$; where p and P are the weights of solute and solution, resp.

[b] $E = \frac{G\ 100}{G + 100}$; [c] g/1 acetone; [d] should be mmol/1 (compiler).

[e] g of acetone required to dissolve 1 g of solute; [f] volume (cm^3) of acetone required to dissolve 1 g of solute.

AUXILIARY INFORMATION

METHOD/APPARATUS/PROCEDURE:

A special all-glass app was constructed enabling the prepn of satd solns, agitation by bubbling a stream of acetone-satd N, filtration, and distn off the solvent without contact with air. Two exchangeable dissoln vessels of 15 and 8 cm^3 working capacity were used depending on the soly of solute. The app was immersed in a thermostat. The vols of acetone used were 15 or 5 cm^3, and the equilibration time was 2-2.5 h. The satd solns were filtered, weighed, the solvent was distd off, the residues were dried at 105°C, weighed, and examd for the presence of solvated acetone.

SOURCE AND PURITY OF MATERIALS:

The source of the materials was not specified. Pure, anhyd acetone was used. The absence of impurities and water was confirmed by procedures of the German Pharmacopeia VI and Spanish Pharmacopeia VIII.

The purity of the solute was not specified.

ESTIMATED ERROR:

Soly: measurements were repeated until 2 values not differing in the second decimal were obtained (author).

Temp: ±0.1°C (author).

REFERENCES:

COMPONENTS:

(1) Benzenesulfonamide, 4-amino-N-[4-[(methylamino)sulfonyl]phenyl]- (Neo-uliron); $C_{13}H_{15}N_3O_4S_2$; [547-53-5]

(2) Phosphoric acid, monopotassium salt; KH_2PO_4; [7778-77-0]

(3) Water; H_2O; [7732-18-5]

ORIGINAL MEASUREMENTS:

Krüger- Thiemer, E.

Arch. Dermatol. Syphilis 1942, *183*, 90-116.

VARIABLES:

One temperature: ca 20°C; one pH: 4.37

PREPARED BY:

R. Piekos

EXPERIMENTAL VALUES:

Solubility of Neo-uliron in a 0.735M (10%) KH_2PO_4 solution of pH 4.37, at room temperature (about 20°C), is 0.0157 g% (4.60×10^{-4} mol dm^{-3} solution, compiler).

AUXILIARY INFORMATION

METHOD/APPARATUS/PROCEDURE:

Neo-uliron (0.5 g) was dissolved in 10 cm^3 of the 0.735N (10%) KH_2PO_4 soln, shaken for 2 h, and filtered. A 1-cm^3 aliquot of the filtrate was then withdrawn, cooled, acidified with 2N HCl, and the sulfonamide content was detd colorimetrically by the method of Marshall modified by Kimmig (1) using an Authenrieth colorimeter. The pH was measured on an ultraionograph using a glass electrode.

SOURCE AND PURITY OF MATERIALS:

Neo-uliron was the product manufd by "Bayer". The source and purity of the remaining materials were not specified.

ESTIMATED ERROR:

Soly: precision ±5% (author).
Temp: not specified.
pH : ±0.05 pH unit (author).

REFERENCES:

1. Kimmig, J. *Arch. Dermatol.* 1938, *176*, 722; *Erg. Hyg.* 1941, *24*, 398.

COMPONENTS:	ORIGINAL MEASUREMENTS:
(1) Benzenesulfonamide, 4-amino-[4-[(methylamino)sulfonyl]phenyl]- (Neo-uliron); $C_{13}H_{15}N_3O_4S_2$; [547-53-5] (2) Phosphoric acid, disodium salt; Na_2HPO_4; [7558-94-4] (3) Water; H_2O; [7732-18-5]	Krüger-Thiemer, E. *Arch. Dermatol. Syphilis* 1942, *183*, 90-116.
VARIABLES:	PREPARED BY:
One temperature: ca 20°C; one pH: 8.74	R. Piekos

EXPERIMENTAL VALUES:

Solubility of Neo-uliron in a 0.705M (10%) Na_2HPO_4 solution of pH 8.74, at room temperature (about 20°C) is 0.130 g% (3.81×10^{-3} mol dm^{-3} solution, compiler).

AUXILIARY INFORMATION

METHOD/APPARATUS/PROCEDURE:

Neo-uliron (0.5 g) was dissolved in 10 cm^3 of the 0.705M (10%) Na_2HPO_4 soln, shaken for 2 h, and filtered. A 1-cm^3 aliquot of the filtrate was then withdrawn, cooled, acidified with 2N HCl, and the sulfonamide content was detd colorimetrically by the method of Marshall modified by Kimmig (1) using an Authenrieth colorimeter. The pH was measured on an ultraionograph using a glass electrode.

SOURCE AND PURITY OF MATERIALS:

Neo-uliron was the product manufd by "Bayer". The source and purity of the remaining materials were not specified.

ESTIMATED ERROR:

Soly: precision ±5% (author).
Temp: not specified.
pH : ±0.05 pH unit (author).

REFERENCES:

1. Kimmig, J. *Arch. Dermatol.* 1938, *176*, 722; *Erg. Hyg.* 1941, *24*, 398.

COMPONENTS:

(1) Benzenesulfonamide, 4-amino-N-[4-[(methylamino)sulfonyl]phenyl]- (Neo-uliron); $C_{13}H_{15}N_3O_4S_2$; [547-53-5]

(2) Phosphoric acid, disodium salt; Na_2HPO_4; [7558-94-4]

(3) Phosphoric acid, monopotassium salt; KH_2PO_4; [7778-77-0]

(4) Water; H_2O; [7732-18-5]

ORIGINAL MEASUREMENTS:

Krüger-Thiemer, E.

Arch. Dermatol. Syphilis 1942, *183*, 90-116.

VARIABLES:

Temperature, pH

PREPARED BY:

R. Piekos

EXPERIMENTAL VALUES:

Composition of 1/15M phosphate buffer solutions			pH	Solubility			
				Room temp (ca 20°C)		37°C	
Na_2HPO_4	KH_2PO_4	%Content		g%	10^4 mol dm^{-3} solution[a]	g%	10^4 mol dm^{-3} solution[a]
1.0	99.0	0.91	4.944	0.028	8.20	-	-
10.0	90.0	0.91	5.906	0.025	7.32	0.046	13.47
61.1	38.9	0.93	7.005	0.027	7.91	0.058	16.99
9.5	0.5	0.733[b]	7.51	0.032	9.37	-	-
94.7	5.3	0.95	8.018	0.078	22.85	-	-

[a] Calculated by compiler

[b] Molar content; 10% buffer solution

AUXILIARY INFORMATION

METHOD/APPARATUS/PROCEDURE:

Neo-uliron (0.5 g) was dissolved in 10 cm^3 of a buffer soln, shaken for 2 h at 20°C (or left for 48 h at 37°C), and filtered at respective temp. A 1-cm^3 aliquot of the filtrate was then withdrawn, cooled (dild for expts at 37°C), acidified with 1 cm^3 of 2N HCl and the sulfonamide content was detd colorimetrically by the method of Marshall modified by Kimmig (1) using an Authenrieth colorimeter. The pH was detd on an ultra-ionograph using a glass electrode.

SOURCE AND PURITY OF MATERIALS:

Neo-uliron was the product manufd by "Bayer". The source and purity of the remaining materials were not specified.

ESTIMATED ERROR:

Soly: precision: ±5% (author).
Temp: not specified.
pH : ±0.05 pH unit (author).

REFERENCES:

1. Kimmig, J. *Arch. Dermatol.* 1938, *176*, 722; *Erg. Hyg.* 1941, *24*, 398.

COMPONENTS:

(1) Benzenesulfonamide, 4-amino-N-[4-[(methylamino)sulfonyl]phenyl]-; $C_{13}H_{15}N_3O_4S_2$; [547-53-5]

(2) 2-Propanone (acetone); C_3H_6O; [67-64-1]

ORIGINAL MEASUREMENTS:

Gutierrez, F. H.

Anales fis. quim. (Madrid) 1945, *41*, 537-60.

VARIABLES:

Temperature

PREPARED BY:

R. Piekos

EXPERIMENTAL VALUES:

t/°C	G^a	E^b	$X_g/1^c$	mol/1^d acetone	mmol/mol acetone	$1{:}X_g^e$	$1 + X_{cc}^f$
0	17.565	14.940	143.084	419.1	29.80	5.79	6.98
5	18.507	15.617	149.685	438.4	31.41	5.40	6.67
10	19.309	16.185	154.530	452.6	32.84	5.17	6.47
15	20.062	16.709	159.914	468.4	34.12	4.98	6.25
20	22.107	18.104	174.911	512.3	37.60	4.52	5.72
25	23.202	18.881	182.182	533.6	39.46	4.32	5.49
30	25.474	20.302	198.507	580.5	43.33	3.92	5.04
35	28.002	21.876	216.539	634.3	47.63	3.55	4.52
40	30.735	23.509	235.887	690.9	52.28	3.25	4.24
45	37.109	27.065	282.631	827.8	63.12	2.69	3.54
50	47.626	32.261	359.814	1053.9	82.71	2.10	2.78

$^aG = \frac{p\ 100}{P - p}$, where p and P are the weights of solute and solution, resp.

$^bE = \frac{G\ 100}{G + 100}$; ^{c}g/1 acetone; ^{d}should be mmol/1 (compiler).

^{e}g of acetone required to dissolve 1 g of solute; ^{f}volume (cm^3) of acetone required to dissolve 1 g of solute.

AUXILIARY INFORMATION

METHOD/APPARATUS/PROCEDURE:

A special all-glass app was contructed enabling the prepn of satd solns, agitation by bubbling a stream of acetone-satd N, fitration, and distn off the solvent without contact with air. Two exhangeable dissoln vessels of 15 and 8 cm^3 working capacity were used depending on the soly of solute. The app was immersed in a thermostat. The vols of acetone used were 15 or 5 cm^3, and the equilibration time was 2-2.5 h. The satd solns were filtered, weighed, the solvent was distd off, the residues were dried at 105°C, weighed, and examd for the presence of solvated acetone.

SOURCE AND PURITY OF MATERIALS:

The source of the materials was not specified. Pure, anhyd acetone was used. The absence of impurities and water was confirmed by procedures of the German Pharmacopeia VI and Spanish Pharmacopeia VIII.

The purity of the solute was not specified.

ESTIMATED ERROR:

Soly: measurements were repeated until 2 values not differing in the second decimal were obtained (author).

Temp: ±0.1°C (author).

REFERENCES:

COMPONENTS:

(1) Acetamide, N-[4-[[[4-[(methylamino)-sulfonyl]phenyl]animo]sulfonyl]phenyl]-(acetyl Neo-uliron); $C_{15}H_{17}N_3O_5S_2$; [71119-14-7]

(2) Phosphoric acid, disodium salt; Na_2HPO_4; [7558-94-4]

(3) Water; H_2O;]7732-18-5]

ORIGINAL MEASUREMENTS:

Krüger-Thiemer, E.

Arch. Dermatol. Syphilis 1942, *183*, 90-116.

VARIABLES:

One temperature: ca 20°C; one pH: 8.74

PREPARED BY:

R. Piekos

EXPERIMENTAL VALUES:

Solubility of acetyl Neo-uliron in a 0.705 M (10%) Na_2HPO_4 solution of pH 8.74 at room temperature (about 20°C) is 0.021 g% (5.5×10^{-4} mol dm^{-3} solution, compiler).

AUXILIARY INFORMATION

METHOD/APPARATUS/PROCEDURE:

Acetyl Neo-uliron (0.5 g) was dissolved in 10 cm^3 of the 0.705 M (10%) Na_2HPO_4 soln of pH 8.74, shaken for 2 h at room temp (about 20°C), and filtered. The filtrate was treated with equal vol of 2N HCl, and refluxed for 15 min. After proper diln, a 1-cm^3 aliquot was withdrawn, acidified, cooled, and the sulfonamide content was detd colorimetrically (as Neo-uliron) by the Marshall method modified by Kimmig (1) using an Authenrieth colorimeter. The pH was detd on an ultra-ionograph using a glass electrode.

SOURCE AND PURITY OF MATERIALS:

Acetyl Neo-uliron (source not specified) gave no coloration upon diazotization of its satd soln, thus showing absence of Neo-uliron.

The source and purity of the remaining materials were not specified.

ESTIMATED ERROR:

Soly: precision ±5% (author)

Temp: not specified

pH : ±0.05 pH unit (author)

REFERENCES:

1. Kimmig, J. *Arch. Dermatol.* 1938, *176*, 722; *Erg. Hyg.* 1941, *24*, 398.

COMPONENTS:

(1) Acetamide, N-[4-[[[4-[(methylamino)-sulfonyl]phenyl]amino]sulfonyl]phenyl]-(acetyl Neo-uliron); $C_{15}H_{17}N_3O_5S_2$; [71119-14-7]

(2) Phosphoric acid, monopotassium salt; KH_2PO_4; [7778-77-0]

(3) Water; H_2O; [7732-18-5]

ORIGINAL MEASUREMENTS:

Krüger-Thiemer, E.

Arch. Dermatol. Syphilis 1942, *183*, 90-116.

VARIABLES:

One temperature: ca 20°; one pH: 4.37

PREPARED BY:

R. Piekos

EXPERIMENTAL VALUES:

Solubility of acetyl Neo-uliron in a 0.735M (10%) KH_2PO_4 solution of pH 4.37 at room temperature (about 20°C) is 0.0020 g% (5.2 x 10^{-5} mol dm^{-3} solution, compiler).

AUXILIARY INFORMATION

METHOD/APPARATUS/PROCEDURE:

Acetyl Neo-uliron (0.5 g) was dissolved in 10 cm^3 of the 0.735M (10%) KH_2PO_4 soln of pH 4.37, shaken for 2 h at room temp (about 20°C), and filtered. The filtrate was treated with equal vol of 2N HCl, and refluxed for 15 min. After proper diln, a 1-cm^3 aliquot was withdrawn, acidified, cooled, and the sulfonamide content was detd colorimetrically (as Neo-uliron) by the Marshall method modified by Kimmig (1) using an Authenrieth colorimeter. The pH was detd on an ultra-ionograph using a glass electrode.

SOURCE AND PURITY OF MATERIALS:

Acetyl Neo-uliron (source not specified) gave no coloration upon diazotization of its sat soln, thus showing absence of Neo-uliron.

The source and purity of the remaining materials were not specified.

ESTIMATED ERROR:

Soly: precision ±5% (author).

Temp: not specified.

pH : ±0.05 pH unit (author).

REFERENCES:

1. Kimmig, J. *Arch. Dermatol.* 1938, *176*, 722; *Erg. Hyg.* 1941, *24*, 398.

COMPONENTS:

(1) Acetamide, N-[4-[[[4-[(methylamino)-sulfonyl[phenyl]amino]sulfonyl]phenyl]-(acetyl Neo-uliron); $C_{15}H_{17}N_3O_5S_2$; [71119-14-7]

(2) Phosphoric acid, disodium salt; Na_2HPO_4; [7558-94-4]

(3) Phosphoric acid, monopotassium salt; KH_2PO_4; [7778-77-0]

(4) Water; H_2O; [7732-18-5]

VARIABLES:

Temperature, pH

ORIGINAL MEASUREMENTS:

Krüger-Thiemer, E.

Arch. Dermatol Syphilis 1942, *183*, 90-116.

PREPARED BY:

R. Piekos

EXPERIMENTAL VALUES:

Composition of 1/15M phosphate buffer solutions				Solubility			
				Room temp (ca 20°C)		37°C	
Na_2HPO_4	KH_2PO_4	%Content	pH	g%	10^4 mol dm^{-3} solution	g%	10^4 mol dm^{-3} solution
1.0	99.0	0.91	4.944	0.0019	0.495	-	-
10.0	90.0	0.91	5.906	0.0022	0.573	0.0024	0.625
61.1	38.9	0.93	7.005	0.0038	0.991	0.0051	1.33
9.5	0.5	0.733[b]	7.51	0.0040	1.043	-	-
94.7	5.3	0.95	8.018	0.0128	3.338	-	-

[a]Calculated by compiler.

[b]Molar content; 10% buffer solution.

AUXILIARY INFORMATION

METHOD/APPARATUS/PROCEDURE:

Acetyl Neo-uliron (0.5 g) was dissolved in 10 cm^3 of a buffer soln, shaken for 2 h at 20°C (or left for 48 h at 37°C), filtered at respective temp. The filtrate was treated with equal vol of 2N HCl and refluxed for 15 min. After proper diln, a 1 cm^3 aliquot was withdrawn, cooled, and the sulfonamide content was detd colorimetrically (as Neo-uliron) by the Marshall method modified by Kimmig (1) using Authenrieth colorimeter. the pH was detd on an ultraionograph using a glass electrode.

SOURCE AND PURITY OF MATERIALS:

Acetyl Neo-uliron (source not specified) gave no coloration upon diazotization of its satd soln, thus showing absence of Neo-uliron.

The source and purity of the remaining materials were not specified.

ESTIMATED ERROR:

Soly: precision ±5% (author).

Temp: not specified.

pH : ±0.05 pH unit (author).

REFERENCES:

1. Kimmig, J. Arch. Dermatol. 1938, *176*, 722; *Erg. Hyg.* 1941, *24*, 398.

COMPONENTS:

(1) Benzenesulfonamide, 4-amino-N-[4-[(dimethylamino)sulfonyl]phenyl]-; $C_{14}H_{17}N_3O_4S_2$; [515-67-3]

(2) 2-Propanone (acetone); C_3H_6O; [67-64-1]

ORIGINAL MEASUREMENTS:

Gutierrez, F. H.
Anales Fis. quim. (Madrid) 1945, *41*, 537-60.

VARIABLES:

Temperature

PREPARED BY:

R. Piekos

EXPERIMENTAL VALUES:

$t/^{\circ}C$	G^a	E^b	X_g/l^c	mol/l[d] acetone	mmol/mol acetone	$1:X_g^e$	$1 + X_{cc}^f$
0	26.007	20.639	211.853	509	42.5	3.84	4.72
5	27.025	21.275	218.575	615	44.1	3.70	4.57
10	28.036	21.897	225.129	633	45.8	3.56	4.44
15	29.064	22.519	231.669	654	47.5	3.44	4.31
20	30.092	23.130	238.072	672	49.0	3.32	4.20
25	31.120	23.733	244.354	689	50.9	3.21	4.09
30	31.898	24.214	248.581	701	52.1	3.13	4.92
35	33.500	25.093	259.055	731	54.7	2.98	3.86
40	35.705	26.311	274.000	745	58.3	2.80	3.65
45	39.169	28.144	298.242	841	64.0	2.55	3.35
50	44.509	30.800	336.265	949	72.7	2.25	2.98

[a] $G = \frac{p\ 100}{P - p}$, where p and P are the weights of solute and solution, resp.

[b] $E = \frac{G\ 100}{G + 100}$; [c] g/l acetone; [d] should be mmol/l (compiler);

[e] g of acetone required to dissolve 1 g of solute; [f] volume (cm^3) of acetone required to dissolve 1 g of solute.

AUXILIARY INFORMATION

METHOD/APPARATUS/PROCEDURE:

A special all-glass app was contructed enabling the prepn of satd solns, agitation by bubbling a stream of acetone-satd N, filtration, distn off the solvent without contact with air. Two exchangeable dissoln vessels of 15 and 8 cm^3 working capacity were used depending on the soly of solute. The app was immersed in a thermostat. The vols of acetone used were 15 or 5 cm^3, and the equilibration time was 2-2.5 h. The satd solns were filtered, weighed, the solvent was distd off, the residues were dried at 105°C, weighed, and examd for the presence of solvated acetone.

SOURCE AND PURITY OF MATERIALS:

The source of the materials was not specified. Pure, anhyd acetone was used. The absence of impurities and water was confirmed by procedures of the German Pharmacopeia VI and Spanish Pharmacopeia VIII.
The purity of the solute was not specified.

ESTIMATED ERROR:

Soly: measurements were repeated until 2 values not differing in the second decimal were obtained (author).

Temp: ±0.1°C (author).

REFERENCES:

COMPONENTS:	EVALUATOR:
(1) Benzenesulfonamide, 4-amino-N-2-pyridinyl- (sulfapyridine); $C_{11}H_{11}N_3O_2S$; [144-83-2] (2) Water or (3) Aqueous phosphate buffers	Anthony N. Paruta Department of Pharmaceutics University of Rhode Island Kingston, Rhode Island, USA and Ryszard Piekos Faculty of Pharmacy, University of Gdansk Gdansk, Poland 1986

CRITICAL EVALUATION:

The solubility of sulfapyridine was studied in water mainly at 310K in eleven papers (1-11) as shown in Table I.

Table I: Solubility of sulfapyridine in water at various temperatures

	10^3 mol dm^{-3} (*indicates mol Kg^{-1})		
Reference	298K	303K/308K	310K
1	-	-	2.2
2	-	2.3(308K)	-
3	-	-	1.98
4	-	-	1.8
5	-	-	8
6	1.07*	-	1.95*
7	-	-	1.89
8	-	-	2.1
9	-	-	1.48
10	-	2.01(303K)	-
11	-	-	2.09 (pH=6)

Lebel, Schroeder and Simesen (2) gave a value at 308K and Yamazaki et al. (10) at 303K, both values are about the right order of magnitude and in line with the other results. Kitao et al. (11) appear to have determined the solubility at 310K where the pH value is the natural pH produced by the solute in water. Neish's (9) value of 1.48 x 10^{-3}mol dm^{-3} is too low with respect to the other values probably due to the poor control of temperature. Tréfouël's (5) value is about four times greater than the apparent average. Sapozhnikova and Postovskii (7) used a one hour equilibration which is felt to be inadequate. Although the value given (7) is in line it was not taken into account in the determination of the recommended value. The remaining values (1,3,4,6,8) all determined with appropriate methods were used to calculate the simple average. The recommended value for sulfapyridine in water at 310K is 2 x 10^{-3}mol dm^{-3}.

Solubilities of sulfapyridine in aqueous buffered solutions at 293K and 310K are shown in Table II.

Table II: Solubility of sulfapyridine at various temperatures and pH values

		10^3 mol dm^{-3}	
Reference	pH	293K	310K
12	5.90	2.808	4.252
13	6.0	2.6	-
14	5.9	-	2.4
12	7.0	2.848	4.252
13	7.0	2.7	-
14	7.0	-	2.3
12	8.0	3.249	-
13	8.0	2.8	-

The values given by Langecker (13) appear to be quite low, it should be noted that the solution was boiled for one hour and then stored for an unspecified period. It is possible that the low value's are due to decomposition. Krüger-Thiermer (11), and Pulver and Suter (12) are in reasonable agreement. Both used 0.067M mol dm^{-3} phosphate buffer solutions. At pH 6,7 and 8 the recommended values are at 293K, 2.7 x 10^{-3} mol dm^{-3}, 2.8 x 10^{-3} mol dm^{-3} and 3.0 x 10^{-3} mol dm^{-3} respectively.

It would be of interest to compare these values with those in pure water. Dissolving sulfapyridine in water results in a near neutral solution (pH=7); and enhanced dissolution at a lower temperature, 310K to 293K, may be due to a salting-in effect. It should be noted that the buffer concentration is 6.6 x 10^{-3} mol dm^{-3} and the solute is a fraction (≅ 2-3 x 10^{-3} mol dm^{-3}) of these highly soluble salts.

REFERENCES:

(1) Hug, E. *Rev. soc. Argentina biol.* 1940, *16*, 662-6.
(2) Lebel, H.; Schroeder, E.; Simesen, M. *Acta Med. Scand.* 1940, *105(4)*, 395-410.
(3) Roblin, R.O., Jr.; Winnek, P.S. *J. Am. Chem. Soc.* 1940, *62*, 1999-2002.
(4) Durel, M.P.; Allinne, M. *Bull. Soc. Med. Hop. Paris* III 1941, 251-9.
(5) Tréfouël, M. *Bull. Acad. Med. Paris* 1941, *124*, 546-54.
(6) Clark, W.G.; Strakosch, E.A.; Levitan, N.I. *J. Lab. Clin. Med.* 1942, *28*, 188-9.
(7) Sapozhnikova, N.V.; Postovskii, I.Ya. *Zh. Prikl. Khim.* 1944, *17*, 427-34.
(8) Langecker, H. *Arch. Exptl. Path. Pharmakol.* 1948, *205*, 291-301.
(9) Neish, W.J.P. *Rec. trav. chim.* 1948, *67*, 361-71.
(10) Yamazaki, M.; Aoki.; Kamada, A.; Yata, N. *Yakuzaigaku* 1967, *27(1)*, 37-40.
(11) Kitao, K.; Kubo, K.; Morishita, T.; Yata, N.; Kamada, A. *Chem. Pharm. Bull.* 1973, *21*, 2417-26.
(12) Krüger-Thiemer, E. *Arch. Dermatol. Syphilis* 1942, *183*, 90-116.
(13) Pulver, R.; Suter, R. *Schweiz. Med. Wochenschr.* 1943, *73(13)*, 403-8.
(14) Langecker, H. *Arch. Exptl. Path. Pharmakol.* 1948, *205*, 291-301.

COMPONENTS:	ORIGINAL MEASUREMENTS:
(1) Benzenesulfonamide, 4-amino-N-2-pyridinyl- (sulfapyridine); $C_{11}H_{11}N_3O_2S$; [144-83-2] (2) Water; H_2O; [7732-18-5]	Hug, E. *Rev. soc. Argentina biol.* 1940, *16*, 662-6.
VARIABLES:	PREPARED BY:
One temperature: 37°C	R. Piekos

EXPERIMENTAL VALUES:

Solubility of sulfapyridine in water at 37°C is 54 mg% (2.2×10^{-3} mol dm^{-3} solution, compiler).

AUXILIARY INFORMATION

METHOD/APPARATUS/PROCEDURE:	SOURCE AND PURITY OF MATERIALS:
A satd soln of sulfapyridine was prepd by heating on a water bath. The soln was then cooled down to 37°C and maintained at this temp for 7 days.	Nothing specified
	ESTIMATED ERROR:
	Nothing specified
	REFERENCES:

COMPONENTS:

(1) Benzenesulfonamide, 4-amino-N-2-pyridinyl- (sulfapyridine); $C_{11}H_{11}N_3O_2S$; [144-83-2]

(2) Water; H_2O; [7732-18-5]

ORIGINAL MEASUREMENTS:

Lebel, H.; Schroeder, E.; Simesen, M. M. *Acta Med. Scand.* 1940, *105(4)*, 395-410.

VARIABLES:

One temperature: 35°C

PREPARED BY:

R. Piekos

EXPERIMENTAL VALUES:

Solubility of sulfapyridine in water at 35°C is 1:1800 corresponding to 56 mg% (2.3 x 10^{-3} mol dm^{-3} solution, compiler).

AUXILIARY INFORMATION

METHOD/APPARATUS/PROCEDURE:

Nothing specified

SOURCE AND PURITY OF MATERIALS:

Nothing specified

ESTIMATED ERROR:

Nothing specified

REFERENCES:

COMPONENTS:

(1) Benzenesulfonamide, 4-amino-N-2-pyridinyl- (sulfapyridine); $C_{11}H_{11}N_3O_2S$; [144-83-2]

(2) Water; [7732-18-5]

ORIGINAL MEASUREMENTS:

Roblin, R. O., Jr.; Winnek, P. S.

J. Am. Chem. Soc. 1940, *62*, 1999-2002.

VARIABLES:

One temperature: 37°C

PREPARED BY:

R. Piekos

EXPERIMENTAL VALUES:

Solubility of sulfapydridine in water at 37°C is 49.5 mg/100 cm^3 solution (1.98×10^{-3} mol dm^{-3}, compiler).

AUXILIARY INFORMATION

METHOD/APPARATUS/PROCEDURE:

Excess sulfonamide in water was heated and stirred on a steam bath for 30 min. The suspension was then agitated for 24 h in a thermostat at 37°C. A sample of the satd soln was withdrawn through a glass filter, dild, and analyzed by the modified Marshall method (1) using a General Electric recording spectrophotometer for comparing the colors with those of the standards.

SOURCE AND PURITY OF MATERIALS:

Sulfapyridine, mp 190-1°C, was probably prepd by the authors.

Purity of the water was not specified.

ESTIMATED ERROR:

Nothing specified

REFERENCES:

1. Bratton, A. C.; Marshall, E. K., Jr. *J. Pharmacol.* 1939, *66*, 4.

COMPONENTS:	ORIGINAL MEASUREMENTS:
(1) Benzenesulfonamide, 4-amino-N-2-pyridinyl- (sulfapyridine); $C_{11}H_{11}N_3O_2S$; [144-83-2] (2) Water; H_2O; [7732-18-5]	Durel, M. P.; Allinne, M. *Bull. Soc. Med. Hop. Paris III* 1941, 251-9.
VARIABLES: One temperature: 37°C	PREPARED BY: R. Piekos

EXPERIMENTAL VALUES:

Solubility of sulfapyridine in water at 37°C is 0.45 g/liter (1.8×10^{-3} mol dm^{-3}, compiler).

AUXILIARY INFORMATION

METHOD/APPARATUS/PROCEDURE:	SOURCE AND PURITY OF MATERIALS:
A mixt of sulfapyridine and water was agitated for 24 hours at 37°C.	Source and purity of sulfapyridine were not specified. Distilled water was used.
	ESTIMATED ERROR: Nothing specified
	REFERENCES:

COMPONENTS:	ORIGINAL MEASUREMENTS:
(1) Benzenesulfonamide, 4-amino-N-2-pyridinyl- (sulfapyridine); $C_{11}H_{11}N_3O_2S$; [144-83-2] (2) Water; H_2O; [7732-18-5]	Tréfouël, M. *Bull. Acad. Med. Paris* 1941, *124*, 546-54.
VARIABLES: One temperature: 37°C	PREPARED BY: R. Piekos

EXPERIMENTAL VALUES:

Solubility of sulfapyridine in water at 37°C is 0.2 part per 100 parts water (8×10^{-3} mol kg^{-1} water, compiler).

AUXILIARY INFORMATION

METHOD/APPARATUS/PROCEDURE:	SOURCE AND PURITY OF MATERIALS:
Sulfapyridine was diazotized, coupled with N-naphthyl-1-N-diethyl-3-propylenediamine and assayed colorimetrically.	Nothing specified
	ESTIMATED ERROR: Nothing specified
	REFERENCES:

COMPONENTS:

(1) Benzenesulfonamide, 4-amino-N-2-pyridinyl- (sulfapyridine); $C_{11}H_{11}N_3O_2S$; [144-83-2]

(2) Water; H_2O; [7732-18-5]

ORIGINAL MEASUREMENTS:

Clark, W. G.; Strakosch, E. A.; Levitan, N. I. *J. Lab. Clin. Med.* 1942, *28*, 188-9.

VARIABLES:

Temperature

PREPARED BY:

R. Piekos

EXPERIMENTAL VALUES:

$t/^{\circ}C$	Solubility	
	g/100 g water	10^3 mol kg^{-1} water[a]
25	0.0268	1.07
37	0.0486	1.95

[a] Calculated by compiler

AUXILIARY INFORMATION

METHOD/APPARATUS/PROCEDURE:

A small tinted glass container contg excess sulfapyridine in water was shaken in a water bath thermostat for 24 h. The satd soln was then filtered by aspiration through a washed and dried asbestos filter stick into a weighed weighing bottle. The entire app was kept at the temp at which the compd was dissolved. The amt dissolved was then detd by the method of Bratton and Marshall (1), using a photoelectric colorimeter.

SOURCE AND PURITY OF MATERIALS:

Neither source nor purity of sulfapyridine was specified.

CO_2-free distd water was used.

ESTIMATED ERROR:

Soly: not specified.

Temp: ±0.1°C (authors).

REFERENCES:

1. Bratton, A. C.; Marshall, E. K., Jr. *J. Biol. Chem.* 1939, *128*, 537.

COMPONENTS:

(1) Benzenesulfonamide, 4-amino-N-2-pyridinyl- (sulfapyridine); $C_{11}H_{11}N_3O_2S$; [144-83-2]

(2) Water; H_2O; [7732-18-5]

ORIGINAL MEASUREMENTS:

Sapozhnikova, N. V.; Postovskii, I. Ya. *Zh. Prikl Khim.* 1944, *17*, 427-34.

VARIABLES:

Temperature

PREPARED BY:

R. Piekos

EXPERIMENTAL VALUES:

$t/^{\circ}C$	Solubility: Weight %	Solubility: 10^3 mol kg^{-1} water[a]
20	0.0194	0.778
37	0.0470	1.89
50	0.094	3.8
75	0.24	9.6
99	0.61[b]	25

[a] Calculated by compiler

[b] Calculated from the heat of dissolution (10,040 cal mol^{-1})

AUXILIARY INFORMATION

METHOD/APPARATUS/PROCEDURE:

Sulfapyridine was dissolved in water to form a satd soln which was occasionally agitated in a glass vessel immersed in a thermostat. The equilibrium was usually attained after 1 h. Five to 100-cm^3 samples of the satd soln were placed in Pt crucibles or dishes and evapd to dryness at temps lower than 110-115°C. The residue was dried to const wt at 105-110°C and weighed.

SOURCE AND PURITY OF MATERIALS:

Pure, recrystd sulfapyridine was used. Its mp conformed to that reported in the literature.

Purity of the water was not specified.

ESTIMATED ERROR: Soly: quite reliable results were obtained over the temp range 20-75°C. At higher temps the accuracy was poor due to evapn of water during sampling (authors).
Temp: ±0.05°C (authors).

REFERENCES:

COMPONENTS:	ORIGINAL MEASUREMENTS:
(1) Benzenesulfonamide, 4-amino-N-2-pyridinyl- (sulfapyridine); $C_{11}H_{11}N_3O_2S$; [144-83-2] (2) Water; H_2O; [7732-18-5]	Langecker, H. *Arch. Exptl. Path. Pharmakol.* 1948, *205*, 291-301.
VARIABLES:	PREPARED BY:
One temperature: 37°C	R. Piekos

EXPERIMENTAL VALUES:

Solubility of sulfapyridine in water at 37°C is 53 mg% (2.1×10^{-3} mol dm^{-3}, compiler).

AUXILIARY INFORMATION

METHOD/APPARATUS/PROCEDURE:

An excess of sulfapyridine in water was boiled and left for 24 h in a vessel protected from access of CO_2. The concn of sulfapyridine was detd colorimetrically by the method of Bratton and Marshall (1) using a Havemann colorimeter (2), as well as by microanal detn of the solid residue.

SOURCE AND PURITY OF MATERIALS:

Source and purity of the materials were not specified.

ESTIMATED ERROR:

Nothing specified

REFERENCES:

1. Bratton, A. G.; Marshall, E. K. *J. Biol. Chem.* 1939, *128*, 537.
2. Havemann, R. *Klin. Wochenschr.* 1940, p. 503.

COMPONENTS:	ORIGINAL MEASUREMENTS:
(1) Benzenesulfonamide, 4-amino-N-2-pyridinyl- (sulfapyridine); $C_{11}H_{11}N_3O_2S$; [144-83-2] (2) Water; H_2O; [7732-18-5]	Neish, W. J. P. *Rec. trav. chim.* 1948, *67*, 361-71.

VARIABLES:	PREPARED BY:
One temperature: 37°C	R. Piekos

EXPERIMENTAL VALUES:

Solubility of sulfapyridine in water at 37°C is 370 γ/ml (1.48×10^{-3} mol dm^{-3}, compiler).

AUXILIARY INFORMATION

METHOD/APPARATUS/PROCEDURE:

A suspension of sulfapyridine (10 mg/5 ml) in water was kept for 5 h at 37°C and 1 h at room temp before fitration. Soly was detd by the Westfall's method (1) based on diazotization of the sulfonamide, coupling with Na 2-napthol-3,6-disulfonate and comparing the color with that of a std soln in a Klett colorimeter.

SOURCE AND PURITY OF MATERIALS:

Sulfapyridine: not specified.
Distd water was used.

ESTIMATED ERROR:

Nothing specified

REFERENCES:

1. Westfall, B. B. *J. Nat. Cancer Inst.* 1945, *6*, 23.

COMPONENTS:	ORIGINAL MEASUREMENTS:
(1) Benzenesulfonamide, 4-amino-N-2-pyridinyl- (sulfapyridine); $C_{11}H_{11}N_3O_2S$; [144-83-2] (2) Water; H_2O; [7732-18-5]	Yamazaki, M.; Aoki, M.; Kamada, A.; Yata, N. *Yakuzaigaku* 1967, *27(1)*, 37-40.
VARIABLES:	PREPARED BY:
One temperature: 30°C	R. Piekos

EXPERIMENTAL VALUES:

Solubility of sulfapyridine in water at 30°C is 2.01 mmol/L (0.501 g dm^{-3}, compiler).

AUXILIARY INFORMATION

METHOD/APPARATUS/PROCEDURE:

Sulfapyridine (0.5 g) was placed in an L-shaped tube together with 20 ml of water. The mixt was shaken in a thermostat until equilibrium was attained. The sulfapyridine was then assayed in the supernatant spectrophotometrically at 545 nm on a Beckmann DU spectrophotometer. The results were taken from a calibration graph.

SOURCE AND PURITY OF MATERIALS:

Nothing specified.

ESTIMATED ERROR:

Soly: not specified.

Temp: ±1°C (authors).

REFERENCES:

COMPONENTS:	ORIGINAL MEASUREMENTS:
(1) Benzenesulfonamide, 4-amino-N-2-pyridinyl- (sulfapyridine); $C_{11}H_{11}N_3O_2S$; [144-83-2] (2) Water; H_2O; [7732-18-5]	Kitao, K.; Kubo, K.; Morishita, T.; Yata, N.; Kamada, A. *Chem. Pharm. Bull.* 1973, *21*, 2417-26.
VARIABLES:	PREPARED BY:
One temperature: 37°C	R. Piekos

EXPERIMENTAL VALUES:

Solubility of sulfapyridine in water at 37°C is 2.09 mmol dm^{-3} solution.

AUXILIARY INFORMATION

METHOD/APPARATUS/PROCEDURE:	SOURCE AND PURITY OF MATERIALS:
The sulfapyridine concn in the aq soln (pH 6) was detd by diazotization. No details were given.	Comm available sulfapyridine was used as supplied. Deionized water was used.
	ESTIMATED ERROR: Soly: not specified. Temp: ±1°C (authors).
	REFERENCES:

COMPONENTS:	ORIGINAL MEASUREMENTS:
(1) Benzenesulfonamide, 4-amino-N-2-pyridinyl- (sulfapyridine); $C_{11}H_{11}N_3O_2S$; [144-83-2] (2) Sodium hydroxide; NaOH; [1310-73-2] (3) Water; H_2O; [7732-18-5]	Rose, F. L.; Martin, A.R.; Bevan, H. G. L. *J. Pharm. Exp. Therap.* 1943, *77*, 127-42.
VARIABLES:	PREPARED BY:
pH	R. Piekos

EXPERIMENTAL VALUES:

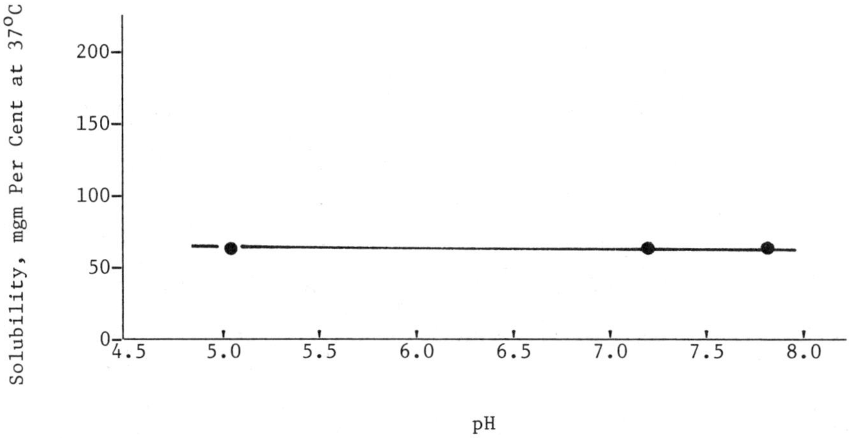

AUXILIARY INFORMATION

METHOD/APPARATUS/PROCEDURE:

An excess of sulfapyridine was stirred in boiling water, the soln was cooled to 37°C, the temp being maintained thermostatically, and 0.1N NaOH was added to increase the pH. The pH was measured by means of a glass electrode - calomel half-cell system, and was permitted to reach equilibrium before a reading was taken. The concn of sulfapyridine in soln was detd colorimetrically by withdrawing a sample through a filter-tip into a preheated micropipet.

SOURCE AND PURITY OF MATERIALS:

The source and purity of sulfapyridine was not specified.
Water was doubly distilled.

ESTIMATED ERROR:

Nothing specified.

REFERENCES:

COMPONENTS:

(1) Benzenesulfonamide, 4-amino-N-2-pyridinyl- (sulfapyridine); $C_{11}H_{11}N_3O_2S$; [144-83-2]

(2) Sodium chloride; NaCl; [7647-14-5]

(3) Water; H_2O; [7732-18-5]

ORIGINAL MEASUREMENTS:

Avico, U.; Cavazutti, G.; di Francesco, R.; Signoretti Ciranni, E.; Zuccaro, P. *Farmaco, Ed. Pratica* 1975, *30(1)*, 40-6.

VARIABLES:

Temperature

PREPARED BY:

R. Piekos

EXPERIMENTAL VALUES:

$t/^{\circ}C$	Solubility of amorphous sulfapyridine in equimolal NaCl solutions	
	g/100 g water	10^3 mol kg^{-1} water[a]
25	0.61	2.4
35	0.75	3.0
40	0.81	3.2

[a]Calculated by compiler

AUXILIARY INFORMATION

METHOD/APPARATUS/PROCEDURE:

A soln of Na salt of sulfapyridine was added to a HCl soln contg stoichiometric quantity of the acid to neutralize the salt. The neutralization was carried out in a thermostat and the pH of the mixt was maintained close to that of a satd sulfapyridine soln. The procedure was repeated using various initial concn of the reagents to find the max concn of sulfapyridine at which no pptn occurred.

SOURCE AND PURITY OF MATERIALS:

Source and purity of sulfapyridine was not specified. The mp of crystalline sulfapyridine was 191-3°C.

Purity of the water was not specified.

ESTIMATED ERROR:

Nothing specified.

REFERENCES:

COMPONENTS:	ORIGINAL MEASUREMENTS:
(1) Benzenesulfonamide, 4-amino-N-2-pyridinyl- (sulfapyridine); $C_{11}H_{11}N_3O_2S$; [144-83-2] (2) Phosphoric acid, disodium salt; Na_2HPO_4; [7558-94-4] (3) Water; H_2O; [7732-18-5]	Krüger-Thiemer, E. *Arch. Dermatol. Syphilis* 1942, *183*, 90-116.
VARIABLES:	**PREPARED BY:**
One temperture: ca 20°C; one pH: 8.74	R. Piekos

EXPERIMENTAL VALUES:

Solubility of sulfapyridine in a 0.705M (10%) Na_2HPO_4 solution of pH 8.74, at room temperature (about 20°C), is 0.075 g% (3.0 x 10^{-3} mol dm^{-3} solution, compiler).

AUXILIARY INFORMATION

METHOD/APPARATUS/PROCEDURE:

Sulfapyridine (0.5 g) was dissolved in the 0.705M (10%) Na_2HPO_4 soln (pH 8.74), at room temp (about 20°C), shaken for 2 h, and filtered. A 1-cm^3 aliquot of the filtrate was withdrawn, cooled, acidified with 2N HCl, and the sulfapyridine content was detd colorimetrically by the method of Marshall modified by Kimmig (1) using an Authenrieth colorimeter. The pH was detd on an ultra-ionograph using a glass electrode.

SOURCE AND PURITY OF MATERIALS:

Sulfapyridine was the product manufd by Nordmark under the name Eubasin. The source and purity of the remaining materials were not specified.

ESTIMATED ERROR:

Soly: precision ±5% (author).
Temp: not specified.
pH : ±0.05 pH unit (author).

REFERENCES:

1. Kimmig, J. *Arch. Dermatol.* *176*, 722; *Erg. Hyg.* 1941, *24*, 398.

COMPONENTS:

(1) Benzenesulfonamide, 4-amino-N-2-pyridinyl- (sulfapyridine); $C_{11}H_{11}N_3O_2S$; [144-83-2]

(2) Phosphoric acid, monopotassium salt; KH_2PO_4; [7778-77-0]

(3) Water; H_2O; [7732-18-5]

ORIGINAL MEASUREMENTS:

Krüger-Thiemer, E.

Arch. Dermatol. Syphilis 1942, *183*, 90-116.

VARIABLES:

One temperature: ca 20°C; one pH: 4.37

PREPARED BY:

R. Piekos

EXPERIMENTAL VALUES:

Solubility of sulfapyridine in a 0.735M (10%) KH_2PO_4 solution of pH 4.37, at room temperature (about 20°C), is 0.054 g% (2.17 mol dm^{-3} solution, compiler).

AUXILIARY INFORMATION

METHOD/APPARATUS/PROCEDURE:

Sulfapyridine was dissolved in a 0.735M (10%) KH_2PO_4 soln in amt of 0.5 g/10 cm^3, shaken for 2 h, and filtered. A 1-cm^3 aliquot of the filtrate was withdrawn, cooled, acidified with 2N HCl, and the sulfapyridine content was detd colorimetrically by the method of Marshall modified by Kimmig (1) using Authenrieth colorimeter. The pH was detd on an ultraionograph using a glass electrode.

SOURCE AND PURITY OF MATERIALS:

Sulfapyridine was the product manufd by Nordmark under the name Eubasin.

The source and purity of the remaining materials were not specified.

ESTIMATED ERROR:

Soly: precision ±5% (author).

Temp: not specified.

pH : ±0.05 pH unit (author).

REFERENCES:

1. Kimmig, J. *Arch. Dermatol.* 1938, *176*, 722; *Erg. Hyg.* 1941, *24*, 398.

COMPONENTS:

(1) Benzenesulfonamide, 4-amino-N-2-pyridinyl- (sulfapyridine); $C_{11}H_{11}N_3O_2S$; [144-83-2]

(2) Phosphoric acid, disodium salt; Na_2HPO_4; [7558-94-4]

(3) Phosphoric acid, monopotassium salt; KH_2PO_4; [7778-77-0]

(4) Water; H_2O; [7732-18-5]

ORIGINAL MEASUREMENTS:

Krüger-Thiemer, E.

Arch. Dermatol. Syphilis 1942, *183*, 90-116.

VARIABLES:

Temperature; pH

PREPARED BY:

R. Piekos

EXPERIMENTAL VALUES:

Composition of 1/15M phosphate buffer solutions				Solubility			
				Room temp (ca 20°C)		37°C	
Na_2HPO_4	KH_2PO_4	%Content	pH	g%	10^3 mol dm^{-3} solution[a]	g%	10^3 mol dm^{-3} solution[a]
1.0	99.0	0.91	4.944	0.070	2.808	-	-
10.0	90.0	0.91	5.906	0.070	2.808	0.106	4.252
61.1	38.9	0.93	7.005	0.071	2.848	0.106	4.252
9.5	0.5	0.733[b]	7.51	0.060	2.407	-	-
94.7	5.3	0.95	8.018	0.081	3.249	-	-

[a] Calculated by compiler.

[b] Molar content; 10% buffer solution.

AUXILIARY INFORMATION

METHOD/APPARATUS/PROCEDURE:

Sulfapyridine (0.5 g) was dissolved in 10 cm^3 of a buffer soln, shaken for 2 h at 20°C (or left for 48 h at 37°C), and filtered at respective temp. A 1-cm^3 aliquot of the filtrate was then withdrawn, cooled (dild for expts at 37°C), acidified with 1 cm^3 of 2N HCl, and the sulfapyridine content was detd colorimetrically by the method of Marshall modified by Kimmig (1) using an Authenrieth colorimeter. The pH was detd on an ultraionograph using a glass electrode.

SOURCE AND PURITY OF MATERIALS:

Sulfapyridine was the product manufd by Nordmark under the name Eubasin.

The source and purity of the remaining materials were not specified.

ESTIMATED ERROR:

Soly: precision ±Z% (author).

Temp: not specified.

pH : ±0.05 pH unit (author).

REFERENCES:

1. Kimmig, J. *Arch. Dermatol.* 1938, *176*; 722; *Erg. Hyg.* 1941, *24*, 398.

COMPONENTS:

(1) Benzenesulfonamide, N-2-pyridinyl- (sulfapyridine); $C_{11}H_{11}N_3O_2S$; [144-83-2]

(2) Phosphoric acid, disodium salt; Na_2HPO_4; [7558-94-4]

(3) Phosphoric acid, monopotassium salt; KH_2PO_4; [7778-77-0]

(4) Water; H_2O; [7732-18-5]

ORIGINAL MEASUREMENTS:

Pulver, R.; Suter, R.

Schweiz. Med. Wochenschr. 1943, *73(13)*, 403-8.

VARIABLES:

pH

PREPARED BY:

R. Piekos

EXPERIMENTAL VALUES:

pH	Solubility of sulfapyridine in M/15 phosphate buffers (according to Sørensen) at 20°C	
	mg%	10^3 mol dm^{-3} [a]
6.0	64	2.6
7.0	67	2.7
8.0	70	2.8

[a] Calculated by compiler

AUXILIARY INFORMATION

METHOD/APPARATUS/PROCEDURE:

Nothing specified.

SOURCE AND PURITY OF MATERIALS:

Nothing specified.

ESTIMATED ERROR:

Nothing specified.

REFERENCES:

COMPONENTS:

(1) Benzenesulfonmaide, 4-amino-N-2-pyridinyl- (sulfapyridine); $C_{11}H_{11}N_3O_2S$; [144-83-2]

(2) Phosphoric acid, disodium salt; Na_2HPO_4; [7558-94-4]

(3) Phosphoric acid, monopotassium salt; KH_2PO_4; [7778-77-0]

(4) Water; H_2O; [7732-18-5]

ORIGINAL MEASUREMENTS:

Langecker, H.

Arch. Exptl. Path. Pharmakol. 1948, *205*, 291-301.

VARIABLES:

pH

PREPARED BY:

R. Piekos

EXPERIMENTAL VALUES:

pH of the 1/15M phosphate buffer	Solubility at 37°C	
	mg%	10^3 mol dm^{-3} [a]
4.9	44	1.8
5.9	60	2.4
7.0	58	2.3
7.5	62	2.5

[a] Calculated by compiler.

AUXILIARY INFORMATION

METHOD/APPARATUS/PROCEDURE:

An excess of sulfapyridine was added to the buffer soln and boiled for 1 h in a sealed ampul followed by keeping the ampul at 37°C. The concn of sulfapyridine was detd colorimetrically by the method of Bratton and Marshall (1) using a Havemann colorimeter (2), as well as by microanal detn of the solid residue.

SOURCE AND PURITY OF MATERIALS:

Source and purity of the materials were not specified.

ESTIMATED ERROR:

Nothing specified.

REFERENCES:

1. Bratton, A. G.; Marshall, E. K., Jr. *J. Biol. Chem.* 1939, *128*, 537.
2. Havemann, R. *Klin. Wochenschr.* 1940, p. 503.

COMPONENTS:

(1) Benzenesulfonamide, 4-amino-N-2-pyridinyl- (sulfapyridine); $C_{11}H_{11}N_3O_2S$; [144-83-2]

(2) Phosphoric acid, disodium salt; Na_2HPO_4; [7558-94-4]

(3) Phosphoric acid, monopotassium salt; KH_2PO_4; [7778-77-0]

(4) Water; H_2O; [7732-18-5]

ORIGINAL MEASUREMENTS:

Yamazaki, M.; Aoki, M.; Kamada, A.; Yata, N. *Yakuzaigaku,* 1967, *27(1),* 37-40.

VARIABLES:

One temperature: 30°C; one pH: 7.4

PREPARED BY:

R. Piekos

EXPERIMENTAL VALUES:

Solubility of sulfapyridine in a phosphate buffer solution of pH 7.4 (μ = 0.17) at 30°C is 1.91 mmol/L (0.476 g dm^{-3}, compiler).

AUXILIARY INFORMATION

METHOD/APPARATUS/PROCEDURE:

Sulfapyridine (0.5 g) was placed in an L-shaped tube together with 20 ml of the buffer soln. The mixt was then shaken in a thermostat until equilibrium was attained. The sulfapyridine was then assayed in the supernatant spectrophotometrically at 545 nm on a Beckmann DU spectrophotometer. The results were taken from a calibration graph.

SOURCE AND PURITY OF MATERIALS:

Nothing specified.

ESTIMATED ERROR:

Soly: not specified.

Temp: ±1°C (authors).

REFERENCES:

COMPONENTS:

(1) Benzenesulfonamide, 4-amino-N-2-pyridinyl- (sulfapyridine); $C_{11}H_{11}N_3O_2S$; [144-83-2]

(2) Phosphoric acid, disodium salt; Na_2HPO_4; [7558-94-4]

(3) Phosphoric acid, monopotassium salt; KH_2PO_4; [7778-77-0]

(4) Water; H_2O; [7732-18-5]

ORIGINAL MEASUREMENTS:

Hekster, Y. A.; Vree, T. B.; Damsma, J.E.; Friesen, W. T. *J. Antimicrob. Chemother.* 1981, *8*, 133-44.

VARIABLES:

pH

PREPARED BY:

R. Piekos

EXPERIMENTAL VALUES:

pH	Solubility at 25°C	
	mg/l	10^4 mol dm^{-3} [a]
5.5	120	4.81
7.5	200	8.02

[a]Calculated by compiler.

AUXILIARY INFORMATION

METHOD/APPARATUS/PROCEDURE:

Satd solns of sulfapyridine were prepd in phosphate buffers of pH 5.5 and 7.5 at room temp (25°C). The concn of the solute was measured by means of a Spectra Physics 3500B high-performance liquid chromatograph equipped with a column oven (Model 748) and a Pye-Unicam LC-UV spectrophotometric detector. The detector was connected to a 1-mV recorder. A stainless steel column (10 cm x 4.6 mm i.d.) was packed with Lichrosorb RPS, 5 μm, obtained from Chrompack. An injection loop of 100 μl was used. The oven temp was 40°C. Detection of sulfapyridine was performed at 260 nm.

SOURCE AND PURITY OF MATERIALS:

The source and purity of the materials were not specified.

ESTIMATED ERROR:

The detection limit of the solute by HPLC was 0.5 mg/l (authors). The error in temperature and pH was not specified.

REFERENCES:

COMPONENTS:	ORIGINAL MEASUREMENTS:
(1) Benzenesulfonamide, 4-amino-N-2-pyridinyl- (sulfapyridine); $C_{11}H_{11}N_3O_2S$; [144-83-2] (2) Calcium chloride; $CaCl_2$; [10043-52-4] (3) Potassium chloride; KCl; [7447-40-7] (4) Sodium chloride; NaCl; [7647-14-5] (5) Water; H_2O; [7732-18-5]	Hawking, F. *Lancet*, 1941, *240*, 786-8.
VARIABLES: Temperature	PREPARED BY: R. Piekos

EXPERIMENTAL VALUES:

t/°C	Solubility of bicarbonate-free Locke's solution[a]	
	mg/100 ml	10^4 mol dm^{-3} [a]
17	17	6.8
36	41	16.5

[a]The solution contained NaCl 9 g, KCl 0.2 g, $CaCl_2$ 0.2 g, water 1 liter, and had a pH of 6.8.

[b]Calculated by compiler.

AUXILIARY INFORMATION

METHOD/APPARATUS/PROCEDURE:

Sulfapyridine was shaken up with the bicarbonate-free Locke's soln for many hours in a tube which was corked to prevent loss of CO_2. The supernatant was filtered through a paper, dild in a hot room to prevent pptn, and sulfapyridine was detd by the method of Marshall and Litchfield (1).

SOURCE AND PURITY OF MATERIALS:

Nothing specified.

ESTIMATED ERROR:

Soly: average of 3 detns has been given (authors).

Temp: not specified.

REFERENCES:

1. Marshall, E. K., Jr.; Litchfield, J. T., Jr. *Science* 1938, *88*, 85.

COMPONENTS:

(1) Benzenesulfonamide, 4-amino-N-2-pyridinyl- (sulfapyridine); $C_{11}H_{11}N_3O_2S$; [144-83-2]

(2) Ethanol; C_2H_6O; [64-17-5]

(3) Water; H_2O; [7732-18-5]

ORIGINAL MEASUREMENTS:

Sapozhnikova, N. V.; Postovskii, I. Ya.

Zh. Prikl. Khim. 1944, *17*, 427-34.

VARIABLES:

Concentration of ethanol, temperature

PREPARED BY:

R. Piekos

EXPERIMENTAL VALUES:

Concentration of ethanol	Solubility			
	at 37°C		at 75°C	
Weight%	Weight%	10^2 mol kg^{-1} solvent[a]	Weight%	10^2 mol kg^{-1} solvent[a]
0.0	0.047	0.19	0.25	1.00
19.2	0.32	1.29	0.91	3.68
38.3	0.37	1.49	1.57	6.40
57.6	0.58	2.34	3.00	12.4
67.2	-	-	3.10	12.8
76.4	0.54	2.18	3.0	12.4
96.0	0.39	1.57	2.3	9.4

[a]Calculated by compiler

AUXILIARY INFORMATION

METHOD/APPARATUS/PROCEDURE:

Sulfapyridine was dissolved in EtOH-water mixts to form satd solns which were occasionally agitated in glass vessels immersed in a thermostat. The equilibrium was usually attained after 1 h. Five- to 100-cm^3 samples of the satd soln were placed in Pt crucibles or dishes and evapd to dryness at temps lower than 110-115°C. The residue was dried to const wt at 105-110°C and weighed.

SOURCE AND PURITY OF MATERIALS:

Pure, recrystd sulfapyridine was used. Its mp conformed to that reported in the literature.

The purity of ethanol and water was not specified.

ESTIMATED ERROR:

Soly: quite reliable results were obtained (authors).

Temp: ±0.05°C (authors).

REFERENCES:

COMPONENTS:

(1) Benzenesulfonamide, 4-amino-N-2-pyridinyl- (sulfapyridine); $C_{11}H_{11}N_3O_2S$; [144-83-2]

(2) Urea; CH_4NO; [57-13-6]

(3) Water; H_2O; [7732-18-5]

ORIGINAL MEASUREMENTS:

Sobin, S. S.

J. Lab. Clin. Med. 1942, *27*, 1657-8.

VARIABLES:

Concentration of urea

PREPARED BY:

R. Piekos

EXPERIMENTAL VALUES:

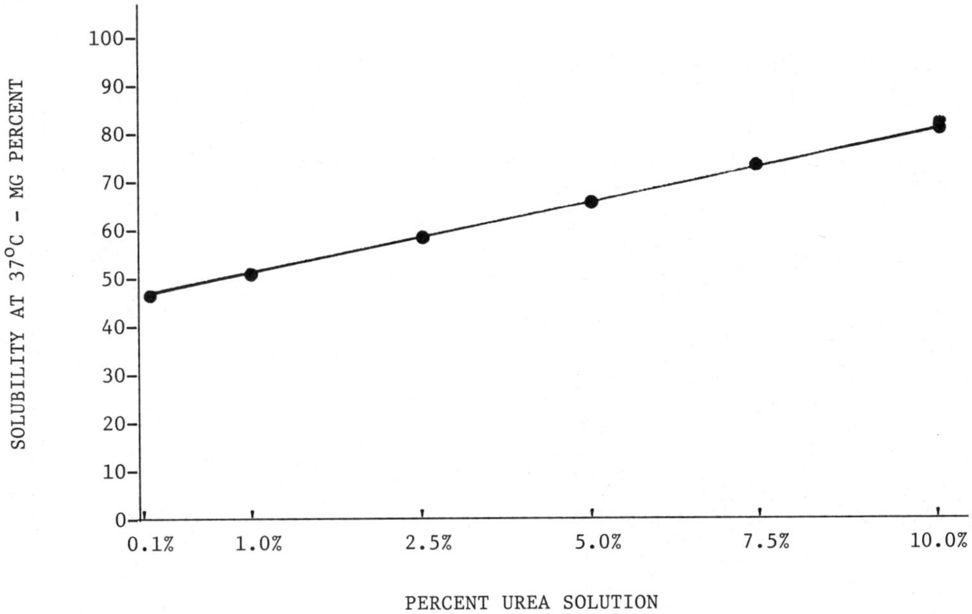

Solubility in a 10 per cent solution at 37°C is 73.3 mg per 100 cm^3 (2.94×10^{-3} mol dm^{-3}, compiler).

AUXILIARY INFORMATION

METHOD/APPARATUS/PROCEDURE:

Urea solns of varying concns from 0.1 to 10% were incubated at 37°C with an excess of sulfapyridine, shaken at intervals, and filtered through two thicknesses of Whatman No. 42 filter paper. After appropriate diln the free sulfonamide was detd by the method of Bratton and Marshall (1) using the Evelyn colorimeter and a No. 540 filter.

SOURCE AND PURITY OF MATERIALS:

Nothing specified.

ESTIMATED ERROR:

Nothing specified.

REFERENCES:

1. Bratton, A. C.; Marshall, E. K., Jr. *J. Biol. Chem.* 1939, *128*, 537.

COMPONENTS:

(1) Benzenesulfonamide, 4-amino-N-2-pyridinyl- (sulfapyridine); $C_{11}H_{11}N_3O_2S$; [144-83-2]

(2) Ethanol; C_2H_6O; [64-17-5]

(3) 1,2,3-Propanetriol; C_3H_8O; [56-81-5]

(4) Water; H_2O; [7732-18-5]

ORIGINAL MEASUREMENTS:

Dolique, R.; Foucault, J.

Trav. soc. pharm. Montpellier 1952, *12*, 145-53.

VARIABLES:

One temperature: 26-28°C

PREPARED BY:

R. Piekos

EXPERIMENTAL VALUES:

Solubility of sulfapyridine in a mixture of 1,2,3-propanetriol and 95% ethanol (2:1 by wt) at 26-28°C is 0.575% (2.32×10^{-2} mol kg^{-1} solvent, compiler).

AUXILIARY INFORMATION

METHOD/APPARATUS/PROCEDURE:

The sulfapyridine content was detd by diazotization of the amine group in a cold acidified 0.1N KNO_2 soln. an excess of KNO_2 was detected by using iodinated starch.

SOURCE AND PURITY OF MATERIALS:

Nothing specified

ESTIMATED ERROR:

Nothing specified

REFERENCES:

COMPONENTS:

(1) Benzenesulfonamide, 4-amino-N-2-pyridinyl- (sulfapyridine); $C_{11}H_{11}N_3O_2S$; [144-83-2]

(2) Ethanol; C_2H_6O; [64-17-5]

(3) 1,2,3-propanetriol; $C_3H_8O_3$; [56-81-5]

(4) Urea; CH_4N_2O; [57-13-6]

(5) Water; H_2O; [7732-18-5]

ORIGINAL MEASUREMENTS:

Dolique, R.; Foucault, J.

Trav. soc. pharm. Montpellier 1952, *12*, 145-53.

VARIABLES:

One temperature: 26-28°C

PREPARED BY:

R. Piekos

EXPERIMENTAL VALUES:

Solubility of sulfapyridine at 26-28°C in a saturated solution of urea in a mixture of 1,2,3-propanetriol and 95° ethanol (2:1 by wt.), containing 54.5 g of urea per 100 g of the mixture, is 0.78% (3.2×10^{-2} mol kg^{-1} solvent, compiler).

AUXILIARY INFORMATION

METHOD/APPARATUS/PROCEDURE:

The sulfapyridine content was detd by diazotization of the amine group in a cold acidified 0.1N KNO_2 soln. An excess of KNO_2 was detected by using iodinated starch.

SOURCE AND PURITY OF MATERIALS:

Nothing specified.

ESTIMATED ERROR:

Nothing specified.

REFERENCES:

COMPONENTS:

(1) Benzenesulfonamide, 4-amino-N-2-pyridinyl- (sulfapyridine); $C_{11}H_{11}N_3O_2S$; [144-83-2]

(2) 1H-Purine-2,6-dione, 3,7-dihydro-1,3,7-trimethyl- (caffeine); $C_8H_{10}N_4O_2$; [58-08-2]

(3) Water; H_2O; [7732-18-5]

ORIGINAL MEASUREMENTS:

Neish, W. J. P. *Rec. trav. chim.* 1948, *67*, 361-71.

VARIABLES:

Concentration of caffeine

PREPARED BY:

R. Piekos

EXPERIMENTAL VALUES:

Concentration of caffeine g/100 ml	Solubility of sulfapyridine at 37°C	
	γ /ml	10^3 mol dm^{-3} [a]
0.5	470	1.88
0.75	480	1.92

AUXILIARY INFORMATION

METHOD/APPARATUS/PROCEDURE:

A suspension of sulfapyridine in caffeine soln was kept for 5 h at 37°C and 1 h at room temp before filtration. Soly was detd by the Westfall's method (1) based on diazotization of the sulfonamide, coupling with Na 2-napthol-3,6-disulfonate and comparing the color with that of a std soln in a Klett colorimeter.

SOURCE AND PURITY OF MATERIALS:

Sulfapyridine: not specified.

Anhydrous caffeine was a good commercial product (source not specified).

Distd water was used.

ESTIMATED ERROR:

Nothing specified.

REFERENCES:

1. Westfall, B. B. *J. Nat. Cancer Inst.* 1945, *6*, 23.

COMPONENTS:

(1) Benzenesulfonamide, 4-amino-N-2-pyridinyl- (sulfapyridine); $C_{11}H_{11}N_3O_2S$; [144-83-2]

(2) Methane, trichloro- (chloroform); $CHCl_3$; [67-66-3]

ORIGINAL MEASUREMENTS:

Yamazaki, M.; Aoki, M.; Kamada, A.; Yata, N. *Yakuzaigaku*, 1967, *27(1)*, 37-40.

VARIABLES:

One temperature: 30°C

PREPARED BY:

R. Piekos

EXPERIMENTAL VALUES:

Solubility of sulfapyridine in chloroform at 30°C is 3.27 mmol/L (0.815 g dm^{-3}, compiler).

AUXILIARY INFORMATION

METHOD/APPARATUS/PROCEDURE:

Sulfapyridine (0.5 g) was placed in an L-shaped tube together with 20 ml of chloroform. The mixt was shaken in a thermostat until equilibrium was attained. The sulfapyridine was then assayed in the supernatant spectrophotometrically at 545 nm on a Beckmann DU spectrophotometer. The results were taken from a calibration graph.

SOURCE AND PURITY OF MATERIALS:

Nothing specified.

ESTIMATED ERROR:

Soly: not specified.

Temp: ±1°C (authors).

REFERENCES:

COMPONENTS:

(1) Benzenesulfonamide, 4-amino-N-2-pyridinyl- (sulfapyridine); $C_{11}N_{11}N_3O_2S$; [144-83-2]

(2) Methane, trichloro-; $CHCl_3$; [67-66-3]

ORIGINAL MEASUREMENTS:

Kitao, K.; Kubo, K.; Morishita, T.; Yata, N.; Kamada, A.

1973, *21*, 2417-26.

VARIABLES:

One temperature: 37°C

PREPARED BY:

R. Piekos

EXPERIMENTAL VALUES:

Solubility of sulfapyridine in $CHCl_3$ at 37°C is 2.86 mmol dm^{-3} solution.

AUXILIARY INFORMATION

METHOD/APPARATUS/PROCEDURE:

One ml of the sulfapyridine soln in $CHCl_3$ at equilibrium was taken into a test tube. After evapn of the solvent, the residue was dissolved in 1N NaOH, the soln was properly dild with deionized water and the concn of sulfapyridine was detd by diazotization.

SOURCE AND PURITY OF MATERIALS:

Comm available sulfapyridine was used as supplied.

Neither source nor purity of $CHCl_3$ was specified.

ESTIMATED ERROR:

Soly: not specified.

Temp: ±1°C (authors).

REFERENCES:

COMPONENTS:	ORIGINAL MEASUREMENTS:
(1) Benzenesulfonamide, 4-amino-N-2-pyridinyl- (sulfapyridine); $C_{11}H_{11}N_3O_2S$; [144-83-2] (2) 2-Propanol; C_3H_8O; [67-63-0]	Burlage, H. M. *J. Am. Pharm. Assoc., Sci. Ed.* 1948, *37*, 345.
VARIABLES:	**PREPARED BY:**
One temperature: 25°C	R. Piekos

EXPERIMENTAL VALUES:

Solubility of sulfapyridine in 2-propanol at 25°C is 0.1750 g/100 cm^3 solution (7.020 x 10^{-3} mol dm^{-3}, compiler).

AUXILIARY INFORMATION

METHOD/APPARATUS/PROCEDURE:

Satd solns of sulfapyridine in 2-propanol were prepd at 25°C and definite vols of the solns were measured into tared dishes by means of standard pipets. The alcohol was allowed to evap at room temp and the residue was dried at 105°C. In the case of losses due to apparent decompn, the residue was dried in a desiccator (1).

SOURCE AND PURITY OF MATERIALS:

The sulfapyridine was manufd by Merck and was of the U.S.P. purity. The source and purity of 2-propanol were not specified.

ESTIMATED ERROR:

Nothing specified.

REFERENCES:

1. Burlage, H. M. *J. Am. Pharm. Assoc., Sci. Ed.* 1947, *36(1)*, 16.

COMPONENTS:	ORIGINAL MEASUREMENTS:
(1) Benzenesulfonamide, 4-amino-N-2-pyridinyl- (sulfapyridine); $C_{11}H_{11}N_3O_2S$; [144-83-2] (2) 2-Propanone (acetone); C_3H_6O; [67-64-1]	Gutierrez, F. H. *Anales fis. quim. (Madrid)* 1945, *41*, 537-60.
VARIABLES: Temperature	**PREPARED BY:** R. Piekos

EXPERIMENTAL VALUES:

t/°C	G^a	E^b	X_g/l^c	mol/l[d] acetone	mmol/mol acetone	$1:X_g^e$	$1 + X_{cc}^f$
0	0.791	0.785	6.443	26	1.8	126.42	155.28
5	0.825	0.818	6.663	27	1.9	121.21	150.15
10	1.058	1.047	8.496	34	2.2	94.52	117.79
15	1.447	1.426	11.534	46	3.4	69.11	86.73
20	1.613	1.587	12.762	51	3.8	62.00	78.37
25	1.902	1.866	14.935	60	4.4	52.63	66.93
30	2.416	2.359	18.828	76	5.6	41.39	53.11
35	2.999	2.717	23.191	91	7.0	33.44	43.12
40	3.552	3.430	27.238	109	8.3	28.15	36.34
45	4.310	4.132	32.816	132	10.0	23.20	30.47
50	5.180	4.924	39.135	156	12.1	19.30	25.55

$^aG = \frac{p\ 100}{P - p}$, where p and P are weights of solute and solution, resp.

Continued on the next page.

AUXILIARY INFORMATION

METHOD/APPARATUS/PROCEDURE:

A special all-glass app was contructed enabling the prepn of satd solns, agitation by bubbling a stream of acetone-satd N, filtration, and distn off the solvent without contact with air. Two exchangeable dissoln vessels of 15 and 8 cm^3 working capacity were used depending on the soly of solute. The app was immersed in a thermostat. The vols of acetone used were 15 or 5 cm^3, and the equilibration time was 2-2.5 h. The satd solns were filtered, weighed, the solvent was distd off, the residues were dried at 105°C, weighed, and examd for the presence of solvated acetone.

SOURCE AND PURITY OF MATERIALS:

The source of the materials was not specified. Pure, anhyd acetone was used. The absence of impurities and water was confirmed by procedures of the German Pharmacopeia VI and Spanish Pharmacopeia VIII.
The purity of sulfapyridine was not specified.

ESTIMATED ERROR:

Soly: measurements were repeated until 2 values not differeing in the second decimal were obtained (author).

Temp: ±0.1°C (author).

REFERENCES:

Continued from previous page.

[b] $E = \frac{G\ 100}{G + 100}$; [c] g/l acetone; [d] should be mmol/l acetone (compiler);

[e] g of acetone required to dissolve 1 g of solute; [f] volume (cm^3) of acetone required to dissolve 1 g of solute.

COMPONENTS:

(1) Benzenesulfonamide, 4-amino-N-2-pyridinyl- (sulfapyridine); $C_{11}H_{11}N_3O_2S$; [144-83-2]

(2) Methylcyclohexanone; $C_7H_{12}O$; [1331-22-2]

ORIGINAL MEASUREMENTS:

Barber, H. J.; Wilkinson, J. H. *Quart. J. Pharm. Pharmacol.* 1946, *19*, 248-55.

VARIABLES:

One temperature: 25°C

PREPARED BY:

R. Piekos

EXPERIMENTAL VALUES:

Approximate solubility of sulfapyridine in methylcyclohexanone at 25°C is 3.5 per cent w/v (0.14 mol dm^{-3} solution, compiler).

AUXILIARY INFORMATION

METHOD/APPARATUS/PROCEDURE:

Nothing specified.

SOURCE AND PURITY OF MATERIALS:

Nothing specified.

ESTIMATED ERROR:

Nothing specified.

REFERENCES:

COMPONENTS:	ORIGINAL MEASUREMENTS:
(1) Benzenesulfonamide, 4-amino-N-2-pyridinyl- (sulfapyridine); $C_{11}H_{11}N_3O_2S$; [114-83-2] (2) Methylcyclohexanone; $C_7H_{12}O$; [1331-22-2]	Barber, H. J.; Wilkinson, J. H. *Pharm. J.* 1946, 105-6.
VARIABLES: One temperature: 25°C	PREPARED BY: R. Piekos

EXPERIMENTAL VALUES:

Approximate solubility of sulfapyridine in methylcyclohexanone at 25°C is 3.5 per cent w/v (0.14 mol dm^{-3} solution, compiler).

AUXILIARY INFORMATION

METHOD/APPARATUS/PROCEDURE:	SOURCE AND PURITY OF MATERIALS:
Nothing specified.	Nothing specified.
	ESTIMATED ERROR: Nothing specified.
	REFERENCES:

COMPONENTS:	ORIGINAL MEASUREMENTS:
(1) Benzenesulfonamide, 4-amino-N-3-pyridinyl-; $C_{11}H_{11}N_3O_2S$; [599-81-5] (2) Water; H_2O; [7732-18-5]	Roblin, R. O., Jr.; Winnek, P. S. *J. Am. Chem. Soc.* 1940, *62*, 1999-2002.
VARIABLES:	**PREPARED BY:**
One temperature: 37°C	R. Piekos

EXPERIMENTAL VALUES:

Solubility of 4-amino-N-3-pyridinylbenzenesulfonamide in water at 37°C is 3.3 mg/100 cm^3 solution (1.3 x 10^{-4} mol dm^{-3}, compiler).

AUXILIARY INFORMATION

METHOD/APPARATUS/PROCEDURE:

Excess sulfonamide in water was heated and stirred on a steam bath for 30 min. The suspension was then agitated for 24 h in a thermostat at 37°C. A sample of the satd soln was withdrawn through a glass filter, dild, and analyzed by the modified Marshall method (1) using a General Electric recording spectrophotometer for comparing the colors with those of the standards.

SOURCE AND PURITY OF MATERIALS:

The sulfonamide, mp 258-9°C (dec), was probably prepd by the authors by reacting 3-aminopyridine with acetylsulfanilyl chloride followed by hydrolysis. Purity of the water was not specified.

ESTIMATED ERROR:

Nothing specified.

REFERENCES:

1. Bratton, A. C.; Marshall, E. K., Jr. *J. Pharmacol.* 1939, *66*, 4.

COMPONENTS:	ORIGINAL MEASUREMENTS:
(1) Hydroxycalcium, (4-amino-N-2-pyridinyl-benzenesulfonamidato)-; $C_{11}H_{11}CaN_3O_3S$; [77400-70-5] (2) 2-Propanone (acetone); C_3H_6O; [67-64-1]	Gutierrez, F. H. *Anales fis. quim. (Madrid)* 1945, *41*, 537-60.
VARIABLES:	**PREPARED BY:**
Temperature	R. Piekos

EXPERIMENTAL VALUES:

$t/^{o}C$	G^a	E^b	X_g/l^c	mol/l[d] acetone	mmol/mol acetone	$1:X_g^e$	$1 + X_{cc}^f$
0	1.554	1.526	12.659	41.5	2.9	64.35	78.99
5	1.256	1.244	10.159	33.0	2.4	79.62	98.45
10	0.843	0.841	6.809	22.6	1.6	117.92	146.86
15	0.754	0.748	6.010	19.7	1.4	132.63	166.39
20	0.654	0.649	5.264	17.2	1.2	152.91	189.96
25	0.559	0.555	4.389	14.4	1.05	179.61	227.84
30	0.543	0.540	4.232	13.9	1.03	184.16	236.29
35	0.480	0.478	3.712	12.2	0.91	208.25	269.39
40	0.455	0.453	3.492	11.4	0.86	229.80	286.36
45	0.420	0.418	3.198	10.5	0.79	238.09	312.69
50	0.383	0.382	2.894	9.5	0.73	261.09	345.51

$^{a}G = \frac{p\ 100}{P - p}$, where p and P are the weights of solute and solution, resp.

$^{b}E = \frac{G\ 100}{G + 100}$; [c] g/l acetone; [d] should be mmol/l acetone (compiler);

[e] g of acetone required to dissolve 1 g of solute; [f] volume (cm^3) of acetone required to dissolve 1 g of solute.

AUXILIARY INFORMATION

METHOD/APPARATUS/PROCEDURE:

A special all-glass app was constructed enabling the prepn of satd solns, agitation by bubbling a stream of acetone-satd N, filtration, and distn off the solvent without contact with air. Two exchangeable dissoln vessels used depending on the soly of solute. The app was immersed in a thermostat. The vols of acetone used were 15 or 8 cm^3, and the equilibration time was 2-2.5 h. The satd solns were filtered, weighed, the solvent was distd off, the residues were dried at 105°C, weighed, and examd for the presence of solvated acetone.

SOURCE AND PURITY OF MATERIALS:

The source of the materials was not specified. Pure, anhyd acetone was used. The absence of impurities and water in it was confirmed by procedures of the German Pharmacopeia VI and Spanish Pharmacopeia VIII. The purity of the solute was not specified.

ESTIMATED ERROR:

Soly: measurements were repeated until 2 values not differing in the second decimal were obtained (author).

Temp: ±0.1°C (author).

REFERENCES:

COMPONENTS:

(1) Hydroxycalcium, (4-amino-N-2-pyridinylbenzenesulfonamidato)-, dihydrate; $C_{11}H_{11}CaN_3O_3S \cdot 2H_2O$; [77400-71-6]

(2) 2-Propanone (acetone); C_3H_6O; [67-64-1]

ORIGINAL MEASUREMENTS:

Gutierrez, F. H.

Anales fis. quim. (Madrid) 1945, *41*, 537-60.

VARIABLES:

Temperature

PREPARED BY:

R. Piekos

EXPERIMENTAL VALUES:

t/°C	G[a]	E[b]	X_g/l[c]	mol/l[d] acetone	mmol/mol acetone	1:X_g[e]	1 + X_{cc}[f]
0	0.526	0.523	4.285	12.6	0.90	190.80	233.10
5	0.531	0.528	4.295	12.7	0.91	188.32	232.49
10	0.541	9.538	4.344	12.8	0.91	184.84	230.41
15	0.648	0.644	5.165	15.2	1.10	154.32	193.61
20	0.621	0.617	4.893	14.4	1.06	157.21	204.37
25	0.588	0.585	4.617	13.6	1.00	170.07	216.57
30	0.550	0.547	4.286	12.6	0.94	181.82	233.32
35	0.520	0.517	4.021	11.8	0.89	192.31	248.69
40	0.479	0.476	3.676	10.7	0.82	208.77	275.30
45	0.469	0.466	3.571	10.5	0.80	213.21	280.03
50	0.392	0.390	2.962	8.8	0.67	255.102	337.95

[a] $G = \frac{p\ 100}{P - p}$, where p and P are the weights of solute and solution, resp.

[b] $E = \frac{G\ 100}{G + 100}$; [c] g/l acetone; [d] should be mmol/l acetone (compiler);

[e] g of acetone required to dissolve 1 g of solute; [f] volume (cm^3) of acetone required to dissolve 1 g of solute.

AUXILIARY INFORMATION

METHOD/APPARATUS/PROCEDURE:

A special all-glass app was constructed enabling the prepn of satd solns, agitation by bubbling a stream of acetone-satd N, filtration, and distn off the solvent without contact with air. Two exchangeable dissoln vessels of 15 and 8 cm^3 working capacity were used depending on the soly of solute. The app was immersed in a thermostat. The vols of acetone used were 15 or 5 cm^3, and the equilibration time was 2-2.5 h. The satd solns were filtered, weighed, the solvent was distd off, the residues were dried at 105°C, weighed, and examd for the presence of solvated acetone.

SOURCE AND PURITY OF MATERIALS:

The source of the materials was not specified. Pure, anhyd acetone was used. The absence of impurities and water in it was confirmed by procedures of the German Pharmacopeia VI and Spanish Pharmacopeia VIII. The purity of the solute was not specified.

ESTIMATED ERROR:

Soly: measurements were repeated until 2 values not differing in the second decimal were obtained (author).

Temp: ±0.1°C (author).

REFERENCES:

COMPONENTS:	ORIGINAL MEASUREMENTS:
(1) Benzenesulfonamide, 4-amino-N-2-pyridinyl-, monosodium salt (sodium sulfapyridine); $C_{11}H_{10}N_3NaO_2S$; [127-57-1] (2) Water; H_2O; [7732-18-5]	Clark, W. G.; Strakosch, E. A.; Levitan, N.I. *J. Lab. Clin. Med.* 1942, *28*, 188-9.

VARIABLES:	PREPARED BY:
Temperature	R. Piekos

EXPERIMENTAL VALUES:

t/°C	Solubility	
	g/100 g water	mol kg^{-1} water[a]
25	52.0	1.92
37	80.0	2.95

AUXILIARY INFORMATION

METHOD/APPARATUS/PROCEDURE:

A small tinted glass container contg excess Na sulfapyridine in water was shaken in a water bath thermostat for 24 h. The satd soln was then filtered by aspiration through a washed and dried asbestos filter stick into a weighed weighing bottle. The entire app was kept at the temp at which the compd was dissolved. The amt dissolved was then detd by the method of Bratton and Marshall (1), using a photoelectric colorimeter.

SOURCE AND PURITY OF MATERIALS:

Neither source nor purity of Na sulfapyridine was specified.

CO_2-free distd water was used.

ESTIMATED ERROR:

Soly: not specified.

Temp: ±0.1°C (authors).

REFERENCES:

1. Bratton, A. C.; Marshall, E. K., Jr. *J. Biol. Chem.* 1939, *128*, 537.

COMPONENTS:

(1) Zinc, (T-4)-bis(4-amino-N-2-pyridinylbenzenesulfonamidato-N^N,O)-Zn(II) sulfapyridine); $C_{22}H_{18}N_6O_4S_2Zn$; [71261-89-7]

(2) Water; H_2O; [7732-18-5]

ORIGINAL MEASUREMENTS:

Fox, Ch. L., Jr.; Modak, S. Standford, J. W.; Fox, P. L. *Scand. J. Plast. Reconstr. Surg.* 1979, *13(1)*, 89-94.

VARIABLES:

One temperture: 28-30°C

PREPARED BY:

R. Piekos

EXPERIMENTAL VALUES:

Solubility of Zn(II) sulfapyridine in water at room temperature (28-30°C)[a] is 44.6 mg% (7.94 x 10^{-4} mol dm^{-3} solution, compiler).

[a]Value given by one of the authors (S. M.) in personal communication.

AUXILIARY INFORMATION

METHOD/APPARATUS/PROCEDURE:

Satd soln of Zn(II) sulfapyridine was prepd in water and after 24 h aliquots from the clear supernatant were assayed for sulfapyridine content using the colorimeter method of Bratton and Marshall (1). The soly value was then calcd from the molecular formula.

SOURCE AND PURITY OF MATERIALS:

The Zn(II) sulfapyridine was prepd by the authors as follows: an inorg Zn salt was reacted with Na salt of sulfapyridine and the ppt was analyzed and characterized. No details were given, however.
Purity of the materials were not specified.

ESTIMATED ERROR:

Nothing specified

REFERENCES:

1. Bratton, A. C.; Marshall, E. K., Jr. *J. Biol. Chem.* 1939, *120*, 537.

COMPONENTS:	ORIGINAL MEASUREMENTS:
(1) Benzenesulfonamide, 4-amino-N-2-pyridinyl-, monosilver salt; (Ag sulfapyridine); $C_{11}H_{10}AgN_3O_2S$ [24342-38-9] (2) Nitric acid; HNO_3; [7697-37-2] (3) Potassium nitrate; KNO_3; [7757-79-1] (4) Water; H_2O; [7732-18-5]	Nesbitt, R. U., Jr.; Sandmann, B. J. *J. Pharm. Sci.* 1978, *67(7)*, 1012-17.
VARIABLES: pH	PREPARED BY: R. Piekos

EXPERIMENTAL VALUES:

Comparison of Total Silver Sulfapyridine Molar Solubility, S, Determined by the Method of Known Subtraction with the Molar Concentration of the Silver Ion Determined by Direct Potentiometry on Identical Samples at 25±0.1°, 0.1M Ionic Strength, in Nitric Acid Buffer

	pH 2.294		pH 2.486	
	$S \times 10^3$	$[Ag^+] \times 10^3$	$S \times 10^3$	$[Ag^+] \times 10^3$
	8.973	8.143	5.472	7.425
	8.889	8.018	5.514	7.514
	9.030	8.143	5.579	7.632
	9.030	8.143	5.304	7.484
	8.889	8.018	5.514	7.514
			5.660	7.454
Mean	8.962	8.093	5.508	7.504

	pH 2.583		pH 2.848	
	$S \times 10^3$	$[Ag^+] \times 10^3$	$S \times 10^3$	$[Ag^+] \times 10^3$
	5.366	4.540	3.031	2.954
	5.336	4.488	2.974	2.874
	5.356	4.401	2.866	2.774
	5.346	4.505	2.979	2.863
	5.366	4.470	2.788	2.678
	5.356	4.436		
Mean	5.354	4.473	2.928	2.839

to be contd.

AUXILIARY INFORMATION

METHOD/APPARATUS/PROCEDURE:

Mixts of 100 g of Ag sulfapyridine and 25 or 27 ml of the nitric acid buffer were placed in paraffin-coated vials, adjusted to an ionic strength 0.1M with KNO_3, and rotated end over end in a thermostated bath until equilibrium soly was obtained (3-7 days). After filtration through 20M glass filtering crucibles, the solns were analyzed at 25±0.1 °C in paraffin coated beakers for Ag^+ ions with a silver-ion selective electrode (No. 94-16, Orion Res. Cambridge, Mass) standardized at the temp indicated and 0.1M ionic strength. The pH was measured with a triple-purpose pH electrode (Corning Sci. Instruments, Medfield, Mass) standardized using buffers meeting NBS requirements. The nitric acid buffers were prepd by diln of 0.1M HNO_3 and were adjusted to an ionic strength of 0.1M with KNO_3.

SOURCE AND PURITY OF MATERIALS:

All reagents were anal or USP grade. Ag sulfapyridine was prepd by the method of Rosenzweig and Fuchs (1) and recrystd from concd ammonia (2).

Water had a sp cond of (1-10) x 10^{-7} ohm^{-1} cm^{-1}.

The source of the reagent was not specified.

ESTIMATED ERROR:

Soly: not specified.

Temp: ±0.1°C (authors)

pH : accuracy ±0.001 pH unit (authors).

REFERENCES:

1. Rosenzweig, S.; Fuchs, W. *U.S. pat.* 2,536,095 (1951)
2. Sandmann, B.J.; Nesbitt, R. U., Jr. Sandmann, R. A.; *J. Pharm. Sci.* 1974, *63*, 948.

COMPONENTS:

(1) Benzenesulfonamide, 4-amino-N-2-pyridinyl-, monosilver salt, (Ag sulfapyridine); $C_{11}H_{10}AgN_3O_2S$; [24342-38-9]

(2) 4-Morpholinepropanesulfonic acid; $C_7H_{15}NO_4S$; [1132-61-2]

(3) 4-Morpholinepropanesulfonic acid, sodium salt; $C_7H_{14}NNaO_4S$; [71119-22-7]

(4) Potassium nitrate; KNO_3; [7757-79-1]

(5) Water; H_2O; [7732-18-5]

ORIGINAL MEASUREMENTS:

Nesbitt, R. U., Jr.; Sandmann, B. J.

J. Pharm. Sci. 1978, *67(7)*, 1012-17.

VARIABLES:

Hydronium-ion concentration

PREPARED BY:

R. Piekos

EXPERIMENTAL VALUES:

Equilibrium values of S^2 (S = total molar solubility) versus $[H_3O^+]$ for Ag sulfapyridine in 0.05M 4-morpholinepropanesulfonic acid buffer at 0.1M ionic strength (KNO_3) and 25±0.1°C.

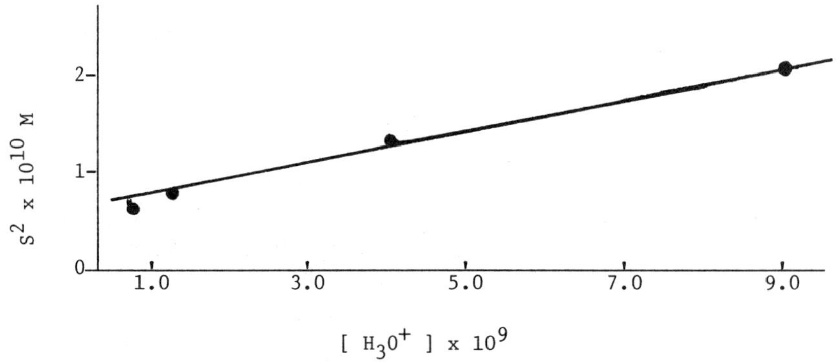

AUXILIARY INFORMATION

METHOD/APPARATUS/PROCEDURE:

Mixts of 100 mg of Ag sulfapyridine and 25 or 27 ml of the 4-morpholinepropanesulfonic acid buffer were placed in paraffin-coated vials, adjusted to an ionic strength 0.1M with KNO_3, and rotated end over end in a thermostated bath until equilibrium soly was obtained (3-7 days). After filtration through 20M glass filtering crucibles, the solns were analyzed at 25±0.1°C in paraffin-coated beakers for Ag^+ ions with a silver-ion selective electrode (No. 94-16, Orion Res., Cambridge, Mass) standardized at the temp indicated and 0.1M ionic strength. The pH was measured with a triple-purpose pH electrode (Corning Sci. Instruments, Medfield, Mass) standardized using buffers meeting NBS requirements. The buffers were prepd with a total molar concn of 0.05M and adjusted to an ionic strength of 0.1M with KNO_3

SOURCE AND PURITY OF MATERIALS:

All reagents used were anal or USP grade.

Ag sulfapyridine was prepd by the method of Rosenzweig and Fuchs (1) and recrystd from concd ammonia (2).

Water had a sp cond of (1-10) x 10^{-7} $ohm^{-1}cm^{-1}$

The buffer soln was from US Biochem. Corp., Cleveland, Ohio (purity not specified).

ESTIMATED ERROR:

Soly: not specified.

Temp: ±0.1°C (authors)

pH : accuracy ±0.001 pH unit (authors)

REFERENCES:

1. Rosenzweig, S.; Fuchs, W. *U.S. pat* 2,536,095 (1951)
2. Sandmann, B.J.; Nesbitt, R.U., Jr.; Sandmann, R. A. *J. Pharm.* 1974, *63*, 948.

COMPONENTS:

Continued from previous page

ORIGINAL MEASUREMENTS:

Nesbitt, R. U., Jr.; Sandmann, B. J.

J. Pharm. Sci. 1978, *67(7)*, 1012-17.

VARIABLES:

pH

PREPARED BY:

R. Piekos

EXPERIMENTAL VALUES:

Calculation of the Solubility Product of Ag Sulfapyridine[a], K_s, at 25°C and 0.1M Ionic Strength

pH	f_o	$[Ag^+]$	S	K_s
2.294	1.418×10^{-7}	8.093×10^{-3}	8.890×10^{-3}	1.02×10^{-11}
2.486	3.066×10^{-7}	5.508×10^{-3}	7.504×10^{-3}	1.27×10^{-11}
2.583	4.480×10^{-7}	4.473×10^{-3}	5.354×10^{-3}	1.07×10^{-11}
2.848	1.184×10^{-6}	2.828 x 10-3	2.937×10^{-3}	9.80×10^{-12}
			Mean	$(1.09 \pm 0.13)10^{-11}$

[a]from eq. $K_s = f_o\,[Ag^+]S$, where

$$f_o = \left(1 + \frac{[H_3O^+]}{K_2} + \frac{[H_3O^+]^2}{K_1\ K_2}\right)^{-1}$$

S is the total molar solubility, and K_1 and K_2 are the apparent dissociation constants of the N^4 - (amino) and N^1 - (amido) hydrogens of sulfapyridine, respectively.

[a]K_s reported as mean ± SD.

to be contd

AUXILIARY INFORMATION

METHOD/APPARATUS/PROCEDURE:

SOURCE AND PURITY OF MATERIALS:

ESTIMATED ERROR:

REFERENCES:

COMPONENTS:

(1) Ag sulfapyridine; $C_{11}H_{10}AgN_3O_2S$; [24342-38-9]

(2) Nitric acid; HNO_3; [7693-37-2]

(3) Potassium nitrate; KNO_3; [7757-79-1]

(4) Water; H_2O; [7732-18-5]

for details see previous page

ORIGINAL MEASUREMENTS:

Nesbitt, R. U., Jr.; Sandmann, B. J.

J. Pharm. Sci. 1978, *67(7)*, 1012-17.

VARIABLES:

pH

PREPARED BY:

R. Piekos

EXPERIMENTAL VALUES:

Molar solubility, S, of Ag sulfapyridine versus pH at 0.1M ionic strength and 25±0.1°C. Key: • calculated from the equation

$$S = \left(\frac{[H_3O^+]^2 K_s}{K_1 K_2} + \frac{K_s}{K_2} [H_3O^+] \right)^{0.5} \left(1 + \frac{f_o S}{K_d}\right),$$

where K_s is the solubility product of Ag sulfapyridine, and K_1, K_2, and K_d are the apparent dissociation constants of the N^4 (amido) hydrogen, N^1 (amino) hydrogens and Ag sulfapyridine, respectively; o , experimental values.

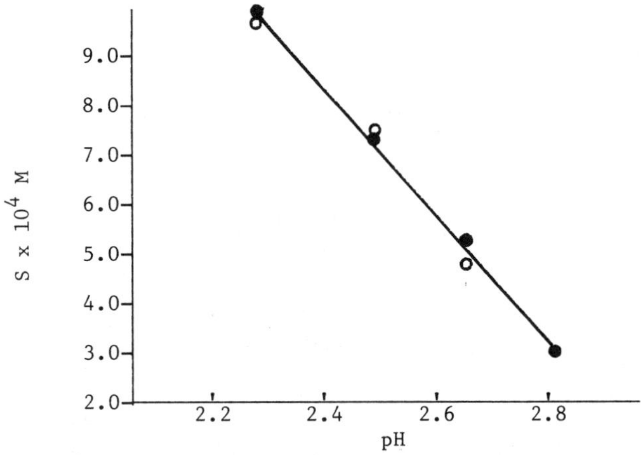

AUXILIARY INFORMATION

METHOD/APPARATUS/PROCEDURE:

SOURCE AND PURITY OF MATERIALS:

ESTIMATED ERROR:

REFERENCES:

COMPONENTS:	EVALUATOR:
(1) Acetamide, N-[4-[(2-pyridinylamino)-sulfonyl]phenyl]- (acetyl sulfapyridine) $C_{13}H_{13}N_3O_3S$; [19077-98-6] (2) Water	Anthony N. Paruta Department of Pharmaceutics University of Rhode Island Kingston, Rhode Island, USA and Ryszard Piekos Faculty of Pharmacy, University of Gdansk Gdansk, Poland 1986

CRITICAL EVALUATION:

Table I: Solubility of Acetyl sulfapyridine in water, 310K

Reference	10^3 mol dm^{-3} 310K
1	2.4
2	0.721
3	1.1
4	0.51
5	0.82 - 1.1 (pH 6.6-7.2)

Solubility values are reported for acetyl sulfapyridine in water at 310K and are shown in Table I. These values (1-6) are quite divergent and should not be averaged, not even to an approximate degree. It would seem that a value of less that 2.0 x 10^{-3} mol dm^{-3} would be expected since the acetyl derivatives should possess a lower solubility that the unacetylated parent compound. An approximate value of 1 x 10^{-3} mol dm^{-3} can be suggested. This is about one-half the value for the parent compound.

In Table II, the solubility of acetyl sulfapyridine at 293K at various pH values are shown.

Table II: Solubility of Acetyl sulfapyridine at various pH levels, 293K

Reference	pH	10^3 mol dm^{-3} 293K
6	5.9	1.06
7	6.0	0.75
6	7.0	1.10
7	7.0	1.0
6	8.0	1.60
7	8.0	1.3

The values of Krüger-Thiemer (7) refer to two hour equilibrium which is considered inadequate. Pulver and Suter (8) do not specify the techniques used, their results, however are consistant. It appears that between a pH of 5 to 7, the solubility is about the same and an approximate value of 1 x 10^{-3} mol dm^{-3} in phosphate buffer 0.067 mol dm^{-3} at 293K is suggested. At a pH of 8, there should be an increase in solubility due to the formation of a greater concentration of water soluble species according to the Henderson-Hasselbach expression. The approximate solubility in phosphate buffer (0.067 mol dm^{-3}) at 293K is about 1.5 x 10^{-3} mol dm^{-3}.

REFERENCES:

(1) Hug, E. *Rev. soc. Argentina biol.* 1940, *16*, 662-6.
(2) Lebel, H.; Schroeder, E.; Simesen, M. *Acta Med. Scand.* 1940, *105(4)*, 395-410
(3) Roblin, R.O., Jr.; Williams, J.H.; Winnek, P.S.; English, J.P. *J. Am. Chem. Soc.* 1940, *62*, 2002-5.
(4) Durel, M.P.; Allinne, M. *Bull. Soc. Med. Hop. Paris* III 1941, 251-9.
(5) Sapozhnikova, N.V.; Postovskii, I.Ya. *Zh. Prikl. Khim.* 1944, *17*, 427-34.
(6) Langecker, H. *Arch. Exptl. Path. Pharmakol.* 1948, *205*, 291-301.
(7) Krüger-Thiemer, E. *Arch. Dermatol. Syphilis* 1942, *183*, 90-116.
(8) Pulver, R.; Suter, R. *Schweiz. Med. Wochenschr.* 1943, *73(13)*, 403-8.

COMPONENTS:

(1) Acetamide, N-[4-[(2-pyridinylamino)sulfonyl]phenyl]- (acetyl sulfapyridine); $C_{13}H_{13}N_3O_3S$; [19077-98-6]

(2) Water; H_2O; [7732-18-5]

ORIGINAL MEASUREMENTS:

Hug, E.; *Rev. soc. Argentina biol.* 1940, *16*, 662-6.

VARIABLES:

One temperature: 37°C

PREPARED BY:

R. Piekos

EXPERIMENTAL VALUES:

Solubility of acetyl sulfapyridine in water at 37°C is 60 mg% based on sulfapyridine (62 mg% as acetyl sulfapyridine = 2.4×10^{-3} mol dm^{-3} solution, compiler).

AUXILIARY INFORMATION

METHOD/APPARATUS/PROCEDURE:

A satd soln of acetyl sulfapyridine was prepd by heating on a water bath. The soln was then cooled down to 37°C and maintained at this temp for 7 days.

SOURCE AND PURITY OF MATERIALS:

Nothing specified.

ESTIMATED ERROR:

Nothing specified.

REFERENCES:

COMPONENTS:	ORIGINAL MEASUREMENTS:
(1) Acetamide, N-[4-[(2-pyridinylamino)sulfonyl]phenyl]- (acetyl sulfapyridine); $C_{13}H_{13}N_3O_3S$; [19077-98-6] (2) Water; H_2O; [7732-18-5]	Lebel, H.; Schroeder, E.; Simesen, M. *Acta. Med. Scand.* 1940, *105(4)*, 395-410.
VARIABLES:	PREPARED BY:
One temperature: 35°C	R. Piekos

EXPERIMENTAL VALUES:

Solubility of acetyl sulfapyridine in water at 35°C is 1:1750 corresponding to 57 mg% (2.1 x 10^{-3} mol dm^{-3} solution, compiler).

AUXILIARY INFORMATION

METHOD/APPARATUS/PROCEDURE:	SOURCE AND PURITY OF MATERIALS:
Nothing specified	Nothing specified
	ESTIMATED ERROR:
	Nothing specified
	REFERENCES:

COMPONENTS:	ORIGINAL MEASUREMENTS:
(1) Acetamide, N-[4-[(2-pyridinylamino) sulfonyl]phenyl]- (acetyl sulfapyridine); $C_{13}H_{13}N_3O_3S$; [19077-98-6] (2) Water; H_2O; [7732-18-5]	Roblin, R. O., Jr.; Williams, J. H.; Winnek, P. S.; English, J. P. *J. Am. Chem. Soc.* 1940, *62*, 2002-5.
VARIABLES:	**PREPARED BY:**
One temperature: 37°C	R. Piekos

EXPERIMENTAL VALUES:

Solubility of acetyl sulfapyridine in water at 37°C is 21.0 mg/100 cm^3 solution (7.21×10^{-4} mol dm^{-3}, compiler).

AUXILIARY INFORMATION

METHOD/APPARATUS/PROCEDURE:

Excess acetyl sulfapyridine in water was heated and stirred on a steam bath for 30 min. The suspension was then agitated for 24 h in a thermostat at 37°C. A sample of the satd soln was withdrawn through a glass filter, dild, and analyzed by the Marshall method (1) using a General Electric spectrophotometer for comparing the colors developed with those of the standards.

SOURCE AND PURITY OF MATERIALS:

Acetyl sulfapyridine was prepd by condensing recrystd acetylsulfanilyl chloride with 2-aminopyridine in a pyridine soln. The crude product was recrystd from EtOH or PrOH (2).
Purity of the water was not specified.

ESTIMATED ERROR:

Nothing specified

REFERENCES:

1. Bratton, A.C.; Marshall, E. K., Jr. *J. Pharmacol* 1939, *66*, 4.
2. Winterbottom, R. *J. Am. Chem. Soc.* 1940, *62*, 160.

COMPONENTS:	ORIGINAL MEASUREMENTS:
(1) Acetamide, N-[4[(2-pyridinylamino)sulfonyl]phenyl]- (acetyl sulfapyridine); $C_{13}H_{13}N_3O_3S$; [19077-98-6] (2) Water; H_2O; [7732-18-5]	Durel, M. P.; Allinne, M. *Bull. Soc. Med. Hop. Paris III* 1941, 251-9.
VARIABLES:	PREPARED BY:
One temperature: 37°C	R. Piekos

EXPERIMENTAL VALUES:

Solubility of acetyl sulfapyridine in water at 37°C is 0.32 g/liter (1.1×10^{-3} mol dm^{-3}, compiler).

AUXILIARY INFORMATION

METHOD/APPARATUS/PROCEDURE:	SOURCE AND PURITY OF MATERIALS:
A mixt of acetyl sulfapyridine and water was agitated for 24 hours at 37°C.	Source and purity of acetyl sulfapyridine was not specified. Distilled water was used.
	ESTIMATED ERROR:
	Nothing specified.
	REFERENCES:

COMPONENTS:

(1) Acetamide, N-[4-[(2-pyridinylamino)sulfonyl]phenyl]- $C_{13}H_{13}N_3O_3S$; [19077-98-6]

(2) Water; H_2O; [7732-18-5]

ORIGINAL MEASUREMENTS:

Sapozhnikova, N. V.; Postovskii, I. Ya.
Zh. Prikl. Khim. 1944, *17*, 427-34.

VARIABLES:

Temperature

PREPARED BY:

R. Piekos

EXPERIMENTAL VALUES:

t/°C	Solubility	
	Weight%	10^3 mol kg^{-1} water[a]
20	0.0056	0.19
37	0.015	0.51
50	0.026[b]	0.89
75	0.067	2.30
99	0.20	6.88

[a] calculated by compiler

[b] calculated from the heat of dissolution (10,480 cal mol^{-1})

AUXILIARY INFORMATION

METHOD/APPARATUS/PROCEDURE:

The sulfonamide was dissolved in water to form a satd soln which was occasionally agitated in a glass vessel immersed in a thermostat. The equilibrium was usually attained after 1 h. Five - to 100-cm^3 samples of the satd soln were placed in Pt crucibles or dishes and evapd to dryness at temps lower than 110-115°C. The residue was dried to const wt at 105-110°C and weighed.

SOURCE AND PURITY OF MATERIALS:

Pure, recrystd sulfonamide was used.

Its mp conformed to that reported in the literature.

Purity of the water was not specified.

ESTIMATED ERROR:

Soly: quite reliable results were obtained over the temp range 20-75°C. At higher temps the accuracy was poor due to evapn of water during sampling (authors). Temp: ±0.05°C (authors)

REFERENCES:

COMPONENTS:

(1) Acetamide, N-[4-[(2-pyridinylamino)sulfonyl]phenyl]- (acetyl sulfapyridine); $C_{13}H_{13}N_3O_3S$; [19077-98-6]

(2) Water; H_2O; [7732-18-5]

ORIGINAL MEASUREMENTS:

Langecker, H.
Arch. Exptl. Path. Pharmakol. 1948, *205*, 291-301.

VARIABLES:

pH

PREPARED BY:

R. Piekos

EXPERIMENTAL VALUES:

pH	Solubility at 37°C	
	mg%	10^3 mol dm^{-3} [a]
7.2	31	1.1
6.6	24	0.82

[a] Calculated by compiler

AUXILIARY INFORMATION

METHOD/APPARATUS/PROCEDURE:

An excess of acetyl sulfapyridine in water was boiled for 1 h in a sealed ampul followed by keeping the soln at 37°C. Before assaying, the solute was treated with 2.6N NaOH soln (1) to cleave the acetyl group and the sulfapyridine was detd colorimetrically by the method of Bratton and Marshall (2) using a Havemann colorimeter (3), as well as by microanal detn of the solid residue.

SOURCE AND PURITY OF MATERIALS:

Source and purity of the materials were not specified.

ESTIMATED ERROR:

Nothing specified

REFERENCES:

1. Scudi, J. V. *J. Lab. Clin. Med.* 1940, *25*, 404.
2. Bratton, A. G.; Marshall, E. K., Jr. *J. Biol. Chem.* 1939, *128*, 537.
3. Havemann, R. *Klin. Wochenschr.* 1940, p. 503.

COMPONENTS:

(1) Acetamide, N-[4-[(2-pyridinylamino)sulfonyl]phenyl]- (acetyl sulfapyridine); $C_{13}H_{13}N_3O_3S$; [19077-98-6]

(2) Sodium hydroxide; NaOH; [1310-73-2]

(3) Water; H_2O; [7732-18-5]

ORIGINAL MEASUREMENTS:

Rose, F. L.; Martin, A. R.; Bevan, H. G. L.

J. Pharm. Exp. Therap. 1943, *77*, 127-42.

VARIABLES:

pH

PREPARED BY:

R. Piekos

EXPERIMENTAL VALUES:

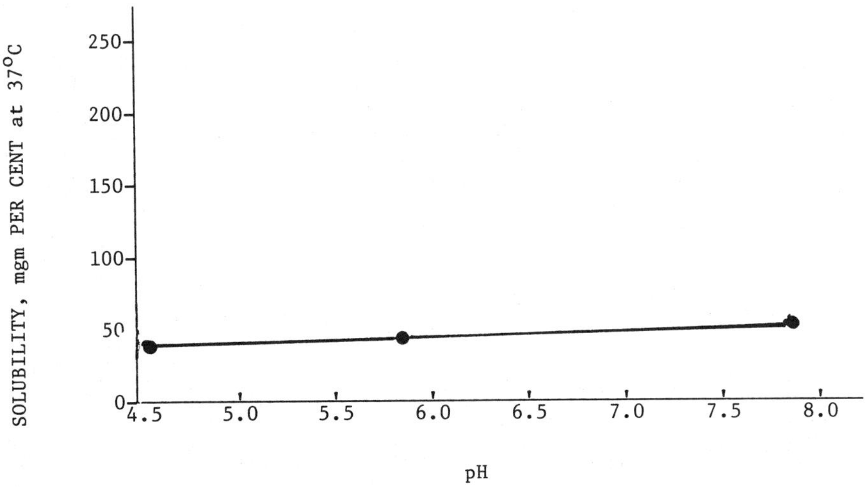

AUXILIARY INFORMATION

METHOD/APPARATUS/PROCEDURE:

An excess of acetyl sulfapyridine was stirred in boiling water, the soln was cooled to 37°C, the temp being maintained thermostatically, and 0.1N NaOH was added to increase the pH. The pH was measured by means of a glass electrode - calomel electrode half-cell system and was permitted to reach equilibrium before a reading was taken. The drug was then de-acetylated and the concn of sulfapyridine in soln was detd colorimetrically by withdrawing a sample through a filter-tip into a preheated micropipet.

SOURCE AND PURITY OF MATERIALS:

The source and purity of acetyl sulfapyridine were not specified.

Water was doubly distilled.

ESTIMATED ERROR:

Nothing specified

REFERENCES:

COMPONENTS:

(1) Acetamide, N-[4-[(2-pyridinylamino)sulfonyl]phenyl]- (acetyl sulfapyridine); $C_{13}H_{13}N_3O_3S$; [19077-98-6]

(2) Phosphoric acid, disodium salt; Na_2HPO_4 [7558-94-4]

(3) Water; H_2O; [7732-18-5]

ORIGINAL MEASUREMENTS:

Krüger-Thiemer, E.

Arch. Dermatol. Syphilis 1942, *183*, 90-116.

VARIABLES:

One temperature: ca 20°C; one pH: 8.74

PREPARED BY:

R. Piekos

EXPERIMENTAL VALUES:

Solubility of acetyl sulfapyridine in a 0.705M (10%) Na_2HPO_4 solution of pH 8.74 at room temperature (about 20°C) is 0.042 g% (1.4 x 10^{-3} mol dm^{-3} solution, compiler).

AUXILIARY INFORMATION

METHOD/APPARATUS/PROCEDURE:

Acetyl sulfapyridine (0.5 g) was dissolved in 10 cm^3 of the 0.705M (10%) Na_2HPO_4 soln of pH 8.74, shaken for 2 h at room temp (about 20°C), and filtered. The filtrate was treated with equal vol of 2N HCl and refluxed for 15 min. After proper diln, a 1-cm^3 aliquot was withdrawn, acidified, cooled, and the sulfonamide content was detd colorimetrically (as sulfapyridine) by the Marshall method modified by Kimmig (1) using an Authenrieth colorimeter. The pH was detd on an ultraionograph using a glass electrode.

SOURCE AND PURITY OF MATERIALS:

Acetyl sulfapyridine (source not specified) gave no coloration upon diazotization of its satd soln, thus showing absence of sulfapyridine. The source and purity of the remaining materials were not specified.

ESTIMATED ERROR:

Soly: precision ±5% (author).

Temp: not specified.

pH : ±0.05 pH unit (author).

REFERENCES:

1. Kimmig, J. *Arch. Dermatol.* 1938, *176*, 722; *Erg. Hyg.* 1941, *24*, 398.

COMPONENTS:

(1) Acetamide, N-[4-[(-pyridinylamino)sulfonyl]phenyl]- (acetyl sulfapyridine); $C_{13}H_{13}N_3O_3S$; [19077-98-6]

(2) Phosphoric acid, monopotassium salt; KH_2PO_4 [7778-77-0]

(3) Water; H_2O; [7732-18-5]

ORIGINAL MEASUREMENTS:

Krüger-Thiemer, E.

Arch. Dermatol. Syphilis 1942, *183*, 90-116.

VARIABLES:

One temperature: ca 20°C; one pH: 4.37

PREPARED BY:

R. Piekos

EXPERIMENTAL VALUES:

Solubility of acetyl sulfapyridine in a 0.735M (10%) KH_2PO_4 solution of pH 4.37 at room temperature (about 20°C) is 0.020 g% (6.9 x 10^{-4} mol dm^{-3} solution, compiler).

AUXILIARY INFORMATION

METHOD/APPARATUS/PROCEDURE:

Acetyl sulfapyridine (0.5 g) was dissolved in 10 cm^3 of the 0.735M (10%) KH_2PO_4 soln, shaken for 2 h at room temp (about 20°C), and filtered. The filtrate was treated with equal vol of 2N HCl, and refluxed for 15 min After proper diln, a 1-cm^3 aliquot was withdrawn, acidified, cooled, and the sulfonamide content was detd colorimetrically by the Marshall method modified by Kimmig (1) using an Authenrieth colorimeter. The pH was detd on an ultraionograph using a glass electrode.

SOURCE AND PURITY OF MATERIALS:

Acetyl sulfapyridine (source not specified) gave no coloration upon diazotization of its satd soln, thus showing absence of sulfapyridine. The source and purity of the remaining materials were not specified.

ESTIMATED ERROR:

Soly: precision ±5% (author).

Temp: not specified.

pH : ±0.05 pH unit (author).

REFERENCES:

1. Kimmig, J. *Arch. Dermatol.* 1938, *176*, 722; *Erg. Hyg.* 1941, *24*, 398.

COMPONENTS:

(1) Acetamide, N-[4-[(2-pyridinylamino)sulfonyl]phenyl]- (acetyl sulfapyridine); $C_{13}H_{13}N_3O_3S$; [19077-98-6]

(2) Phosphoric acid, disodium salt; Na_2HPO_4; [7558-94-4]

(3) Phosphoric acid, monopotassium salt; KH_2PO_4; [7778-77-0]

(4) Water; H_2O; [7732-18-5]

ORIGINAL MEASUREMENTS:

Krüger-Thiemer E.

Arch. Dermtol. Syphilis 1942, *183*, 90-116.

VARIABLES:

Temperature; pH

PREPARED BY:

R. Piekos

EXPERIMENTAL VALUES:

Composition of 1/15M phosphate buffer solutions			pH	Solubility			
				Room temp (ca 20°C)		37°C	
Na_2HPO_4	KH_2PO_4	%Content		g%	10^3 mol dm^{-3} solution[a]	g%	10^3 mol dm^{-3} solution[a]
1.0	99.0	0.91	4.944	0.030	1.03	-	-
10.0	90.0	0.91	5.906	0.031	1.06	0.030	1.03
61.1	38.9	0.93	7.005	0.032	1.10	0.033	1.13
9.5	0.5	0.733[b]	7.51	0.026	0.89	-	-
94.7	5.3	0.95	8.018	0.047	1.60	-	-

[a]Calculated by compiler.

[b]Molar content; 10% buffer solution.

AUXILIARY INFORMATION

METHOD/APPARATUS/PROCEDURE:

Acetyl sulfapyridine (0.5 g) was dissolved in 10 cm^3 of a buffer soln, shaken for 2 h at 20°C (or left for 48 h at 37°C), and filtered at respective temp. The filtrate was treated with equal vol of 2N HCl and refluxed for 15 min. After proper diln, a 1-cm^3 aliquot was withdrawn, acidified, cooled, and the sulfonamide content was detd (as sulfapyridine) colorimetrically by the Marshall method modified by Kimmig (1) using an Authenrieth colorimeter. The pH was detd on an ultraionograph using a glass electrode.

SOURCE AND PURITY OF MATERIALS:

Acetyl sulfapyridine (source not specified) gave no coloration upon diazotization of its satd soln, thus showing absence of sulfapyridine. The source and purity of the remaining materials were not specified.

ESTIMATED ERROR:

Soly: precision ±5% (author).

Temp: not specified.

pH : ±0.05 pH unit (author).

REFERENCES:

1. Kimmig, J. *Arch. Dermatol.* 1938, *176*, 722; *Erg. Hyg.* 1941, *24*, 398.

COMPONENTS:

(1) Acetami e, N-[4-[(2-pyridinylamino)sulfonyl]phenyl]- (acetyl sulfapyridine); $C_{13}H_{13}N_3O_3S$; [19077-98-6]

(2) Phosphoric acid, disodium salt; Na_2HPO_4; [7558-94-4]

(3) Phosphoric acid, monopotassium salt; KH_2PO_4; [7778-77-0]

(4) Water; H_2O; [7732-18-5]

ORIGINAL MEASUREMENTS:

Pulver, R.; Suter, R.

Schweiz. Med. Wochenschr. 1943, *73(13)*, 403-8.

VARIABLES:

pH

PREPARED BY:

R. Piekos

EXPERIMENTAL VALUES:

pH	Solubility of acetyl sulfapyridine in M/15 phosphate buffers (according to Sørensen) at 20°C	
	mg%	10^3 mol dm^{-3} [a]
6.0	22	0.75
7.0	30	1.0
8.0	39	1.3

[a]Calculated by compiler

AUXILIARY INFORMATION

METHOD/APPARATUS/PROCEDURE:

Nothing specified

SOURCE AND PURITY OF MATERIALS:

Nothing specified

ESTIMATED ERROR:

Nothing specified

REFERENCES:

COMPONENTS:

(1) Acetamide, N-[4-[(2-pyridinylamino)-sulfonyl]phenyl]- ($\underline{N}^4$-acetyl sulfapyridine); $C_{13}H_{13}N_3O_3S$; [19077-98-6]

(2) Phosphoric acid, disodium salt; Na_2HPO_4; [7558-94-4]

(3) Phosphoric acid, monopotassium salt; KH_2PO_4; [7778-77-0]

(4) Water; H_2O; [7732-18-5]

ORIGINAL MEASUREMENTS:

Hekster, Y. A.; Vree, T. B.
Damsma, J. E.; Friesen, W. T.
J. Antimicrob. Chemother. 1981, *8*, 133-44.

VARIABLE:

pH

PREPARED BY:

R. Piekos

EXPERIMENTAL VALUES:

pH	Solubility at 25°C mg/l	10^3 mol dm^{-3} [a]
5.5	313	1.07
7.5	440	1.51

[a]Calculated by compiler

AUXILIARY INFORMATION

METHOD/APPARATUS/PROCEDURE:

Satd solns of $\underline{N}^4$-acetylsulfapyridine were prepd in phosphate buffers of pH 5.5 and 7.5 at room temp (25°C). The concn of the solute was measured by means of a Spectra Physics 3500B high-performance liquid chromatograph equipped with a column oven (Model 748) and a Pye-Unicam LC-UV spectrophotometric detector. The detector was connected to a 1-mV recorder. A stainless steel column (10 cm x 4.6 mm i.d.) was packed with Lichrosorb RPS 5 μm, obtained from Chrompack. An injection loop of 100μl was used. The oven temp was 40° C. Detection of $\underline{N}^4$-acetylsulfapyridine was performed at 260 nm.

SOURCE AND PURITY OF MATERIALS:

The source and purity of the materials were not specified.

ESTIMATED ERROR:

The detection limit of the solute was 0.5 mg/l (authors).
The error in temperature and pH was not specified.

REFERENCES:

COMPONENTS:	ORIGINAL MEASUREMENTS:
(1) Benzenesulfonamide, 4-amino-N-(5-bromo-2-pyridinyl)-; $C_{11}H_{10}BrN_3O_2S$; [16805-99-5] (2) Water; H_2O; [7732-18-5]	Roblin, R. O., Jr.; Winnek, P. S. *J. Am. Chem. Soc.* 1940, *62*, 1999-2002.
VARIABLES:	**PREPARED BY:**
One temperature: 37°C	R. Piekos

EXPERIMENTAL VALUES:

Solubility of 4-amino-N-(5-bromo-2-pyridinyl)benzenesulfonamide in water at 37°C is 3.8 mg/100 cm^3 solution (1.2 x 10^{-4} mol dm^{-3}, compiler).

AUXILIARY INFORMATION

METHOD/APPARATUS/PROCEDURE:

Excess sulfonamide in water was heated and stirred on a steam bath for 30 min. The suspension was then agitated for 24 h in a thermostat at 37°C. A sample of the satd soln was withdrawn through a glass filter, dild, and analyzed by the modified Marshall method (1) using a General Electric recording spectrophotometer for comparing the colors with those of the standards.

SOURCE AND PURITY OF MATERIALS:

The sulfonamide, mp 199-200°C (cor), was prepd by the authors. Anal: %C 40.2 (calcd 40.2); %H 3.0 (3.0).
Purity of the water was not specified.

ESTIMATED ERROR:

Nothing specified

REFERENCES:

1. Bratton, A. C.; Marshall, E. K., Jr. *J. Pharmacol.* 1939, *66*, 4.

COMPONENTS:	ORIGINAL MEASUREMENTS:
(1) Benzenesulfonamide, 4-amino-N-(5-iodo-2-pyridinyl); $C_{11}H_{10}N_3O_2S$; [71119-21-6] (2) Water; H_2O; [7732-18-5]	Roblin, R. O., Jr.; Winnek P. S. *J. Am. Chem. Soc.* 1940, *62*, 1990-2002.
VARIABLES: One temperature: 37°C	PREPARED BY: R. Piekos

EXPERIMENTAL VALUES:

Solubility of 4-amino-N(5-iodo-2-pyridinyl)benzenesulfonamide in water at 37°C is 1.3 mg/100 cm^3 solution (3.5×10^{-5} mol dm^{-3}, compiler).

AUXILIARY INFORMATION

METHOD/APPARATUS/PROCEDURE:

Excess sulfonamide in water was heated and stirred on a steam bath for 30 min. The suspension was then agitated for 24 h in a thermostat at 37°C. A sample of the satd soln was withdrawn through a glass filter, dild, and analyzed by the modified Marshall method (1) using a General Electric recording spectrophotometer for comparing the colors with those of the standards.

SOURCE AND PURITY OF MATERIALS:

The sulfonamide, mp 220-1°C (cor), was prepd by the authors. Anal: %C 35.2 (calcd 35.2); %H 2.5 (2.7). Purity of the water was not specified.

ESTIMATED ERROR:

Nothing specified

REFERENCES:

1. Bratton, A. C.; Marshall, E. K., Jr. *J. Pharmacol.* 1939, *66*, 4.

COMPONENTS:

(1) Benzenesulfonamide, 4-amino-N-(5-nitro-2-pyridinyl)-; $C_{11}H_{10}N_4O_4S$; [39588-36-8]

(2) Water; H_2O; [7732-18-5]

ORIGINAL MEASUREMENTS:

Roblin, R. O., Jr.; Winnek, P. S.

J. Am. Chem. Soc. 1940, *62*, 1999-2002.

VARIABLES:

One temperature: 37°C

PREPARED BY:

R. Piekos

EXPERIMENTAL VALUES:

Solubility of 4-amino-N-(5-nitro-2-pyridinyl)benzenesulfonamide in water at 37°C is 3.7 mg/100 cm^3 solution (1.3 x 10^{-4} mol dm^{-3}, compiler).

AUXILIARY INFORMATION

METHOD/APPARATUS/PROCEDURE:

Excess sulfonamide in water was heated and stirred on a steam bath for 30 min. The suspension was then agitated for 24 h in a thermostat at 37°C. A sample of the satd soln was withdrawn through a glass filter, dild, and analyzed by the modified Marshall method (1) using a General Electric recording spectrophotometer for comparing the colors with those of the standards.

SOURCE AND PURITY OF MATERIALS:

The sulfonamide, mp 220-1°C (cor) was prepd by the authors. Anal: %C 45.0 (calcd 44.9); %H 3.2 (3.4).

Purity of the water was not specified.

ESTIMATED ERROR:

Nothing specified

REFERENCES:

1. Bratton, A. C.; Marshall, E. K., Jr. *J. Pharmacol.* 1939, *66*, 4.

COMPONENTS:

(1) Benzenesulfonamide, 4-amino-N-(5-amino-2-pyridinyl)-; $C_{11}H_{12}N_4O_2S$; [16840-28-1]

(2) Water; H_2O; [7732-18-5]

ORIGINAL MEASUREMENTS:

Roblin, R. O., Jr.; Winnek, P. S.

J. Am. Chem. Soc. 1940, *62*, 1999-2002.

VARIABLES:

One temperature: 37°C

PREPARED BY:

R. Piekos

EXPERIMENTAL VALUES:

Solubility of 4-amino-N-(5-amino-2-pyridinyl)benzenesulfonamide in water at 37°C is 418 mg/100 cm^3 solution (1.58×10^{-2} mol dm^{-3}, compiler).

AUXILIARY INFORMATION

METHOD/APPARATUS/PROCEDURE:

Excess sulfonamide in water was heated and stirred on a steam bath for 30 min. The suspension was then agitated for 24 h in a thermostat at 37°C. A sample of the satd soln was withdrawn through a glass filter, dild, and analyzed by the modified Marshall method (1) using a General Electric recording spectrophotometer for comparing the colors with those of the standards.

SOURCE AND PURITY OF MATERIALS:

The sulfonamide, mp 157-8°C (cor), was prepd by the authors. Anal: %C 49.8 (calcd 50.0); %H 4.5 (4.5).

Purity of the water was not specified.

ESTIMATED ERROR:

Nothing specified

REFERENCES:

1. Bratton, A. C.; Marshall, E. K., Jr. *J. Pharmacol.* 1939, *66*, 4.

COMPONENTS:

(1) Benzenesulfonamide, 4-amino-N-(3-ethoxy-2-pyridinyl)-; $C_{13}H_{15}N_3O_3S$; [71119-19-2]

(2) Water; H_2O; [7732-18-5]

ORIGINAL MEASUREMENTS:

Roblin, R. O., Jr.; Winnek, P. S.

J. Am. Chem. Soc. 1940, *62*, 1999-2002.

VARIABLES:

One temperature: 37°C

PREPARED BY:

R. Piekos

EXPERIMENTAL VALUES:

Solubility of 4-amino-N-(3-ethoxy-2-pyridinyl)benzenesulfonamide in water at 37°C is 23.5 mg/100 cm^3 (8.47×10^{-4} mol dm^{-3}, compiler).

AUXILIARY INFORMATION

METHOD/APPARATUS/PROCEDURE:

Excesss sulfonamide in water was heated and stirred on a steam bath for 30 min. The suspension was then agitated for 24 h in a thermostat at 37°C. A sample of the satd soln was withdrawn through a glass filter, dild, and analyzed by the modified Marshall method (1) using a General Electric recording spectrophotometer for comparing the colors with those of the standards.

SOURCE AND PURITY OF MATERIALS:

The sulfonamide, mp 198-200°C (cor), was prepd by the authors. Anal: %C 53.0 (calcd 53.2); %H 5.0 (5.1).

Purity of the water was not specified.

ESTIMATED ERROR:

Nothing specified

REFERENCES:

1. Bratton, A. C.; Marshall, E. K., Jr. *J. Pharmacol.* 1939, *66*, 4.

COMPONENTS:

(1) Benzenesulfonamide, 4-amino-N-[1(2-hydroxyethyl)-1,2-dihydro-2-pyridinyl]-; $C_{13}H_{17}N_3O_3S$; [71119-27-2]

(2) Water; H_2O; [7732-18-5]

ORIGINAL MEASUREMENTS:

Shepherd, R. G.; Bratton, A. C.; Blanchard, K. C. *J. Am. Chem. Soc.* 1942, *64*, 2532-7.

VARIABLES:

One temperature: 37°C

PREPARED BY:

R. Piekos

EXPERIMENTAL VALUES:

Solubility of 4-amino-N-[1-(2-hydroxyethyl)-1,2-dihydro-2-pyridinyl]-benzenesulfonamide at 37°C is 440 mg% (1.5×10^{-2} mol dm^{-3} solution, compiler).

AUXILIARY INFORMATION

METHOD/APPARATUS/PROCEDURE:

The sulfonamide was assayed colorimetrically (1). No details were reported.

SOURCE AND PURITY OF MATERIALS:

The sulfonamide, m.p. 184-5°C, was synthesized by the authors. Analysis: %C 53.41 (calcd 53.23); %H 5.18 (5.15), %N 14.35 (14.33). Colorimetric factor 0.630 (calcd 0.587).
Purity of the water was not specified.

ESTIMATED ERROR:

Nothing specified.

REFERENCES:

1. Bratton, A. C.; Marshall, E. K., Jr. *J. Biol. Chem.* 1939, *128*, 537.

COMPONENTS:	ORIGINAL MEASUREMENTS:
(1) Benzenesulfonamide, 4-amino-N-(1-carboxymethyl-1,2-dihydro-2-pyridinyl)-; $C_{13}H_{15}N_3O_4S$; [71119-28-3] (2) Water; H_2O; [7732-18-5]	Shepherd, R. G.; Bratton, A. C.; Blanchard, K. C., *J. Am. Chem. Soc.* 1942, *64*, 2532-7.
VARIABLES:	PREPARED BY:
One temperature: 37°C	R. Piekos

EXPERIMENTAL VALUES:

Solubility of 4-amino-N-(1-carboxymethyl-1,2-dihydro-2-pyridinyl)benzenesulfonamide in water at 37°C is 754 mg% (2.45 x 10^{-2} mol dm^{-3} solution, compiler).

AUXILIARY INFORMATION

METHOD/APPARATUS/PROCEDURE:	SOURCE AND PURITY OF MATERIALS:
The sulfonamide was assayed colorimetrically (1). No details were reported.	The sulfonamide, m.p. 165°C (dec) was synthesized by the authors. Analysis: %C 50.75 (calcd 50.81), %H 4.37 (4.26), %N 13.59 (13.67). Colorimetric factor 0.570 (calcd 0.560). Purity of the water was not specified.
	ESTIMATED ERROR:
	Nothing specified
	REFERENCES:
	1. Bratton, A. C.; Marshall, E. K., Jr. *J. Biol. Chem.* 1939, *128*, 537.

COMPONENTS:

(1) Benzenesulfonamide, 4-amino-N-(2-chloro-5-pyridinyl)-; $C_{11}H_{10}ClN_3O_2S$; [34392-82-0]

(2) Water; H_2O; [7732-18-5]

ORIGINAL MEASUREMENTS:

Roblin, R. O., Jr.; Winnek, P. S.

J. Am. Chem. Soc. 1940, *62*, 1999-2002.

VARIABLES:

One temperature: 37°C

PREPARED BY:

R. Piekos

EXPERIMENTAL VALUES:

Solubility of 4-amino-N-(2-chloro-5-pyridinyl)benzenesulfonamide in water at 37°C is 18.0 mg/100 cm^3 (6.34×10^{-4} mol dm^{-3}, compiler).

AUXILIARY INFORMATION

METHOD/APPARATUS/PROCEDURE:

An excess of the sulfonamide was heated and stirred on a steam bath for 30 min. The suspension was then agitated for 24 h at 37°C. A sample of the satd. soln. was withdrawn through a sintered glass filter into a bottle held at the same temp. An aliquot of the satd. soln. was dild. and analyzed by the Marshall method (1). A General Electric recording spectrophotometer was used in comparing the colors developed with those of the standards.

SOURCE AND PURITY OF MATERIALS:

The sulfonamide: m.p. (cor.) 186-7°C.

Purity of the water was not specified.

ESTIMATED ERROR:

None specified

REFERENCES:

1. Bratton, A. C.; Marshall, E. K., Jr. *J. Pharmacol.* 1939, *66*, 4.

COMPONENTS:

(1) Benzenesulfonamide, 4-amino-N-(2-bromo-5-pyridinyl)-; $C_{11}H_{10}BrN_3O_2S$; [17103-43-4]

(2) Water; H_2O; [7732-18-5]

ORIGINAL MEASUREMENTS:

Roblin, R. O., Jr.; Winnek, P. S.

J. Am. Chem. Soc. 1940, *62*, 1999-2002.

VARIABLES:

One temperature: 37°C

PREPARED BY:

R. Piekos

EXPERIMENTAL VALUES:

Solubility of 4-amino-N-(2-bromo-5-pyridinyl)benzenesulfonamide in water at 37°C is 12.2 mg/100 cm^3 (3.72×10^{-4} mol dm^{-3} - compiler).

AUXILIARY INFORMATION

METHOD/APPARATUS/PROCEDURE:

An excess of the sulfonamide was heated and stirred on a steam bath for 30 min. The suspension was then agitated for 24 h at 37°C. A sample of the satd. soln. was withdrawn through a sintered glass filter into a bottle held at the same temp. An aliquot of the satd soln was dild and analyzed by the Marshall method (1). A General Electric recording spectrophotometer was used in comparing the colors developed with those of the standards.

SOURCE AND PURITY OF MATERIALS:

The sulfonamide: m.p. (cor.) 196-7°C.

Purity of the water was not specified.

ESTIMATED ERROR:

None specified

REFERENCES:

1. Bratton, A. C.; Marshall, E. K., Jr. *J. Pharmacol.* 1939, *66*, 4.

COMPONENTS:

(1) Benzenesulfonamide, 4-amino-N-(2-hydroxy-5-pyridinyl)-; $C_{11}H_{11}N_3O_3S$; [71119-20-5]

(2) Water; H_2O; [7732-18-5]

ORIGINAL MEASUREMENTS:

Roblin, R. O., Jr.; Winnek, P. S.
J. Am. Chem. Soc. 1940, *62*, 1999-2002.

VARIABLES:

One temperature: 37°C

PREPARED BY:

R. Piekos

EXPERIMENTAL VALUES:

Solubility of 4-amino-N-(2-hydroxy-5-pyridinyl)benzenesulfonamide in water at 37°C is 258 mg/100 cm^3 solution (9.72×10^{-3} mol dm^{-3}, compiler).

AUXILIARY INFORMATION

METHOD/APPARATUS/PROCEDURE:

Excess sulfonamide in water was heated and stirred on a steam bath for 30 min. The suspension was then agitated for 24 h in a thermostat at 37°C. A sample of the satd soln was withdrawn through a glass filter, dild, and analyzed by the modified Marshall method (1) using a General Electric recording spectrophotometer for comparing colors with those of the standards.

SOURCE AND PURITY OF MATERIALS:

The sulfonamide, mp 243-4°C (dec, cor) was prepd by the authors. Anal: %C 49.8 (calcd 49.8); %H 4.2 (4.2).
Purity of the water was not specified.

ESTIMATED ERROR:

Nothing specified

REFERENCES:

1. Bratton, A. C.; Marshall, E. K., Jr. *J. Pharmacol.* 1939, *66*, 4.

COMPONENTS:

(1) Benzenesulfonamide, 4-amino-N-(2-amino-5-pyridinyl)-; $C_{11}H_{12}N_4O_2S$; [17103-45-6]

(2) Water; H_2O; [7732-18-5]

ORIGINAL MEASUREMENTS:

Roblin, R. O., Jr.; Winnek, P. S.
J. Am. Chem. Soc. 1940, *62*, 1999-2002.

VARIABLES:

One temperature: 37°C

PREPARED BY:

R. Piekos

EXPERIMENTAL VALUES:

Solubility of 4-amino-N-(2-amino-5-pyridinyl)benzenesulfonamide in water at 37°C is 129 mg/100 cm^3 solution (4.88×10^{-3} mol dm^{-3}, compiler).

AUXILIARY INFORMATION

METHOD/APPARATUS/PROCEDURE:

Excess sulfonamide in water was heated and stirred on a steam bath for 30 min. The suspension was then agitated for 24 h in a thermostat at 37°C. A sample of the satd soln was withdrawn through glass filter, dild, and analyzed by the modified Marshall method (1) using a General Electric recording spectrophotometer for comparing the colors with those of the standards.

SOURCE AND PURITY OF MATERIALS:

The sulfonamide, mp 207-8°C (cor), was probably that synthesized by Winterbottom (2). Purity of the water was not specified.

ESTIMATED ERROR:

Nothing specified

REFERENCES:

1. Bratton, A. C.; Marshall, E. K., Jr. *J. Pharmacol.* 1939, *66*, 4.
2. Winterbottom, R. *J. Am. Chem. Soc.* 1940, *62*, 160.

COMPONENTS:

(1) Benzenesulfonamide, 4-amino-N-(2-ethoxy-5-pyridinyl)-; $C_{13}H_{15}N_3O_3S$; [71720-65-5]

(2) Water; H_2O; [7732-18-5]

ORIGINAL MEASUREMENTS:

Roblin, R. O., Jr.; Winnek, P. S.

J. Am. Chem. Soc. 1940, *62*, 1999-2002.

VARIABLES:

One temperature: 37°C

PREPARED BY:

R. Piekos

EXPERIMENTAL VALUES:

Solubility of 4-amino-N-(2-ethoxy-5-pyridinyl)benzenesulfonamide in water at 37°C is 3.6 mg/100 cm^3 solution (1.3 x 10^{-4} mol dm^{-3}, compiler).

AUXILIARY INFORMATION

METHOD/APPARATUS/PROCEDURE:

Excess sulfonamide in water was heated and stirred on a steam bath for 30 min. The suspension was then agitated for 24 h in a thermostat at 37°C. A sample of the satd soln was withdrawn through a glass filter, dild, and analyzed by the modified Marshall method (1) using a General Electric recording spectrophotometer for comparing colors with those of the standards.

SOURCE AND PURITY OF MATERIALS:

The sulfonamide, mp 207-8°C (cor), was prepd by the authors. Anal: %C 53.2 (calcd 53.2); %H 5.1 (5.1).

Purity of the water was not specified.

ESTIMATED ERROR:

Nothing specified

REFERENCES:

1. Bratton, A. C.; Marshall, E. K., Jr. *J. Pharmacol.* 1939, *66*, 4.

COMPONENTS:	EVALUATOR:
(1) Benzenesulfonamide, 4-amino-N-methyl-N-2-pyridinyl-; $C_{12}H_{13}N_3O_2S$; [51543-29-4] (2) Water	Anthony N. Paruta Department of Pharmaceutics University of Rhode Island Kingston, Rhode Island, USA and Ryszard Piekos Faculty of Pharmacy, University of Gdansk Gdansk, Poland 1986

CRITICAL EVALUATION:

The solubility of the compound was reported by Kitao et al. (2) in 1973, and 31 years earlier by Shepherd et al. (1). It is conceivable that this low-melting solid be rather insoluble in water and the average of the two reported values should be considered as the recommended value at 310K in water, 4.96×10^{-3} mol dm^{-3}.

REFERENCES:

(1) Shepherd, R. G.; Bratton, A. C.; Blanchard, K. C. *J. Am. Chem. Soc.* 1942, *64*, 2532-7.

(2) Kitao, K.; Kubo, K.; Morishita, T.; Yata, N.; Kamada, A. *Chem. Pharm. Bull.* 1973, *21*, 2417-26.

COMPONENTS:	ORIGINAL MEASUREMENTS:
(1) Benzenesulfonamide, 4-amino-N-methyl-N-2-pyridinyl-; $C_{12}H_{13}N_3O_2S$; [51543-29-4] (2) Water; H_2O; [7732-18-5]	Shepherd, R. G.; Bratton, A. C.; Blanchard, K. C.; *J. Am. Chem. Soc.* 1942, *64*, 2532-7.
VARIABLES:	PREPARED BY:
One temperature: 37°C	R. Piekos

EXPERIMENTAL VALUES:

Solubilty of 4-amino-N-methyl-N-2-pyridinylbenzenesulfonamide in water at 37°C is 136 mg% (5.17 x 10^{-3} mol dm^{-3} solution, compiler).

AUXILIARY INFORMATION

METHOD/APPARATUS/PROCEDURE:

The sulfonamide was assayed colorimetrically (1). No details were given.

SOURCE AND PURITY OF MATERIALS:

The sulfonamide, mp 86.5-7.0°C, was synthesized by the authors. Anal: %C 54.76 (54.74); %H 4.82 (4.98); %N 15.94 (15.96). Colorimetric factor: 0.658 (calcd 0.654). Purity of the water was not specified.

ESTIMATED ERROR:

Nothing specified

REFERENCES:

1. Bratton, A. C.; Marshall, E. K., Jr. *J. Biol. Chem.* 1939, *128*, 537.

COMPONENTS:	ORIGINAL MEASUREMENTS:
(1) Benzenesulfonamide, 4-amino-N-methyl-N-2-pyridinyl-; $C_{12}H_{13}N_3O_2S$; [51543-29-4] (2) Water; H_2O; [7732-18-5]	Kitao, K.; Kubo, K.; Morishita, T.; Yata, N.; Kamada, A. *Chem. Pharm. Bull.* 1973, *21*, 2417-26.
VARIABLES: One temperature: 37°C	PREPARED BY: R. Piekos

EXPERIMENTAL VALUES:

Solubility of 4-amino-N-methyl-N-2-pyridinylbenzenesulfonamide in water at 37°C is 4.74 mmol dm^{-3} solution.

AUXILIARY INFORMATION

METHOD/APPARATUS/PROCEDURE:	SOURCE AND PURITY OF MATERIALS:
The sulfonamide was assayed by diazotization. No details were given.	The sulfonamide was synthesized by the authors. Its purity was not specified. Deionized water was used.
	ESTIMATED ERROR: Soly: not specified. Temp: ±1°C (authors).
	REFERENCES:

COMPONENTS:	ORIGINAL MEASUREMENTS:
(1) Benzenesulfonamide, 4-amino-N-methyl-N-2-pyridinyl-; $C_{12}H_{13}N_3O_2S$; [51543-29-4] (2) Methane, trichloro-; $CHCl_3$; [67-66-3]	Kitao, K.; Kubo, K.; Morishita, T.; Yata, N.; Kamada, A. *Chem. Pharm. Bull.* 1973, *21*, 2417-26.
VARIABLES:	PREPARED BY:
One temperature: 37°C	R. Piekos

EXPERIMENTAL VALUES:

Solubility of 4-amino-N-methyl-N-2-pyridinylbenzenesulfonamide in $CHCl_3$ at 37°C is more than 2630 mmol dm^{-3} solution.

AUXILIARY INFORMATION

METHOD/APPARATUS/PROCEDURE:

One ml of the sulfonamide soln in $CHCl_3$ at equilibrium was taken into a test tube. After evapn of the solvent, the residue was dissolved in 1N HCl, the soln was properly dild with deionized water, and the concn of the sulfonamide was detd by diazotization.

SOURCE AND PURITY OF MATERIALS:

The sulfonamide was prepd by the authors. Its purity was not specified. Neither source nor purity of the $CHCl_3$ was specified.

ESTIMATED ERROR:

Soly: not specified.

Temp: ±1°C (authors).

REFERENCES:

COMPONENTS:	EVALUATOR:
(1) Benzenesulfonamide, 4-amino-N-(1-methyl-1,2-dihydro-2-pyridinyl)-; $C_{12}H_{13}N_3O_2S$; [51543-30-7] (2) Water	Anthony N. Paruta Department of Pharmaceutics University of Rhode Island Kingston, Rhode Island, USA and Ryszard Piekos Faculty of Pharmacy, University of Gdansk Gdansk, Poland 1986

CRITICAL EVALUATION:

There were two reports on the solubility of this compound in water at 310K (1,2). Shepherd et al. (1) gave a solubility value of 4.25×10^{-3} mol dm^{-3} in 1942. Kitao et al. (2) gave a value of 3.69×10^{-3} mol dm^{-3}. These values are about 15% in difference, and can be used to indicate a tentative value of 3.97×10^{-3} mol dm^{-3} in water at 310K.

REFERENCES:

(1) Shepherd, R. G.; Bratton, A. C.; Blanchard, K. C. *J. Am. Chem. Soc.* 1942, *64*, 2532-7.

(2) Kitao, K.; Kubo, K.; Morishita, T.; Yata, N.; Kamada, A. *Chem. Pharm. Bull.* 1973, *21*, 2417-26.

COMPONENTS:

(1) Benzenesulfonamide, 4-amino-N-(1-methyl-1,2-dihydro-2-pyridinyl)-; $C_{12}H_{13}N_3O_2S$; [51543-30-7]

(2) Water; H_2O; [7732-18-5]

ORIGINAL MEASUREMENTS:

Shepherd, R. G.; Bratton, A. C.; Blanchard, K.C. *J. Am. Chem. Soc.* 1942, *64*, 2532-7.

VARIABLES:

One temperature: 37 °C

PREPARED BY:

R. Piekos

EXPERIMENTAL VALUES:

Solubility of 4-amino-N-(1-methyl-1,2-dihydro-2-pyridinyl)benzenesulfonamide in water at 37°C is 112 mg% (4.25 x 10^{-3} mol dm^{-3}, compiler).

AUXILIARY INFORMATION

METHOD/APPARATUS/PROCEDURE:

The sulfonamide was assayed colorimetrically (1). No details were given.

SOURCE AND PURITY OF MATERIALS:

The sulfonamide , mp 232-3°C, was synthesized by the authors. Analysis: %C 54.68 (calcd 54.74); %H 4.85 (4.98); %N 15.90 (15.96). Colorimetric factor: 0.670 (calcd 0.654).

Purity of the water was not specified.

ESTIMATED ERROR:

Nothing specified

REFERENCES:

1. Bratton, A. C.; Marshall, E. K., Jr. *J. Biol. Chem.* 1939, *128*, 537.

COMPONENTS:	ORIGINAL MEASUREMENTS:
(1) Benzenesulfonamide, 4-amino-N-(1-methyl-1,2-dihydro-2-pyridinyl)-; $C_{12}H_{13}N_3O_2S$; [51543-30-7] (2) Water; H_2O; [7732-18-5]	Kitao, K.; Kubo, K.; Morishita, T.; Yata, N.; Kamada, A. *Chem. Pharm. Bull.* 1973, *21*, 2417-26.
VARIABLES:	**PREPARED BY:**
One temperature: 37°C	R. Piekos

EXPERIMENTAL VALUES:

Solubility of 4-amino-N-(1-methyl-1,2-dihydro-2-pyridinyl)benzenesulfonamide in water at 37°C is 3.69 mmol dm^{-3} solution.

AUXILIARY INFORMATION

METHOD/APPARATUS/PROCEDURE:

The sulfonamide was assayed by diazotization. No details were given.

SOURCE AND PURITY OF MATERIALS:

The sulfonamide was synthesized by the authors. Its purity was not specified. Deionized water was used.

ESTIMATED ERROR:

Soly: not specified.

Temp: ±1°C (authors).

REFERENCES:

COMPONENTS:	ORIGINAL MEASUREMENTS:
(1) Benzenesulfonamide, 4-amino-N-(1-methyl-1,2-dihydro-2-pyridinyl)-; $C_{12}H_{13}N_3O_2S$; [51543-30-7] (2) Methane, trichloro-; $CHCl_3$; [67-66-3]	Kitao, K.; Kubo, K.; Morishita, T.; Yata, N.; Kamada, A. *Chem. Pharm. Bull.* 1973, *21*, 2417-26.
VARIABLES:	**PREPARED BY:**
One temperature: 37°C	R. Piekos

EXPERIMENTAL VALUES:

Solubility of 4-amino-N-(2-methyl-1,2-dihydro-2-pyridinyl)benzenesulfonamide in $CHCl_3$ at 37°C is 5.53 mmol dm^{-3} solution.

AUXILIARY INFORMATION

METHOD/APPARATUS/PROCEDURE:

One ml of the sulfonamide soln in $CHCl_3$ at equilibrium was taken into a test tube. After evapn of the solvent the residue was dissolved in 1N HCl, the soln was properly dild with deionized water, and the concn of the sulfonamide was detd by diazotization.

SOURCE AND PURITY OF MATERIALS:

The sulfonamide was prepd by the authors. Its purity was not specified. Neither source nor purity of the $CHCl_3$ was specified.

ESTIMATED ERROR:

Soly: not specified.

Temp: ±1°C (authors).

REFERENCES:

COMPONENTS:

(1) Pyridine, 2,5-bis[[(4-aminophenyl)-sulfonyl]amino]-; $C_{17}H_{17}N_5O_4S_2$; [71119-18-1]

(2) Water; H_2O; [7732-18-5]

ORIGINAL MEASUREMENTS:

Roblin, R. O., Jr.; Winnek, P. S.
J. Am. Chem. Soc. 1940, *62*, 1999-2002.

VARIABLES:

One temperature: 37°C

PREPARED BY:

R. Piekos

EXPERIMENTAL VALUES:

Solubility of 2,5-bis[[(4-aminophenyl)sulfonyl]amino]pyridine in water at 37°C is 49.5 mg/100 cm^3 solution (1.18×10^{-3} mol dm^{-3}, compiler).

AUXILIARY INFORMATION

METHOD/APPARATUS/PROCEDURE:

Excess sulfonamide in water was heated and stirred on a steam bath for 30 min. The suspension was then agitated for 24 h in a thermostat at 37°C. A sample of the satd soln was withdrawn through a glass filter, dild, and analyzed by the modified Marshall method (1) using a General Electric recording spectrophotometer for comparing colors with those of the standards.

SOURCE AND PURITY OF MATERIALS:

The sulfonamide, mp 215-6°C (cor), was prepd by the authors. Anal: %C 48.8 (calcd 48.7); %H 4.1 (4.1).
Purity of the water was not specified.

ESTIMATED ERROR:

Nothing specified

REFERENCES:

1. Bratton, A. C.; Marshall, E. K., Jr. *J. Pharmacol.* 1939, *66*, 4.

COMPONENTS:

(1) Benzenesulfonamide, 4-amino-N-3-pyridazinyl-; $C_{10}H_{10}N_4O_2S$; [515-62-8]

(2) Water; H_2O; [7732-18-5]

ORIGINAL MEASUREMENTS:

Anderson, G. W.; Faith, H. E.; Marson, H.W. Winnek, P. S.; Roblin, R. O., Jr.

J. Am. Chem. Soc. 1942, *64*, 2902-5.

VARIABLES:

One temperature: 37°C

PREPARED BY:

R. Piekos

EXPERIMENTAL VALUES:

Solubility of 4-amino-N-3-pyridazinylbenzenesulfonamide in water at 37°C is 221 mg/100 cm^3 solution (8.83×10^{-3} mol dm^{-3}, compiler).

AUXILIARY INFORMATION

METHOD/APPARATUS/PROCEDURE:

Excess sulfonamide in water was heated and stirred on a steam bath for 30 min. The suspension was then agitated for 24 h in a thermostat. A sample of the satd soln was withdrawn through a glass filter, dild, and analyzed by the Marshall method (1) using a General Electric recording spectrophotometer for comparing the colors developed with those of the standards.

SOURCE AND PURITY OF MATERIALS:

The sulfonamide, mp 189-90°C, was prepd by the authors. Anal: %C 47.7 (calcd 48.0); %H 4.0 (4.0); %N 22.8 (22.4). Purity of the water was not specified.

ESTIMATED ERROR:

Nothing specified

REFERENCES:

1. Bratton, A. C.; Marshall, E. K., Jr. *J. Pharmacol.* 1939, *66*, 4.

COMPONENTS:

(1) Benzenesulfonamide, 4-amino-N-(6-chloro-3-pyridazinyl)- (sulfachlorpyridazine); $C_{10}H_9ClN_4O_2S$; [80-32-0]

(2) Phosphoric acid, disodium salt; Na_2HPO_4; [7558-94-4]

(3) Phosphoric acid, monopotassium salt; KH_2PO_4; [7778-77-0]

(4) Water; H_2O; [7732-18-5]

ORIGINAL MEASUREMENTS:

Riess, W.

Intern. Congr. Chemotheraphy, Proc. 3rd, Stuttgart 1963, *1*, 627-32.

VARIABLES:

One temperature: 20°C; one pH: 7.4

PREPARED BY:

R. Piekos

EXPERIMENTAL VALUES:

Solubility of sulfachlorpyridazine in a M/15 Sörensen buffer solution (pH 7.4) at 20°C is 1200 mg% (4.215×10^{-2} mol dm^{-3} solution, compiler).

AUXILIARY INFORMATION

METHOD/APPARATUS/PROCEDURE:

Sörensen buffer solns of pH varying between 7 and 8 were prepd, satd with sulfachlorpyridazine at 20°C, their pH was measured at equilibrium, and the sulfachlorpyridazine was assayed colorimetrically. The measured pH values were then plotted against concn, and the soly at pH 7.4 was detd by interpolation (personal comunication).

SOURCE AND PURITY OF MATERIALS:

Nothing specified

ESTIMATED ERROR:

Nothing specified

REFERENCES:

COMPONENTS:	ORIGINAL MEASUREMENTS:
(1) Benzenesulfonamide, 4-amino-N-(6-chloro-3-pyridazinyl)- (sulfachlorpyridazine); $C_{10}H_9ClN_4O_2S$; [80-32-0] (2) Methane, trichloro- (chloroform); $CHCl_3$; [67-66-3]	Riess, W. *Intern. Congr. Chemotherapy, Proc. 3rd, Stuttgart* 1963, *1*, 627-32.
VARIABLES:	**PREPARED BY:**
One temperature: 20°C	R. Piekos

EXPERIMENTAL VALUES:

Solubility of sulfachlorpyridazine in chloroform at 20°C is 59 mg%

(2.1 x 10^{-3} mol dm^{-3} solution, compiler).

AUXILIARY INFORMATION

METHOD/APPARATUS/PROCEDURE:	SOURCE AND PURITY OF MATERIALS:
Nothing specified	Nothing specified
	ESTIMATED ERROR:
	Nothing specified
	REFERENCES:

COMPONENTS:	ORIGINAL MEASUREMENTS:
(1) Benzenesulfonamide, 4-amino-N-(6-methyl-3-pyridazinyl)-(sulfamethylpyridazine); $C_{11}H_{12}N_4O_2S$; [5433-63-6] (2) Phosphoric acid, disodium salt; Na_2HPO_4; [7558-77-0] (3) Phosphoric acid, monopotassium salt; KH_2PO_4; [7778-77-0] (4) Water; H_2O; [7732-18-5]	Riess, W. *Intern. Congr. Chemotherapy, Proc. 3rd, Stuttgart* 1963, *1*, 627-32.
VARIABLES: One temperature: 20°C; one pH: 7.4	PREPARED BY: R. Piekos

EXPERIMENTAL VALUES:

Solubility of sulfamethylpyridazine in M/15 Sörensen buffer solution (pH 7.4) at 20°C is 140 mg% (5.30 x 10^{-3} dm^{-3} solution, compiler).

AUXILIARY INFORMATION

METHOD/APPARATUS/PROCEDURE:	SOURCE AND PURITY OF MATERIALS:
Sörensen buffer solns of pH varying between 7 and 8 were prepd, satd with sulfamethylpyridazine at 20°C, their pH was measured at equilibrium, and the sulfamethylpyridazine was assayed colorimetrically. The measured pH values were then plotted against concn and the soly at pH 7.4 was detd by interpolation (personal communication).	Nothing specified ESTIMATED ERROR: Nothing specified REFERENCES:

COMPONENTS:	ORIGINAL MEASUREMENTS:
(1) Benzenesulfonamide, 4-amino-N-(6-methyl-3-pyridazinyl)- (sulfamethylpyridazine); $C_{11}H_{12}N_4O_2S$; [5433-63-6] (2) Methane, trichloro- (chloroform); $CHCl_3$; [67-66-3]	Reiss, W. *Intern. Congr. Chemotherapy, Proc. 3rd, Stuttgart* 1963, *1*, 627-32.
VARIABLES:	**PREPARED BY:**
One temperature: 20°C	R. Piekos

EXPERIMENTAL VALUES:

Solubility of sulfamethylpyridazine in chloroform at 20°C is 144 mg%

(5.45×10^{-3} mol dm^{-3} solution, compiler).

AUXILIARY INFORMATION

METHOD/APPARATUS/PROCEDURE:

Nothing specified

SOURCE AND PURITY OF MATERIALS:

Nothing specified

ESTIMATED ERROR:

Nothing specified

REFERENCES:

COMPONENTS:	ORIGINAL MEASUREMENTS:
(1) Benzenesulfonamide, 4-amino-N-(6-methoxy-3-pyridazinyl)- (sulfamethoxypyridazine); $C_{11}H_{12}N_4O_3S$; [80-35-3] (2) Water; H_2O; [7732-18-5]	Yamazaki, M.; Aoki, M.; Kamada, A.; Yata, N. *Yakuzaigaku* 1967, *27(1)*, 37-40.
VARIABLES:	**PREPARED BY:**
One temperature: 30°C	R. Piekos

EXPERIMENTAL VALUES:

Solubility of sulfamethoxypyridazine in water at 30°C is 3.36 mmol/L (0.942 g dm^{-3}, compiler).

AUXILIARY INFORMATION

METHOD/APPARATUS/PROCEDURE:	SOURCE AND PURITY OF MATERIALS:
Sulfamethoxypyridazine (0.5 g) was placed in an L-shaped tube together with 20 ml of water. The mixt was shaken in a thermostat until equilibrium was attained. The sulfamethoxypyridazine was assayed in the supernatant spectrophotometrically at 545 nm on a Beckmann DU spectrophotometer. The results were taken from a calibration graph.	Nothing specified
	ESTIMATED ERROR: Soly: not specified Temp: ±1°C (authors)
	REFERENCES:

COMPONENTS:	ORIGINAL MEASUREMENTS:
(1) Benzenesulfonamide, 4-amino-N-(6-methoxy-3-pyridazinyl)- (sulfamethoxypyridazine); $C_{11}H_{12}N_4O_3S$; [80-35-3] (2) Hydrochloric acid; HCl; [7647-01-0] (3) Water; H_2O; [7732-18-5]	Ogata, H.; Shibazaki, T.; Inoue, T.; Ejima, A. *Chem. Pharm. Bull.* 1979, *27(6)*, 1281-6.
VARIABLES:	PREPARED BY:
One temperature: 37°C	R. Piekos

EXPERIMENTAL VALUES:

Solubility of sulfamethoxypyridazine in 0.1N HCl at 37°C is 17.243 mg/ml (6.1516×10^{-2} mol dm^{-3}, compiler).

AUXILIARY INFORMATION

METHOD/APPARATUS/PROCEDURE:	SOURCE AND PURITY OF MATERIALS:
A centrifuge tube contg 30 ml of 0.1N HCl and 0.5-3.0 g of the sulfamethoxypyridazine powder was tightly sealed and shaken at 37°C. The concn of the dissolved drug was detd spectrophotometrically following filtration through a Millipore filter (type EH, pore size 0.5 μm), and the procedure was repeated every 24 h until a const concn was obtained.	Comm available 250-mg uncoated tablets of sulfamethoxypyridazine were used. Hydrochloric acid was of reagent grade. ESTIMATED ERROR: Nothing specified REFERENCES:

COMPONENTS:	ORIGINAL MEASUREMENTS:
(1) Benzenesulfonamide, 4-amino-N-(6-methoxy-3-pyridazinyl)- (sulfamethoxypyridazine); $C_{11}H_{12}N_4O_3S$; [80-35-3] (2) Phosphoric acid, disodium salt; Na_2HPO_4; [7558-94-4] (3) 1,2,3-Propanetricarboxylic acid, 2-hydroxy- (citric acid); $C_6H_8O_7$; [77-92-9] (4) Water; H_2O; [7732-18-5]	Bertazzoli, C.; Buogo, A.; Ciceri, C.; Ghione, M.; Turolla, E.; Zavaglio, V. *Minerva Med.* 1961, *52(40)*, 1789-96.
VARIABLES: pH	PREPARED BY: R. Piekos

EXPERIMENTAL VALUES:

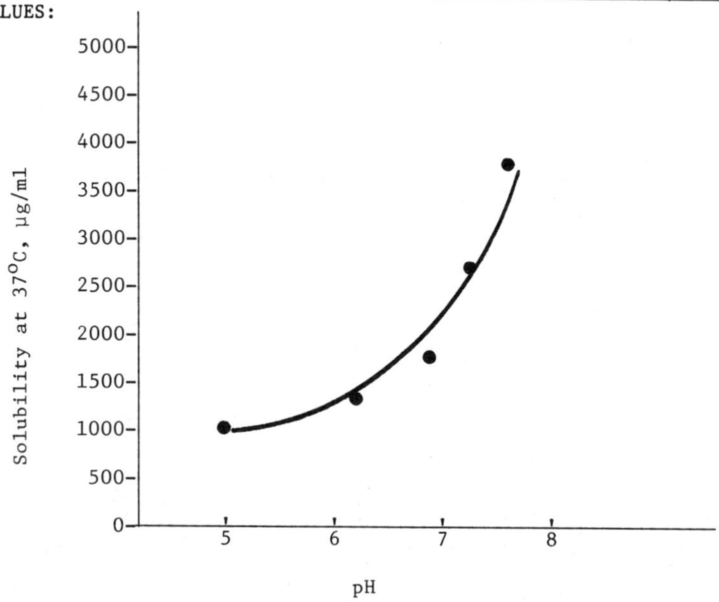

AUXILIARY INFORMATION

METHOD/APPARATUS/PROCEDURE:	SOURCE AND PURITY OF MATERIALS:
The soly of sulfamethoxypyridazine in McIlvaine's Na_2HPO_4 - citric acid buffer solns was detd under agitation at 37°C. No details were given.	Nothing specified
	ESTIMATED ERROR: Nothing specified
	REFERENCES:

COMPONENTS:

(1) Benzenesulfonamide, 4-amino-N-(6-methoxy-3-pyridazinyl)- (sulfamethoxypyridazine); $C_{11}H_{12}N_4O_3S$; [80-35-3]

(2) Phosphoric acid, disodium salt; Na_2HPO_4; [7558-94-4]

(3) Phosphoric acid, monopotassium salt; KH_2PO_4; [7778-77-0]

(4) Water; H_2O; [7732-18-5]

ORIGINAL MEASUREMENTS:

Bandelin, F. J.; Malesh, W.

J. Am. Pharm. Assoc., Sci. Ed. 1959, *48*, 177-81.

VARIABLES:

pH

PREPARED BY:

R. Piekos

EXPERIMENTAL VALUES:

Solubility of sulfamethoxypyridazine in buffers of varying mixtures of $Na_2HPO_4 \cdot 7H_2O$ (71.6 g/l distilled water; 0.27 mol dm^{-3}, compiler) and KH_2PO_4 (36.3 g/l distilled water; 0.27 mol dm^{-3} , compiler) at 37°C.

Initial pH	Solubility	
	mg/100 ml	10^2 mol dm^{-3} [a]
4.5	720	2.57
5.0	740	2.64
5.5	770	2.75
6.0	800	2.85
6.5	920	3.28
7.0	1380	4.923

[a] calculated by compiler

AUXILIARY INFORMATION

METHOD/APPARATUS/PROCEDURE:

Solns were prepd by adding an excess of sulfamethoxypyridazine to 10 ml of buffer soln at each pH level in 18 x 150-mm test tubes, stoppering the tubes and placing them in a water bath at 37°C with gentle agitation for 24 h. The mixt was then filtered and a 1-ml aliquot was accurately pipetted into a volumetric flask for diln and analysis. The balance was retained for pH detn to ascertain any change in pH value. The sulfonamide was assayed colorimetrically by the method of Bratton and Marshall as described in detail by Biamonte and Schneller (1). A standard curve was prepd using accurately prepd standard solutions.

SOURCE AND PURITY OF MATERIALS:

Neither source nor purity of the reagents were specified. Distilled water was used.

ESTIMATED ERROR:

Soly: av values of duplicate runs are reported (authors).

Temp and pH: not specified.

REFERENCES:

1. Biamonte, A. R.; Schneller, G. E. *J. Am. Pharm. Assoc., Sci. Ed.*, 1952, *41*, 341.

COMPONENTS:

(1) Benzenesulfonamide, 4-amino-N-(6-methoxy-3-pyridazinyl)- (sulfamethoxypyridazine); $C_{11}H_{12}N_4O_3S$; [80-35-3]

(2) Phosphoric acid, disodium salt; Na_2HPO_4; [7558-94-4]

(3) Phosphoric acid, monopotassium salt; KH_2PO_4; [7778-77-0]

(4) Water; H_2O; [7732-18-5]

ORIGINAL MEASUREMENTS:

Riess, W.

Intern. Congr. Chemotherapy, Proc. 3rd, Stuttgart 1963, *1*, 627-32.

VARIABLES:

One temperature: 20°C; one pH: 7.4

PREPARED BY:

R. Piekos

EXPERIMENTAL VALUES:

Solubility of sulfamethoxypyridazine in a M/15 Sörensen buffer solution (pH 7.4) at 20°C is 200 mg% (7.13×10^{-3} mol dm^{-3} solution, compiler).

AUXILIARY INFORMATION

METHOD/APPARATUS/PROCEDURE:

Sörensen buffer solutions of pH varying between 7 and 8 were prepd, satd with sulfamethoxypyridazine at 20°C, their pH was measured at equilibrium, and the sulfamethoxypyridazine was assayed colorimetrically. The measured pH values were plotted against concn, and the soly at pH 7.4 was detd by interpolation (personal communication).

SOURCE AND PURITY OF MATERIALS:

Nothing specified

ESTIMATED ERROR:

Nothing specified

REFERENCES:

COMPONENTS:

(1) Benzenesulfonamide, 4-amino-N-(6-methoxy-3-pyridazinyl)- (sulfamethoxypyridazine); $C_{11}H_{12}N_4O_3S$; [80-35-3]

(2) Phosphoric acid, disodium salt; Na_2HPO_4; [7558-94-4]

(3) Phosphoric acid, monopotassium salt; KH_2PO_4; [7778-77-0]

(4) Water; H_2O; [7732-18-5]

ORIGINAL MEASUREMENTS:

Yamazaki, M.; Aoki, M.; Kamada, A.; Yata, N. *Yakuzaigaku* 1967, *27(1)*, 37-40

VARIABLES:

One temperature: 30°C; one pH: 7.4

PREPARED BY:

R. Piekos

EXPERIMENTAL VALUES:

Solubility of sulfamethoxypyridazine in a phosphate buffer solution of pH 7.4 (μ = 0.17) at 30°C is 6.12 mmol/L (1.71 g dm^{-3}, compiler).

AUXILIARY INFORMATION

METHOD/APPARATUS/PROCEDURE:

Sulfamethoxypyridazine (0.5 g) was placed in an L-shaped tube together with 20 ml of the buffer soln. The mixt was shaken in a thermostat until equilibrium was attained. The sulfamethoxypyridazine was assayed in the supernatant spectrophotometrically at 545 nm on a Beckmann DU spectrophotometer. The results were taken from a calibration curve.

SOURCE AND PURITY OF MATERIALS:

Nothing specified.

ESTIMATED ERROR:

Soly and pH: not specified.

Temp: ±1°C (authors)

REFERENCES:

COMPONENTS:

(1) Benzenesulfonamide, 4-amino-N-(6-methoxy-3-pyridazinyl)- (sulfamethoxypyridazine); $C_{11}H_{12}N_4O_3S$; [80-35-3]

(2) Calcium chloride; $CaCl_2$; [10043-52-4]

(3) Magnesium chloride; $MgCl_2$; [7786-30-3]

(4) Phosphoric acid, monoammonium salt; $NH_4H_2PO_4$; [7722-76-1]

(5) Potassium chloride; KCl; [7447-40-7]

(6) Sodium chloride; NaCl; [7647-14-5]

(7) Urea; CH_4N_2O; [57-13-6]

(8) Water; H_2O; [7732-18-5]

ORIGINAL MEASUREMENTS:

Bandelin, F. J.; Malesh, W.

J. Am. Pharm. Assoc., Sci. Ed. 1959, *48*, 177-81.

VARIABLES:

pH at 37°C

PREPARED BY:

R. Piekos

EXPERIMENTAL VALUES:

Solubility of sulfamethoxypyridazine in a solution containing $CaCl_2$ 0.1143, $MgCl_2$ 0.121, $NH_4H_2PO_4$ 0.300, KCl 1.660, NaCl 2.950 and urea 20 g/dm^3 (synthetic urine, Mosher Vehicle) at 37°

Equilibrium pH	Solubility	
	mg/100 ml	10^2 mol/dm^3 [a]
4.5	460	1.64
5.0	466	1.66
5.5	475	1.69
6.0	488	1.74
6.5	552	1.97
7.0	862	3.07

[a] calculated by compiler

AUXILIARY INFORMATION

METHOD/APPARATUS/PROCEDURE:

Excess sulfamethoxypyridazine was added to aliquots of synthetic urine solns and 1% H_3PO_4 or 1% NaOH solns were used to adjust the pH to the required value. The solns were agitated for 24 h with addn of acid or base to keep them at the desired pH level until equilibrium was attained. Then the solns were filtered and in aliquots the sulfonamide was assayed spectrophotometrically by the method described by Biamonte and Schneller (1).

SOURCE AND PURITY OF MATERIALS:

Nothing specified

ESTIMATED ERROR:

Soly: average values of 2 detns were given.

Temp: not specified.

pH ; not specified

REFERENCES:

1. Biamonte, A. R.; Schneller, G. E. *J. Am. Pharm. Assoc., Sci. Ed.* 1952, *41*, 341.

COMPONENTS:

(1) Benzenesulfonamide, 4-amino-N-(6-methoxy-3-pyridazinyl)- (sulfamethoxypyridazine); $C_{11}H_{12}N_4O_3S$; [80-35-3]

(2) Ethanol, 2,2'-iminodi-(diethanolamine); $C_4H_{11}NO_2$; [111-42-2]

(3) 1,3 -Propanediol; $C_3H_8O_2$; [504-63-2]

(4) Water; H_2O; [7732-18-5]

ORIGINAL MEASUREMENTS:

Lombardi, R. B.

Rev. Farm. (Buenos Aires) 1968, *110(7-8)*, 154-9.

VARIABLES:

One temperature: 20°C

PREPARED BY:

R. Piekos

EXPERIMENTAL VALUES:

Maximum concentration of sulfamethoxypyridazine in a mixture of diethanolamine 1 ml, 1,3-propanediol 60 ml, and water 39 ml, at 20°C, is 14% (0.50 mol kg^{-1} solvent, compiler).

AUXILIARY INFORMATION

METHOD/APPARATUS/PROCEDURE:

To the diethanolamine solution in water warmed up to 40°C, sulfamethoxypyridazine was added followed by 1,3-propanediol previously warmed to 50°C, and the mixt was agitated. The pH of the mixt was 7.9 at 20°C.

SOURCE AND PURITY OF MATERIALS:

The source and purity of the materials, with the exception of water, was not specified.

Distd water was used.

ESTIMATED ERROR:

Nothing specified.

REFERENCES:

COMPONENTS:

(1) Benzenesulfonamide, 4-amino-N-(6-methoxy-3-pyridazinyl)- (sulfamethoxypyridazine); $C_{11}H_{12}N_4O_3S$; [80-35-3]

(2) Ethanol; C_2H_6O; [64-17-5]

(3) Piperazine; $C_4H_{10}N_2$; [110-85-0]

(4) Poly(oxy-1,2-ethanediyl), α-[(tetrahydro-2-furanyl)methyl]-ω-hydroxy-(glycofurol); $(C_2H_4O)_nC_5H_{10}O_2$; [31692-85-0]

(5) Water; H_2O; [7732-18-5]

ORIGINAL MEASUREMENTS:

Lombardi, R. B.

Rev. Farm. (Buenos Aires) 1968, *110(7-8)*, 154-9.

VARIABLES: One temperature: 20°C

PREPARED BY:

R. Piekos

EXPERIMENTAL VALUES:

Maximum concentration of sulfamethoxypyridazine in a mixture of ethanol 96°GL 10 ml, piperazine 3.750 g, glycofurol 70 ml, and water 20 ml, at 20°C, is 18-19% (0.64 - 0.68 mol kg^{-1} solvent, compiler).

AUXILIARY INFORMATION

METHOD/APPARATUS/PROCEDURE:

To glycofurol, warmed to 60°C, through which a stream of nitrogen was bubbled, sulfamethoxypyridazine was added, followed by an aq piperazine soln and EtOH. The stream of nitrogen was bubbled throughout. The pH of the soln was 7.9 at 20°C.

SOURCE AND PURITY OF MATERIALS:

The source and purity of the materials were not specified, with the exception of water. Water for injection was used.

ESTIMATED ERROR:

Nothing specified.

REFERENCES:

COMPONENTS:

(1) Benzenesulfonamide,4-amino-N-(6-methoxy-3-pyridazinyl)- (sulfamethoxypyridazine); $C_{11}H_{12}N_4O_3S$; [80-35-3]

(2) Ethanol; C_2H_6O; [64-17-5]

(3) Piperazine; $C_4H_{10}N_2$; [110-85-0]

(4) Poly(oxy-1,2-ethanediyl), α-hydro -ω-hydroxy- (PEG 300); $(C_2H_4O)_nH_2O$; [25322-68-3] 300

(5) Water; H_2O; [7732-18-5]

ORIGINAL MEASUREMENTS:

Lombardi, R. B.

REv. Farm. (Buenos Aires) 1968, *110(7-8),* 154-9.

VARIABLES:

One temperature: 20°C

PREPARED BY:

R. Piekos

EXPERIMENTAL VALUES:

Maximum concentration of sulfamethoxypyridazine in a mixture of ethanol 96°GL 10 ml, piperazine 2 g, PEG 300 50 ml, and water 40 ml, at 20°C , is 18% (0.64 mol kg^{-1} solvent, compiler).

AUXILIARY INFORMATION

METHOD/APPARATUS/PROCEDURE:

To a soln of piperazine in water, EtOH was added, followed by sulfamethoxypyridazine, and PEG 300 previously warmed to 50-55°C. Nitrogen was bubbled through the mixt to ensure agitation. The pH of the mixt was 7.8 at 20°C.

SOURCE AND PURITY OF MATERIALS:

The source and purity of the materials, with the exception of water, were not specified. Water for injections was used.

ESTIMATED ERROR:

Nothing specified

REFERENCES:

COMPONENTS:	ORIGINAL MEASUREMENTS:
(1) Benzenesulfonamide, 4-amino-N-(6-methoxy-3-pyridazinyl)-(sulfamethoxypyridazine); $C_{11}H_{12}N_4O_3S$; [80-35-3] (2) Piperazine; $C_4H_{10}N_2$; [110-85-0] (3) Poly(oxy-1,2-ethanediyl), α-hydro- ω -hydroxy- (PEG 300); $(C_2H_4O)_nH_2O$; [25322-68-3] 300 (4) 1,3-Propanediol; C_3H_8O; [504-63-2] (5) Water; H_2O; [7732-18-5]	Lombardi, R. B. *Rev. Farm. (Buenos Aires)* 1968, *110(7-8)*, 154-9.
VARIABLES: One temperature: 20°C	PREPARED BY: R. Piekos

EXPERIMENTAL VALUES:

Maximum concentration of sulfamethoxypyridazine in a mixture of piperazine 3.5 g, PEG 300 40 ml, 1,3-propanediol 40 ml, and water 27.5 ml, at 20°C, is 15% (0.54 mol kg^{-1} solvent, compiler).

AUXILIARY INFORMATION

METHOD/APPARATUS/PROCEDURE:	SOURCE AND PURITY OF MATERIALS:
To a soln of piperazine in water, a mixt of PEG 300 and 1,3-propanediol, previously heated to 50°C, was added, followed by sulfamethoxypyridazine. Nitrogen was bubbled through the mixt to ensure agitation. The pH of the mixt was 7.8 at 20°C.	The source and purity of the materials, with the exception of water, were not specified. Water for injections was used. ESTIMATED ERROR: Nothing specified. REFERENCES:

COMPONENTS:	ORIGINAL MEASUREMENTS:
(1) Benzenesulfonamide,4-amino-N-(6-methoxy-3-pyridazinyl)-(sulfamethoxypyridazine); $C_{11}H_{12}N_4O_3S$; [80-35-3] (2) Methane, trichloro- (chloroform); $CHCl_3$; [67-66-3]	Riess, W. *Intern. Congr. Chemotherapy, Proc. 3rd, Stuttgart* 1963, *1*, 627-32.
VARIABLES:	PREPARED BY:
One temperature: 20°C	R. Piekos

EXPERIMENTAL VALUES:

Solubility of sulfamethoxypyridazine in chloroform at 20°C is 390 mg%

(1.39×10^{-2} mol dm^{-3} solution, compiler).

AUXILIARY INFORMATION

METHOD/APPARATUS/PROCEDURE:	SOURCE AND PURITY OF MATERIALS:
Nothing specified	Nothing specified
	ESTIMATED ERROR:
	Nothing specified
	REFERENCES:

COMPONENTS:	ORIGINAL MEASUREMENTS:
(1) Benzenesulfonamide, 4-amino-N-(6-methoxy-3-pyridazinyl)- (sulfamethoxypyridazine); $C_{11}H_{12}N_4O_3S$; [80-35-5] (2) Methane, trichloro- (chloroform); $CHCl_3$; [67-66-3]	Yamazaki, M.; Aoki, M.; Kamada, A.; Yata, N. *Yakuzaigaku* 1967, *27(1)*, 37-40.
VARIABLES:	PREPARED BY:
One temperature: 30°C	R. Piekos

EXPERIMENTAL VALUES:

Solubility of sulfamethoxypyridazine in chloroform at 30°C is 14.03 mmol/L (3.933 g dm^{-3}, compiler).

AUXILIARY INFORMATION

METHOD/APPARATUS/PROCEDURE:	SOURCE AND PURITY OF MATERIALS:
Sulfamethoxypyridazine (0.5 g) was placed in an L-shaped tube together with 20 ml of chloroform. The mixt was shaken in a thermostat until equilibrium was attained. The sulfamethoxypyridazine was assayed in the supernatant spectrophotometrically at 545 nm on a Beckmann DU spectrophotometer. The results were taken from a calibration graph.	Nothing specified
	ESTIMATED ERROR: Soly: not specified Temp: ±1°C (authors)
	REFERENCES:

COMPONENTS:	ORIGINAL MEASUREMENTS:
(1) Cobalt, bis [4-amino-N-(6-methoxy-3-pyridazinyl)benzenesulfonamidato-]diaqua ($CoR_2 \cdot 2H_2O$); $C_{22}H_2CoN_8O_6S_2 \cdot 2H_2O$; [72384-11-3] (2) Hydrochloric acid; HCl; [7647-01-0] (3) Water; H_2O; [7732-18-5]	Gogorishvili, P. V.; Tskitishvili, M. G.; Chrelashvili, M. V. *Izv. Akad. Nauk Gruz. SSR, Ser. Khim* 1979, *5(3)*, 199-204.
VARIABLES:	PREPARED BY:
Concentration of HCl (pH)	R. Piekos

EXPERIMENTAL VALUES:

Concentration of HCl	pH	Solubility (mol/l) at 25°C based on		$10^{11} \times K_{so}$
		$[Co^{2+}]$	$[RH_2^+]$	
2.5×10^{-2}	6.15	1.3×10^{-2}	8.09×10^{-3}	-
1.0×10^{-2}	6.25	6.6×10^{-3}	5.15×10^{-3}	1.51
5.0×10^{-3}	6.62	3.6×10^{-3}	4.36×10^{-3}	1.56
2.5×10^{-3}	6.92	2.0×10^{-3}	2.80×10^{-3}	1.50
1.0×10^{-3}	7.31	1.28×10^{-3}	8.45×10^{-4}	1.53
5.0×10^{-4}	7.54	1.05×10^{-3}	8.32×10^{-4}	1.51
2.5×10^{-4}	7.78	5.84×10^{-4}	8.23×10^{-4}	1.55
1.0×10^{-4}	7.92	5.67×10^{-4}	7.90×10^{-4}	1.50
5.0×10^{-5}	8.12	5.51×10^{-4}	7.87×10^{-4}	1.50
2.5×10^{-5}	8.23	5.48×10^{-4}	7.15×10^{-4}	1.52
			Mean	1.52

AUXILIARY INFORMATION

METHOD/APPARATUS/PROCEDURE:	SOURCE AND PURITY OF MATERIALS:
$CoR_2 \cdot 2H_2O$ was placed in a glass vessel, 100 ml of aq HCl was added, and the mixt was shaken for 6 h in a thermostat at 25°C. After attaining equilibrium, the contents of Co^{2+} and S were detd, and the pH measured.	Nothing specified
	ESTIMATED ERROR: Soly: not specified K_{so} : standard deviation 5.64×10^{-13} (compiler). Temp: not specified.
	REFERENCES:

COMPONENTS:

(1) Cobalt, bis[4-amino-N-(6-methoxy-3-pyridazinyl)benzenesulfonamidato]diaqua- ($CoR_2 \cdot 2H_2O$); $C_{22}H_{22}CoN_8O_6S_2 \cdot 2H_2O$; [72384-11-3]

(2) Acetic acid, cobalt(2+) salt; $C_4H_6CoO_4$; [71-48-7]

(3) Benzenesulfonamide, 4-amino-N-(6-methoxy-3-pyridazinyl)-, monosodium salt, (NaR); $C_{11}H_{11}NaN_4O_3S$; [2577-32-4]

(4) Water; H_2O; [7732-18-5]

VARIABLES:

Co^{2+}/R^- ratio; pH

ORIGINAL MEASUREMENTS:

Gogorishvili, P. V.; Tskitishvili, M. G.; Chrelashvili, M. V.

Izv. Akad. Nauk Gruz. SSR, Ser. Khim. 1979, 5(3), 199-204.

PREPARED BY:

R. Piekos

EXPERIMENTAL VALUES:

Volume ratio of 0.1M solns of Co^{2+}acetate and NaR	pH	$[Co^{2+}]$ mol/l	Excess of NaR or Co^{2+} acetate mol/l	$[R^-]$ mol/l	pLi[a] based on $[Co^{2+}]$	pLi[a] based on $[R^-]$
10:90	9.25	7.09×10^{-6}	7.0×10^{-3}	7.02×10^{-3}	5.16	5.45
20:80	9.09	2.12×10^{-5}	4.0×10^{-3}	4.04×10^{-3}	4.69	4.71
25:75	8.75	3.90×10^{-5}	2.5×10^{-3}	2.58×10^{-3}	4.44	4.43
30:70	8.14	2.06×10^{-4}	1.0×10^{-3}	1.43×10^{-3}	3.80	3.79
1:20	7.95	7.52×10^{-4}	-	1.49×10^{-3}	3.32	3.30
35:65	7.85	4.76×10^{-4}	2.5×10^{-4}	4.83×10^{-4}	3.91	3.82
40:60	7.75	1.37×10^{-3}	1.0×10^{-3}	7.59×10^{-4}	3.65	3.64
50:50	7.62	4.04×10^{-3}	2.5×10^{-3}	1.47×10^{-3}	3.10	3.09
60:40	7.49	7.67×10^{-3}	4.0×10^{-3}	7.48×10^{-3}	2.78	2.44
70:30	7.46	1.14×10^{-2}	5.5×10^{-3}	1.19×10^{-2}	2.59	2.59
80:20	7.40	1.51×10^{-2}	7.0×10^{-3}	1.50×10^{-2}	2.48	2.46
90:10	7.33	1.88×10^{-2}	8.5×10^{-3}	2.02×10^{-2}	2.41	2.41

[a]pLi = -log (solubility), solubility being expressed in mol/l and measured at 25°C.

AUXILIARY INFORMATION

METHOD/APPARATUS/PROCEDURE:

The effect of the concns of Co^{2+} and R^- on the soly of $CoR_2 \cdot H_2O$ was detd by the method of isomolar series of solns. The ppt of CoR_2 was obtained by mixing together 0.1M solns of Co^{2+} acetate and NaR taken in proportions specified in the Table (cf. Experimental Values box). The total vol after mixing was always 100 ml. After attaining equilibrium by shaking for 6 h, the pH, and the Co^{2+} and S content were detd in soln.

SOURCE AND PURITY OF MATERIALS:

$CoR_2 \cdot H_2O$ was obtained by mixing together 0.1M solns of Co^{2+} acetate and NaR. Neither source nor purity of the reagents was specified.

ESTIMATED ERROR:

Nothing specified

REFERENCES:

COMPONENTS:	ORIGINAL MEASUREMENTS:
(1) Magnesium, (T-4)-bis[4-amino-N-(6-methoxy-3-pyridazinyl)benzenesulfonamidato]-hydrate; $C_{22}H_{20}MgN_8O_6S_2 \cdot nH_2O$; [84812-83-9] (2) Hydrochloric acid; HCl; [7647-01-0] (3) Water; H_2O; [7732-18-5]	Tskitishvili, M. G.; Shvelashvili, A. E.; Mikadze, I. I.; Zhorzholiani, N. B.; Chrelashvili, M. V. *Izv. Akad. Nauk Gruz. SSR, Ser. Khim.* 1981, *7(4)*, 300-4.
VARIABLES:	PREPARED BY:
pH	R. Piekos

EXPERIMENTAL VALUES:

K_{so} over the HCl concentration range 5.0×10^{-3} to 1.5×10^{-5} mol dm^{-3}, at 25°C, is 2.00×10^{-3}.

AUXILIARY INFORMATION

METHOD/APPARATUS/PROCEDURE:

The earlier described apparatus and method was used (1): in a glass vessel , a mixt of 100 ml of HCl of appropriate concn and the solute were placed and shaken for 6 h in a water thermostat at 25°C. After attaining equilibrium, the pH of the soln was measured and the Mg^{2+} and S content was determined to calculate K_{so} The pH was measured on a pH-673 pH meter.

SOURCE AND PURITY OF MATERIALS:

0.1M solns of chem pure $Mg(OAc)_2$, monosodium salt of sulfapyridazine , and HCl as well as doubly distd water were used. The source of the materials was not specified.

ESTIMATED ERROR:

Nothing specified

REFERENCES:

1. Tskitishvili, M. G.; Mikadze, I. I. *Soobshch. Akad. Nauk Gruz. SSR* 1978, *89(3)*, 589.

COMPONENTS:

(1) Manganese, bis[4-amino-N-(6-methoxy-3-pyridazinyl)benzenesulfonamidato)-hydrate; $C_{22}H_{20}MnN_8O_6S_2 \cdot nH_2O$; [84812-82-8]

(2) Hydrochloric acid; HCl; [7647-01-0]

(3) Water; H_2O; [7732-18-5]

ORIGINAL MEASUREMENTS:

Tskitishvili, M. G.; Shvelashvili, A. E.; Mikadze, I. I.; Zhorzholiani, N. B.; Chrelashvili, M. V. *Izv. Akad. Nauk Gruz. SSR, Ser. Khim* 1981, *7(4)*, 300-4.

VARIABLES:

pH

PREPARED BY:

R. Peikos

EXPERIMENTAL VALUES:

Concentration of HCl (mol/l)	pH	10^6 K_{so} at 25°C
1.0×10^{-2}	7.65	1.99
5.0×10^{-3}	7.71	1.97
2.5×10^{-3}	7.73	1.94
1.0×10^{-3}	7.75	1.99
5.0×10^{-4}	7.76	2.00
2.5×10^{-4}	7.78	1.97
1.0×10^{-4}	7.80	2.02
5.0×10^{-5}	7.82	2.04
1.5×10^{-5}	7.83	1.99
	Mean	1.99

AUXILIARY INFORMATION

METHOD/APPARATUS/PROCEDURE:

The earlier described apparatus and method was used (1): in a glass vessel, a mixt of 100 ml of HCl of appropriate concn and the solute were placed and shaken for 6 h in a water thermostat at 25°C. After attaining equilibrium, the pH of the soln was measured and the Mn^{2+} and S content was determined to calculate K_{so}. The pH was measured on a pH-673 pH meter.

SOURCE AND PURITY OF MATERIALS:

0.1M solns of chem pure $Mn(OAc)_2$, monosodium salt of sulfapyridazine, and HCl as well as doubly distd water were used. The source of the materials was not specified.

ESTIMATED ERROR:

K_{so} : std deviation 3×10^{-8} (compiler).

Temp and pH: not specified.

REFERENCES:

1. Tskitishvili, M. G.; Mikadze, I. I. *Soobshch. Akad. Nauk Gruz. SSR* 1978, *89(3)*, 589.

COMPONENTS:

(1) Nickel, bis[4-amino-N-6-methoxy-3-pyridazinyl)benzenesulfonamidato]-hydrate; $C_{22}H_{20}N_8NiO_6S_2 \cdot nH_2O$; [84825-01-4]

(2) Hydrochloric acid; HCl; [7647-01-0]

(3) Water; H_2O; [7732-18-5]

ORIGINAL MEASUREMENTS:

Tskitishvili, M. G.; Shvelashvili, A. E.; Mikadze, I. I.; Zhorzholiani, N. B.; Chrelashvili, M. V.;

Izv. Akad. Nauk Gruz. SSR, Ser. Khim. 1981, *7(4)*, 300-4.

VARIABLES:

pH

PREPARED BY:

R. Piekos

EXPERIMENTAL VALUES:

Concentration of HCl (mol/l)	pH	10^7 K_{so} at 25^oC
1.0×10^{-2}	7.15	1.88
5.0×10^{-3}	7.38	1.87
2.5×10^{-3}	7.65	1.86
1.0×10^{-3}	7.90	1.87
5.0×10^{-4}	8.01	1.84
2.5×10^{-4}	8.04	1.83
1.0×10^{-4}	8.07	1.79
5.0×10^{-5}	8.07	1.80
2.5×10^{-5}	8.07	1.80
	Mean	1.84

AUXILIARY INFORMATION

METHOD/APPARATUS/PROCEDURE:

The earlier described apparatus and method was used (1): in a glass vessel, a mixt of 100 ml of HCl of appropriate concn and the solute were placed and shaken for 6 h in a water thermostat at 25^oC. After attaining equilibrium, the pH of the soln was measured and the Ni^{2+} and S content was determined to calculate K_{so} The pH was measured on a pH-673 pH meter.

SOURCE AND PURITY OF MATERIALS:

0.1M solns of chem pure $Ni(OAc)_2$, monosodium salt of sulfapyridazine, and HCl as well as doubly distd water were used. The source of the materials was not specified.

ESTIMATED ERROR:

K_{so}: std deviation 3.5×10^{-9} (compiler).

Temp and pH: not specified.

REFERENCES:

1. Tskitishvili, M. G.; Mikadze, I. I. *Soobshch. Akad. Nauk Gruz. SSR* 1978, *89(3)*, 589.

COMPONENTS:

(1) Acetamide, N-[4-[[(6-methoxy-2-pyridazinyl)amino]sulfonyl]phenyl]- (acetyl sulfamethoxypyridazine); $C_{13}H_{14}N_4O_4S$; [127-75-3]

(2) Phosphoric acid, disodium salt; Na_2HPO_4; [7558-94-4]

(3) Phosphoric acid, monopotassium salt; KH_2PO_4; [7889-77-0]

(4) Water; H_2O; [7732-18-5]

ORIGINAL MEASUREMENTS:

Bandelin, F. J.; Malesh, W.

J. Am. Pharm. Assoc., Sci. Ed. 1959, *48*, 177-81.

VARIABLES:

pH

PREPARED BY:

R. Piekos

EXPERIMENTAL VALUES:

Solubility of acetyl sulfamethoxypyridazine in buffers of varying mixtures of $Na_2HPO_4 \cdot 7H_2O$ (71.6 g/l distilled water; 0.27 mol dm^{-3}, compiler) and KH_2PO_4 (36.3 g/l distilled water; 0.27 mol dm^{-3}, compiler) at 37°C

Equilibrium pH	Solubility (based on sulfamethoxypyridazine)	
	mg/100 ml	10^3 mol dm^{-3} [a]
4.5	22	0.78
4.8	24	0.86
5.4	26	0.93
5.9	28	1.0
6.2	30	1.1
7.0	41	1.5

AUXILIARY INFORMATION

METHOD/APPARATUS/PROCEDURE:

Solns were prepd by adding an excess of acetyl sulfamethoxypyridazine to 10 ml of buffer soln at each pH level in 18 x 150 mm test tubes, stoppering the tubes, and placing them in water bath at 37°C with gentle agitation for 24 h. The solute was then hydrolyzed with 5% H_2SO_4 for 1 h to liberate the free sulfonamide. One-ml aliquot of the hydrolyzate was accurately pipetted into a volumetric flask for diln and analysis. The sulfonamide was assayed colorimetrically by the method of Bratton and Marshall as described in detail by Biamonte and Schneller (1). A standard curve was prepd using accurately prepd standard solutions.

SOURCE AND PURITY OF MATERIALS:

Neither source nor purity of the reagents were specified. Distilled water was used.

ESTIMATED ERROR:

Soly: av values of duplicate runs are reported (authors).

Temp and pH: not specified.

REFERENCES:

1. Biamonte, A. R.; Schneller, G. E. *J. Am. Pharm. Assoc., Sci. Ed.* 1952, *41*, 341.

COMPONENTS:

(1) Acetamide, N-[4[[(6-methoxy-2-pyridazinyl)amino]sulfonyl]phenyl]-(acetyl sulfamethoxypyridazine); $C_{13}H_{14}N_4O_4S$; [127-75-3]

(2) Calcium chloride; $CaCl_2$; [10043-52-4]

(3) Magnesium chloride; $MgCl_2$; [7786-30-3]

(4) Phosphoric acid, monoammonium salt; $NH_4H_2PO_4$; [7722-76-1]

(5) Potassium chloride; KCl; [7447-40-7]

(6) Sodium chloride; NaCl; [7647-14-5]

(7) Urea; CH_4N_2O; [57-13-6]

(8) Water; H_2O; [7732-18-5]

ORIGINAL MEASUREMENTS:

Bandelin, F. J.; Malesh, W.

J. Am. Pharm. Assoc., Sci. Ed.

1959, *48*, 177-81.

VARIABLES:

pH at 37°C

PREPARED BY:

R. Piekos

EXPERIMENTAL VALUES:

Solubility of acetyl sulfamethoxypyridazine in a solution containing $CaCl_2$ 0.143, $MgCl_2$ 0.121, $NH_4H_2PO_4$ 0.300, KCl 1.660, NaCl 2.950 and urea 20 g/dm^3 (synthetic urine, Mosher vehicle) at 37°C

Equilibrium pH	Solubility	
	mg/100 ml	10^2 mol/dm^3 [a]
	as sulfamethoxypyridazine	
4.5	165	0.51
5.0	168	0.52
5.5	174	0.54
6.0	182	0.56
6.5	212	0.66
7.0	290	0.90

[a] calculated by compiler

AUXILIARY INFORMATION

METHOD/APPARATUS/PROCEDURE:

Excess acetyl sulfamethoxypyridazine was added to aliquots of synthetic urine solns and 1% H_3PO_4 or 1% NaOH solns were used to adjust the pH to the required value. The solns were agitated for 24 h with addn of acid or base to keep them at the desired pH level until equilibrium was attatined. Then the solns were filtered and in aliquots the acetylsulfonamide was assayed spectrophotometrically by the method described by Biamonte and Schneller (1). Before detn the soln was refluxed with 5% H_2SO_4 for 1 h to liberate the free amino compound.

SOURCE AND PURITY OF MATERIALS:

Nothing specified.

ESTIMATED ERROR:

Soly: average values of 2 detns were given.

Temp: not specified .

pH : not specified.

REFERENCES:

1. Biamonte, A. R.; Schneller, G. E. *J. Am. Pharm. Assoc., Sci. Ed.* 1952, *41*, 341.

COMPONENTS:

(1) Benzenesulfonamide, 4-amino-N-[6-(methylthio)-3-pyridazinyl]- (sulfamethylmercaptopyridazine); $C_{11}H_{12}N_4O_2S_2$; [7758-81-8]

(2) Phosphoric acid, disodium salt; Na_2HPO_4; [7558-94-4]

(3) Phosphoric acid, monopotassium salt; KH_2PO_4; [7778-77-0]

(4) Water; H_2O; [7732-18-5]

ORIGINAL MEASUREMENTS:

Riess, W.

Intern. Congr. Chemotherapy, Proc. 3rd, Stuttgart 1963, *1*, 627-32.

VARIABLES:

One temperature: 20°C; one pH: 7.4

PREPARED BY:

R. Piekos

EXPERIMENTAL VALUES:

Solubility of sulfamethylmercaptopyridazine in a M/15 Sörensen buffer solution (pH 7.4) at 20°C is 29 mg% (9.8×10^{-4} mol dm^{-3} solution, compiler).

AUXILIARY INFORMATION

METHOD/APPARATUS/PROCEDURE:

Sörensen buffer solns of pH varying between 7 and 8 were prepd, satd with sulfamethylmercaptopyridazine at 20°C, their pH was measured at equilibrium, and the sulfamethylmercaptopyridazine was assayed colorimetrically. The measured pH values were plotted against concn, and the soly at pH 7.4 was detd by interpolation (personal communication).

SOURCE AND PURITY OF MATERIALS:

Nothing specified

ESTIMATED ERROR:

Nothing specified

REFERENCES:

COMPONENTS:	ORIGINAL MEASUREMENTS:
(1) Benzenesulfonamide, 4-amino-N-(6-methylthio)-3-pyridazinyl]- (sulfamethylmercaptopyridazine); $C_{11}H_{12}N_4O_2S_2$; [7758-81-8] (2) Methane, trichloro- (chloroform); $CHCl_3$; [67-66-3]	Reiss, W. *Intern. Congr. Chemotherapy, Proc. 3rd, Stuttgart* 1963, *1*, 627-32.
VARIABLES:	PREPARED BY:
One temperature: 20°C	R. Piekos

EXPERIMENTAL VALUES:

Solubility of sulfamethylmercaptopyridazine in chloroform at 20°C is 150 mg% (5.06×10^{-3} mol dm^{-3} solution, compiler).

AUXILIARY INFORMATION

METHOD/APPARATUS/PROCEDURE:	SOURCE AND PURITY OF MATERIALS:
Nothing specified	Nothing specified
	ESTIMATED ERROR:
	Nothing specified
	REFERENCES:

COMPONENTS:	EVALUATOR:
(1) Benzenesulfonamide, 4-amino-N-2-pyrimidinyl- (sulfapyrimidine); $C_{10}H_{10}N_4O_2S$; [68-35-9] (2) Water	Anthony N. Paruta Department of Pharmaceutics University of Rhode Island Kingston, Rhode Island, USA and Ryszard Piekos Faculty of Pharmacy, University of Gdansk Gdansk, Poland 1986

CRITICAL EVALUATION:

There were twelve reports (1-12) giving values for the aqueous solubility of this compound at four temperatures as shown in Table I.

Table I: Solubility of Sulfapyrimidine in water at various temperatures

	10^4 mol dm^{-3} (*indicates mol kg^{-1})			
Reference	293K	298K	303K	310K
1	-	-	-	4.91
2	-	3.08*	-	5.07*
3	-	-	-	21.4
4	-	-	-	4.8
5	-	-	6.0	-
6	-	-	3.46	-
7	-	-	5.4	-
8	1.81*	-	3.04*	5.15*
9	2.36	-	-	-
10	-	-	-	5.07
11	-	-	-	5.11
12	-	-	-	5.11

The solubility at 298 was reported (2) only once and seems to be correct. At 293K, Elworthy and Worthington (8) and Corby and Elworthy (9), in two studies three years apart, used the same technique of percolation over a period of 5-14 days. The values differ by some 30% possibly due to the reported technique. While no recommended value can be given, an approximate solubility value of 2 x 10^{-4} mol dm^{-3} for sulfapyrimidine can be suggested. At 303K, there were four values (5-8) reported. The only reasonable values appear to be those of Elworthy and Worthington (8), and Higuchi and Lach (6). Those given by Bhattacharyya (5) and Yamazaki et al. (7) are much larger and not considered further. The other values (6,8) do not agree well; the lower value being 84% of the higher value (6) given by Higuchi. Therefore, a recommended value cannot be given but a tentative solubility value of about 3.3 x 10^{-4} mol dm^{-3} can be suggested at 303K and seems to be in accord with other temperature data. At 310K, there were eight values available (1-4,8,10-12). That given by Kikuth (3) was obviously out of line at about four times greater than the rest. The remaining values (1,2,4,8,10-12) were quite close, and the equilibrium time was at least 24 hours (1,2,4), up to 3 to 5 days (10-12), or 7-14 days (8). The average of these values, 5.03 x 10^{-4} mol dm^{-3} is the recommended solubility of sulfapyrimidine in water at 310K.

REFERENCES:

(1) Roblin, R.O., Jr.; Williams, J. H.; Winnek, P. S.; English, J.P. *J. Am. Chem. Soc.* 1940, *62*, 2002-5.
(2) Clark, W. G.; Strakosch, E. A.; Levitan, N. I.; *J. Lab. Clin. Med.* 1942, *28*, 188-9.
(3) Kikuth, W. *Med. Welt.* 1943, *17(26/27)*, 483-6.
(4) Langecker, H. *ARch. Exptl. Path. Pharmakol* 1948, *205*, 291-301.
(5) Bhattacharyya, R. Basu, U. P. *Indian Pharmacist* 1950, *6(3)*, 77-8,86.
(6) Higuchi, T. Lach, J. L. *J. Amer. Pharm. Assoc., Sci. Ed.* 1954, *43*, 349-54.
(7) Yamazaki, M.; Aoki, M.; Kamada, A.; Yata, N. *Yakuzaigaku* 1967, *27(1)*, 37-40.
(8) Elworthy, P. H.; Worthington, H. E. C. *J. Pharm. Pharmac.* 1968, *20*, 830-5.
(9) Corby, T. C.; Elworthy, P. H. *J. Pharm. Pharmac.* 1971, *23*, Suppl. 39S-48S.
(10) Watari, N.; Kaneniwa, N. *Chem. Pharm. Bull.* 1976, *24(11)*, 2577-84.
(11) Kaneniwa, N.; Watari, N. *Chem. Pharm. Bull.* 1978, *26(3)* 813-26.
(12) Watari, N.; Kaneniwa, N.; Hanano, M. *Int. J. Pharm.* 1980, *6(2)*, 155-66.

COMPONENTS:

(1) Benzenesulfonamide, 4-amino-N-2-pyrimidinyl- (sulfapyrimidine); $C_{10}H_{10}N_4O_2S$; [68-35-9]

(2) Water; H_2O; [7732-18-5]

ORIGINAL MEASUREMENTS:

Roblin, R. O., Jr.; Williams, J. H.; Winnek, P. S.; English, J. P.

J. Am. Chem. Soc. 1940, *62*, 2002-5.

VARIABLES:

One temperature: 37°C

PREPARED BY:

R. Piekos

EXPERIMENTAL VALUES:

Solubility of sulfapyrimidine in water at 37°C is 12.3 mg/100 cm^3 solution (4.91×10^{-4} mol dm^{-3}, compiler).

AUXILIARY INFORMATION

METHOD/APPARATUS/PROCEDURE:

Excess sulfonamide in water was heated and stirred on a steam bath for 30 min. The suspension was then agitated for 24 h in a thermostat at 37°C. A sample of the satd soln was withdrawn through a glass filter, dild, and analyzed by the Marshall method (1) using a General Electric recording spectrophotometer for comparing the colors developed with those of the standards.

SOURCE AND PURITY OF MATERIALS:

Sulfapyrimidine, mp 255-6°C (dec, cor), was prepd by the authors. Anal: %C 48.1 (calcd 48.0); %H 4.0 (4.0); %N 21.7 (22.4).

Purity of the water was not specified.

ESTIMATED ERROR:

Nothing specified

REFERENCES:

1. Bratton, A. C.; Marshall, E. K., Jr. *J. Pharmacol.* 1939, *66*, 4.

COMPONENTS:	ORIGINAL MEASUREMENTS:
(1) Benzenesulfonamide, 4-amino-N-2-pyrimidinyl- (sulfadiazine); $C_{10}H_{10}N_4O_2S$; [68-35-9] (2) Water; H_2O; [7732-18-5]	Clark, W. G.; Strakosch, E. A.; Levitan, N. I. *J. Lab. Clin. Med.* 1942, *28*, 188-9.
VARIABLES:	PREPARED BY:
Temperature	R. Piekos

EXPERIMENTAL VALUES:

t/°C	Solubility	
	g/100 g water	10^4 mol kg^{-1} water[a]
25	0.0077	3.08
37	0.0127	5.07

[a]Calculated by compiler

AUXILIARY INFORMATION

METHOD/APPARATUS/PROCEDURE:

A small tinted glass container contg excess sulfadiazine in water was shaken in a water bath thermostat for 24 h. The satd soln was then filtered by aspiration through a washed and dried asbestos filter stick into a weighed weighing bottle. The entire app was kept at the temp at which the compd was dissolved. The amt dissolved was then detd by the method of Bratton and Marshall (1) using a photoelectric colorimeter.

SOURCE AND PURITY OF MATERIALS:

Neither source nor purity of sulfadiazine was specified.

CO_2- free distd water was used.

ESTIMATED ERROR:

Soly: not specified.

Temp: ±0.1°C (authors).

REFERENCES:

1. Bratton, a. C.; Marshall, E. K., Jr. *J. Biol. Chem.* 1939, *128*, 537.

COMPONENTS:

(1) Benzenesulfonamide, 4-amino-N-2-pyrimidinyl- (sulfapyrimidine); $C_{10}H_{10}N_4O_2S$; [68-35-9]

(2) Water; H_2O; [7732-18-5]

ORIGINAL MEASUREMENTS:

Kikuth, W.

Med. Welt 1943, *17(26/27)*, 483-6.

VARIABLES:

One temperature: 37°C

PREPARED BY:

R. Piekos

EXPERIMENTAL VALUES:

Solubility of sulfapyrimidine in water at 37°C is 53.5 mg/100 cm^3 solution (2.14×10^{-3} mol dm^{-3}, compiler).

AUXILIARY INFORMATION

METHOD/APPARATUS/PROCEDURE:

Nothing specified

SOURCE AND PURITY OF MATERIALS:

Sulfapyrimidine was a product of Bayer.
The pH of the water was 7.0.

ESTIMATED ERROR:

Nothing specified.

REFERENCES:

COMPONENTS:	ORIGINAL MEASUREMENTS:
(1) Benzenesulfonamide, 4-amino-N-2-pyrimidinyl- (sulfadiazine); $C_{10}H_{10}N_4O_2S$; [68-35-9] (2) Water; H_2O; [7732-18-5]	Langecker, H. *Arch. Exptl. Path. Pharmakol.* 1948, *205*, 291-301.
VARIABLES:	**PREPARED BY:**
One temperature: 37°C	R. Piekos

EXPERIMENTAL VALUES:

Solubility of sulfadiazine in water at 37°C is 12 mg% (4.8×10^{-4} mol dm^{-3}, compiler).

AUXILIARY INFORMATION

METHOD/APPARATUS/PROCEDURE:

An excess of sulfadiazine was boiled with water and left for 24 h in a vessel protected from access of CO_2. The concn of sulfadiazine was detd colorimetrically by the method of Bratton and Marshall (1) using a Havemann colorimeter (2), as well as by microanal detn of the solid residue.

SOURCE AND PURITY OF MATERIALS:

Source and purity of the materials were not specified.

ESTIMATED ERROR:

Nothing specified

REFERENCES:

1. Bratton, A. G.; Marshall, E. K., Jr. *J. Biol. Chem.* 1939, *128*, 537.
2. Havemann, R. *Klin. Wochenschr.* 1940, p. 503.

COMPONENTS:

(1) Benzenesulfonamide, 4-amino-N-2-pyrimidinyl- (sulfadiazine); $C_{10}H_{10}N_4O_2S$; [68-35-9]

(2) Water; H_2O; [7732-18-5]

ORIGINAL MEASUREMENTS:

Bhattacharyya, R.; Basu, U. P.

Indian Pharmacist 1950, *6(3)*, 77-8, 86.

VARIABLES:

One temperature: 30°C

PREPARED BY:

R. Piekos

EXPERIMENTAL VALUES:

Solubility of sulfadiazine in water at 30°C is 15 mg per 100 ml (6.0×10^{-4} mol dm^{-3}, compiler).

AUXILIARY INFORMATION

METHOD/APPARATUS/PROCEDURE:

A weighed sample of sulfadiazine was placed in a clean reagent bottle and a known vol of water was added. The mixt was shaken in a mech shaker at 80-100 strokes/min. After at least 24 h the mixt was filtered through a clean, dried and weighed sintered-glass crucible. At the end of the filtration the crucible was washed with about 1 ml of water, dried at 105°C for 2-3 h, cooled, and weighed to const wt.

SOURCE AND PURITY OF MATERIALS:

Neither source nor purity of the sulfadiazine was specified.

Doubly distd water was used.

ESTIMATED ERROR:

Soly: not specified.

Temp: ±0.2°C (authors).

REFERENCES:

COMPONENTS:

(1) Benzenesulfonamide, 4-amino-N-2-pyrimidinyl- (sulfadiazine); $C_{10}H_{10}N_4O_2S$; [68-35-9]

(2) Water; H_2O; [7732-18-5]

ORIGINAL MEASUREMENTS:

Higuchi, T.; Lach, J. L.
J. Amer. Pharm. Assoc., Sci. Ed. 1954, *43*, 349-54.

VARIABLES:

One temperature: 30°C

PREPARED BY:

R. Piekos

EXPERIMENTAL VALUES:

Solubility of sulfadiazine in water at 30°C is 3.46×10^{-4} mol dm^{-3} solution (9.11×10^{-2} g dm^{-3}, compiler).

AUXILIARY INFORMATION

METHOD/APPARATUS/PROCEDURE:

Excess sulfadiazine (25 mg) was placed in a 125-ml glass-stoppered bottle together with 50 ml of water. The bottle was placed in a mech shaker in a const temp bath and equilibrated for 8 h at 30°C. Aliquot of the supernatant liquid was analyzed for sulfadiazine by the method of Bratton and Marshall (1).

SOURCE AND PURITY OF MATERIALS:

Recrystd sulfadiazine (U.S.P.), mp 255-6°C, and distilled water were used.

ESTIMATED ERROR:

Nothing specified.

REFERENCES:

1. Bratton, A. C.; Marshall, E. K., Jr. *J. Biol. Chem.* 1939, *128*, 537.

COMPONENTS:

(1) Benzenesulfonamide, 4-amino-N-2-pyrimidinyl- (sulfadiazine); $C_{10}H_{10}N_4O_2S$; [68-35-9]

(2) Water; H_2O; [7732-18-5]

ORIGINAL MEASUREMENTS:

Yamazaki, M.; Aoki, M.; Kamada, A.; Yata, N. *Yakuzaigaku,* 1967, *27(1),* 37-40.

VARIABLES:

One temperature: 30°C

PREPARED BY:

R. Piekos

EXPERIMENTAL VALUES:

Solubility of sulfadiazine in water at 30°C is 0.54 mmol/L (0.14 g dm^{-3}, compiler).

AUXILIARY INFORMATION

METHOD/APPARATUS/PROCEDURE:

Sulfadiazine (0.5 g) was placed in an L-shaped tube together with 20 ml of water. The mixt was shaken in a thermostat until equilibrium was attained. The sulfadiazine content was assayed in the supernatant spectrophotometrically at 545 nm on a Beckmann DU spectrophotometer. The results were taken from a calibration graph.

SOURCE AND PURITY OF MATERIALS:

Nothing specified.

ESTIMATED ERROR:

Soly: not specified.

Temp: ±1°C (authors).

REFERENCES:

COMPONENTS:

(1) Benzenesulfonamide, 4-amino-N-2-pyrimidinyl- (sulfadiazine); $C_{10}H_{10}N_4O_2S$; [68-35-9]

(2) Water; H_2O; [7732-18-5]

ORIGINAL MEASUREMENTS:

Elworthy, P. H.; Worthington, H. E. C.

J. Pharm. Pharmac. 1968, *20*, 830-5.

VARIABLES:

Temperature

PREPARED BY:

R. Piekos

EXPERIMENTAL VALUES:

t/°C	Solubility		
	Weight %	10^6 mole fraction	10^4 mol kg^{-1} water[a]
20	0.00454	3.27	1.81
30	0.00760	5.47	3.04
40	0.01290	9.29	5.15

[a]Calculated by compiler

AUXILIARY INFORMATION

METHOD/APPARATUS/PROCEDURE:

Solns were presatd by shaking with powd sulfadiazine for 24 h and transferred to a soly app which was of the percolation type and a modification of that used by Davies and Griffiths (1). The soln was recycled in the app through a sintered-glass filter until satd (7 - 14 days). Samples were dild with water and assayed spectrophotometrically at 270 nm. Suitable calibration lines were prepd.

SOURCE AND PURITY OF MATERIALS:

Sulfadiazine (B. P. quality) was twice recrystd from an EtOH - DMF mixt (3:1 by vol) and dried over P_2O_5. Its mp was 255°C. Assay by the Pharmacopeial method gave 100.0% purity calcd with reference to the material dried at 105°C.

Purity of the water was not specified.

ESTIMATED ERROR:

Soly: mean values of duplicate runs are given (authors).

Temp: ±0.05°C (authors).

REFERENCES:

1. Davies, M; Griffiths, D. M. L. *Trans. Faraday Soc.* 1953, *49*, 1405.

COMPONENTS:	ORIGINAL MEASUREMENTS:
(1) Benzenesulfonamide, 4-amino-N-2-pyrimidinyl- (sulfadiazine); $C_{10}H_{10}N_4O_2S$; [68-35-9] (2) Water; H_2O; [7732-18-5]	Corby, T. C.; Elworthy, P. H. *J. Pharm. Pharmac.* 1971, *23*, Suppl. 39S-48S.
VARIABLES: One temperature: 20°C	PREPARED BY: R. Piekos

EXPERIMENTAL VALUES:

Concentration of sulfadiazine in a saturated aqueous solution at 20°C is 0.236 millimolal (5.91×10^{-3}% by wt, compiler).

AUXILIARY INFORMATION

METHOD/APPARATUS/PROCEDURE:

The soly was detd in an app where water percolated through a plug of sulfadiazine supported on a 5/3 sintered glass disc. The percolation was continued until the soln was satd (5-6 days). The concn of sulfadiazine was detd spectrophotometrically from a calibration curve using a Uvispek spectrophotometer (1).

SOURCE AND PURITY OF MATERIALS:

Sulfadiazine (Macarthy's Ltd, Romford) was a British Pharmacopeia product. It was recrystd twice from a DMF-EtOH mixt (1:3) and dried at 40°C over P_2O_5. Its mp was 254° (decompn).

Tap water once distd from glass was used.

ESTIMATED ERROR:

Soly: not specified.

Temp: ±0.05°C (authors).

REFERENCES:

1. Elworthy, P. H.; Lipscomb, F. J. *J. Pharm. Pharmac.* 1968, *20*, 790.

COMPONENTS:	ORIGINAL MEASUREMENTS:
(1) Benzenesulfonamide, 4-amino-N-2-pyrimidinyl- (sulfadiazine); $C_{10}H_{10}N_4O_2S$; [68-35-9] (2) Water; H_2O; [7732-18-5]	Watari, N.; Kaneniwa, N. *Chem. Pharm. Bull.* 1976, *24(11)*, 2577-84.
VARIABLES:	PREPARED BY:
One temperature: 37°C	R. Piekos

EXPERIMENTAL VALUES:

Total solubility of sulfadiazine in water at 37°C is 0.127 mg/ml solution (5.07×10^{-4} mol dm^{-3}, compiler.)

AUXILIARY INFORMATION

METHOD/APPARATUS/PROCEDURE:

An excess of sulfadiazine, required to saturate water, was placed in a flask contg 25 ml of water. The flask was shaken (2 strokes/s) at the amplitude of 3 cm in a thermostatically controlled water bath at 37°C. One-ml sample was removed every 6 h (total equilibration period 3-5 days) using a warmed Millipore filter syringe with a filter pore size of 0.45 μ (Millipore HAWP 01300) and the filtrate was dild with water and assayed spectrophotometrically (1).

SOURCE AND PURITY OF MATERIALS:

Commercial sulfadiazine of the Japanese Pharmacopeia grade and distd water were used.

ESTIMATED ERROR:

Temp: ±0.05°C (authors).

Soly: not specified.

REFERENCES:

1. Kaneniwa, N.; Watari, N. *Chem. Pharm. Bull.* 1974, *22*, 1699.

COMPONENTS:	ORIGINAL MEASUREMENTS:
(1) Benzenesulfonamide, 4-amino-N-2-pyrimidinyl- (sulfadiazine); $C_{10}H_{10}N_4O_2S$; [68-35-9] (2) Water; H_2O; [7732-18-5]	Kaneniwa, N.; Watari, N. *Chem. Pharm. Bull.* 1978, *26(3)*, 813-26.
VARIABLES:	**PREPARED BY:**
One temperature: 37°C	R. Piekos

EXPERIMENTAL VALUES:

Solubility of sulfadiazine in water at 37°C is 0.128 mg/ml solution (5.11×10^{-4} mol dm^{-3}, compiler).

AUXILIARY INFORMATION

METHOD/APPARATUS/PROCEDURE:

An excess of sulfadiazine was placed in a flask contg 25 ml of water. The flask was shaken (2 strokes/s at the amplitude of 3 cm) in a thermostatically controlled water bath at 37°C. One-ml sample was withdrawn every 6 h (total equilibration period was 3-5 days) using a warmed Millipore filter syringe with a filter pore size of 0.45 μ (Millipore HAWP 01300) and the filtrate was dild with water and assayed spectrophotometrically (1).

SOURCE AND PURITY OF MATERIALS:

Commercial sulfadiazine of the Japanese Pharmacopeia grade and distd water were used.

ESTIMATED ERROR:

Soly: not specified.

Temp: ±0.05°C (authors).

REFERENCES:

1. Kaneniwa, N.; Watari, N. *Chem. Pharm. Bull.* 1974, *22*, 1699.

COMPONENTS:

(1) Benzenesulfonamide, 4-amino-N-2-pyrimidinyl- (sulfadiazine); $C_{10}H_{10}N_4O_2S$; [68-35-9]

(2) Water; H_2O; [7732-18-5]

ORIGINAL MEASUREMENTS:

Watari, N.; Kaneniwa, N.; Hanano, M.
Int. J. Pharm. 1980, *6(2)*, 155-66.

VARIABLES:

One temperature: 37°C

PREPARED BY:

R. Piekos

EXPERIMENTAL VALUES:

Solubility of sulfadiazine in water at 37°C is 12.8 mg/100 ml

(5.11×10^{-4} mol dm^{-3}, compiler).

AUXILIARY INFORMATION

METHOD/APPARATUS/PROCEDURE:

The earlier developed method was employed (1), whereby an excess of sulfadiazine, required to saturate medium, was placed in a flask contg 25 ml of water. The flask was stroked (2 strokes/s) at an amplitude of 3 cm, in a thermostatically controlled bath. One-ml sample was removed every 6 h (total equilibration time was 3-5 days) using a warmed Millipore filter syringe with a filter pore size of 0.45 μ (Millipore HAWP 01300) and the filtrate was dild with water and assayed spectrophotometrically).

SOURCE AND PURITY OF MATERIALS:

Sulfadiazine was of the Japanese Pharmacopeia grade.

Distilled water was used.

ESTIMATED ERROR:

Soly: not specified.

Temp: ±0.05°C (authors).

REFERENCES:

1. Kaneniwa, N.; Watari, N.
Chem. Pharm. Bull. 1974, *22*, 1699.

COMPONENTS:	ORIGINAL MEASUREMENTS:
(1) Benzenesulfonamide, 4-amino-N-2-pyrimidinyl- (sulfadiazine); $C_{10}H_{10}N_4O_2S$; [68-35-9] (2) Hydrochloric acid; HCl; [7647-01-0] (3) Water; H_2O; [7732-18-5]	Nogami, H.; Nagai, T.; Suzuki, A. *Chem. Pharm. Bull.* 1966, *14(4)*, 339-50.
VARIABLES:	PREPARED BY:
Concentration of HCl	R. Piekos

EXPERIMENTAL VALUES:

Concentration of HCl	Solubility of sulfadiazine at 37°C	
mol dm^{-3}	mg/100 ml	10^4 mol dm^{-3} [a]
10^{-1}	92.0	36.8
10^{-2}	18.0	7.19
10^{-3}	11.1	4.43

[a] Calculated by compiler

AUXILIARY INFORMATION

METHOD/APPARATUS/PROCEDURE:

Soly of sulfadiazine was detd from dissoln rate data obtained by the rotating disk method.

SOURCE AND PURITY OF MATERIALS:

Commercial sulfadiazine J. P. was used. Purity of the remaining materials was not specified.

ESTIMATED ERROR:

Nothing specified.

REFERENCES:

COMPONENTS:

(1) Benzenesulfonamide, 4-amino-N-2-pyrimidinyl- (sulfadiazine); $C_{10}H_{10}N_4O_2S$; [68-35-9]

(2) Hydrochloric acid; HCl; [7647-01-0]

(3) Sodium chloride; NaCl; [7647-14-5]

(4) Water; H_2O; [7732-18-5]

ORIGINAL MEASUREMENTS:

Krebs, H. A. ; Speakman, J. C.
J. Chem. Soc. 1945, 593-5.

VARIABLES:

pH

PREPARED BY:

R. Piekos

EXPERIMENTAL VALUES:

pH at saturation	Solubility of sulfadiazine at 25.0°C in solution of ionic strength 0.1M maintained by NaCl	
	mg/100 ml	10^3 mol dm^{-3} [a]
1.00[b]	68	2.7
1.26	66	2.6
1.55	25.2	1.01
1.89	16.5	0.659

[a] Calculated by compiler

[b] in 0.15N HCl

AUXILIARY INFORMATION

METHOD/APPARATUS/PROCEDURE:

The HCl solns were satd by shaking them with an excess of solid sulfadiazine for at least 3 h. They were then quickly filtered. The sulfadiazine concn was detd by the method of Bratton and Marshall (1), and the pH values were measured with the glass electrode, which was standardized in terms of the buffer solns recommended by Hitchcock and Taylor (2).

SOURCE AND PURITY OF MATERIALS:

The sulfadiazine, mp 256°C, was obtained by recrystg a comm specimen (source not specified) from water.
Purity of the remaining materials was not specified.

ESTIMATED ERROR:

Nothing specified.

REFERENCES:

1. Bratton, A. C.; Marshall, E. K., Jr. *J. Biol. Chem.* 1939, *128*, 537.
2. Hitchcock, D. I.; Taylor, A. C. *J. Am. Chem. Soc.* 1937, *59*, 1812.

COMPONENTS:	ORIGINAL MEASUREMENTS:
(1) Benzenesulfonamide, 4-amino-N-2-pyrimidinyl- (sulfadiazine); $C_{10}H_{10}N_4O_2S$; [68-35-9] (2) Sodium hydroxide; NaOH; [1310-73-2] (3) Water; H_2O; [7732-18-5]	Rose, F. L.; Martin, A. R.; Bevan, H. G. L. *J. Pharm. Exp. Therap.* 1943, *77*, 127-42.

VARIABLES:	PREPARED BY:
pH	R. Piekos

EXPERIMENTAL VALUES:

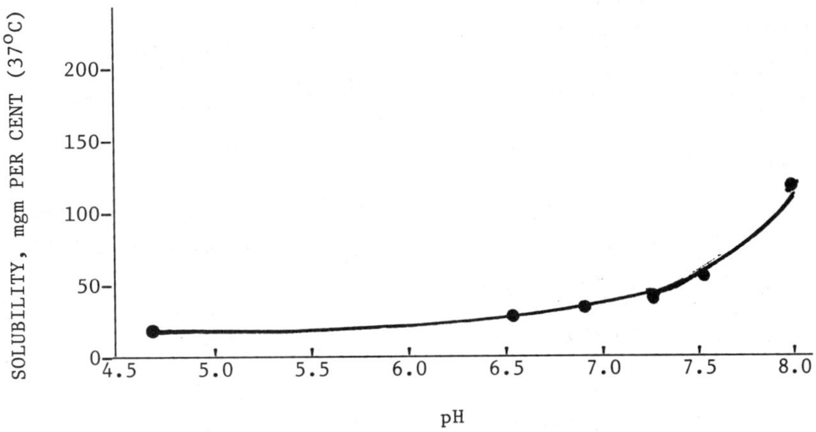

AUXILIARY INFORMATION

METHOD/APPARATUS/PROCEDURE:

An excess of sulfadiazine was stirred in boiling water, the soln was cooled to 37°C, the temp being maintained thermostatically, and 0.1N NaOH was added to increase the pH. The pH was measured by means of a glass electrode - calomel half-cell system, and was permitted to reach equilibrium before a reading was taken. The concn of sulfadiazine in soln was detd colorimetrically by withdrawing a sample through a filter-tip into a preheated micropipet.

SOURCE AND PURITY OF MATERIALS:

The source and purity of sulfadiazine were not specified.

Water was doubly distilled.

ESTIMATED ERROR:

Nothing specified.

REFERENCES:

COMPONENTS:

(1) Benzenesulfonamide, 4-amino-N-2-pyrimidinyl- (sulfadiazine); $C_{10}H_{10}N_4O_2S$; [68-35-9]

(2) Sodium hydroxide; NaOH; [1310-73-2]

(3) Water; H_2O; [7732-18-5]

ORIGINAL MEASUREMENTS:

Holz, E.; Garcia Onandia, A.; Holz, S.

Acta Cient. Venezolana 1955, *6(2)*, 68-73.

VARIABLES:

Concentration of NaOH

PREPARED BY:

R. Piekos

EXPERIMENTAL VALUES:

Concentration of NaOH soln	Volume of NaOH soln required to dissolve 1 g of sulfadiazine at 26°C	Solubility of sulfadiazine at 26°C
N	cm^3	mol dm^{-3} NaOH soln[a]
1/10	43.3	0.0923
1/4	17.3	0.231
1/2	8.35	0.478
1	4.5	0.89
1.5	2.8	1.4
2	2.1	1.9
2.5	1.65	2.42
3	1.4	2.8
3.5	1.2	3.3
3.75	1.2	3.3
4	101	0.0396

[a] Calculated by compiler

AUXILIARY INFORMATION

METHOD/APPARATUS/PROCEDURE:

Nothing specified.

SOURCE AND PURITY OF MATERIALS:

Nothing specified. Distd water was used.

ESTIMATED ERROR:

Nothing specified.

REFERENCES:

COMPONENTS:

(1) Benzenesulfonamide, 4-amino-N-2-pyrimidinyl- (sulfadiazine); $C_{10}H_{10}N_4O_2S$; [68-35-9]

(2) Sodium hydroxide; NaOH; [1310-73-2]

(3) Water; H_2O; [7732-18-5]

ORIGINAL MEASUREMENTS:

Nogami, H.; Nagai, T.; Suzuki, A.

Chem. Pharm. Bull. 1966, *14(4)*, 339-50.

VARIABLES:

Concentration of sodium hydroxide

PREPARED BY:

R. Piekos

EXPERIMENTAL VALUES:

Concentration of NaOH	Solubility of sulfadiazine at 37°C	
10^3 mol dm^{-3}	mg/100 ml	10^3 mol dm^{-3}
1	35.2	1.41
1.4	46.9	1.88
2	68.1	2.72
3	96.0	3.84
4	118	4.72
5	153	6.12
10	272	10.9
30	760	30.4
50	1270	50.8
100	2520	100.8

AUXILIARY INFORMATION

METHOD/APPARATUS/PROCEDURE:

Soly of sulfadiazine was detd from dissoln rate data obtained by the rotating disk method.

SOURCE AND PURITY OF MATERIALS:

Commercial sulfadiazine J. P. was used. Purity of the remaining materials was not specified.

ESTIMATED ERROR:

Nothing specified.

REFERENCES:

COMPONENTS:

(1) Benzenesulfonamide, 4-amino-N-2-pyrimidinyl- (sulfadiazine); $C_{10}H_{10}N_4O_2S$; [68-35-9]

(2) Potassium hydroxide; KOH; [1310-58-3]

(3) Water; H_2O; [7732-18-5]

ORIGINAL MEASUREMENTS:

Nogami, H.; Nagai, T.; Suzuki, A.
Chem. Pharm. Bull. 1966, *14(4)*, 339-50.

VARIABLES:

Concentration of KOH

PREPARED BY:

R. Piekos

EXPERIMENTAL VALUES:

Concentration of KOH	Solubility of sulfadiazine at 37°C	
10^2 mol dm^{-3}	mg/100 ml	10^2 mol dm^{-3}
0.1	35	0.141
0.5	131	0.524
1.0	326	1.30
5.0	1350	5.40
10.0	2710	1.08[a]

[a] should be 10.8 - compiler

AUXILIARY INFORMATION

METHOD/APPARATUS/PROCEDURE:

Soly of sulfadiazine was detd from dissoln rate data obtained by the rotating disk method.

SOURCE AND PURITY OF MATERIALS:

Commercial sulfadiazine J. P. was used. Purity of the remaining materials was not specified.

ESTIMATED ERROR:

Nothing specified.

REFERENCES:

COMPONENTS:	ORIGINAL MEASUREMENTS:
(1) Benzenesulfonamide, 4-amino-N-2-pyrimidinyl- (sulfadiazine); $C_{10}H_{10}N_4O_2S$; [68-35-9] (2) Carbonic acid, disodium salt; Na_2CO_3; [497-19-8] (3) Water; H_2O; [7732-18-5]	Takubo, T.; Matsumaru, H.; Tsuchiya, S.; Hiura, M. *Chem. Pharm. Bull.* <u>1973</u>, *21(7)*, 1440-5.
VARIABLES:	PREPARED BY:
One temperature: 37°C; one pH: 11.3	R. Piekos

EXPERIMENTAL VALUES:

Solubility of sulfadiazine in a Na_2CO_3 solution (2.120 g Na_2CO_3/100 ml water) of pH 11.3 at 37°C is 29.12 mg/ml solution[a] (0.1163 mol dm^{-3} solution, compiler).

[a]Numerical value to the graphical data given by one of the authors (S.T.) in personal communication.

AUXILIARY INFORMATION

METHOD/APPARATUS/PROCEDURE:

Aliquots of the Na_2CO_3 soln were placed in glass-stoppered flasks with excess of sulfadiazine. The flasks were allowed to stand at 37±1°C and shaken vigorously for 4 h until equilibrium was established. One ml of the supernatant was removed by means of a filter pipet and sulfadiazine was assayed by the previously reported method (1).

SOURCE AND PURITY OF MATERIALS:

The sulfadiazine was of pharmaceutical purity. The source and purity of Na_2CO_3 were not specified.
Distd water was used.

ESTIMATED ERROR:

Soly and pH: not specified.
Temp: ±1°C (authors).

REFERENCES:

1. Takubo, T.; Tsuchiya, S.; Hiura, M. *Yakuzaigaku* <u>1971</u>, *31*, 298.

COMPONENTS:	ORIGINAL MEASUREMENTS:
(1) Benzenesulfonamide, 4-amino-N-2-pyrimidinyl- (sulfadiazine); $C_{10}H_{10}N_4O_2S$; [68-35-9] (2) Carbonic acid, monosodium salt; $NaHCO_3$; [144-55-8] (3) Water; H_2O; [7732-18-5]	Takubo, T.; Matsumaru, H.; Tsuchiya, S.; Hiura, M. *Chem. Pharm. Bull.* 1973, *21(7)*, 1440-5.
VARIABLES:	PREPARED BY:
One temperature: 37°C; one pH: 8.4	R. Piekos

EXPERIMENTAL VALUES:

Solubility of sulfadiazine in a $NaHCO_3$ solution (1.680 g $NaHCO_3$/100 ml water) of pH 8.4 at 37°C is 8.06 mg/ml solution[a] (3.22×10^{-2} mol dm^{-3} solution, compiler).

AUXILIARY INFORMATION

METHOD/APPARATUS/PROCEDURE:

Aliquots of the $NaHCO_3$ soln were placed in glass-stoppered flasks with excess of sulfadiazine. The flasks were allowed to stand at 37±1°C and shaken vigorously for 4 h until equilibrium was established. One ml of the supernatant was removed by means of a filter pipet and sulfadiazine was assayed by the previously reported method (1).

SOURCE AND PURITY OF MATERIALS:

The sulfadiazine was of pharmaceutical grade. The source and purity of $NaHCO_3$ was not specified.

Distd. water was used.

ESTIMATED ERROR:

Soly and pH: not specified.

Temp: ±1°C (authors).

REFERENCES:

1. Takubo, T.; Tsuchiya, S.; Hiura, M. *Yakuzaigaku*, 1971, *31*, 298.

COMPONENTS:

(1) Benzenesulfonamide, 4-amino-N-2-pyrimidinyl- (sulfadiazine); $C_{10}H_{10}N_4O_2S$; [68-35-9]

(2) Sodium chloride; NaCl; [7647-14-5]

(3) Water; H_2O; [7732-18-5]

ORIGINAL MEASUREMENTS:

Langecker, H.

Arch. Exptl. Path. Pharmakol. 1948, *205*, 291-301.

VARIABLES:

One temperature: 37°C

PREPARED BY:

R. Piekos

EXPERIMENTAL VALUES:

Solubility of sulfadiazine in a 0.9 % w/w NaCl solution at 37°C is 16 mg%

(6.4×10^{-4} mol dm^{-3}, compiler).

AUXILIARY INFORMATION

METHOD/APPARATUS/PROCEDURE:

An excess of sulfadiazine in a 0.9% w/w NaCl soln was boiled for 1 h in a sealed ampul followed by keeping the ampul at 37°C. The sulfadiazine concn was detd colorimetrically by the method of Bratton and Marsahll (1) using a Havemann colorimeter (2), and by microanal detn of the solid residue.

SOURCE AND PURITY OF MATERIALS:

Source and purity of the materials were not specified.

ESTIMATED ERROR:

Nothing specified.

REFERENCES:

1. Bratton, A. G.; Marshall, E. K., Jr. *J. Biol. Chem.* 1939, *128*, 537.
2. Havemann, R. *Klin. Wochenschr.* 1940, p. 503.

COMPONENTS:	ORIGINAL MEASUREMENTS:
(1) Benzenesulfonamide, 4-amino-N-2-pyrimidinyl- (sulfadiazine); $C_{10}H_{10}N_4O_2S$; [68-35-9] (2) Sodium chloride: NaCl; [7647-14-5] (3) Water; H_2O; [7732-18-5]	Nogami, H.; Nagai, T.; Suzuki, A. *Chem. Pharm. Bull.* 1966, *14(4)*, 339-50.
VARIABLES:	PREPARED BY:
Concentration of sodium chloride	R. Piekos

EXPERIMENTAL VALUES:

Concentration of NaCl	Solubility of sulfadiazine at 37°C	
mol dm^{-3}	mg/100 ml	10^4 mol dm^{-3} [a]
10^{-3}	12.1	4.83
10^{-2}	11.3	4.51
10^{-1}	10.2	4.07
1	9.0	3.6

[a] Calculated by compiler

AUXILIARY INFORMATION

METHOD/APPARATUS/PROCEDURE:

Soly of sulfadiazine was detd from dissoln rate data obtained by the rotating disk method.

SOURCE AND PURITY OF MATERIALS:

Commercial sulfadiazine J. P. was used. Purity of the remaining materials was not specified.

ESTIMATED ERROR:

Nothing specified.

REFERENCES:

COMPONENTS:	ORIGINAL MEASUREMENTS:
(1) Benzenesulfonamide, 4-amino-N-2-pyrimidinyl- (sulfadiazine); $C_{10}H_{10}N_4O_2S$; [68-35-9] (2) Sodium chloride; NaCl; [7647-14-5] (3) Water; H_2O; [7732-18-5]	Avico, U.; Cavazuti, G.; di Francesco, R.; Signoretti Ciranni, E.; Zuccaro, P. *Farmaco, Ed. Practica* 1975, *30(1)*, 40-6.
VARIABLES:	PREPARED BY:
Temperature	R. Piekos

EXPERIMENTAL VALUES:

t/°C	Solubility of amorphous sulfadiazine in equimolal NaCl solutions	
	g/100 g water	10^3 mol kg^{-1} water[a]
25	0.151	6.03
35	0.151	6.03
40	0.151	6.03

[a] Calculated by compiler

AUXILIARY INFORMATION

METHOD/APPARATUS/PROCEDURE:

A soln of Na salt of sulfadiazine was added to a HCl soln contg stoichiometric quantity of the acid to neutralize the salt. The neutralization was carried out in a thermostat and the pH of the mixt was maintained close to that of a satd sulfadiazine soln. The procedure was repeated using various initial concns of the reagents to find the max concn of sulfadiazine at which no pptn occurred.

SOURCE AND PURITY OF MATERIALS:

Source and purity of sulfadiazine was not specified. The mp of crystalline sulfadiazine was 253-6°C.

Purity of the water was not specified.

ESTIMATED ERROR:

Nothing specified.

REFERENCES:

COMPONENTS:

(1) Benzenesulfonamide, 4-amino-N-2-pyrimidinyl- (sulfapyrimidine); $C_{10}H_{10}N_4O_2S$; [68-35-9]

(2) Phosphoric acid, monopotassium salt; KH_2PO_4; [7778-77-0]

(3) Water; H_2O; [7732-18-5]

ORIGINAL MEASUREMENTS:

Krüger-Thiemer, E.

Arch. Dermatol. Syphilis 1942, *183*, 90-116.

VARIABLES:

One temperature: ca 20°C; one pH: 4.37

PREPARED BY:

R. Piekos

EXPERIMENTAL VALUES:

Solubility of sulfapyrimidine in a 0.735M (10%) KH_2PO_4 solution of pH 4.37 at room temperature (about 20°C) is 0.0070 g% (2.8 x 10^{-4} mol dm^{-3} solution, compiler).

AUXILIARY INFORMATION

METHOD/APPARATUS/PROCEDURE:

Sulfapyrimidine (0.5 g) was dissolved in 10 cm^3 of a 0.735M (10%) KH_2PO_4 soln of pH 4.37, shaken for 2 h at room temp (about 20°C), and filtered. A 1-cm^3 aliquot of the filtrate was withdrawn, cooled, acifified with 1 cm^3 of 2N HCl, and the sulfapyrimidine content was detd colorimetrically by the method of Marshall modified by Kimmig (1) using an Authenrieth colorimeter. The pH was detd on an ultraionograph using a glass electrode.

SOURCE AND PURITY OF MATERIALS:

Sulfapyrimidine was manufd by Schering (purity not specified). The source and purity of the remaining materials were not specified.

ESTIMATED ERROR:

Soly: precision ±5% (author).

Temp: not specified.

pH : ±0.05 pH unit (author).

REFERENCES:

1. Kimmig, J. *Arch. Dermatol.* 1938, *176*, 722; *Erg. Hyg.* 1941, *24*, 398.

COMPONENTS:

(1) Benzenesulfonamide, 4-amino-N-2-pyrimidinyl- (sulfadiazine); $C_{10}H_{10}N_4O_2S$; [68-35-9]

(2) Phosphoric acid; disodium salt; Na_2HPO_4; [7558-94-4]

(3) Water; H_2O; [7732-18-5]

ORIGINAL MEASUREMENTS:

Nogami, H.; Nagai, T.; Suzuki, A. *Chem. Pharm. Bull.* 1966, *14(4)*, 339-50.

VARIABLES:

Concentration of Na_2HPO_4

PREPARED BY:

R. Piekos

EXPERIMENTAL VALUES:

Concentration of Na_2HPO_4	Solubility of sulfadiazine at 37°C	
mol dm^{-3}	mg/100 ml	10^{-3} mol dm^{-3}
10^{-4}	15.8	0.63
10^{-3}	33.1	1.3
10^{-2}	104	4.2
10^{-1}	242	9.7

AUXILIARY INFORMATION

METHOD/APPARATUS/PROCEDURE:

Soly of sulfadiazine was detd from dissoln rate data obtained by the rotating disk method.

SOURCE AND PURITY OF MATERIALS:

Commercial sulfadiazine J. P. was used. Purity of the remaining materials was not specified.

ESTIMATED ERROR:

Nothing specified.

REFERENCES:

COMPONENTS:

(1) Benzenesulfonamide, 4-amino-N-2-pyrimidinyl- (sulfapyrimidine); $C_{10}H_{10}N_4O_2S$; [68-35-9]

(2) Hydrochloric acid; HCl; [7647-01-0]

(3) Phosphoric acid, trisodium salt; Na_3PO_4; [7601-54-9]

(4) Water; H_2O; [7732-18-5]

ORIGINAL MEASUREMENTS:

Stricker, H.

Pharm. Ind. 1971, *33(7)*, 446-54.

VARIABLES:

pH

PREPARED BY:

R. Piekos

EXPERIMENTAL VALUES:

pH	Solubility at 37°C	
	g/100 ml	10^3 mol dm^{-3} [a]
5.7	ca 0.03	1.2
7.5	ca 0.11	4.4

[a] Calculated by compiler

AUXILIARY INFORMATION

METHOD/APPARATUS/PROCEDURE:

Soly was detd in a Sartorius-Loesemodell SM 16 751 soly apparatus contg 100 ml of solvent. The solvent consisted of 0.06N HCl whose pH was adjusted to the desired value by addn of a concd Na_3PO_4 soln.
Aliquots of solns were acidifed, diazotized and coupled with N-(1-naphthyl)-ethylenediammonium chloride. After subsequent addn of an acetate buffer soln of pH 5 and acetone, extinction of the solns was measured at 500 nm.

SOURCE AND PURITY OF MATERIALS:

Neither source nor the purity of the materials was specified.

ESTIMATED ERROR:

Soly: not specified.

Temp: ±0.5°C (authors).

REFERENCES:

COMPONENTS:

(1) Benzenesulfonamide, 4-amino-N-2-pyrimidinyl- (sulfadiazine); $C_{10}H_{10}N_4O_2S$; [68-35-9]

(2) Carbonic acid, disodium salt; Na_2CO_3; [497-19-8]

(3) Carbonic acid, monosodium salt; $NaHCO_3$; [144-55-8]

(4) Water; H_2O; [7732-18-5]

ORIGINAL MEASUREMENTS:

Takubo, T.; Matsumaru, H.; Tsuchiya, S.; Hiura, M.

Chem. Pharm. Bull. 1973, *21(7)*, 1440-5.

VARIABLES:

pH

PREPARED BY:

R. Piekos

EXPERIMENTAL VALUES:

Na_2CO_3	$NaHCO_3$	pH	Solubility at 37°C	
g/100 ml water	g/100 ml water		mg/ml soln[a]	10^2 mol dm^{-3} soln[b]
0.212	1.512	9.1	9.32	3.724
0.848	1.008	9.8	18.82	7.520
1.908	0.168	10.7	26.94	10.764

[a] Numerical values to the graphical ones were given by one of the authors (S. T.) in personal communication.

[b] Calculated by compiler

AUXILIARY INFORMATION

METHOD/APPARATUS/PROCEDURE:

Aliquots of the carbonate buffer solns were placed in glass-stoppered flasks with excess of sulfadiazine. The flasks were allowed to stand at 37±1°C and shaken vigorously for 4 h until equilibrium was established. One ml of the supernatant was removed by means of a filter pipet and sulfadiazine was assayed by the previously reported method (1).

SOURCE AND PURITY OF MATERIALS:

The sulfadiazine was of pharmaceutical grade. The source and purity of Na_2CO_3 and $NaHCO_3$ were not specified.

Distd water was used.

ESTIMATED ERROR:

Soly and pH: not specified.

Temp: ±1°C

REFERENCES:

1. Takubo, T.; Tsuchiya, S.; Hiura, M. *Yakuzaigaku* 1971, *31*, 298.

COMPONENTS:

(1) Benzenesulfonamide, 4-amino-N-2-pyrimidinyl- (sulfapyrimidine); $C_{10}H_{10}N_4O_2S$; [68-35-9]

(2) Phosphoric acid, disodium salt; Na_2HPO_4; [7558-94-4]

(3) Phosphoric acid, monopotassium salt; KH_2PO_4; [7778-77-0]

(4) Water; H_2O; [7732-18-5]

ORIGINAL MEASUREMENTS:

Krüger-Thiemer, E.

Arch. Dermatol. Syphilis 1942, *183*, 90-116.

VARIABLES:

pH

PREPARED BY:

R. Piekos

EXPERIMENTAL VALUES:

Composition of 1/15M phosphate buffer solutions			pH	Solubility at room temperature (about 20°C)	
Na_2HPO_4	KH_2PO_4	%Content		g%	10^3 mol dm^{-3} solution[a]
1.0	99.0	0.91	4.944	0.0087	0.35
10.0	90.0	0.91	5.906	0.0101	0.403
61.1	38.9	0.93	7.005	0.033	1.3
9.5	0.5	0.733[b]	7.51	0.065	2.6
94.7	5.3	0.95	8.018	0.127	5.07

[a] Calculated by compiler

[b] Molar content; 10% buffer solution

AUXILIARY INFORMATION

METHOD/APPARATUS/PROCEDURE:

Sulfapyrimidine (0.5 g) was dissolved in 10 cm^3 of the buffer soln, shaken for 2 h at room temp (about 20°C), and filtered. A 1-cm^3 aliquot of the filtrate was withdrawn, cooled, acidified with 1 cm^3 of 2N HCl, and the sulfapyrimidine content was detd colorimetrically by the method of Marshall modified by Kimmig (1) using an Authenrieth colorimeter. The pH was detd on an ultraionograph using a glass electrode.

SOURCE AND PURITY OF MATERIALS:

Sulfapyrimidine was manufd by Schering (purity not specified). The source and purity of the remaining materials were not specified.

ESTIMATED ERROR:

Soly: precision ±5% (author).

Temp: not specified.

pH : ±0.05 pH unit (author).

REFERENCES:

1. Kimmig, J. *Arch. Dermatol.* *176*, 722; *Erg. Hyg.* 1941, *24*, 398.

COMPONENTS:

(1) Benzenesulfonamide, 4-amino-N-2-pyrimidinyl- (sulfadiazine); $C_{10}H_{10}N_4O_2S$; [68-35-9]

(2) Phosphoric acid, disodium salt; Na_2HPO_4; [7558-94-4]

(3) Phosphoric acid, monopotassium salt; KH_2PO_4; [7778-77-0]

(4) Water; H_2O; [7732-18-5]

ORIGINAL MEASUREMENTS:

Pulver, R.; Suter, R.

Schweiz. Med. Wochenschr. 1943, *73(13)*, 403-8.

VARIABLES:

pH

PREPARED BY:

R. Piekos

EXPERIMENTAL VALUES:

pH	Solubility of sulfadiazine in M/15 phosphate buffers (according to Sørensen) at 20°C	
	mg%	10^3 mol dm^{-3} [a]
6.0	13	0.52
7.0	45	1.80
8.0	191	7.63

[a] Calculated by compiler

AUXILIARY INFORMATION

METHOD/APPARATUS/PROCEDURE:

Nothing specified.

SOURCE AND PURITY OF MATERIALS:

Nothing specified.

ESTIMATED ERROR:

Nothing specified.

REFERENCES:

COMPONENTS:

(1) Benzenesulfonamide, 4-amino-N-2-pyrimidinyl- (sulfadiazine); $C_{10}H_{10}N_4O_2S$; [68-35-9]

(2) Phosphoric acid, disodium salt; Na_2HPO_4; [7558-94-4]

(3) Phosphoric acid, monopotassium salt; KH_2PO_4; [7778-77-0]

(4) Water; H_2O; [7732-18-5]

ORIGINAL MEASUREMENTS:

Gilligan, D. R.; Garb, S.; Plummer, M. N. *Proc. Soc. Exp. Biol. Med.* 1943, *52*, 248-50.

VARIABLES:

pH

PREPARED BY:

R. Piekos

EXPERIMENTAL VALUES:

The curve of solubility of sulfadiazine in M/15 phosphate buffer solutions of varying pH's at 37°C

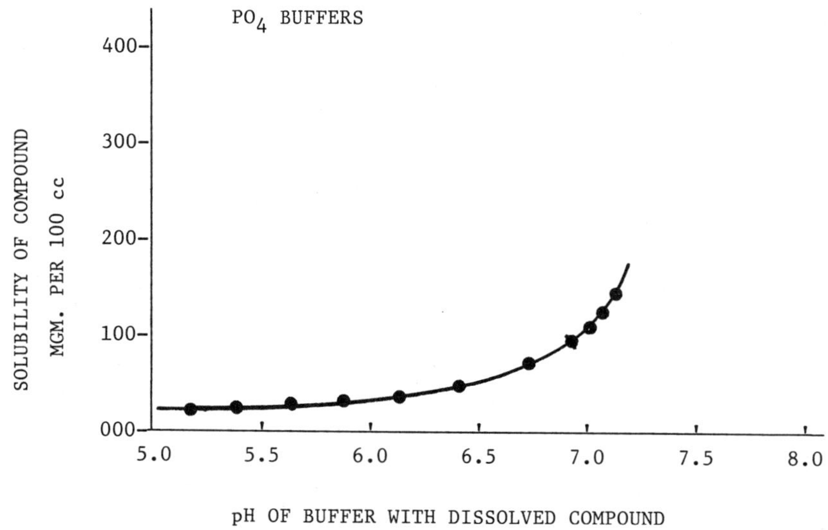

AUXILIARY INFORMATION

METHOD/APPARATUS/PROCEDURE:

An amt of sulfadiazine in large excess of that to be dissolved was added to M/15 phosphate buffer solns, shaken for 18 h in a water bath at 37°C and filtered at the same temp. The pH of the filtrate was measured immediately with a Beckmann pH meter and appropriate temp corrections were made. Sulfadiazine was assayed by the method of Bratton and Marshall.

SOURCE AND PURITY OF MATERIALS:

Sulfadiazine (purity not specified) was supplied by Lederle Lab, Inc.

The source and purity of the remaining materials were not specified.

ESTIMATED ERROR:

Nothing specified.

REFERENCES:

1. Bratton, A. C.; Marshall, E. K., Jr. *J. Biol. Chem.* 1939, *128*, 537.

COMPONENTS:	ORIGINAL MEASUREMENTS:
(1) Benzenesulfonamide, 4-amino-N-2-pyrimidinyl- (sulfadiazine); $C_{10}H_{10}N_4O_2S$; [68-35-9] (2) Phosphoric acid, disodium salt; Na_2HPO_4; [7558-94-4] (3) Phosphoric acid, monopotassium salt; KH_2PO_4; [7778-77-0] (4) Water; H_2O; [7732-18-5]	Gilligan, D. R.; Plummer, M. N. *Proc. Soc. Exp. Biol. Med.* 1943, *53*, 142-5.
VARIABLES:	PREPARED BY:
pH	R. Piekos

EXPERIMENTAL VALUES:

Solubility of sulfadiazine in M/15 phosphate buffer solutions at 37°C

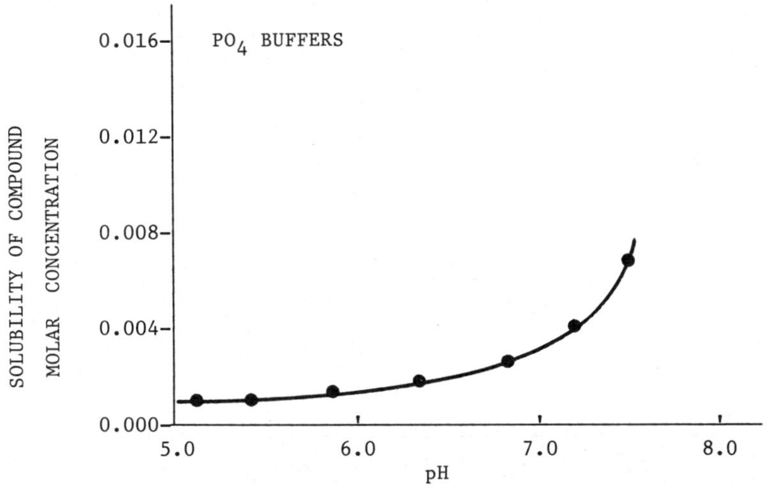

AUXILIARY INFORMATION

METHOD/APPARATUS/PROCEDURE:

An excess of sulfadiazine was shaken in M/15 phosphate buffer solns of various pH values for 18 h in a water bath at 37°C, and filtered in an incubator room at this temp. The pH of the filtrate was measured immediately at room temp with a Beckmann electrode pH meter and appropriate corrections for the difference between room temp and 37°C were applied. The amt of dissolved compd was measured by the method of Bratton and Marshall (1).

SOURCE AND PURITY OF MATERIALS:

Sulfadiazine was supplied by Lederle Labs, Inc. The source and purity of the remaining materials were not specified.

ESTIMATED ERROR:

Nothing specified.

REFERENCES:

1. Bratton, A. C.; Marshall, E. K., Jr. *J. Biol. Chem.* 1939, *128*, 537.

COMPONENTS:

(1) Benzenesulfonamide, 4-amino-N-2-pyrimidinyl- (sulfapyrimidine); $C_{10}H_{10}N_4O_2S$; [68-35-9]

(2) Phosphoric acid, disodium salt; Na_2HPO_4; [7558-94-4]

(3) Phosphoric acid, monopotassium salt; KH_2PO_4; [7778-77-0]

(4) Water; H_2O; [7732-18-5]

ORIGINAL MEASUREMENTS:

Frisk, A. R.; Hagerman, G.; Helander, S.; Sjögren, B. *Hygiea* 1946, *108(12)*, 639-51.

VARIABLES:

pH

PREPARED BY:

R. Piekos

EXPERIMENTAL VALUES:

Solubility of sulfapyrimidine at 37°C

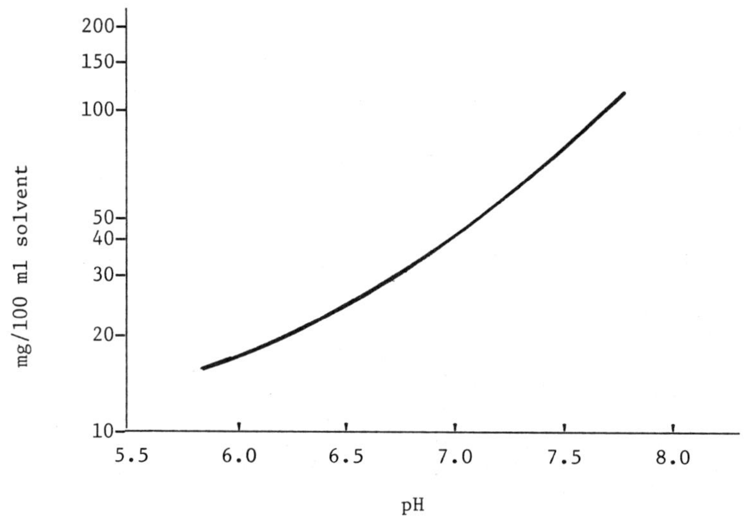

AUXILIARY INFORMATION

METHOD/APPARATUS/PROCEDURE:

An excess of sulfapyrimidine in the phosphate buffer was shaken at 37°C for 24 h. The concn of the solute was detd by the Bratton and Marshall method (1) using a photoelec colorimeter.

SOURCE AND PURITY OF MATERIALS:

Neither source nor purity of the materials was specified.

ESTIMATED ERROR:

Soly: precision ±2 mg/100 ml (authors).

Temp and pH: not specified.

REFERENCES:

1. Bratton, A. C.; Marshall, E. K., Jr. *J. Biol. Chem.* 1939, *128*, 537.

COMPONENTS:

(1) Benzenesulfonamide, 4-amino-N-2-pyrimidinyl- (sulfadiazine); $C_{10}H_{10}N_4O_2S$; [68-35-9]

(2) Phosphoric acid, disodium salt; Na_2HPO_4; [7558-94-4]

(3) Phosphoric acid, monopotassium salt; KH_2PO_4; [7778-77-0]

(4) Water; H_2O; [7732-18-5]

ORIGINAL MEASUREMENTS:

Langecker, H.

Arch. Exptl. Path. Pharmakol. 1948, *205*, 291-301.

VARIABLES:

pH

PREPARED BY:

R. Piekos

EXPERIMENTAL VALUES:

pH of the 1/15M phosphate buffer	Solubility at 37°C mg%	10^4 mol dm^{-3} [a]
4.9	16	6.4
5.9	21	8.4
6.9	29	12
7.0	54	22
7.5	81	32

[a] Calculated by compiler

AUXILIARY INFORMATION

METHOD/APPARATUS/PROCEDURE:

An excess of sulfadiazine was added to a buffer soln and boiled for 1 h in a sealed ampul followed by keeping the ampul at 37°C. The concn of sulfadiazine was detd colorimetrically by the method of Bratton and Marshall (1) using a Havemann colorimeter (2), as well as by microanal detn of the solid residue.

SOURCE AND PURITY OF MATERIALS:

Source and purity of the materials were not specified.

ESTIMATED ERROR:

Nothing specified.

REFERENCES:

1. Bratton, A. G.; Marshall, E. K, Jr. *J. Biol. Chem.* 1939, *128*, 537.
2. Havemann, R. *Klin. Wochenschr.* 1940, p. 503.

COMPONENTS:

(1) Benzenesulfonamide, 4-amino-N-2-pyrimidinyl- (sulfadiazine); $C_{10}H_{10}N_4O_2S$; [68-35-9]

(2) Phosphoric acid, disodium salt; Na_2HPO_4; [7558-94-4]

(3) Phosphoric acid, monopotassium salt; KH_2PO_4; [7778-77-0]

(4) Water; H_2O; [7732-18-5]

ORIGINAL MEASUREMENTS:

Riess, W.

Intern. Congr. Chemotherapy, Proc., 3rd, Stuttgart 1963, *1*, 627-32.

VARIABLES:

One temperature: 20°C; one pH: 7.4

PREPARED BY:

R. Piekos

EXPERIMENTAL VALUES:

Solubility of sulfadiazine in a M/15 Sörensen buffer solution (pH 7.4) at 20°C is 65 mg% (2.6×10^{-3} mol dm^{-3} solutions, compiler).

AUXILIARY INFORMATION

METHOD/APPARATUS/PROCEDURE:

Sörensen buffer solns of pH varying between 7 and 8 were prepd, satd with sulfadiazine at 20°C, their pH was measured at equilibrium, and the sulfadiazine was assayed colorimetrically. The measured pH values were then plotted against concn and the soly at pH 7.4 was detd by interpolation (personal communication).

SOURCE AND PURITY OF MATERIALS:

Nothing specified.

ESTIMATED ERROR:

Nothing specified.

REFERENCES:

COMPONENTS:

(1) Benzensulfonamide, 4-amino-N-2-pyrimidinyl- (sulfadiazine); $C_{10}H_{10}N_4O_2S$; [68-35-9]

(2) Phosphoric acid, disodium salt; Na_2HPO_4; [7558-94-4]

(3) Phosphoric acid, monopotassium salt; KH_2PO_4; [7778-77-0]

(4) Water; H_2O; [7732-18-5]

ORIGINAL MEASUREMENTS:

Yamazaki, M.; Aoki, M.; Kamada, A.; Yata, N. *Yakuzaigaku* 1967, *27(1)*, 37-40.

VARIABLES:

One temperature: 30°C; one pH: 7.4

PREPARED BY:

R. Piekos

EXPERIMENTAL VALUES:

Solubility of sulfadiazine in a phosphate buffer solution of pH 7.4 (μ = 0.17) at 30°C is 2.70 mmol/L (0.676 g dm^{-3}, compiler).

AUXILIARY INFORMATION

METHOD/APPARATUS/PROCEDURE:

Sulfadiazine (0.5 g) was placed in an L-shaped tube together with 20 ml of the buffer soln. The mixt was shaken in a thermostat until equilibrium was attained. The sulfadiazine was assayed in the supernatant spectrophotometrically at 545 nm on a Beckmann DU spectrophotometer. The results were taken from a calibration graph.

SOURCE AND PURITY OF MATERIALS:

Nothing specified.

ESTIMATED ERROR:

Soly and pH: not specified.

Temp: ±1°C (authors).

REFERENCES:

COMPONENTS:

(1) Benzenesulfonamide, 4-amino-N-2-pyrimidinyl- (sulfadiazine); $C_{10}H_{10}N_4O_2S$; [68-35-9]

(2) Phosphoric acid, disodium salt; Na_2HPO_4; [7558-94-4]

(3) Phosphoric acid, monopotassium salt; KH_2PO_4; [7778-77-0]

(4) Water; H_2O; [7732-18-5]

ORIGINAL MEASUREMENTS:

Hekster, Y. A.; Vree, T. B.; Damsma, J. E.; Friesen, W. T.

J. Antimicrob. Chemother. 1981, *8*, 133-44.

VARIABLES:

pH

PREPARED BY:

R. Piekos

EXPERIMENTAL VALUES:

pH	Solubility at 25°C mg/l	10^3 mol dm^{-3} [a]
5.5	265	1.06
7.5	950	3.80

[a]Calculated by compiler

AUXILIARY INFORMATION

METHOD/APPARATUS/PROCEDURE:

Satd solns of sulfadiazine were prepd in phosphate buffers of pH 5.5 and 7.5 at room temp (25°C). The concn of the solute was measured by means of a Spectra Physics 3500B high-performance liquid chromatograph equipped with a column oven (Model 748) and a Pye-Unicam LC-UV spectrophotometric detector. The detector was connected to a 1-mV recorder. A stainless steel column (10 cm x 4.6 mm i.d.) was packed with Lichrosorb RPS, 5 μm, obtained from Chrompack. An injection loop of 100 μl was used. The oven temp was 40°C. Detection of sulfadiazine was performed at 260 nm.

SOURCE AND PURITY OF MATERIALS:

The source and purity of the materials were not specified.

ESTIMATED ERROR:

The detection limit of the solute by HPLC was 0.5 mg/l (authors).
The error in temperature and pH was not specified.

REFERENCES:

COMPONENTS:

(1) Benzenesulfonamide, 4-amino-N-2-pyrimidinyl- (sulfadiazine); $C_{10}H_{10}N_4O_2S$; [68-35-9]

(2) Calcium chloride; $CaCl_2$; [10043-52-4]

(3) Potassium chloride; KCl; [7447-40-7]

(4) Sodium chloride: NaCl; [7647-14-5]

(5) Water; H_2O; [7732-18-5]

ORIGINAL MEASUREMENTS:

Hawking, F.

Lancet 1941, *240*, 786-8.

VARIABLES:

Temperature

PREPARED BY:

R. Piekos

EXPERIMENTAL VALUES:

t/°C	Solubility in bicarbonate-free Locke's solution[a] mg/100 ml	10^4 mol dm^{-3} [b]
17	7.8	3.12
36	18.0	7.19

[a]The solution contained NaCl 9 g, KCl 0.2 g, $CaCl_2$ 0.2 g, water 1 liter, and had a pH of 6.8.

[b]Calculated by compiler

AUXILIARY INFORMATION

METHOD/APPARATUS/PROCEDURE:

Sulfadiazine was shaken up with the bicarbonate-free Locke's soln for many hours in a tube which was corked to prevent loss of CO_2. The supernatant was filtered through a paper, dild in a hot room to prevent pptn, and sulfadiazine was detd by the method of Marshall and Litchfield (1).

SOURCE AND PURITY OF MATERIALS:

Nothing specified.

ESTIMATED ERROR:

Soly: average of 3 detns has been given (authors).

Temp: not specified.

REFERENCES:

1. Marshall, E. K., Jr.; Litchfield, J. T., Jr. *Science* 1938, *88*, 85.

COMPONENTS:

(1) Benzenesulfonamide, 4-amino-N-2-pyrimidinyl- (sulfadiazine); $C_{10}H_{10}N_4O_2S$; [68-35-9]

(2) Boric acid, disodium salt; $Na_2B_4O_7$; [1330-43-4]

(3) Phosphoric acid; H_3PO_4; [7664-38-2]

(4) Phosphoric acid; monosodium salt; NaH_2PO_4; [7558-80-7]

(5) Sodium chloride; NaCl; [7647-14-5]

(6) Water; H_2O; [7732-18-5]

ORIGINAL MEASUREMENTS:

Krebs, H. A.; Speakman, J. C.

J. Chem. Soc. 1945, *42*, 593-5.

PREPARED BY:

R. Piekos

VARIABLES:

pH

EXPERIMENTAL VALUES:

pH at saturation	Solubility of sulfadiazine at 25.0°C in phosphate - borate buffer solutions[a] mg/100 ml	10^3 mol dm^{-3} [b]
6.01	8.5	0.34
6.35	11.1	0.443
6.82	19.4	0.775
7.23	43.5	1.74
7.56	86	3.4
7.67	114	4.56
8.00	229	9.15

[a] The buffer solutions were prepared by adding NaOH to a solution of 0.0375M H_3PO_4 + 0.00938M Na tetraborate (i.e. equivalent to 0.0188M H_3PO_4 + 0.0188M NaH_2PO_4 + 0.0375M HBO_2) chosen so as to give an approx uniform buffering capacity over the range of pH, diluting and adding the appropriate quantity of N-NaCl to maintain the ionic strength at 0.1M. After dilution, the final total concentration of each acid was 0.0268M.

[b] Calculated by compiler.

AUXILIARY INFORMATION

METHOD/APPARATUS/PROCEDURE:

The buffer solns were satd by shaking them vigorously with an excess of solid sulfadiazine for at least 3 h. They were then quickly filtered. The sulfadiazine concn was detd by the method of Bratton and Marshall (1), and the pH values were measured with the glass electrode, which was standardized in terms of the buffer solns recommended by Hitchcock and Taylor (2).

SOURCE AND PURITY OF MATERIALS:

The sulfadiazine, mp 256°C, was obtained by recrystg a common specimen from water. Neither source nor purity of the remaining materials was specified.

ESTIMATED ERROR:

Nothing specified.

REFERENCES:

1. Bratton, A. C.; Marshall, E. K., Jr. *J. Biol. Chem.* 1939, *128*, 537.

2. Hitchcock, D. I.; Taylor, A. C. *J. Am. Chem. Soc.* 1937, *59*, 1812.

COMPONENTS:

(1) Benzenesulfonamide, 4-amino-N-2-pyrimidinyl- (sulfadiazine); $C_{10}H_{10}N_4O_2S$; [68-35-9]

(2) Acetic acid; $C_2H_4O_2$; [64-19-7]

(3) Acetic acid, sodium salt; $C_2H_3NaO_2$; [127-09-3]

(4) Phosphoric acid; H_3PO_4; [7664-38-2]

(5) Phosphoric acid, monosodium salt; NaH_2PO_4; [7558-80-7]

(6) Sodium chloride; NaCl; [7647-14-5]

(7) Water; H_2O; [7732-18-5]

ORIGINAL MEASUREMENTS:

Krebs, H. A.; Speackman, J. C.

J. Chem. Soc. 1945, *42*, 593-5.

PREPARED BY:

R. Piekos

VARIABLES:

pH

EXPERIMENTAL VALUES:

pH at saturation	Solubility of sulfadiazine at 25.0°C in acetate - phosphate buffer solutions[a]	
	mg/100 ml	10^4 mol dm^{-3} [a]
2.31	9.3	3.7
2.69	7.5	3.0
3.06	6.9	2.8
4.89	6.3	2.5

[a]The buffer solutions were prepared by adding NaOH to a solution of 0.0375M H_3PO_4 + 0.0375M CH_3COOH chosen so as to give an approx uniform buffering capacity over the range of pH, diluting, and adding the appropriate quantity of N-NaCl to maintain the ionic strength at 0.1M.

[b] Calculated by compiler.

AUXILIARY INFORMATION

METHOD/APPARATUS/PROCEDURE:

The buffer solns were satd by shaking them vigorously with an excess of solid sulfadiazine for at least 3 h. They were then quickly filtered. The sulfadiazine concn was detd by the method of Bratton and Marshall (1), and the pH values were measured with the glass electrode, which was standardized in terms of the buffer solns recommended by Hitchcock and Taylor (2).

SOURCE AND PURITY OF MATERIALS:

The sulfadiazine, mp 256°C, was obtained by recrystg a common specimen from water. Neither source nor purity of the remaining materials was specified.

ESTIMATED ERROR:

Nothing specified.

REFERENCES:

1. Bratton, A. C.; Marshall, E.K., Jr. *J. Biol. Chem.* 1939, *128*, 537.

2. Hitchcock, D. I.; Taylor, A. C. *J. Am. Chem. Soc.* 1937, *59*, 1812.

COMPONENTS:

(1) Benzenesulfonamide, 4-amino-N-2-pyrimidinyl- (sulfadiazine); $C_{10}H_{10}N_4O_2S$; [68-35-9]

(2) 1,2,3-Propanetricarboxylic acid, 2-hydroxy, disodium salt; (Na citrate); $C_6H_6Na_2O_7$; [144-33-2]

(3) Sodium hydroxide; NaOH; [1310-73-2]

(4) Water; H_2O; [7732-18-5]

ORIGINAL MEASUREMENTS:

Gilligan, D. R.; Garb, S.; Plummer, M. N. *Proc. Soc. Exp. Biol. Med.* 1943, *52*, 248-50.

VARIABLES:

pH

PREPARED BY:

R. Piekos

EXPERIMENTAL VALUES:

The curve of solubility of sulfadiazine in M/10 citrate plus NaOH buffer solutions at 37°C.

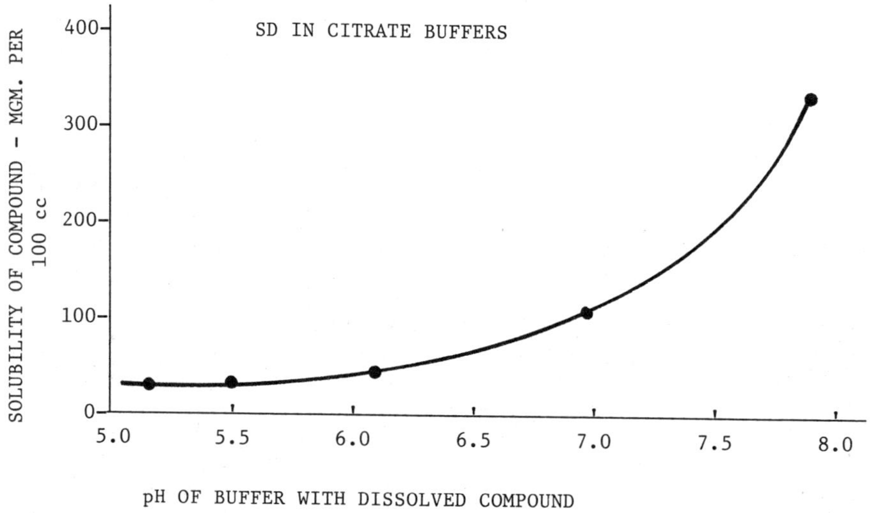

AUXILIARY INFORMATION

METHOD/APPARATUS/PROCEDURE:

An amt of sulfadiazine in large excess of that to be dissolved was added to M/10 citrate plus NaOH buffers, shaken for 18 h in a water bath at 37°C and filtered at the same temp. The pH of the filtrate was measured immediately with a Beckmann pH meter and appropriate corrections were made. Sulfadiazine was assayed by the method of Bratton and Marshall (1).

SOURCE AND PURITY OF MATERIALS:

Sulfadiazine (purity not specified) was supplied by Lederle Labs, Inc.
The source and purity of the remaining materials were not specified.

ESTIMATED ERROR:

Nothing specified.

REFERENCES:

1. Bratton, A. C.; Marshall, E. K., Jr. *J. Biol. Chem.* 1939, *128*, 537.

COMPONENTS:

(1) Benzenesulfonamide, 4-amino-N-2-pyrimidinyl- (sulfadiazine); $C_{10}H_{10}N_4O_2S$; [68-35-9]

(2) 1,2,3-Propanetricarboxylic acid, 2-hydroxy-, disodium salt; (Na citrate); $C_6H_6Na_2O_7$; [144-33-2]

(3) Sodium hydroxide; NaOH; [1310-73-2]

(4) Water; H_2O; [7732-18-5]

ORIGINAL MEASUREMENTS:

Gilligan, D. R.; Plummer, M. N.

Proc. Soc. Exp. Biol. Med. 1943, *53*, 142-5.

VARIABLES:

pH

PREPARED BY:

R. Piekos

EXPERIMENTAL VALUES:

Solubility of sulfadiazine in M/10 Na citrate + NaOH solutions at 37°C

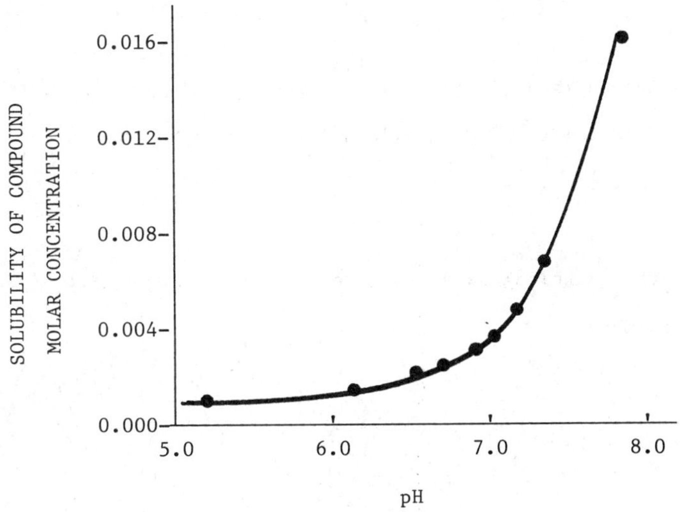

AUXILIARY INFORMATION

METHOD/APPARATUS/PROCEDURE:

An excess of sulfadiazine was shaken in M/10 Na citrate + NaOH solns of various pH values for 18 h in a water bath at 37°C, and filtered in an incubator room at this temp. The pH of the filtrate was measured immediately at room temp with a Beckmann glass electrode pH meter and appropriate corrections for the difference between room temp and 37°C were applied. The amt of dissolved compd was measured by the method of Bratton and Marshall (1).

SOURCE AND PURITY OF MATERIALS:

Sulfadiazine was supplied by Lederle Labs, Inc. The source and purity of the remaining materials were not specified.

ESTIMATED ERROR:

Nothing specified.

REFERENCES:

1. Bratton, A. C.; Marshall, E. K., Jr. *J. Biol. Chem.* 1939, *128*, 537.

COMPONENTS:

(1) Benzenesulfonamide, 4-amino-N-2-pyrimidinyl- (sulfadiazine); $C_{10}H_{10}N_4O_2S$; [68-35-9]

(2) 1,2,3-Propanetricarboxylic acid, 2-hydroxy- (citric acid); $C_6H_8O_7$; [77-92-9]

(3) Water; H_2O; [7732-18-5]

ORIGINAL MEASUREMENTS:

Takubo, T.; Matsumaru, H.; Tsuchiya, S.; Hiura, M.
Chem. Pharm. Bull. 1973, *21(7)*, 1440-5.

VARIABLES:

One temperature: 37°C; one pH: 2.1

PREPARED BY:

R. Piekos

EXPERIMENTAL VALUES:

Solubility of sulfadiazine in a citric acid solution (2.100 g citric acid per 100 ml water) of pH 2.1 at 37°C is 0.20 mg/ml solution[a] (8.0 x 10^{-4} mol dm^{-3} solution, compiler).

[a]Numerical value to the graphical data was given by one of the authors (S. T.) in personal communication.

AUXILIARY INFORMATION

METHOD/APPARATUS/PROCEDURE:

Aliquots of the citric acid soln were placed in glass-stoppered flasks with excess of sulfadiazine. The flasks were allowed to stand at 37±1°C and shaken vigorously for 4 h until equilibrium was established. One ml of the supernatant was removed by means of a filter pipet and sulfadiazine was assayed by the previously reported method (1).

SOURCE AND PURITY OF MATERIALS:

The sulfadiazine was of pharmaceutical grade. The source and purity of the citric acid was not specified.
Distd water was used.

ESTIMATED ERROR:

Soly and pH: not specified.
Temp: ±1°C (authors).

REFERENCES:

1. Takubo, T.; Tsuchiya, S.; Hiura, M. *Yakuzaigaku* 1971, *31*, 298.

COMPONENTS:

(1) Benzenesulfonamide, 4-amino-N-2-pyrimidinyl- (sulfadiazine); $C_{10}H_{10}N_4O_2S$; [68-35-9]

(2) Phosphoric acid, disodium salt; Na_2HPO_4; [7558-94-4]

(3) 1,2,3-Propanetricarboxylic acid, 2-hydroxy- (citric acid); $C_6H_8O_7$; [77-92-9]

(4) Water; H_2O; [7732-18-5]

ORIGINAL MEASUREMENTS:

Nogami, H.; Nagai, T.; Suzuki, A. *Chem. Pharm. Bull.* 1966, *14(4)*, 339-50.

VARIABLES:

pH

PREPARED BY:

R. Piekos

EXPERIMENTAL VALUES:

pH of Mc'Ilvaine buffer solutions[a] at constant ionic strength of 0.35.		Solubility of sulfadiazine at 37°C	
at the beginning of the run	at the end of the run	mg/100 ml	10^3 mol dm^{-3} [b]
4.17	4.19	10.7	0.4275
5.06	5.09	11.0	0.4395
6.56	6.59	93.0	3.716
7.23	7.10	175	6.992
8.14	7.30	242	9.669

[a]Obtained by mixing together Na_2HPO_4 and citric acid solutions of proper concentrations.

[b]Calculated by compiler.

AUXILIARY INFORMATION

METHOD/APPARATUS/PROCEDURE:

Soly of sulfadiazine was detd from dissoln rate data obtained by the rotating disk method.

SOURCE AND PURITY OF MATERIALS:

Commercial sulfadiazine J. P. was used. Purity of the remaining materials was not specified.

ESTIMATED ERROR:

Nothing specified.

REFERENCES:

COMPONENTS:

(1) Benzenesulfonamide, 4-amino-N-2-pyrimidinyl- (sulfadiazine); $C_{10}H_{10}N_4O_2S$; [68-35-9]

(2) Phosphoric acid, disodium salt; Na_2HPO_4; [7558-94-4]

(3) 1,2,3-Propanetricarboxylic acid, 2-hydroxy- (citric acid); $C_6H_8O_7$; [77-92-9]

(4) Water; H_2O; [7732-18-5]

ORIGINAL MEASUREMENTS:

Takubo, T.; Matsumaru, H.; Tsuchiya, S.; Hiura, M. *Chem. Pharm. Bull.* 1973, *21(7)*, 1440-5.

VARIABLES:

pH

PREPARED BY:

R. Piekos

EXPERIMENTAL VALUES:

Citric acid g/100 ml water	Na_2HPO_4 g/100 ml water	pH	Solubility at 37°C mg/ml soln[a]	Solubility at 37°C 10^3 mol dm^{-3} soln[b]
1.680	0.572	3.1	0.13	0.519
1.260	1.144	4.2	0.11	0.439
0.840	1.716	5.8	0.16	0.639
0.420	2.228	6.8	0.49	1.96

[a] Numerical values to the graphical data were given by one of the authors (S. T.) in personal communication.

[b] Calculated by compiler.

AUXILIARY INFORMATION

METHOD/APPARATUS/PROCEDURE:

Aliquots of the buffer solns were placed in glass-stoppered flasks with excess of sulfadiazine. The flasks were allowed to stand at 37±1°C and shaken vigorously for 4 h until equilibrium was established. One ml of the supernatant was removed by means of a filter pipet and sulfadiazine was assayed by the previously reported method (1).

SOURCE AND PURITY OF MATERIALS:

The sulfadiazine was of pharmaceutical grade. The source and purity of Na_2HPO_4 and citric acid was not specified.
Distd water was used.

ESTIMATED ERROR:

Soly and pH: not specified.

Temp: ±1°C (authors).

REFERENCES:

1. Takubo, T.; Tsuchiya, S.; Hiura, M. *Yakuzaigaku* 1971, *31*, 298.

COMPONENTS:

(1) Benzenesulfonamide, 4-amino-N-2-pyrimidinyl- (sulfadiazine); $C_{10}H_{10}N_4O_2S$; [68-35-9]

(2) Ethanesulfonic acid, 2-[[(3α,5β,7α,12α)-3,7,12-trihydroxy-24-oxocholan-24-yl]-amino]-, monosodium salt (Na taurocholate); $C_{26}H_{45}NO_7S \cdot Na$; [145-42-6]

(3) Phosphoric acid, disodium salt; Na_2HPO_4; [7558-94-4]

(4) Phosphoric acid, monosodium salt; NaH_2PO_4; [7558-80-7]

(5) Water; H_2O; [7732-18-5]

ORIGINAL MEASUREMENTS:

Gasco, M. R.; Aimonetto, S.

Atti Accad. Sci. Torino, Cl. Sci. Fis., Mat. Nat. 1979, *113(1-2)*, 119-22.

VARIABLES:

Concentration of Na taurocholate; pH

PREPARED BY:

R. Piekos

EXPERIMENTAL VALUES:

Concentration of Na taurocholate mM/l solution[a]	Solubility of sulfadiazine at 25°C μM/ml solution[a] pH 6.3	pH 7.2
2.25	0.52	1.43
4.50	0.47	1.34
6.00	0.47	1.39
8.00	0.47	1.35
12.00	0.47	1.43
16.00	0.47	1.43
20.00	0.47	1.43

[a] Numerical values given by the first author in personal communication.

AUXILIARY INFORMATION

METHOD/APPARATUS/PROCEDURE:

The soly of sulfadiazine was detd by the method of Hofmann (1). In a series of 15-ml glass cylinders with ground-in stoppers, 50 mg of sulfadiazine was placed in 15 ml of phosphate buffer solns of increasing Na taurocholate concn. The suspensions were agitated for 20 h at 25°C and filtered. The quantity of sulfadiazine dissolved was detd by measuring surface tension by means of a Dognon-Abribat (Prolabo) tensiometer and spectrophotometrically on a Perkin Elmer EPS-35 spectrophotometer.

SOURCE AND PURITY OF MATERIALS:

Neither source nor purity of the materials was specified.

The phosphate buffer was 0.3M in respect of the Na^+ ion concentration.

ESTIMATED ERROR:

Soly: precision ±2% (authors).

pH : precision ±0.02 pH unit (authors).

Temp: ±0.5°C (authors).

REFERENCES:

1. Hofmann, A. F., *Biochem. J.* 1963, *89*, 57.

COMPONENTS:

(1) Benzenesulfonamide, 4-amino-N-2-pyrimidinyl- (sulfadiazine); $C_{10}H_{10}N_4O_2S$; [68-35-9]

(2) Ethanesulfonic acid, 2-{[[[(3α,5β,7α,12α)-3,7,12-trihydroxy-24-oxacholan-24-yl]amino]acetyl]amino]-, sodium salt (Na tauroglycocholate); $C_{28}H_{48}N_2O_8S \cdot Na$ [11006-55-6]

(3) Phosphoric acid, disodium salt; Na_2HPO_4; [7558-94-4]

(4) Phosphoric acid, monosodium salt; NaH_2PO_4; [7558-80-7]

(5) Water; H_2O; [7732-18-5]

ORIGINAL MEASUREMENTS:

Gasco, M. R.; Aimonetto, S.

Atti Accad. Sci. Torino, Cl. Sci. Fis., Mat. Nat. 1979, *113(1-2)*, 119-22.

VARIABLES:

Concentration of Na tauroglycochlolate; pH

PREPARED BY:

R. Piekos

EXPERIMENTAL VALUES:

Concentration of Na tauroglycocholate mM/l solution	Solubility of sulfadiazine at 25°C µml solution[a]	
	pH 6.3	pH 7.2
2.25	0.43	1.34
4.50	0.39	1.21
6.00	0.41	1.26
8.00	0.41	1.26
12.00	0.43	1.26
16.00	0.43	1.28
20.00	0.47	1.30

[a]Numerical values given by the first author in personal communication.

AUXILIARY INFORMATION

METHOD/APPARATUS/PROCEDURE:

The soly of sulfadiazine was detd by the method of Hofmann (1). In a series of 15-ml glass cylinders with ground-in stoppers, 50 mg of sulfadiazine was placed in 15 ml of phosphate buffer solns of increasing Na tauroglycocholate concn. The suspensions were agitated for 20 h at 25°C and filtered. The quantity of dissolved sulfadiazine was detd by measuring surface tension by means of a Dognon-Abribat (Prolabo) tensiometer and spectrophotometrically on a Perkin Elmer EPS-35 spectrophotometer.

SOURCE AND PURITY OF MATERIALS:

Neither source nor purity of the materials was specified.

The phosphate buffer was 0.3M in respect of the Na^+ ion concn.

ESTIMATED ERROR:

Soly: precision ±2% (authors).

pH : precision ±0.02 pH unit (authors).

Temp: ±0.5°C (authors).

REFERENCES:

1. Hofmann, A. F.; *Biochem. J.* 1963, *89*, 57.

COMPONENTS:

(1) Benzenesulfonamide, 4-amino-N-2-pyrimidinyl- (sulfadiazine); $C_{10}H_{10}N_4O_2S$; [68-35-9]

(2) Sulfuric acid monododecyl ester, sodium salt (Na lauryl sulfate); $C_{12}H_{25}NaO_4S$; [151-21-3]

(3) Water; H_2O; [7732-18-5]

ORIGINAL MEASUREMENTS:

Watari, N.; Kaneniwa, N.

Chem. Pharm. Bull. 1976, *24(11)*, 2577-84.

VARIABLES:

Concentration of Na lauryl sulfate

PREPARED BY:

R. Piekos

EXPERIMENTAL VALUES:

Concentration of Na lauryl sulfate	Total solubility of sulfadiazine at 37°C	
% w/v	mg/ml solution	10^4 mol dm^{-3} [a]
0.01	0.124	4.954
0.05	0.132	5.274
0.10	0.151	6.033
0.25	0.165	6.592
0.50	0.179	7.152
1.00	0.207	8.270
2.00	0.258	10.31
3.00	0.309	12.35
4.00	0.361	14.42
5.00	0.410	16.38
6.00	0.463	18.50

[a] Calculated by compiler.

AUXILIARY INFORMATION

METHOD/APPARATUS/PROCEDURE:

An excess of sulfadiazine was added to 15 ml of the Na lauryl sulfate soln contained in a 50-ml flask and the flask was shaken (2 strokes/s at the amplitude of 3 cm) in a thermostatically controlled water bath at 37°C. One-ml sample was removed every 6 h (total equilibration period was 3-5 days) using a warmed Millipore filter syringe with a filter pore size of 0.45 μ (Millipore HAWP 01300) and the filtrate was dild with water and assayed spectrophotometrically (1).

SOURCE AND PURITY OF MATERIALS:

Commercial sulfadiazine of the Japanese Pharmacopeia grade and distd water were used.

Na lauryl sulfate was of the reagent grade (Wako Pure Chemical Industries Ltd, lot No. PA10233) and used without further purification.

ESTIMATED ERROR:

Soly: not specified.

Temp: ±0.05°C (authors).

REFERENCES:

1. Kaneniwa, N.; Watari, N. *Chem. Pharm. Bull.* 1974, *22*, 1699.

COMPONENTS:

(1) Benzenesulfonamide, 4-amino-N-2-pyrimidinyl- (sulfadiazine); $C_{10}H_{10}N_4O_2S$; [68-35-9]

(2) Ethanol; C_2H_6O; [64-17-5]

(3) 1,2,3-Propanetriol; $C_3H_8O_3$; [56-81-5]

(4) Water; H_2O; [7732-18-5]

ORIGINAL MEASUREMENTS:

Dolique, R.; Foucault, J.

Trav. soc. pharm. Montpellier 1952, *12*, 145-53.

VARIABLES:

One temperature: 26-28°C

PREPARED BY:

R. Piekos

EXPERIMENTAL VALUES:

Solubility of sulfadiazine in a mixture of 1,2,3-propanetriol and 95° ethanol (2:1 by wt) at 26-28°C is 0.27% (1.1×10^{-2} mol kg^{-1} solvent, compiler).

AUXILIARY INFORMATION

METHOD/APPARATUS/PROCEDURE:

The sulfadiazine content was detd by diazotization of the amine group in a cold acidified 0.1N KNO_2 soln. An excess of KNO_3 was detected by using iodinated starch.

SOURCE AND PURITY OF MATERIALS:

Nothing specified.

ESTIMATED ERROR:

Nothing specified.

REFERENCES:

COMPONENTS:
(1) Benzenesulfonamide, 4-amino-N-2-pyrimidinyl- (sulfadiazine); $C_{10}H_{10}N_4O_2S$; [68-35-9]
(2) Ethanol; C_2H_6O; [64-17-5]
(3) 1,2,3-Propanetriol; $C_3H_8O_3$; [56-81-5]
(4) Urea; CH_4N_2O; [57-13-6]
(5) Water; H_2O; [7732-18-5]

ORIGINAL MEASUREMENTS:
Dolique, R.; Foucault, J.
Trav. soc. pharm. Montpellier 1952, *12,* 145-53.

VARIABLES:
One temperature: 26-28°C

PREPARED BY:
R. Piekos

EXPERIMENTAL VALUES:

Solubility of sulfadiazine at 26-28°C in a saturated solution of urea in a mixture of 1,2,3-propanetriol and 95°C ethanol (2:1 by wt), containing 54.5 g of urea per 100 g of the mixture, is 0.26% (1.1 x 10^{-2} mol kg^{-1} solvent, compiler).

AUXILIARY INFORMATION

METHOD/APPARATUS/PROCEDURE:
The sulfadiazine content was detd by diazotization of the amine group in a cold acidified 0.1N KNO_2 soln. An excess of KNO_3 was detected by using iodinated starch.

SOURCE AND PURITY OF MATERIALS:
Nothing specified.

ESTIMATED ERROR:
Nothing specified.

REFERENCES:

COMPONENTS:

(1) Benzenesulfonamide, 4-amino-N-2-pyrimidinyl- (sulfadiazine); $C_{10}H_{10}N_4O_2S$; [68-35-9]

(2) Ethanol; 2,2'-iminobis- (diethanolamine); $C_4H_{11}NO_2$; [111-42-2]

(3) Water; H_2O; [7732-18-5]

ORIGINAL MEASUREMENTS:

Nogami, H.; Nagai, T.; Suzuki, A.

Chem. Pharm. Bull. 1966, *14(4)*, 339-50.

VARIABLES:

Concentration of diaethanolamine

PREPARED BY:

R. Piekos

EXPERIMENTAL VALUES:

Concentration of diethanolamine mol dm^{-3}	Solubility of sulfadiazine at 37°C mg/100 ml	10^2 mol dm^{-3}
10^{-3}	33.8	0.135
10^{-2}	266.0	1.06
10^{-1}	2510.0	10.0

AUXILIARY INFORMATION

METHOD/APPARATUS/PROCEDURE:

Soly of sulfadiazine was detd from dissoln rate data obtained by the rotating disk method.

SOURCE AND PURITY OF MATERIALS:

Commercial sulfadiazine J. P. was used. Purity of the remaining materials was not specified.

ESTIMATED ERROR:

Nothing specified.

REFERENCES:

COMPONENTS:	ORIGINAL MEASUREMENTS:
(1) Benzenesulfonamide, 4-amino-N-2-pyrimidinyl; (sulfadiazine); $C_{10}H_{10}N_4O_2S$; [68-35-9] (2) Ethanol, 2,2',2''-nitrilotris-(triethanolamine); $C_6H_{15}NO_3$; [102-71-6] (3) Water; H_2O; [7732-18-5]	Nogami, H.; Nagai, T.; Suzuki, A. *Chem. Pharm. Bull.* 1966, *14(4)*, 339-50.
VARIABLES:	**PREPARED BY:**
Concentration of triethanolamine	R. Piekos

EXPERIMENTAL VALUES:

Concentration of triethanolamine	Solubility of sulfadiazine at 37°C	
mol dm^{-3}	mg/100 ml	10^2 mol dm^{-3}
10^{-3}	31.4	0.15
10^{-2}	286.0	1.10
10^{-1}	2550.0	10.0

AUXILIARY INFORMATION

METHOD/APPARATUS/PROCEDURE:

Soly of sulfadiazine was detd from dissoln rate data obtained by the rotating disk method.

SOURCE AND PURITY OF MATERIALS:

Commercial sulfadiazine J. P. was used. Purity of the remaining materials was not specified.

ESTIMATED ERROR:

Nothing specified.

REFERENCES:

COMPONENTS:

(1) Benzenesulfonamide, 4-amino-N-2-pyrimidinyl-; (sulfadiazine); $C_{10}H_{10}N_4O_2S$; [68-35-9]

(2) Formamide, N,N-dimethyl- (DMF); C_3H_7NO; [68-12-2]

(3) Water; H_2O; [7732-18-5]

ORIGINAL MEASUREMENTS:

Elworthy, P. H.; Worthington, H. E. C.
J. Pharm. Pharmac. 1968, *20*, 830-5.

VARIABLES:

Temperature: Concentration of DMF

PREPARED BY:

R. Piekos

EXPERIMENTAL VALUES:

% w/w DMF	Sulfadiazine % w/w saturated solution		
	20°	30°C	40°C
0.5	0.00490	0.00828	0.0138
1.0	0.00520	0.00881	0.0147
2.0	0.00598	0.00987	0.0166
3.0	0.00679	0.01110	0.0187
5.0	0.00861	0.01410	0.0233
10.0	0.01700	0.02520	0.0410
20.0	0.03950	0.05790	0.0968
30.0	0.08500	0.12300	0.1880
50.0	0.35200	0.50200	0.7580
70.0	1.90000	2.40000	3.5000
78.0	4.28000	4.85000	6.2000
89.0	9.80000	10.90000	12.0000

AUXILIARY INFORMATION

METHOD/APPARATUS/PROCEDURE:

Solns were presatd by shaking with powd sulfadiazine for 24 h, and transferred to a soly app which was of the percolation type and a modification of that used by Davies and Griffiths (1). The soln was recycled in the app until satd (7-14 days). The solute concn was detd by one of the three methods: (a) concn 1.5%, by evapn of solvent and drying the residue to const wt; (b) concn 0.02 to 1.5%, samples were dild to give a 70% DMF solvent mixt and assayed spectrophotometrically at 270 nm; (c) concn < 0.02%, samples were dild with water and assayed as in (b). In all spectrophotometric assays suitable calibration lines were prepd.

SOURCE AND PURITY OF MATERIALS:

Sulfadiazine (B. P. quality) was twice recrystd from an EtOH-DMF mixt (3:1 by vol) and dried over P_2O_5. Its mp was 255°C. Assay by the Pharmacopeial method gave 100.0% purity calcd with reference to the material dried at 105°C. DMF (May and Baker Ltd) was distd under reduced pressure and gave n_D^{25} = 1.4283. Purity of the water was not specified.

ESTIMATED ERROR:

Soly: mean values of duplicate runs are given (authors).

Temp: ±0.05°C (authors).

REFERENCES:

1. Davies, M.; Griffiths, D. M. L. *Trans. Faraday Soc.* 1953, *49*, 1405.

COMPONENTS:	ORIGINAL MEASUREMENTS:
Continued from previous page.	Elworthy, P. H.; Worthington, H. E. C. *J. Pharm. Pharmac.* 1968, *20*, 830-5.
VARIABLES: Temperature: Concentration of DMF	**PREPARED BY:** R. Piekos

EXPERIMENTAL VALUES:

% w/w DMF	10° mol kg^{-1} water[a]		
	$20^{o}C$	$30^{o}C$	$40^{o}C$
0.5	0.0196	0.0331	0.0551
1.0	0.0208	0.0352	0.0587
2.0	0.0239	0.0394	0.0663
3.0	0.0271	0.0443	0.0747
5.0	0.0344	0.0563	0.0931
10.0	0.0679	0.1010	0.1640
20.0	0.1580	0.2310	0.3870
30.0	0.3400	0.4920	0.7520
50.0	1.4100	2.0200	3.0500
70.0	7.7400	9.8200	14.5000
78.0	17.900	20.4000	26.4000
89.0	42.900	48.9000	54.5000

[a]Calculated by compiler.

AUXILIARY INFORMATION

METHOD/APPARATUS/PROCEDURE:

SOURCE AND PURITY OF MATERIALS:

ESTIMATED ERROR:

REFERENCES:

COMPONENTS:	ORIGINAL MEASUREMENTS:
Continued from previous page.	Elworthy, P. H.; Worthington, H. E. C. *J. Pharm. Pharmac.* 1968, *20*, 830-5.
VARIABLES: Temperature: Concentration of DMF	PREPARED BY: R. Piekos

EXPERIMENTAL VALUES:

% w/w DMF	Mole fraction sulfadiazine in solution		
	20°C	30°	40°C
0.5	3.54×10^{-6}	5.47×10^{-6}	9.97×10^{-6}
1.0	3.78×10^{-6}	6.39×10^{-6}	1.07×10^{-5}
2.0	4.37×10^{-6}	7.21×10^{-6}	1.21×10^{-5}
3.0	5.00×10^{-6}	8.18×10^{-6}	1.38×10^{-5}
5.0	6.44×10^{-6}	1.06×10^{-5}	1.74×10^{-5}
10.0	1.33×10^{-5}	1.97×10^{-5}	3.19×10^{-5}
20.0	3.35×10^{-5}	4.87×10^{-5}	8.21×10^{-5}
30.0	7.91×10^{-5}	1.14×10^{-4}	1.75×10^{-4}
50.0	4.08×10^{-4}	5.84×10^{-4}	8.81×10^{-4}
70.0	2.94×10^{-3}	3.73×10^{-3}	5.50×10^{-3}
78.0	7.75×10^{-3}	8.84×10^{-3}	1.14×10^{-2}
89.0	2.32×10^{-2}	2.60×10^{-2}	2.90×10^{-2}

[a] Calculated by compiler

AUXILIARY INFORMATION

METHOD/APPARATUS/PROCEDURE:	SOURCE AND PURITY OF MATERIALS:
	ESTIMATED ERROR:
	REFERENCES:

COMPONENTS:	ORIGINAL MEASUREMENTS:
(1) Benzenesulfonamide, 4-amino-N-2-pyrimidinyl- (sulfadiazine); $C_{10}H_{10}N_4O_2S$; [68-35-9] (2) Urea; CH_4NO; [57-13-6] (3) Water; H_2O; [7732-18-5]	Sobin, S. S. *J. Lab. Clin. Med.* 1942, *27*, 1657-8.

VARIABLES:	PREPARED BY:
Concentration of urea	R. Piekos

EXPERIMENTAL VALUES:

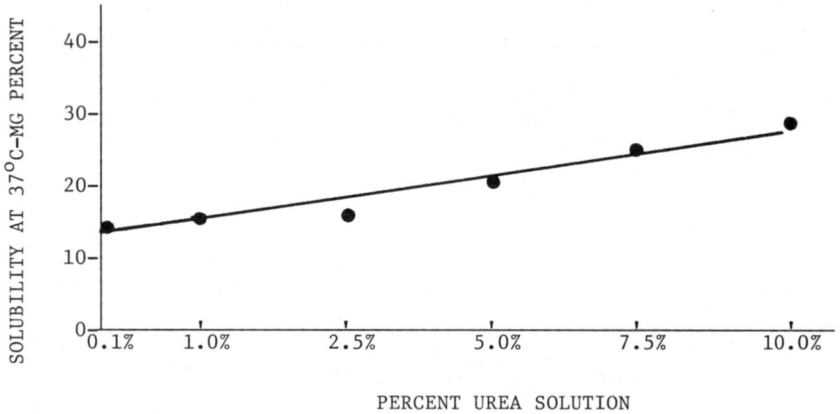

Solubility in a 10 percent urea solution at 37°C is 31.5 mg per 100 cm^3 (1.26×10^{-3} mol dm^{-3}, compiler).

AUXILIARY INFORMATION

METHOD/APPARATUS/PROCEDURE:

Urea solns of varying concns from 0.1 to 10% were incubated at 37°C with an excess of sulfadiazine, shaken at intervals, and filtered through two thicknesses of Whatman No. 42 filter paper. After appropriate diln the free sulfonamide was detd by the method of Bratton and Marshall (1) using the Evelyn colorimeter and a No. 540 filter.

SOURCE AND PURITY OF MATERIALS:

Nothing specified.

ESTIMATED ERROR:

Nothing specified.

REFERENCES:

1. Bratton, A. C.; Marshall, E. K., Jr. *J. Biol. Chem.* 1939, *128*, 537.

COMPONENTS:

(1) Benzenesulfonamide, 4-amino-N-2-pyrimidinyl- (sulfadiazine); $C_{10}H_{10}N_4O_2S$; [68-35-9]

(2) Urea; CH_4N_2O; [57-13-6]

(3) Water; H_2O; [7732-18-5]

ORIGINAL MEASUREMENTS:

Nogami, H.; Nagai, T.; Suzuki, A.

Chem. Pharm. Bull. 1966, *14(4)*, 339-50.

VARIABLES:

Concentration of urea; temperature

PREPARED BY:

R. Piekos

EXPERIMENTAL VALUES:

Concentration of urea	Solubility of sulfadiazine					
	25°C		37°C		50°C	
mol dm^{-3}	mg/100 ml	concn[a]	mg/100 ml	concn[b]	mg/100ml	concn[b]
10^{-3}			15.9	0.635		
10^{-2}			20.2	0.807		
10^{-1}	10.2	4.07	28.8	1.150	34.4	1.37
5×10^{-1}			31.4	1.250		
1			42.3	1.690		

Note: molar concentrations of the solute calculated by compiler.

a - 10^4 mol dm^{-3}

b - 10^3 mol dm^{-3}

AUXILIARY INFORMATION

METHOD/APPARATUS/PROCEDURE:

Soly of sulfadiazine was detd from dissoln rate data obtained by the rotating disk method.

SOURCE AND PURITY OF MATERIALS:

Commercial sulfadiazine J. P. was used. Purity of the remaining materials was not specified.

ESTIMATED ERROR:

Nothing specified.

REFERENCES:

COMPONENTS:	ORIGINAL MEASUREMENTS:
(1) Benzenesulfonamide, 4-amino-N-2-pyrimidinyl- (sulfadiazine); $C_{10}H_{10}N_4O_2S$; [68-35-9] (2) Dextrin; $(C_6H_{10}O_5)_n$; [90004-53-9] (3) Water; H_2O; [7732-18-5]	Lach, J. L.; Cohen, J. *J. Pharm. Sci.* 1963, *52*, 137-42.
VARIABLES:	**PREPARED BY:**
Concentration of dextrin	R. Piekos

EXPERIMENTAL VALUES:

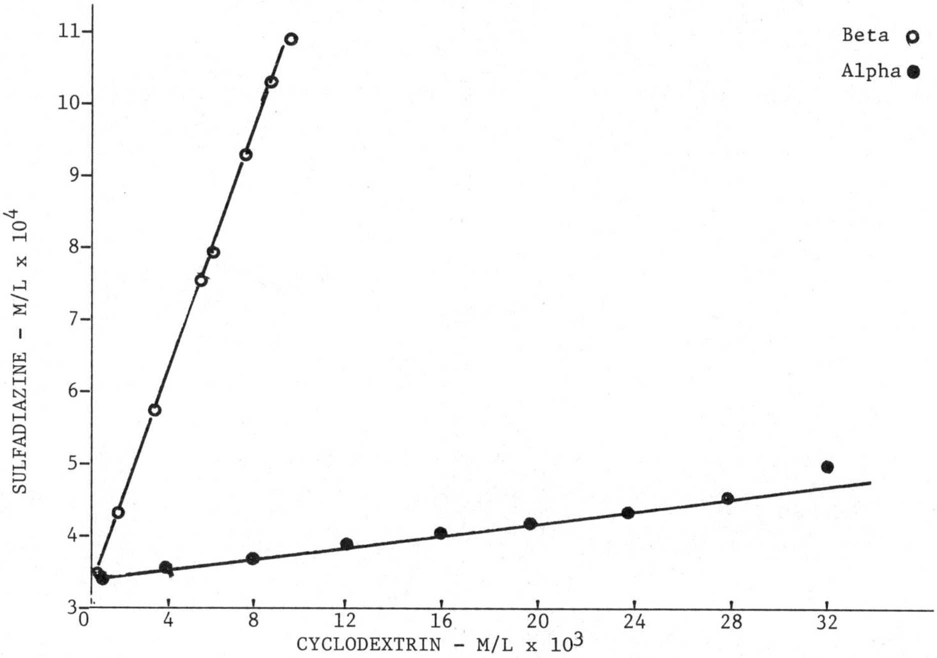

Solubility of sulfadiazine in aqueous solutions of cyclodextrins at 30°C

AUXILIARY INFORMATION

METHOD/APPARATUS/PROCEDURE:

The soly method of Higuchi and Lach was employed (1). Procedure: excess quantity of sulfadiazine was placed in 125-ml glass-stoppered bottles together with varying but accurately weighed amts of dextrin and 50 ml of water. The bottles were placed in a mech shaker in a const temp bath and equilibrated for 8 h at 30°C. Aliquot portions of the supernatant liquid were removed and analyzed for the sulfadiazine concn at the end of this time. Sulfadiazine was assayed spectrophotometrically at 266 nm.

SOURCE AND PURITY OF MATERIALS:

Sulfadiazine, m.p. 255-6°C. Source not specified.

The Schardinger dextrins were prepd by accepted methods (2,3).

Purity of the water was not specified.

ESTIMATED ERROR:

Nothing specified.

REFERENCES:

1. Higuchi, T.; Lach, J. L. *J. Pharm. Sci.* 1954, *43*, 349.
2. French, D.; Lavine, M. L.; Pazur, J. H.; Norberg, E., *J. Am. Chem. Soc.* 1949, *71*, 353.
3. Tilden, E. B.; Hudson, C. S., *J. Bacteriol.* 1942, *43*, 527.

COMPONENTS:

(1) Benzenesulfonamide, 4-amino-N-2-pyrimidinyl- (sulfadiazine); $C_{10}H_{10}N_4O_2S$; [68-35-9]

(2) Poly(oxy-1,2-ethanediyl), α - hexadecyl- ω -hydroxy-, mixt with α- octadecyl- ω -hydroxypoly(oxy-1,2-ethanediyl) (cetomacrogol) [8065-80-3]

(3) Water; H_2O; [7732-18-5]

ORIGINAL MEASUREMENTS:

Corby, T. C.; Elworthy, P. H.

J. Pharm. Pharmac. 1971, *23*, Suppl. 39S-48S.

VARIABLES:

Concentration of cetomacrogol

PREPARED BY:

R. Piekos

EXPERIMENTAL VALUES:

Concentration of cetomacrogol % w/v	Maximum additive concentration of sulfadiazine at 20°C	
	mmol kg^{-1} solvent	% by wt [a]
1	0.272	0.0680
8	0.535	0.134
10	1.01	0.252

[a] Calculated by compiler

AUXILIARY INFORMATION

METHOD/APPARATUS/PROCEDURE:

The soly was detd in an app where the aqueous soln of cetomacrogol was percolated through a plug of sulfadiazine supported on a 5/3 sintered glass disc. The percolation was continued until the soln was satd (5-6 days). The concn of sulfadiazine was detd spectrophotometrically from a calibration curve using a Uvispek spectrophotometer (1).

SOURCE AND PURITY OF MATERIALS:

Sulfadiazine (Macarthy's Ltd, Romford) was a British Pharmacopeia product. It was recrystd twice from a DMF-EtOH mixt (1:3) and dried at 40°C over P_2O_5. Its mp was 254°C (decompn). Cetomacrogol 1000 B.P.C. was of the same origin and was used as received. Tap water once distd from glass was used.

ESTIMATED ERROR:

Soly: nothing specified.

Temp: ±0.05°C (authors).

REFERENCES:

1. Elworthy, P. H.; Lipscomb, F. J. *J. Pharm. Pharmac.* 1968, *20*, 790.

COMPONENTS:

(1) Benzenesulfonamide, 4-amino-N-2-pyrimidinyl- (sulfadiazine); $C_{10}H_{10}N_4O_2S$; [68-35-9]

(2) Poly(oxy-1,2-ethanediyl), α-hydro- ω-hydroxy- (PEG 1000); $(C_2H_4O)_nH_2O$; [25322-68-3] 1000

(3) Water; H_2O; [7732-18-5]

ORIGINAL MEASUREMENTS:

Corby, T. C.; Elworthy, P. H.

J. Pharm. Pharmac. 1971, *23*, Suppl. 39S-48S.

VARIABLES:

PREPARED BY:

R. Piekos

EXPERIMENTAL VALUES:

Concentration of PEG 1000	Solubility of sulfadiazine at 20°C	
% w/v	mmol kg^{-1} solvent	% by wt [a]
10	0.719	0.180
30	2.11	0.525
50	4.20	1.04

[a] Calculated by compiler

AUXILIARY INFORMATION

METHOD/APPARATUS/PROCEDURE:

The soly was detd in an app where the aqueous soln of PEG 1000 was percolated through a plug of sulfadiazine supported on a 5/3 sintered glass disc. The percolation was continued until the soln was satd (5 - 6 days). The concn of sulfadiazine was detd spectrophotometrically from a calibration curve using a Uvispek spectrophotometer (1).

SOURCE AND PURITY OF MATERIALS:

Sulfadiazine (MaCarthy's Ltd, Romford) was a British Pharmacopeia product. It was recrystd twice from a DMF-EtOH mixt (1:3) and dried at 40°C over P_2O_5. Its mp was 254°C (decompn). PEG 1000 was a B.D.H. Laboratory Reagent and was used as received. Tap water once distd from glass was used.

ESTIMATED ERROR:

Soly: nothing specified.

Temp: ±0.05°C (authors).

REFERENCES:

1. Elworthy, P. H.; Lipscomb, F. J. *J. Pharm. Pharmac.* 1968, *20*, 790.

COMPONENTS:

(1) Benzenesulfonamide, 4-amino-N-2-pyrimidinyl- (sulfadiazine); $C_{10}H_{10}N_4O_2S$; [68-35-9]

(2) 1H-Purine-2,6-dione, 3,7-dihydro-1,3,7-trimethyl- (caffeine); $C_8H_{10}N_4O_2$; [58-08-2]

(3) Water; H_2O; [7732-18-5]

ORIGINAL MEASUREMENTS:

Higuchi, T.; Lach, J. L.

J. Amer. Pharm. Assoc., Sci. Ed. 1954, *43*, 349-54.

VARIABLES:

Concentration of caffeine

PREPARED BY:

R. Piekos

EXPERIMENTAL VALUES:

Total solubility of sulfadiazine in water containing caffeine at 30°C

Caffeine	Sulfadiazine		Caffeine	Sulfadiazine	
10^2 mol dm^{-3}	10^4 mol dm^{-3}	g dm^{-3} [a]	10^2 mol dm^{-3}	10^4 mol dm^{-3}	g dm^{-3} [a]
0.000	3.64	0.091	10.470	6.47	0.162
1.454	3.80	0.095	10.742	6.36	0.159
1.467	3.77	0.094	12.410	6.64	0.166
3.716	4.51	0.113	12.865	7.04	0.176
3.728	4.50	0.113	14.556	7.69	0.192
5.285	4.84	0.121	14.957	7.51	0.188
5.964	5.04	0.126			
7.356	5.32	0.133			
7.703	5.55	0.139			

[a] Calculated by compiler

AUXILIARY INFORMATION

METHOD/APPARATUS/PROCEDURE:

Excess quantities of sulfadiazine (25 mg) were placed in 125-ml glass-stoppered bottles together with varying but accurately weighed amts of caffeine and 50-ml portion of water. The bottles were placed in a mech shaker in a const temp bath and equilibrated for 8 h at 30°C. Aliquots of the supernatant liquid were analyzed for the sulfadiazine content by the method of Bratton and Marshall (1).

SOURCE AND PURITY OF MATERIALS:

Recrystd sulfadiazine (U.S.P.), mp 255-6°C, recrystd caffeine (U.S.P.), mp 235-7°C and distd water were used.

ESTIMATED ERROR:

Nothing specified.

REFERENCES:

1. Bratton, A. C.; Marshall, E. K., Jr. *J. Biol. Chem.* 1939, *128*, 537.

COMPONENTS:

(1) Benzenesulfonamide, 4-amino-N-2-pyrimidinyl- (sulfadiazine); $C_{10}H_{10}N_4O_2S$; [68-35-9]

(2) α-D-Glucopyranoside, β-D-fructofuranosyl- (sucrose); $C_{12}H_{22}O_{11}$; [57-50-1]

(3) Water; H_2O; [7732-18-5]

ORIGINAL MEASUREMENTS:

Nogami, H.; Nagai, T.; Suzuki, A. *Chem. Pharm. Bull.* 1966, *14(4)*, 339-50.

VARIABLES:

Concentration of sucrose

PREPARED BY:

R. Piekos

EXPERIMENTAL VALUES:

Concentration of sucrose	Solubility of sulfadiazine at 37°C	
mol dm^{-3}	mg/100 ml	10^4 mol dm^{-3} [a]
10^{-3}	10.9	4.35
10^{-2}	11.0	4.39
10^{-1}	12.0	4.79
1	12.9	5.15

[a] Calculated by compiler

AUXILIARY INFORMATION

METHOD/APPARATUS/PROCEDURE:

Soly of sulfadiazine was detd from dissoln rate data obtained by the rotating disk method.

SOURCE AND PURITY OF MATERIALS:

Commercial sulfadiazine J. P. was used. Purity of the remaining materials was not specified.

ESTIMATED ERROR:

Nothing specified.

REFERENCES:

COMPONENTS:	ORIGINAL MEASUREMENTS:
(1) Benzenesulfonamide, 4-amino-N-2-pyrimidinyl- (sulfadiazine); $C_{10}H_{10}N_4O_2S$; [68-35-9] (2) Methanol; CH_4O; [67-56-1]	Mauger, J. W.; Paruta, A. N.; Gerraughty, R. J. *J. Pharm. Sci.* 1972, 61(1), 97-7.
VARIABLES:	**PREPARED BY:**
Temperature	R. Piekos

EXPERIMENTAL VALUES:

t/°C	Mole fraction solubility (x 10^5)
25	19.3
30	22.9
37	29.9

AUXILIARY INFORMATION

METHOD/APPARATUS/PROCEDURE:

A const temp bath contg screw-capped bottles with sulfadiazine in excess and methanol was rotated for 24 h. Samples were withdrawn through a pledget of glass wool into a pipet, which was wiped clean and allowed to drain into a volumetric flask. Solute concns were detd by spectrophotometric assay at predetd wavelengths using a Cary model 16 spectrophotometer (1).

SOURCE AND PURITY OF MATERIALS:

Sulfadiazine: lot W02235 from Eli Lilly and Co.

Methanol was a spectrophotometric grade solvent from the Mallinckrodt Chemical Works.

ESTIMATED ERROR:

Soly: av values of 3 runs are given (authors).

Temp: ±0.1°C (authors).

REFERENCES:

1. Paruta, A. N.; Mauger, J. W. *J. Pharm. Sci.* 1971, *60*, 432.

COMPONENTS:

(1) Benzenesulfonamide, 4-amino-N-2-pyrimidinyl- (sulfadiazine); $C_{10}H_{10}N_4O_2S$; [68-35-9]

(2) Methanol; CH_4O; [67-56-1]

ORIGINAL MEASUREMENTS:

Mauger, J. W.; Petersen, H., Jr.
Alexander, K. S.; Paruta, A. N.
Drug Dev. Ind. Pharm. 1977, *3(2)*, 163-83.

VARIABLES:

Temperature

PREPARED BY:

R. Piekos

EXPERIMENTAL VALUES:

$t/^{\circ}C$	Solubility		
	mg/ml	10^5 X [a]	10^3 mol dm^{-3} [b]
25	1.18	19.3	4.71
30	1.40	22.9	5.59
37	1.82	29.9	7.27

[a] X = mole fraction

[b] Calculated by compiler

AUXILIARY INFORMATION

METHOD/APPARATUS/PROCEDURE:

A const temp bath contg screw-capped bottles with sulfadiazine in excess and methanol was rotated for 24 h. Samples were withdrawn through a pledget of glass wool into a pipet, which was wiped clean and allowed to drain into a volumetric flask. Soly was detd from absorbance and previously ascertained Beer's law plots detd on a Cary model 6 spectrophotometer (1).

SOURCE AND PURITY OF MATERIALS:

Sulfadiazine: lot W02235, Eli Lilly and Co. Its mp agreed with the literature value. Methanol was a spectrograde solvent from Mallinckrodt Chemical Works.

ESTIMATED ERROR:

Soly: av of at least 3 detns is reported (authors).

Temp: $\pm 0.1^{\circ}C$ (authors).

REFERENCES:

1. Mauger, J.W.; Paruta, A. N.; Gerraughty, R. J. *J. Pharm. Sci.* 1972, *61(1)*, 94.

COMPONENTS:	ORIGINAL MEASUREMENTS:
(1) Benzenesulfonamide, 4-amino-N-2-pyrimidinyl- (sulfadiazine); $C_{10}H_{10}N_4O_2S$; [68-35-9] (2) Ethanol; C_2H_6O; [64-17-5]	Mauger, J. W.; Paruta, A. N.; Gerraughty, R. J. *J. Pharm. Sci.* 1972, *61(1)*, 94-7.
VARIABLES: Temperature	PREPARED BY: R. Piekos

EXPERIMENTAL VALUES:

t/°C	Mole fraction solubility (x 10^5)
25	7.68
30	9.36
37	12.4

AUXILIARY INFORMATION

METHOD/APPARATUS/PROCEDURE:

A const temp bath contg screw-capped bottles with sulfadiazine in excess and ethanol was rotated for 24 h. Samples were withdrawn through a pledget of glass wool into a pipet, which was wiped clean and allowed to drain into a volumetric flask. Solute concns were detd by spectrophotometric assay at predetd wavelengths using a Cary model 16 spectrophotometer (1).

SOURCE AND PURITY OF MATERIALS:

Sulfadiazine: lot W02235 from Eli Lilly and Co.

Ethanol was from the U. S. Industrial Chemicals Co.

ESTIMATED ERROR:

Soly: av values of 3 runs are given (authors).

Temp: ±0.1°C (authors).

REFERENCES:

1. Paruta, A. N.; Mauger, J. W. *J. Pharm. Sci.* 1971, *60*, 432.

COMPONENTS:

(1) Benzenesulfonamide, 4-amino-N-2-pyrimidinyl- (sulfadiazine); $C_{10}H_{10}N_4O_2S$; [68-35-9]

(2) Ethanol, C_2H_6O; [64-17-5]

ORIGINAL MEASUREMENTS:

Mauger, J. W.; Petersen, H., Jr.; Alexander, K. S.; Paruta, A. N. *Drug Dev. Ind. Pharm.* 1977, *3(2)*, 163-83.

VARIABLES:

Temperature

PREPARED BY:

R. Piekos

EXPERIMENTAL VALUES:

t/°C	Solubility		
	mg/ml	10^5 X [a]	10^3 mol dm^{-3} [b]
25	0.33	7.68	1.32
30	0.40	9.36	1.60
37	0.53	12.40	2.12

[a] X = mole fraction

[b] Calculated by compiler

AUXILIARY INFORMATION

METHOD/APPARATUS/PROCEDURE:

A const temp bath contg screw-capped bottles with sulfadiazine in excess and ethanol was rotated for 24 h. Samples were withdrawn through a pledget of glass wool into a pipet, which was wiped clean and allowed to drain into a volumetric flask. Soly was detd from absorbance and previously ascertained Beer's law plots detd on a Cary Model 16 spectrophotometer (1).

SOURCE AND PURITY OF MATERIALS:

Sulfadiazine: lot W02235, Eli Lilly and Co. Its mp agreed with the literature value. Ethanol was from the U. S. Industrial Chemicals Co. Its refractive index value and density agreed with literature values.

ESTIMATED ERROR:

Soly: av of at least 3 detns is reported (authors).

Temp: ±0.1°C (authors).

REFERENCES:

1. Mauger, J. W.; Paruta, A. N. Gerraughty, R. J. *J. Pharm. Sci.* 1972, *61(1)*, 94.

COMPONENTS:

(1) Benzenesulfonamide, 4-amino-N-2-pyrimidinyl- (sulfadiazine); $C_{10}H_{10}N_4O_2S$; [68-35-9]

(2) 1-Propanol; C_3H_8O; [71-23-8]

ORIGINAL MEASUREMENTS:

Mauger, J. W.; Paruta, A. N.; Gerraughty, R. J. *J. Pharm. Sci.* 1972, *61(1)*, 94-7.

VARIABLES:

Temperature

PREPARED BY:

R. Piekos

EXPERIMENTAL VALUES:

t/°C	Mole fraction solubility (X 10^5)
25	4.32
30	5.45
37	7.44

AUXILIARY INFORMATION

METHOD/APPARATUS/PROCEDURE:

A const temp bath contg screw-capped bottles with excess and 1-propanol was rotated for 24 h. Samples were withdrawn through a pledget of glass wool into a pipet, which was wiped clean and allowed to drain into a volumetric flask. Solute concns were detd by spectrophotometric assay at predetd wavelengths using a Cary model 16 spectrophotometer (1).

SOURCE AND PURITY OF MATERIALS:

Sulfadiazine: lot WO2235 from Eli Lilly and Co.

1-Propanol was a Baker Analyzed Reagent from J. T. Baker Chemical Co.

ESTIMATED ERROR:

Soly: av values of 3 runs are given (authors).

Temp: ±0.1°C (authors).

REFERENCES:

1. Paruta, A. N.; Mauger, J. W. *J. Pharm. Sci.* 1971, *60*, 432.

COMPONENTS:

(1) Benzenesulfonamide, 4-amino-N-2-pyrimidinyl- (sulfadiazine); $C_{10}H_{10}N_4O_2S$; [68-35-9]

(2) 1-Propanol; C_3H_8O; [71-23-8]

ORIGINAL MEASUREMENTS:

Mauger, J. W.; Petersen, H., Jr., Alexander, K. S.; Paruta, A. N.

Drug Dev. Ind. Pharm. 1977, *3(2)*, 163-83.

VARIABLES:

Temperature

PREPARED BY:

R. Piekos

EXPERIMENTAL VALUES:

t/°C	Solubility		
	mg/ml	10^5 X [a]	10^4 mol dm^{-3} [b]
25	0.14	4.32	5.59
30	0.18	5.45	7.19
37	0.25	7.44	9.99

[a] X = mole fraction

[b] Calculated by compiler

AUXILIARY INFORMATION

METHOD/APPARATUS/PROCEDURE:

A const temp bath contg screw-capped bottles with sulfadiazine in excess and 1-propanol was rotated for 24 h. Samples were withdrawn through a pledget of glass wool into a pipet, which was wiped clean and allowed to drain into a volumetric flask. Soly was detd from absorbance and previously ascertained Beer's law plots detd on a Cary Model 16 spectrophotometer.

SOURCE AND PURITY OF MATERIALS:

Sulfadiazine: lot W02235, Eli Lilly and Co. Its mp agreed with the literature value. 1-Propanol was a Baker Analyzed Reagent (J. T. Baker Chemical Co.). Its refractive index value and density agreed with literature values.

ESTIMATED ERROR:

Soly: av of at least 3 detns is reported (authors).

Temp: ±0.1°C (authors).

REFERENCES:

1. Mauger, J. W.; Paruta, A. N.; Gerraughty, R. J. *J. Pharm. Sci.* 1972, *61(1)*, 94.

COMPONENTS:	ORIGINAL MEASUREMENTS:
(1) Benzenesulfonamide, 4-amino-N-2-pyrimidinyl- (sulfadiazine); $C_{10}H_{10}N_4O_2S$; [68-35-9] (2) 2-Propanol; C_3H_8O; [67-63-0]	Burlage, H. M. *J. Am. Pharm. Assoc., Sci. Ed.* 1948, *37*, 345.
VARIABLES:	**PREPARED BY:**
One temperature: 25°C	R. Piekos

EXPERIMENTAL VALUES:

Solubility of sulfadiazine in 2-propanol at 25°C is 0.0430 g/100 cm^3 solution (1.72 x 10^{-3} mol dm^{-3}, compiler).

AUXILIARY INFORMATION

METHOD/APPARATUS/PROCEDURE:

Satd solns of sulfadiazine in 2-propanol were prepd at 25°C and definite vols of the solns were measured into tared dishes by means of standard pipets. The alcohol was allowed to evap at room temp and the residue was dried at 105°C. In the case of losses due to apparent decompn, the residue was dried in a desiccator (1).

SOURCE AND PURITY OF MATERIALS:

The sulfadiazine was manufd by Squibb and was of the U.S.P. purity. The source and purity of 2-propanol were not specified.

ESTIMATED ERROR:

Nothing specified.

REFERENCES:

1. Burlage, H. M. *J. Am. Pharm. Assoc., Sci. Ed.* 1947, *36(1)*, 16.

COMPONENTS:

(1) Benzenesulfonamide, 4-amino-N-2-pyrimidinyl- (sulfadiazine); $C_{10}H_{10}N_4O_2S$; [68-35-9]

(2) 1-Butanol; $C_4H_{10}O$; [71-36-3]

ORIGINAL MEASUREMENTS:

Mauger, J. W.; Paruta, A. N.; Gerraughty, R. J. *J. Pharm. Sci.* 1972, *61(1)*, 94-7.

VARIABLES:

Temperature

PREPARED BY:

R. Piekos

EXPERIMENTAL VALUES:

t/°C	Mole fraction solubility (x 10^5)
25	3.18
30	4.09
37	5.66

AUXILIARY INFORMATION

METHOD/APPARATUS/PROCEDURE:

A const temp bath contg screw-capped bottles with sulfadiazine in excess and 1-butanol was rotated for 24 h. Samples were withdrawn through a pledget of glass wool into a pipet, which was wiped clean and allowed to drain into a volumetric flask. Solute concns were detd by spectrophotometric assay at predetd wavelengths using a Cary model 16 spectrophotometer (1).

SOURCE AND PURITY OF MATERIALS:

Sulfadiazine: lot WO2235 from Eli Lilly and Co.

1-Butanol was from the Mallinckrodt Chemical Works.

ESTIMATED ERROR:

Soly: av values of 3 runs are given (authors).

Temp: ±0.1°C (authors).

REFERENCES:

1. Paruta, A. N.; Mauger, J. W. *J. Pharm. Sci.* 1971, *60*, 432.

COMPONENTS:

(1) Benzenesulfonamide, 4-amino-N-2-pyrimidinyl- (sulfadiazine); $C_{10}H_{10}N_4O_2S$; [68-35-9]

(2) 1-Butanol; $C_4H_{10}O$; [71-36-3]

ORIGINAL MEASUREMENTS:

Mauger, J. W.; Petersen, H., Jr.; Alexander, K. S.; Paruta, A. N.

Drug Dev. Ind. Pharm. 1977, *3(2)*, 163-83.

VARIABLES:

Temperature

PREPARED BY:

R. Piekos

EXPERIMENTAL VALUES:

t/°C	Solubility		
	mg/ml	10^5 X [a]	10^4 mol dm^{-3} [a]
25	0.087	3.18	3.48
30	0.111	4.09	4.43
37	0.153	5.66	6.11

[a] X = mole fraction

[b] Calculated by compiler

AUXILIARY INFORMATION

METHOD/APPARATUS/PROCEDURE:

A const temp bath contg screw-capped bottles with sulfadiazine in excess and 1-butanol was rotated for 24 h. Samples were withdrawn through a pledget of glass wool into a pipet, which was wiped clean and allowed to drain into a volumetric flask. Soly was detd from absorbance and previously ascertained Beer's law plots detd on a Cary model 16 spectrophotometer (1).

SOURCE AND PURITY OF MATERIALS:

Sulfadiazine: lot W02235, Eli Lilly and Co. Its mp agreed with the literature value. 1-Butanol was from the Mallinckrodt Chemical Works. Its refractive index value and density agreed with literature values.

ESTIMATED ERROR:

Soly: av of at least 3 detns is reported (authors).

Temp: ±0.1°C (authors).

REFERENCES:

1. Mauger, J. W.; Paruta, A. N.; Gerraughty, R. J. *J. Pharm. Sci.* 1972, *61(1)*, 94.

COMPONENTS:

(1) Benzenesulfonamide, 4-amino-N-2-pyrimidinyl- (sulfadiazine); $C_{10}H_{10}N_4O_2S$; [68-35-9]

(2) 1-Pentanol; $C_5H_{12}O$; [71-41-0]

ORIGINAL MEASUREMENTS:

Mauger, J. W.; Paruta, A. N.; Gerraughty, R. J. *J. Pharm. Sci.* 1972, *61(1)*, 94-7.

VARIABLES:

Temperature

PREPARED BY:

R. Piekos

EXPERIMENTAL VALUES:

t/°C	Mole fraction solubility (x 10^5)
25	2.63
30	3.31
37	4.61

AUXILIARY INFORMATION

METHOD/APPARATUS/PROCEDURE:

A const temp bath contg screw-capped bottles with sulfadiazine in excess and 1-pentanol was rotated for 24 h. Samples were withdrawn through a pledget of glass wool into a pipet, which was wiped clean and allowed to drain into a volumetric flask. Solute concns were detd by spectrophotometric assay at predetd wavelengths using a Cary model 16 spectrophotometer (1).

SOURCE AND PURITY OF MATERIALS:

Sulfadiazine: lot W02235 from Eli Lilly and Co.

1-Pentanol was from Fisher Scientific Co.

ESTIMATED ERROR:

Soly: av values of 3 runs are given (authors).

Temp: ±0.1°C (authors).

REFERENCES:

1. Paruta, A. N.; Mauger, J. W. *J. Pharm. Sci.* 1971, *60*, 432.

COMPONENTS:

(1) Benzenesulfonamide, 4-amino-N-2-pyrimidinyl- (sulfadiazine); $C_{10}H_{10}N_4O_2S$; [68-35-9]

(2) 1-Pentanol; $C_5H_{12}O$; [71-41-0]

ORIGINAL MEASUREMENTS:

Mauger, J. W.; Petersen, H., Jr.; Alexander, K. S.; Paruta, A. N. *Drug Dev. Ind. Pharm.* 1977, *3(2)*, 163-83.

VARIABLES:

Temperature

PREPARED BY:

R. Piekos

EXPERIMENTAL VALUES:

t/°C	Solubility mg/ml	10^5 X [a]	10^4 mol dm^{-3} [b]
25	0.061	2.63	2.437
30	0.076	3.31	3.036
37	0.106	4.61	4.235

[a] X = mole fraction

[b] Calculated by compiler

AUXILIARY INFORMATION

METHOD/APPARATUS/PROCEDURE:

A const temp bath contg screw-capped bottles with sulfadiazine in excess and 1-pentanol was rotated for 24 h. Samples were withdrawn through a pledget of glass wool into a pipet, which was wiped clean and allowed to drain into a volumetric flask. Soly was detd from absorbance and previously ascertained Beer's law plots detd on a Cary model 16 spectrophotometer (1)

SOURCE AND PURITY OF MATERIALS:

Sulfadiazine: lot W02235, Eli Lilly and Co. Its mp agreed with the literature value. 1-Pentanol was from Fisher Scientific Co. Its refractive index value and density agreed with literature values.

ESTIMATED ERROR:

Soly: av of at least 3 detns is reported (authors).

Temp: 0.1°C (authors).

REFERENCES:

1. Mauger, J. W.; Paruta, A. N.; Gerraughty, R. J. *J. Pharm. Sci.* 1972, 61(1), 94..

COMPONENTS:

(1) Benzenesulfonamide, 4-amino-N-2-pyrimidinyl- (sulfadiazine); $C_{10}H_{10}N_4O_2S$; [68-35-9]

(2) 1-Octanol; $C_8H_{18}O$; [111-87-5]

ORIGINAL MEASUREMENTS:

Mauger, J. W.; Paruta, A. N.; Gerraughty, R. J. *J. Pharm. Sci.* 1972, *61(1)*, 94-7.

VARIABLES:

Temperature

PREPARED BY:

R. Piekos

EXPERIMENTAL VALUES:

t/°C	Mole fraction solubility (x 10^5)
25	1.41
30	1.76
37	2.65

AUXILIARY INFORMATION

METHOD/APPARATUS/PROCEDURE:

A const temp bath contg screw-capped bottles with sulfadiazine in excess and 1-octanol was rotated for 24 h. Samples were withdrawn through a pledget of glass wool into a pipet , which was wiped clean and allowed to drain into a volumetric flask. Solute concns were detd by spectrophotometric assay at predetd wavelengths using a Cary model 16 spectrophotometer (1).

SOURCE AND PURITY OF MATERIALS:

Sulfadiazine: lot W02235 from Eli Lilly and Co.

1-Octanol was from Mallinckrodt Chemical Works.

ESTIMATED ERROR:

Soly: av values of 3 runs are given (authors).

Temp: ±0.1°C (authors).

REFERENCES:

1. Paruta, A. N.; Mauger, J. W. *J. Pharm. Sci.* 1971, *60*, 432.

COMPONENTS:

(1) Benzenesulfonamide, 4-amino-N-2-pyrimidinyl- (sulfadiazine); $C_{10}H_{10}N_4O_2S$; [68-35-9]

(2) 1-Octanol; $C_8H_{18}O$; [111-87-5]

ORIGINAL MEASUREMENTS:

Mauger, J. W.; Petersen, H., Jr.; Alexander, K. S.; Paruta, A. N.

Drug Dev. Ind. Pharm. 1977, *3(2)*, 163-83.

VARIABLES:

Temperature

PREPARED BY:

R. Piekos

EXPERIMENTAL VALUES:

t/°C	Solubility		
	mg/ml	10^5 X [a]	10^4 mol dm^{-3} [a]
25	0.022	1.41	0.879
30	0.028	1.76	1.118
37	0.042	2.65	1.678

[a] X = mole fraction

[b] Calculated by compiler

AUXILIARY INFORMATION

METHOD/APPARATUS/PROCEDURE:

A const temp bath contg screw-capped bottles with sulfadiazine in excess and 1-octanol was rotated for 24 h. Samples were withdrawn through a pledget of glass wool into a pipet, which was wiped clean and allowed to drain into a volumetric flask. Soly was detd from absorbance and previously ascertained Beer's law plots detd on a Cary model 16 spectrophotometer (1).

SOURCE AND PURITY OF MATERIALS:

Sulfadiazine: lot W02235, Eli Lilly and Co. Its mp agreed with the literature value. 1-Octanol was from Fisher Scientific Co. Its refractive index and density agreed with literature values.

ESTIMATED ERROR:

Soly: av of at least 3 detns is reported (authors).

Temp: ±0.1°C (authors).

REFERENCES:

1. Mauger, J. W.; Paruta, A. N.; Gerraughty, R. J. *J. Pharm. Sci.* 1972, *61(1)*, 94.

COMPONENTS:

(1) Benzenesulfonamide, 4-amino-N-2-pyrimidinyl- (sulfadiazine); $C_{10}H_{10}N_4O_2S$; [68-35-9]

(2) 1-Decanol; $C_{10}H_{2}O$; [112-30-1]

ORIGINAL MEASUREMENTS:

Mauger, J. W.; Paruta, A. N.; Gerraughty, R. J. *J. Pharm. Sci.* 1972, *61(1)*, 94-7.

VARIABLES:

Temperature

PREPARED BY:

R. Piekos

EXPERIMENTAL VALUES:

t/°C	Mole fraction solubility (x 10^5)
25	7.40
30	8.04
37	9.47

AUXILIARY INFORMATION

METHOD/APPARATUS/PROCEDURE:

A const temp bath contg screw-capped bottles with sulfadiazine in excess and 1-decanol was rotated for 24 h. Samples were withdrawn through a pledget of glass wool into a pipet, which was wiped clean and allowed to drain into a volumetric flask. Solute concns were detd by spectrophotometric assay at predetd wavelengths using a Cary model 16 spectrophotometer (1).

SOURCE AND PURITY OF MATERIALS:

Sulfadiazine: lot W02235 from Eli Lilly and Co.

1-Decanol was from Matheson, Coleman and Bell.

ESTIMATED ERROR:

Soly: av values of 3 runs are given (authors).

Temp: ±0.1°C (authors).

REFERENCES:

1. Paruta, A. N.; Mauger, J. W. *J. Pharm. Sci.* 1971, *60*, 432.

COMPONENTS:	ORIGINAL MEASUREMENTS:
(1) Benzenesulfonamide, 4-amino-N-2-pyrimidinyl- (sulfadiazine); $C_{10}H_{10}N_4O_2S$; [68-35-9] (2) 1-Decanol; $C_{10}H_{22}O$; [112-30-1]	Mauger, J. W.; Petersen, H., Jr. Alexander, K. S.; Paruta, A. N. *Drug Dev. Ind. Pharm.* 1977, *3(2)*, 163-83.
VARIABLES:	**PREPARED BY:**
Temperature	R. Piekos

EXPERIMENTAL VALUES:

t/°C	Solubility		
	mg/ml	10^5 X [a]	10^4 mol dm^{-3} [a]
25	0.097	7.40	3.875
30	0.105	8.04	4.195
37	0.123	9.47	4.914

[a] X = mole fraction

[b] Calculated by compiler

AUXILIARY INFORMATION

METHOD/APPARATUS/PROCEDURE:

A const temp bath contg screw-capped bottles with sulfadiazine in excess and 1-decanol was rotated for 24 h. Samples were withdrawn through a pledget of glass wool into a pipet, which was wiped clean and allowed to drain into a volumetric flask. Soly was detd from absorbance and previously ascertained Beer's law plots detd on a Cary model 16 spectrophotometer (1).

SOURCE AND PURITY OF MATERIALS:

Sulfadiazine: lot W02235, Eli Lilly and Co. Its mp agreed with the literature value. 1-Decanol was from Matheson, Coleman and Bell. Its refractive index and density agreed with literature values.

ESTIMATED ERROR:

Soly: av of at least 3 detns is reported (authors).

Temp: ±0.1°C (authors).

REFERENCES:

1. Mauger, J. W.; Paruta, A. N.; Gerraughty, R. J. *J. Pharm. Sci.* 1972, *61(1)*, 94.

COMPONENTS:

(1) Benzenesulfonamide, 4-amino-N-2-pyrimidinyl- (sulfadiazine); $C_{10}H_{10}N_4O_2S$; [68-35-9]

(2) Acetamide, N,N-dimethyl-; C_4H_9NO; [127-19-5]

ORIGINAL MEASUREMENTS:

Sunwoo, C.; Eisen, H.

J. Pharm. Sci. 1971, *60*, 238-44.

VARIABLES:

One temperature: 25°C

PREPARED BY:

R. Piekos

EXPERIMENTAL VALUES:

The mole fraction solubility of sulfadiazine in N,N-dimethylacetamide at 25°C is 0.0239 (6.57 g/100 g solution, compiler).

AUXILIARY INFORMATION

METHOD/APPARATUS/PROCEDURE:

Soly was detd by the method reported by Restaino and Martin (1). Sulfadiazine was assayed on a Coleman-Hitachi 124 double-beam spectrophotometer at 270 nm after diln of a sample with 95% alcohol or water.

SOURCE AND PURITY OF MATERIALS:

A USP sulfadiazine (American Cyanamide) Co., Pearl River, N. Y.) was recrystd from warm acetone. N,N-Dimethylacetamide was of technical grade.

ESTIMATED ERROR:

Temp: ±1.0° (authors).

Soly: the mean of 3 runs was given (authors).

REFERENCES:

1. Restaino, F. A.; Martin, A. N. *J. Pharm. Sci.* 1964, *53*, 636.

COMPONENTS:

(1) Benzenesulfonamide, 4-amino-N-2-pyrimidinyl- (sulfadiazine); $C_{10}H_{10}N_4O_2S$; [68-35-9]

(2) Formamide, N,N-dimethyl-; C_3H_7NO; [68-12-2]

ORIGINAL MEASUREMENTS:

Elworthy, P. H.; Worthington, H. E. C.

J. Pharm. Pharmac. 1968, *20*, 830-5.

VARIABLES:

Temperature

PREPARED BY:

R. Piekos

EXPERIMENTAL VALUES:

t/°C	Solubility		
	Weight %	10^2 mole fraction	mole kg^{-1} water [a]
20	18.0	6.02	0.782
30	18.7	6.29	0.818
40	19.4	6.57	0.962

[a] Calculated by compiler

AUXILIARY INFORMATION

METHOD/APPARATUS/PROCEDURE:

Solns were prepd by shaking powd sulfadiazine with DMF for 24 h and transferred to a soly app which was of the percolation type and a modification of that used by Davies and Griffiths (1). The soln was recycled in the app through a sintered-glass filter until satd (7-14 days). Samples were evapd and the residues dried to const wt.

SOURCE AND PURITY OF MATERIALS:

Sulfadiazine (B.P. quality) was twice recrystd from an EtOH-DMF mixt (3:1 by vol) and dried over P_2O_5. Its mp was 255°C. Assay by the Pharmacopeial method gave 100.0% purity calcd with reference to the material dried at 105°C.

DMF (May and Baker Ltd) was distd under reduced pressure and gave n_D^{25} = 1.4283.

ESTIMATED ERROR:

Soly: mean values of duplicate runs are given (authors).

Temp: ±0.05°C (authors).

REFERENCES:

1. Davies, M.; Griffiths, D. M. L. *Trans. Faraday Soc.* 1953, *49*, 1405.

COMPONENTS:	ORIGINAL MEASUREMENTS:
(1) Benzenesulfonamide, 4-amino-N-2-pyrimidinyl- (sulfadiazine); $C_{10}H_{10}N_4O_2S$; [68-35-9] (2) n-Hexane; C_6H_{14}; [110-54-3]	Corby, T. C.; Elworthy, P. H. *J. Pharm. Pharmac.* 1971, *23*, Suppl. 39S-48S.
VARIABLES:	PREPARED BY:
One temperature: 20°C	R. Piekos

EXPERIMENTAL VALUES:

Concentration of sulfadiazine in a saturated n-hexane solution at 20°C is 0 millimolal.

AUXILIARY INFORMATION

METHOD/APPARATUS/PROCEDURE:

The soly was detd in an app where n-hexane percolated through a plug of sulfadiazine supported on a 5/3 sintered glass disc. The percolation was continued until the soln was satd (5-6 days). The concn of sulfadiazine was detd spectrophotometrically from a calibration curve using a Uvispek spectrophotometer.

SOURCE AND PURITY OF MATERIALS:

Sulfadiazine (MaCarthy's Ltd, Romford) was a British Pharmacopeia product. It was recrystd twice from a DMF-EtOH mixt (1:3) and dried at 40°C over P_2O_5. Its mp was 254°C (decompn). A spectroscopic grade n-hexane (B.D.H.) was used as received.

ESTIMATED ERROR:

Soly: not specified.

Temp: ±0.05°C (authors).

REFERENCES:

1. Elworthy, P. H.; Lipscomb, F. J. *J. Pharm. Pharmac.* 1968, *20*, 790.

COMPONENTS:	ORIGINAL MEASUREMENTS:
(1) Benzenesulfonamide, 4-amino-N-2-pyrimidinyl- (sulfadiazine); $C_{10}H_{10}N_4O_2S$; [68-35-9] (2) Methane, trichloro- (chloroform); $CHCl_3$; [67-66-3]	Riess, W. *Intern. Congr. Chemotherapy, Proc. 3rd, Stuttgart* <u>1963</u>, *1*, 627-32.
VARIABLES:	PREPARED BY:
One temperature: 20°C	R. Piekos

EXPERIMENTAL VALUES:

Solubility of sulfadiazine in chloroform at 20°C is 15 mg% (6.0×10^{-4} mol dm^{-3} solution, compiler).

AUXILIARY INFORMATION

METHOD/APPARATUS/PROCEDURE:

Nothing specified.

SOURCE AND PURITY OF MATERIALS:

Nothing specified.

ESTIMATED ERROR:

Nothing specified.

REFERENCES:

COMPONENTS:	ORIGINAL MEASUREMENTS:
(1) Benzenesulfonamide, 4-amino-N-2-pyrimidinyl- (sulfadiazine); $C_{10}H_{10}N_4O_2S$; [68-35-9] (2) Methane, trichloro- (chloroform); $CHCl_3$; [67-66-3]	Yamazaki, M.; Aoki, M.; Kamada, A.; Yata, N. *Yakuzaigaku* 1967, *27(1)*, 37-40.
VARIABLES:	**PREPARED BY:**
One temperature: 30°C	R. Piekos

EXPERIMENTAL VALUES:

Solubility of sulfadiazine in chloroform at 30°C is 0.49 mmol/L ($0.12\ g\ dm^{-3}$, compiler).

AUXILIARY INFORMATION

METHOD/APPARATUS/PROCEDURE:

Sulfadiazine (0.5 g) was placed in an L-shaped tube together with 20 ml of chloroform. The mixt was shaken in a thermostat until equilibrium was attained. The sulfadiazine was assayed in the supernatant spectrophotometrically at 545 nm on a Beckmann DU spectrophotometer. The results were taken from a calibration graph.

SOURCE AND PURITY OF MATERIALS:

Nothing specified.

ESTIMATED ERROR:

Soly: not specified.

Temp: ±1°C (authors).

REFERENCES:

COMPONENTS:

(1) Benzenesulfonamide, 4-amino-N-4-pyrimidinyl- (4- sulfapyrimidine); $C_{10}H_{10}N_4O_2S$; [599-82-6]

(2) Water; H_2O; [7732-18-5]

ORIGINAL MEASUREMENTS:

Roblin, R. O., Jr.; Williams, J. H.; Winnek, P. S.; English, J. P.

J. Am. Chem. Soc. 1940, *62*, 2002-5.

VARIABLES:

One temperature: 37°C

PREPARED BY:

R. Piekos

EXPERIMENTAL VALUES:

Solubility of 4-sulfapyrimidine in water at 37°C is 354 mg/100 cm^3 solution (1.41×10^{-2} mol dm^{-3}. compiler).

AUXILIARY INFORMATION

METHOD/APPARATUS/PROCEDURE:

Excess sulfonamide in water was heated and stirred on a steam bath for 30 min. The suspension was then agitated for 24 h in a thermostat at 37°C. A sample of the satd soln was withdrawn through a glass filter, dild, and analyzed by the Marshall method (1) using a General Electric recording spectrophotometer for comparing the colors developed with those of the standards.

SOURCE AND PURITY OF MATERIALS:

The sulfonamide, mp 231-2°C (dec, cor), was prepd by the authors. Anal: %C 48.0 (calcd 48.0); %H 3.4 (3.5); %N 20.2 (19.9). Purity of the water was not specified.

ESTIMATED ERROR:

Nothing specified.

REFERENCES:

1. Bratton, A. C.; Marshall, E. K., Jr. *J. Pharmacol.* 1939, *66*, 4.

COMPONENTS:

(1) Benzenesulfonamide, 4-amino-N-5-pyrimidinyl-; $C_{10}H_{10}N_4O_2S$; [17103-48-9]

(2) Water; H_2O; [7732-18-5]

ORIGINAL MEASUREMENTS:

Roblin, R. O., Jr.; Winnek, P. S.; English, J. P.

J. Am. Chem. Soc. 1942, *64*, 567-70.

VARIABLES:

One temperature: 37°C

PREPARED BY:

R. Piekos

EXPERIMENTAL VALUES:

Solubility of 4-amino-N-5-pyrimidinylbenzenesulfonamide in water at 37°C is 9.8 mg/100 cm^3 solution (3.9 x 10^{-4} mol dm^{-3}, compiler).

AUXILIARY INFORMATION

METHOD/APPARATUS/PROCEDURE:

Excess sulfonamide in water was heated and stirred on a steam bath for 30 min. The suspension was then agitated for 24 h in a thermostat at 37°C. A sample of the satd soln was withdrawn through a glass filter, dild, and analyzed by the Marshall method (1) using a General Electric recording spectrophotometer for comparing the colors developed with those of the standards.

SOURCE AND PURITY OF MATERIALS:

The sulfonamide, mp 260-1°C (cor), was prepd by the authors. Anal: %C 48.1 (calcd 48.0); %H 4.2 (4.0); %N 22.5 (22.4). Purity of the water was not specified.

ESTIMATED ERROR:

Nothing specified.

REFERENCES:

1. Bratton, A. C.; Marshall, E. K., Jr. *J. Pharmacol.* 1939, *66*, 4.

COMPONENTS:

(1) Aluminum, tris(4-amino-*N*-2-pyrimidinylbenzenesulfonamidato-N^N, *O*)- (Al sulfadiazine); $C_{30}H_{27}AlN_{12}O_6S_3$; [71280-76-7]

(2) Water; H_2O; [7732-18-5]

ORIGINAL MEASUREMENTS:

Fox, Ch. L., Jr.; Modak, S. Stanford, J. W.; Fox, P. L.

Scand. J. Plast. Reconstr. Surg. 1979, *13(1)*, 89-94.

VARIABLES:

One temperature: 28-30°C

PREPARED BY:

R. Piekos

EXPERIMENTAL VALUES:

Solubility of Al(III) sulfadiazine in water at room temperature (28-30°C)[a] is 6.6 mg% (8.5×10^{-5} mol dm^{-3} solution, compiler).

[a]Value given by one of the authors (S. M.) in personal communication.

AUXILIARY INFORMATION

METHOD/APPARATUS/PROCEDURE:

Satd soln of Al sulfadiazine was prepared in water and after 24 h aliquots from the clear supernatant were assayed for sulfadiazine content using the colorimetric method of Bratton and Marshall (1). The soly value was then calculated from the molecular formula.

SOURCE AND PURITY OF MATERIALS:

The Al (III) sulfadiazine was prepd by the authors and follows: an inorg Al salt was reacted with Na salt of sulfadiazine and the ppt was analyzed and characterized. No details were given, however. Purity of the materials was not specified.

ESTIMATED ERROR:

Nothing specified.

REFERENCES:

1. Bratton, A. C.; Marshall, E. K., Jr. *J. Biol. Chem.* 1939, *120*, 537.

COMPONENTS:

(1) Cerium, tris(4-amino-N-2-pyrimidinyl-benzenesulfonamidato-N^N,O)- (Ce sulfadiazine); $C_{30}H_{27}CeN_{12}O_6S_3$; [66269-03-2]

(2) Water; H_2O; [7732-18-5]

ORIGINAL MEASUREMENTS:

Fox, Ch. L., Jr.; Modak, S.; Stanford, J. W.; Fox, P. L.; *Scand. J. Plast. Reconstr. Surg.* 1979, *13(1)*, 89-94.

VARIABLES:

One temperature: 28-30°C

PREPARED BY:

R. Piekos

EXPERIMENTAL VALUES:

Solubility of Ce sulfadiazine in water at room temperature (28-30°C)[a] is 83.0 mg% (9.35×10^{-4} mol dm^{-3} solution, compiler).

[a]Value given by one of the authors (S.M.) in personal communication.

AUXILIARY INFORMATION

METHOD/APPARATUS/PROCEDURE:

Satd soln of Ce sulfadiazine was prepd in water and after 24 h aliquots from the clear supernatant were assayed for sulfadiazine content using the colorimetric method of Bratton and Marshall (1). The soly value was then calculated from the molecular formula.

SOURCE AND PURITY OF MATERIALS:

Neither source nor purity of the materials was specified.

The identity of the Ce sulfadiazine was questioned by Bult and associates (2) (compiler).

ESTIMATED ERROR:

Nothing specified.

REFERENCES:

1. Bratton, A. C.; Marshall, E. K., Jr. *J. Biol. Chem.* 1939, *120*, 537.

2. Bult, A.; Hulsing, N.; Weyland, J. W. *Pharm. Weekblad, Sci. Ed.* 1980, *2*, 190.
J. Pharm. Pharmacol. 1981, *33*, 171.

COMPONENTS:

(1) Chromium, bis(4-amino-N-2-pyrimidinylbenzenesulfonamidato-N^N,O)- (Cr(II) sulfadiazine); $C_{20}H_{18}CrN_8O_4S_2$; [71261-84-2]

(2) Water; H_2O; [7732-18-5]

ORIGINAL MEASUREMENTS:

Fox, Ch. L, Jr.; Modak, S.; Stanford, J. W.; Fox, P. L.

Scand. J. Plast. Reconstr. Surg. 1979, *13(1)*, 89-94.

VARIABLES:

One temperature: 28-30°C

PREPARED BY:

R. Piekos

EXPERIMENTAL VALUES:

Solubility of Cr(II) sulfadiazine in water at room temperature (28-30°C)[a] is 15.5 mg% (2.81×10^{-4} mol dm^{-3} solution, compiler).

[a]Value given by one of the authors (S. M.) in personal communication.

AUXILIARY INFORMATION

METHOD/APPARATUS/PROCEDURE:

Satd soln of Cr(II) sulfadiazine was prepd in water and after 24 h aliquots from the clear supernatant were assayed for sulfadiazine content using the colorimetric method of Bratton and Marshall (1). The soly value was then calculated from the molecular formula.

SOURCE AND PURITY OF MATERIALS:

The Cr(II) sulfadiazine was prepd by the authors as follows: an inorg. salt of CR(II) was reacted with Na salt of sulfadiazine and the ppt was analyzed and characterized. No details were given, however.

Purity of the materials was not specified.

ESTIMATED ERROR:

Nothing specified.

REFERENCES:

1. Bratton, A. C.; Marshall, E. K., Jr. *J. Biol. Chem.* 1939, *120*, 537.

COMPONENTS:

(1) Copper, (4-amino-N-2-pyrimidinylbenzenesulfonamidato-N^N,O)- (Cu(I) sulfadiazine); $C_{10}H_9CuN_4O_2S$; [71261-85-3]

(2) Water; H_2O; [7732-18-5]

ORIGINAL MEASUREMENTS:

Fox, Ch. L., Jr.; Modak, S.; Stanford, J. W.; Fox, P. L. *Scand. J. Plast. Reconstr. Surg.* 1979, *13(1)*, 89-94.

VARIABLES:

One temperature: 28-30°C

PREPARED BY:

R. Piekos

EXPERIMENTAL VALUES:

Solubility of Cu(I) sulfadiazine in water at room temperature (28-30°C)[a] is 5.8 mg% (1.8×10^{-4} mol dm^{-3} solution, compiler).

[a]Value given by one of the authors (S.M.) in personal communication.

AUXILIARY INFORMATION

METHOD/APPARATUS/PROCEDURE:

Satd soln of Cu(I) sulfadiazine was prepd in water and after 24 h aliquots from the clear supernatant were assayed for sulfadiazine content using the colorimetric method of Bratton and Marshall (1). The soly value was then calculated from the molecular formula.

SOURCE AND PURITY OF MATERIALS:

The Cu(I) sulfadiazine was prepd by the authors as follows: an inorg Cu(I) salt was reacted with Na salt of sulfadiazine and the ppt was analyzed and characterized. No details were given, however. Purity of the materials was not specified.

ESTIMATED ERROR:

Nothing specified.

REFERENCES:

1. Bratton, A. C.; Marshall, E. K, Jr. *J. Biol. Chem.* 1939, *120*, 537.

COMPONENTS:

(1) Copper, bis(4-amino-N-2-pyrimidinyl-benzenesulfonamidato-N^N,O)- (Cu(II) sulfadiazine); $C_{20}H_{18}CuN_8O_4S_2$; [12171-53-8]

(2) Water; H_2O; [7732-18-5]

ORIGINAL MEASUREMENTS:

Fox, Ch. L., Jr.; Modak, S.; Stanford, J. W.; Fox, P.L.
Scand. J. Plast. Reconstr. Surg. 1979, *13(1)*, 89-94.

VARIABLES:

One temperature: 28-30°C

PREPARED BY:

R. Piekos

EXPERIMENTAL VALUES:

Solubility of Cu(II) sulfadiazine in water at room temperature (28-30°C)[a] is 3.7 mg% (6.6×10^{-5} mol dm^{-3} solution, compiler).

[a]Value given by one of the authors (S. M.) in personal communication.

AUXILIARY INFORMATION

METHOD/APPARATUS/PROCEDURE:

Satd soln of Cu(II) sulfadiazine was prepd in water and after 24 h aliquots from the clear supernatant were assayed for sulfadiazine content using the colorimetric method of Bratton and Marshall (1). The soly value was then calculated from the molecular formula.

SOURCE AND PURITY OF MATERIALS:

Cu(II) sulfadiazine was prepd by the authors as follows: an inorg Cu(II) salt was reacted with Na salt of sulfadiazine and the ppt was analyzed and characterized. No details were given, however.
Purity of the materials was not specified.

ESTIMATED ERROR:

Nothing specified.

REFERENCES:

1. Bratton, A. C.; Marshall, E. K., Jr. *J. Biol. Chem.* 1939, *120*, 537.

COMPONENTS:

(1) Cobalt, bis(4-amino-N-2-pyrimidinyl-benzenesulfonamidato-N^N, O)-
(Co sulfadiazine); $C_{20}H_{18}CoN_8O_4S_2$;
[71280-79-0]

(2) Water; H_2O; [7732-18-5]

ORIGINAL MEASUREMENTS:

Fox, Ch. L., Jr.; Modak, S;
Stanford, J. W.; Fox, P. L.

Scand. J. Plast. Reconstr. Surg. 1979, *13(1)*, 89-94.

VARIABLES:

One temperature: 28-30°C

PREPARED BY:

R. Piekos

EXPERIMENTAL VALUES:

Solubility of Co(II) sulfadiazine in water at room temperature (28-30°C)[a] is 86.9 mg% (1.56×10^{-3} mol dm^{-3} solution, compiler).

[a]Value given by one of the authors (S. M.) in personal communication.

AUXILIARY INFORMATION

METHOD/APPARATUS/PROCEDURE:

Satd soln of Co sulfadiazine was prepd in water and after 24 h aliquots from the clear supernatant were assayed for sulfadiazine content using the colorimetric method of Bratton and Marshall (1). The soly value was then calculated from the molecular formula.

SOURCE AND PURITY OF MATERIALS:

The Co(II) sulfadiazine was prepd by the authors as follows: an inorg Co salt was reacted with Na salt of sulfadiazine and the ppt was analyzed and characterized. No details were given, however.
Purity of the materials was not specified.

ESTIMATED ERROR:

Nothing specified.

REFERENCES:

1. Bratton, A. C.; Marshall, E. K., Jr. *J. Biol. Chem.* 1939, *120*, 537.

COMPONENTS:

(1) Iron, tris(4-amino-N-2-pyrimidinyl-benzenesulfonamidato-N^N,O)- (Fe(III) sulfadiazine); $C_{30}H_{27}FeN_{12}O_6S_3$; [71261-86-4]

(2) Water; H_2O; [7732-18-5]

ORIGINAL MEASUREMENTS:

Fox, Ch. L; Modak, S.; Stanford, J. W.; Fox, P. L.
Scand. J. Plast. Reconstr. Surg. 1979, *13(1)*, 89-94.

VARIABLES:

One temperature: 28-30°C

PREPARED BY:

R. Piekos

EXPERIMENTAL VALUES:

Solubility of Fe(III) sulfadiazine in water at room temperature (28-30°C)[a] is 6.0 mg% (7.5×10^{-5} mol dm^{-3} solution, compiler).

[a]Value given by one of the authors (S. M.) in personal communication.

AUXILIARY INFORMATION

METHOD/APPARATUS/PROCEDURE:

Satd soln of Fe(III) sulfadiazine was prepd in water and after 24 h aliquots from the clear supernatant were assayed for sulfadiazine content using the colorimetric method of Bratton and Marshall (1). The soly value was then calculated from the molecular formula.

SOURCE AND PURITY OF MATERIALS:

The Fe(III) sulfadiazine was prepd by the authors as follows: an inorg Fe(III) salt was reacted with Na salt of sulfadiazine and the ppt was analyzed and characterized. No details were given, however. Purity of the materials was not specified.

ESTIMATED ERROR:

Nothing specified.

REFERENCES:

1. Bratton, A. C.; Marshall, E. K., Jr. *J. Biol. Chem.* 1939, *120*, 537.

COMPONENTS:

(1) Benzenesulfonamide, 4-amino-N-2-pyrimidinyl-, monosilver (+) salt; (AgSD); $C_{10}H_9AgN_4O_2S$; [22199-08-2]

(2) Water; H_2O; [7732-18-5]

ORIGINAL MEASUREMENTS:

Modak, S. M.; Fox, Ch. L. Jr.

Biochem. Pharmacol. 1973, *22*, 2391-404.

VARIABLES:

One temperature: 28°C

PREPARED BY:

R. Piekos

EXPERIMENTAL VALUES:

Solubility of AgSD in water at room temperature (28°C) is 0.5 μ mole per 100 ml water (5×10^{-6} mol dm^{-3} water, compiler).

AUXILIARY INFORMATION

METHOD/APPARATUS/PROCEDURE:

According to personal communication of one of the authors (C.L.F.) the following procefure was employed: A satd soln of ^{110m}Ag sulfadiazine of known sp activity was made and left at room temp (28°C) overnight. The supernatant was then taken out and centrifuged at higher speed to remove small floating particles of the solid. Radioactive measurement was done on the aliquot of the clear supernatant and the soly was calcd using the sp activity of ^{110m}AgSD.

SOURCE AND PURITY OF MATERIALS:

$^{110m}AgNO_3$ obtained from the International Chem and Nuclear Corp was reacted with Na sulfadiazine (source and purity not specified) to give ^{110m}AgSD with a sp activity of 1.14 mCi/mmol.

Purity of the water was not specified.

ESTIMATED ERROR:

Nothing specified.

REFERENCES:

COMPONENTS:	ORIGINAL MEASUREMENTS:
(1) Benzenesulfonamide, 4-amino-N-2-pyrimidinyl-, monosilver (+) salt; (Ag sulfadiazine); $C_{10}H_9AgN_4O_2S$; [22199-08-2] (2) Water; H_2O; [7732-18-5]	Fox, Ch. L., Jr. Modak, S.; Stanford, J. W.; Fox, P. L. *Scand. J. Plast. Reconstr. Surg.* 1979, *13(1)*, 89-94.
VARIABLES:	PREPARED BY:
One temperature: 28-30°C	R. Piekos

EXPERIMENTAL VALUES:

Solubility of Ag sulfadiazine in water at room temperature (28-30°C)[a] is 0.2 mg% (6×10^{-6} mol dm^{-3} solution, compiler).

[a]Values given by one of the authors (S. M.) in personal communication.

AUXILIARY INFORMATION

METHOD/APPARATUS/PROCEDURE:

Satd soln of Ag sulfadiazine was prepd in water and after 24 h aliquots from the clear supernatant were assayed for sulfadiazine content using the colorimetric method of Bratton and Marshall (1). The soly value was then calcd from the molecular formula.

SOURCE AND PURITY OF MATERIALS:

Nothing specified.

ESTIMATED ERROR:

Nothing specified.

REFERENCES:

1. Bratton, A. C.; Marshall, E. K., Jr. *J. Biol. Chem.* 1939, *120*, 537.

COMPONENTS:	ORIGINAL MEASUREMENTS:
(1) Benzenesulfonamide, 4-amino-N-2-pyrimidinyl-, monosilver salt (Ag sulfadiazine); $C_{10}H_9AgN_4O_2S$; [22199-08-2] (2) Water; H_2O; [7732-18-5]	Bult, A.; Klassen, H. B. *Arch. Pharm. (Weinheim)* 1980, *313(12)*, 1016-20.
VARIABLES:	PREPARED BY:
One temperature: 25°C	R. Piekos

EXPERIMENTAL VALUES:

Solubility of Ag sulfadiazine in water at 25±0.1°C is 0.34 mg/100 ml (9.5×10^{-6} mol dm^{-3}, compiler).

AUXILIARY INFORMATION

METHOD/APPARATUS/PROCEDURE:

Ag sulfadiazine was equilibrated with water at 25±0.1°C for one week in a vial with a parafilm-sheet covered rubber closure and wrapped with Al foil. The vial was rotated during equilibration in a thermostated bath. The satd soln was filtered through 25 mm in diam and 1.2 μm av pore size filters (Selectron-filter, type ST69), Schleicher and Schüll, Dassel, Germany). The filtrate was analyzed for sulfadiazine by uv spectrophotometry (Perkin Elmer 124) and for Ag by atomic absorption spectrophotometry.

SOURCE AND PURITY OF MATERIALS:

Ag sulfadiazine was prepd by the earlier reported method (1). Its purity was not specified.

Doubly distd water was used.

ESTIMATED ERROR:

Soly: not specifed.

Temp: ±0.1°C (authors).

REFERENCES:

1. Fox, Ch. L., Jr.; Modak, S. M. *Antimicrob. Agents Chemother.* 1974, *6*, 562.

COMPONENTS:

(1) Benzenesulfonamide, 4-amino-N-2-pyrimidinyl-, monosilver salt (Ag sulfadiazine); $C_{10}H_9AgN_4O_2S$; [22199-08-2]

(2) Nitric acid; HNO_3; [7697-37-2]

(3) Potassium nitrate; KNO_3; [7757-79-1]

(4) Water; H_2O; [7732-18-5]

ORIGINAL MEASUREMENTS:

Nesbitt, R. U., Jr.; Sandmann, B. J.

J. Pharm. Sci. 1977, *66(4)*, 519-22.

VARIABLES:

pH

PREPARED BY:

R. Piekos

EXPERIMENTAL VALUES:

Table I

Total molar solubility, S, of Ag sulfadiazine determined by the method of known subtraction and the molar concentration of the silver ion determined by direct potentiometry on identical samples at 25±0.1°C, ionic strength 0.1M, in nitric acid buffers

	pH 2.128		pH 3.851	
	$S \times 10^5$	$[Ag^+] \times 10^5$	$S \times 10^5$	$[Ag^+] \times 10^5$
	59.18	59.11	6.690	6.455
	58.06	57.97	6.690	6.517
	59.35	59.34	6.434	6.517
	58.53	58.42	6.768	6.475
	59.19	60.00	6.586	6.375
	57.30	57.97	6.612	6.455
Mean	58.60	58.80	6.466	6.629

AUXILIARY INFORMATION

METHOD/APPARATUS/PROCEDURE:

Mixts of 100 mg of Ag sulfadiazine and 25 ml of the HNO_3-KNO_3 buffer with an ionic strength adjusted to 0.1M with KNO_3 were placed in paraffin-coated vials and rotated end-over-end in a thermostated bath until equilibrium soly was obtained. After filtration, the solns were analyzed at 25±0.1°C in paraffin-coated beakers for the Ag-ion concn using a Ag^+-ion selective electrode (No.94-16, Orion Res., Cambridge, Mass) standardized at 25±0.1 °C and an ionic strength of 0.1M. The electrode displayed a Nernstian response through out the concn range of $1 \times 10^{-2} - 1.5 \times 10^{-6}$M for the Ag^+ion. The pH was measured with a pH electrode (Corning Sci Instruments, Medfield, Mass) standardized using standard buffers meeting NBS requirements. The total Ag^+ concn was detd by the method of known subtraction (1,2) in the HNO_3-buffered soln to which a sufficient amt of a std soln of KI was added to precipitate approx one-half of the free Ag^+ ion.

SOURCE AND PURITY OF MATERIALS:

All reagents were of anal grade. Ag sulfadiazine was prepd by the method of Braun and Towle (3) and recrystd. Water had a sp cond of (1-10) x 10^{-7} ohm^{-1} cm^{-1}.

The source of HNO_3 and KNO_3 was not specified.

ESTIMATED ERROR: Soly: the means in Table I are not statistically different at the 5% level (authors). Std deviation of the mean K_s value in Table II is ±0.12 x 10^{-12} (authors). To be contd.

REFERENCES:

1. Durst, R. A., in *"Ion Selective Electrodes"*, R.A. Durst, Ed., NBS Special Publ No 314, US Govt Printing Office, Washington, DC, 1969, p. 381.
2. *Orion Research Inc Newsletter* 1969, *1*, 25.
3. Braun, E. E.; Towle, J. L. *J. Am. Chem. Soc.* 1941, *63*, 3523.

COMPONENTS:	ORIGINAL MEASUREMENTS:
Continued from previous page.	Nesbitt, R. U., Jr.; Sandmann, B. J. *J. Pharm. Sci.* 1977, *66(4)*, 519-22.

VARIABLES:	PREPARED BY:
pH	R. Piekos

EXPERIMENTAL VALUES:

Table II

Calculation of solubility product of Ag sulfadiazine[a], K_s, at 25±0.1°C and ionic strength 0.1M.

pH	f_o^b	$[Ag^+]^2$	K_s
2.122	2.688×10^{-5}	2.980×10^{-7}	8.04×10^{-12}
2.373	6.024×10^{-5}	1.352×10^{-7}	8.16×10^{-12}
2.630	1.279×10^{-4}	6.165×10^{-8}	7.90×10^{-12}
2.891	2.583×10^{-4}	3.139×10^{-8}	8.14×10^{-12}
		Mean	$8.06 \times 10^{-12} \pm 0.12 \times 10^{-12}$

AUXILIARY INFORMATION

METHOD/APPARATUS/PROCEDURE:

SOURCE AND PURITY OF MATERIALS:

ESTIMATED ERROR:

REFERENCES:

COMPONENTS:

Continued from previous page

ORIGINAL MEASUREMENTS:

Nesbitt, R. U., Jr.; Sandmann, B. J.

J. Pharm. Sci. 1977, *66(4)*, 519-22.

VARIABLES:

pH

PREPARED BY:

R. Piekos

EXPERIMENTAL VALUES:

[a]from equation $K_s = f_o S^2$, where $f_o = \left(1 + \frac{[H_3O^+]}{K_2} + \frac{[H_3O^+]^2}{K_1 K_2}\right)^{-1}$

S is the total molar solubility, and K_1 and K_2 are the ionization constants of the N^4- (amine) and N^1- (amide) hydrogens of sulfadiazine, resp.

[b]$K_1 = 8.59 \times 10^{-3}$, and $K_2 = 3.82 \times 10^{-7}$.

Figure

Molar solubility, S, of Ag sulfadiazine versus pH at 0.1M ionic strength at 25±0.1°C. Solid circles denote experimental values and open circles denote those calculated from equation $S^2 = [Ag^+]^2 = \frac{[H_3O^+]K_s}{K_1 K_2} + \frac{[H_3O^+]K_s}{K_2}$

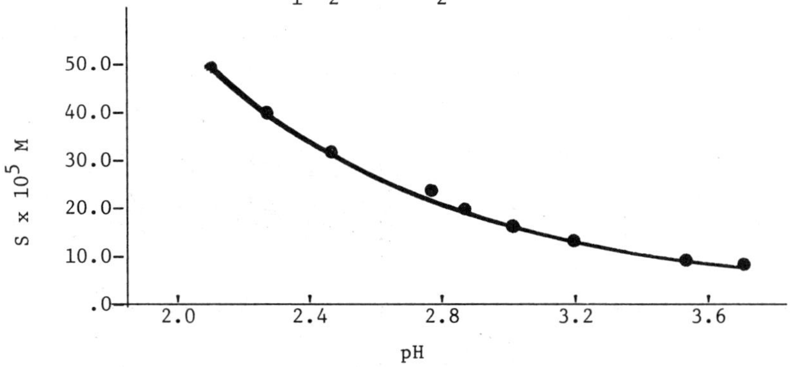

AUXILIARY INFORMATION

METHOD/APPARATUS/PROCEDURE:

SOURCE AND PURITY OF MATERIALS:

ESTIMATED ERROR:

Temp: ±0.1°C (authors).

pH : accuracy ±0.001 pH unit (authors).

REFERENCES:

COMPONENTS:

(1) Benzenesulfonamide, 4-amino-N-2-pyrimidinyl-, monosilver salt (Ag sulfadiazine); $C_{10}H_9AgN_4O_2S$; [22199-08-2]

(2) 4-Morpholineethanesulfonic acid; $C_6H_{13}NO_4S$; [4432-31-9]

(3) 4-Morpholineethanesulfonic acid, sodium salt; $C_6H_{12}NNaO_4S$; [71119-23-8]

(4) Potassium nitrate; KNO_3; [7757-79-1]

(5) Water; H_2O; [7732-18-5]

ORIGINAL MEASUREMENTS:

Nesbitt, R. U., Jr.; Sandmann, B. J.

J. Pharm. Sci. 1977, *66(4)*, 519-22.

VARIABLES:

pH

PREPARED BY:

R. Piekos

EXPERIMENTAL VALUES:

pH	Molar solubility[a], S, in a 0.05M 4-morpholine-ethanesulfonic acid buffer at 0.1M ionic strength (KNO_3) at 25±0.1°C
6	5.40×10^{-6}
7	3.19×10^{-6}

[a]Estimated by equation, $S^2 = [Ag^+]^2 = \frac{[H_3O^+]K_s}{K_2} + K_s$, where K_s is the solubility product contant, (8.06±0.12) x 10^{-12}, and K_2 is the ionization constant of the N^1 (amido)- hydrogen of sulfadiazine (3.82 x 10^{-7}).

AUXILIARY INFORMATION

METHOD/APPARATUS/PROCEDURE:

Mixts of 100 mg of Ag sulfadiazine and 25 ml of 0.05M 4-morpholineethanesulfonic acid buffer were adjusted to an ionic strength of 0.1 M with KNO_3, placed in paraffin-coated vials and rotated in a thermostated bath until equilibrium was obtained. After filtration the solns were analyzed at 25±0.1°C and an ionic stregnth of 0.1M in paraffin-coated beakers for the Ag^+-ion concn using a Ag^+-ion selective electrode standardized using std buffers meeting NBS requirements. The total Ag^+-ion concn was detd by the method of known addn (1); the added reagent was a std soln of $AgNO_3$ representing a 100-fold increase in the free Ag^+ ion present in the sample solns. The pH was measured with triple-purpose pH electrode.

SOURCE AND PURITY OF MATERIALS:

All reagents were of anal grade (source not specified). Ag sulfadiazine was prepd by the method of Braun and Towle(2) and recrystd. The buffer soln was prepd by titrn of the acid with a std NaOH soln. Water had a sp cond.of (1-10) x 10^{-7} ohm^{-1} cm^{-1}.

ESTIMATED ERROR:

Soly: not specified.

Temp: ±0.1°C (authors).

pH : accuracy ±0.001 pH unit (authors).

REFERENCES:

1. *Orion Research Inc Newsletter* 1969, *1*, 25.
2. Braun, C. E.; Towle, J. L. *J. Am. Chem. Soc.* 1941, *63*, 3523.

COMPONENTS:	ORIGINAL MEASUREMENTS:
Continued from previous page.	Nesbitt, R. U., Jr.; Sandmann, B. J. *J. Pharm. Sci.* 1977, *66(4)* 519-22.
VARIABLES: pH	PREPARED BY: R. Piekos

EXPERIMENTAL VALUES:

Figure

Equilibrium values of S^2 versus $[H_3O^+]$ in 0.05M 4-morpholineethanesulfonic acid buffer at 0.1M ionic strength (KNO_3) at 25±0.1°C

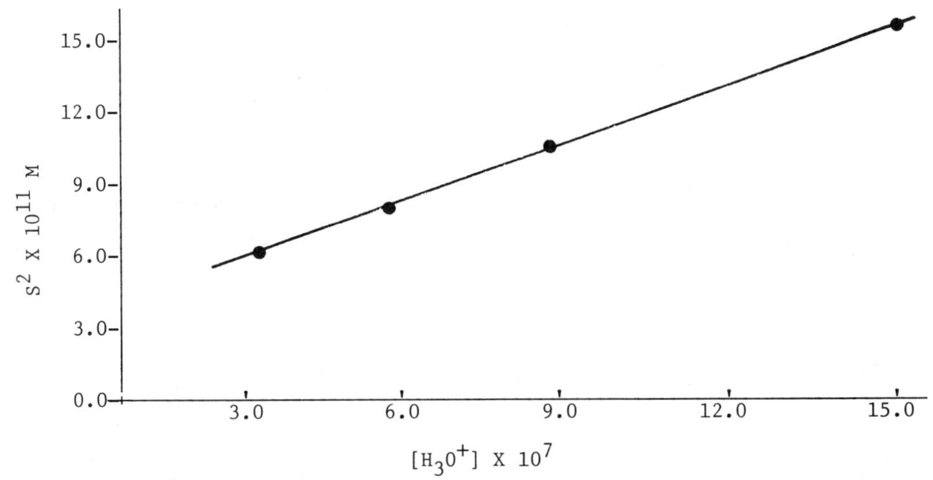

AUXILIARY INFORMATION

METHOD/APPARATUS/PROCEDURE:	SOURCE AND PURITY OF MATERIALS:
	ESTIMATED ERROR:
	REFERENCES:

COMPONENTS:	ORIGINAL MEASUREMENTS:
(1) (T-4)-(4-amino-N-2-pyrimidinylbenzene-sulfonamidato-N^N,O)bis(1H-imidazole-N^3)-silver (Ag sulfadiazine imidazole); $C_{10}H_9AgN_4O_2S \cdot 2C_3H_4N_2$; [76634-39-4] (2) Water; H_2O; [7732-18-5]	Bult, A.; Klasen, H. B. *Arch. Pharm. (Weinheim, Ger.)* 1980, *313(12)*, 1016-20.
VARIABLES:	PREPARED BY:
One temperature: 25°C	R. Piekos

EXPERIMENTAL VALUES:

Solubility of Ag sulfadiazine imidazole in water at 25°C is 3.4 mg/100 ml (6.9×10^{-5} mol dm^{-3}, compiler).

AUXILIARY INFORMATION

METHOD/APPARATUS/PROCEDURE:

Ag sulfadiazine imidazole was equilibrated with doubly distd water at 25±0.1°C for one week in a vial with a paraffin coated sheet rubber closure and wrapped with Al foil. The vial was rotated during equilibration in a thermostated bath. The satd soln was filtered through 25 mm in diam and 1.2 μm av pore size filters (Selectron filter type ST69, Schleicher and Schuell, Dassel, Ger.). The filtrate was analyzed for the sulfonamide by uv spectrophotometry (Perkin Elmer 124) and for Ag by atomic absorption spectrophotometry.

SOURCE AND PURITY OF MATERIALS:

The Ag sulfadiazine imidazole was synthesized by the authors. Analysis- Calcd: C 38.9, H 3.47, N 22.7, S 6.5, Ag 21.9% Found: C 39.1, H 3.5, N 23.2, S 6.5, Ag 21.7%. Sulfadiazine by amperometric titrn: 50.3 (lit 50.52).
Doubly distilled water was used.

ESTIMATED ERROR:

Soly: not specified.
Temp: ±0.1°C (authors).

REFERENCES:

COMPONENTS:	ORIGINAL MEASUREMENTS:
(1) Butanoic acid, 4-oxo-4-[[[4-(2-pyrimidinylamino)sulfonyl]phenyl]amino]-, disilver(1+)salt (Ag succinylsulfadiazine); $C_{14}H_{12}Ag_2N_4O_5S$; [76619-75-5] (2) Water; H_2O; [7732-18-5]	Bult, A.; Klasen, H. B. *Arch. Pharm. (Weinheim, Ger.)* 1980, *313(12)*, 1016-20.
VARIABLES:	PREPARED BY:
One temperature: 25°C	R. Piekos

EXPERIMENTAL VALUES:

Solubility of Ag succinylsulfadiazine in water at 25°C is 0.55 mg/100 ml (9.7×10^{-6} mol dm^{-3}, compiler).

AUXILIARY INFORMATION

METHOD/APPARATUS/PROCEDURE:

Ag succinylsulfadiazine was equilibrated with doubly distd water at 25±0.1°C for one week in a vial with a paraffin coated sheet rubber closure and wrapped with Al foil. The vial was rotated during equilibration in a thermostated bath. The satd soln was filtered through 25 mm in diam and 1.2μm av pore size filters (Selectron filter type ST69, Schleicher and Schuell, Dassel, Ger.) The filtrate was analyzed for the sulfonamide by uv spectrophotometry (Perkin Elmer 124) and for Ag by atomic absorption spectrophotometry.

SOURCE AND PURITY OF MATERIALS:

The Ag succinylsulfadiazine was synthesized by the authors. Analysis - Calcd: C 29.8, H 2.32, N 9.9, S 5.7, Ag 38.3%. Found: C 29.0, H 2.1, N 9.7, S 5.5, Ag 37.0%

Doubly distilled water was used.

ESTIMATED ERROR:

Soly: not specified.

Temp: ±0.1°C (authors).

REFERENCES:

COMPONENTS:

(1) Benzenesulfonamide, 4-amino-N-2-pyrimidinyl, monosodium salt (sodium sulfadiazine); $C_{10}H_9N_4NaO_2S$; [547-32-0]

(2) Water; H_2O; [7732-18-5]

ORIGINAL MEASUREMENTS:

Clark, W. G.; Strakosch, E. A.; Levitan, N. I. *J. Lab. Clin. Med.* 1942, *28*, 188-9.

VARIABLES:

Temperature

PREPARED BY:

R. Piekos

EXPERIMENTAL VALUES:

t/°C	Solubility	
	g/100 g water	mol kg^{-1} water[a]
25	50.0	1.84
37	65.0	2.39

[a] Calculated by compiler.

AUXILIARY INFORMATION

METHOD/APPARATUS/PROCEDURE:

A small tinted glass container contg excess Na sulfadiazine in water was shaken in a water bath thermostat for 24 h. The satd soln was then filtered by aspiration through a washed and dried asbestos filter stick into a weighed weighing bottle. The entire app was kept at the temp at which the compd was dissolved. The amt dissolved was then detd by the method of Bratton and Marshall (1), using a photoelectric colorimeter.

SOURCE AND PURITY OF MATERIALS:

Neither source nor purity of Na sulfadiazine was specified.

CO_2-free distd water was used.

ESTIMATED ERROR:

Soly: not specified.

Temp: ±0.1°C (authors).

REFERENCES:

1. Bratton, A. C.; Marshall, E. K., Jr. *J. Biol. Chem.* 1939, *128*, 537.

COMPONENTS:

(1) Benzenesulfonamide, 4-amino-N-2-pyrimidinyl-, monosodium salt (sodium sulfadiazine); $C_{10}H_{10}N_4O_2S{\cdot}Na$; [547-32-0]

(2) 2-Propanol; C_3H_8O; [67-63-0]

ORIGINAL MEASUREMENTS:

Burlage, H. M.

J. Am. Pharm. Assoc., Sci. Ed. 1948, *37*, 345.

VARIABLES:

One temperature: 25°C

PREPARED BY:

R. Piekos

EXPERIMENTAL VALUES:

Solubility of sodium sulfadiazine in 2-propanol at 25°C is 0.1860 g/100 cm^3 solution (6.832×10^{-3} mol dm^{-3}, compiler).

AUXILIARY INFORMATION

METHOD/APPARATUS/PROCEDURE:

Satd soln of sodium sulfadiazine in 2-propanol were prepd at 25°C and definite vols of the solns were measured into tared dishes by means of standard pipets. The alcohol was allowed to evap at room temp and the residue was dried at 105°C. In the case of losses due to apparent decompn, the residue was dried in a desiccator (1).

SOURCE AND PURITY OF MATERIALS:

The sodium sulfadiazine was of the U. S. P. purity, manufd by Squibb. The source and purity of 2-propanol was not specified.

ESTIMATED ERROR:

Nothing specified.

REFERENCES:

1. Burlage, H. M. *J. Am. Pharm. Assoc., Sci. Ed.* 1947, *36(1)*, 16.

COMPONENTS:	ORIGINAL MEASUREMENTS:
(1) Benzenesulfonamide, 4-amino-N-2-pyrimidinyl-, zinc(2+) salt (zinc sulfadiazine); $C_{20}H_{18}N_8O_4S_2Zn$; [66219-86-1] (2) Water; H_2O; [7732-18-5]	Fox, Ch. L., Jr.; Modak, S.; Stanford, J. W.; Fox, P. L. *Scand. J. Plast. Reconstr. Surg.* 1979, *13(1)*, 89-94.
VARIABLES:	PREPARED BY:
One temperature: 28-30°C	R. Piekos

EXPERIMENTAL VALUES:

Solubility of Zn sulfadiazine in water at room temperature (28-30°C)[a] is 56.0 mg% (9.93×10^{-4} mol dm^{-3} solution, compiler).

[a]Value given by one of the authors (S.M.) in personal communication.

AUXILIARY INFORMATION

METHOD/APPARATUS/PROCEDURE:

Satd soln of Zn sulfadiazine was prepd in water and after 24 h aliquots from the clear supernatant were assayed for sulfadiazine content using the colorimetric method of Bratton and Marshall (1). The soly value was then calculated from the molecular formula.

SOURCE AND PURITY OF MATERIALS:

Neither source nor purity of the materials was specified.
The identity of the Zn sulfadiazine was questioned by Bult and associates (2) (compiler).

ESTIMATED ERROR:

Nothing specified.

REFERENCES:

1. Bratton, A. C.; Marshall, E. K., Jr. *J. Biol. Chem.* 1939, *120*, 537.
2. Bult, A.; Hulsing, N.; Weyland, J. W. *Pharm. Weekblad, Sci. Ed.* 1980, *2*, 190; *J. Pharm. Pharmacol.* 1981, *33*, 171.

COMPONENTS:	EVALUATOR:
(1) Acetamide, N-[4-[(2-pyrimidinyl-amino)-sulfonyl]phenyl]- (acetyl sulfapyrimidine) $C_{12}H_{12}N_4O_3S$; [127-74-2]	Anthony N. Paruta, Department of Pharmaceutics, University of Rhode Island, Kingston, Rhode Island, USA
(2) Water	and Ryszard Piekos, Faculty of Pharmacy, University of Gdansk, Gdansk, Poland 1986

CRITICAL EVALUATION:

The solubility of the acetyl derivative of sulfadiazine in water at 310K has been reported by Roblin et al. (1) and Kikuth (2). The values given were 5.13×10^{-4} mol dm^{-3} and 5.75×10^{-3}, respectively. Only the former (1) supplied sufficient experimental details. Because of the uncertainty in Kikuth's value only a tentative average value of the solubility in water of 5.44×10^{-4} mol dm^{-3} at 310K can be recommended. It should be noted that in this case the acetyl group causes about a ten fold decrease in the equilibrium solubility with respect to the parent compound.

The data reported for buffer solutions are compiled in Table I.

Table I: Solubility of Acetyl sulfapyrimidine at various pH levels in buffer solution, 293K and 310K

Reference	pH	10^4 mol dm^{-3} 293K	310K
3	5.9	4.52	7.2
4	6.0	7.9	-
3	7.0	28.7	56.8
4	7.0	24.6	-
3	8.0	86.5	-
4	8.0	71.2	-

There is a considerable discrepancy in the values. At pH = 7 an approximate value can be suggested, though that of Pulver and Suter (4) is about 86% of Krüger-Thiemer's (3). The simple average of about 26×10^{-4} mol dm^{-3} must be contrasted to the value of the parent compound. Sulfadiazine possesses an aqueous solubility at 293K of about 2×10^{-4} mol dm^{-3}. Since the acetyl derivative usually decreases solubility, it is unexpected to evidence a 13-fold enhancement of solubility.

REFERENCES:

(1) Roblin, R. O., Jr.; Williams, J. H.; Winnek, P. S.; English, J. P. *J. Am. Chem. Soc.* 1940, *62*, 2002-5.

(2) Kikuth, W. *Med. Welt* 1943, *17(26/27)*, 483-6.

(3) Krüger-Thiemer, E. *Arch. Dermatol. Syphilis* 1942, *183* 90-116.

(4) Pulver, R.; Suter, R. *Schweiz. Med. Wochenschr.* 1943, *73(13)*, 403-8.

COMPONENTS:

(1) Acetamide, N-[4-[(2-pyrimidinylamino) sulfonyl]phenyl]- (acetyl sulfapyrimidine); $C_{12}H_{12}N_4O_3S$; [127-74-2]

(2) Water; H_2O; [7732-18-5]

ORIGINAL MEASUREMENTS:

Roblin, R. O., Jr.; Williams, J. H.; Winnek, P. S.; English, J. P.
J. Am. Chem. Soc. <u>1940</u>, *62*, 2002-5.

VARIABLES:

One temperature: 37°C

PREPARED BY:

R. Piekos

EXPERIMENTAL VALUES:

Solubility fo acetyl sulfapyrimidine in water at 37°C is 15.0 mg/100 cm^3 solution (5.13×10^{-4} mol dm^{-3}, compiler).

AUXILIARY INFORMATION

METHOD/APPARATUS/PROCEDURE:

Excess sulfonamide in water was heated and stirred on a steam bath for 30 min. The suspension was then agitated for 24 h in a thermostat at 37°C. A sample of the satd soln was withdrawn through a glass filter, dild, and analyzed by the Marshall method (1) using a General Electric recording spectrophotometer for comparing the colors developed with those of the standards.

SOURCE AND PURITY OF MATERIALS:

The acetyl sulfapyrimidine, mp 258-9°C, (cor), was prepd by the authors. Anal: %C 49.2 (calcd 49.4); %H 4.1 (4.1); %N 19.2 (19.2). Purity of the water was not specified.

ESTIMATED ERROR:

Nothing specified.

REFERENCES:

1. Bratton, A. C.; Marshall, E. K., Jr. *J. Pharmacol.* <u>1939</u>, *66*, 4.

COMPONENTS:	ORIGINAL MEASUREMENTS:
(1) Acetamide, N-[4-[(2-pyrimidinylamino) sulfonyl]phenyl]- (acetyl sulfapyrimidine); $C_{12}H_{12}N_4O_3S$; [127-74-2] (2) Water; H_2O; [7732-18-5]	Kikuth, W. *Med. Welt* 1943, *17(26/27)*, 483-6.
VARIABLES:	**PREPARED BY:**
One temperature: 37°C	R. Piekos

EXPERIMENTAL VALUES:

Solubility of acetyl sulfapyrimidine in water at 37°C is 168 mg/100 cm^3 solution (5.75×10^{-3} mol dm^{-3}, compiler).

AUXILIARY INFORMATION

METHOD/APPARATUS/PROCEDURE:

Nothing specified.

SOURCE AND PURITY OF MATERIALS:

Acetyl sulfapyrimidine: not specified.
The pH of the water was 7.0.

ESTIMATED ERROR:

Nothing specified.

REFERENCES:

COMPONENTS:

(1) Acetamide, N-[4-[(2-pyrimidinylamino)sulfonyl]phenyl]- (acetyl sulfadiazine); $C_{12}H_{12}N_4O_3S$; [127-74-2]

(2) Water; H_2O; [7732-18-5]

ORIGINAL MEASUREMENTS:

Langecker, H.
Arch. Exptl. Path. Pharmakol. 1948, *205*, 291-301.

VARIABLES:

pH

PREPARED BY:

R. Piekos

EXPERIMENTAL VALUES:

pH	Solubility at 37°C	
	mg%	10^4 mol dm^{-3} [a]
5.7	23	7.9
6.3	24	8.2

[a] Calculated by compiler

AUXILIARY INFORMATION

METHOD/APPARATUS/PROCEDURE:

An excess of acetyl sulfadiazine was boiled with water for 1 h in a sealed ampul followed by keeping the ampul at 37°C. Before the assaying, the solute was treated with 2.6N NaOH soln (1) to cleave the acetyl group and sulfadiazine was detd colorimetrically by the method of Bratton and Marshall (2) using a Havemann colorimeter (3), as well as by microanal detn of the solid residue.

SOURCE AND PURITY OF MATERIALS:

Source and purity of the materials were not specified.

ESTIMATED ERROR:

Nothing specified.

REFERENCES:

1. Scudi, J. V. *J. Lab. Clin. Med.* 1940, *25*, 404.
2. Bratton, A. G.; Marshall, E. K., Jr. *J. Biol. Chem.* 1939, *128*, 537.
3. Havemann, R. *Klin. Wochenschr.* 1940, p. 503.

COMPONENTS:

(1) Acetamide, N-[4-[(2-pyrimidinylamino)-sulfonyl]phenyl]- (acetyl sulfadiazine); $C_{12}H_{12}N_4O_3S$; [127-74-2]

(2) Sodium hydroxide; NaOH; [1310-73-2]

(3) Water; H_2O; [7732-18-5]

ORIGINAL MEASUREMENTS:

Rose, F. L.; Martin, A. R.; Bevan, H.G.L. *J. Pharm. Exp. Therap.* 1943, *77*, 127-42.

VARIABLES:

pH

PREPARED BY:

R. Piekos

EXPERIMENTAL VALUES:

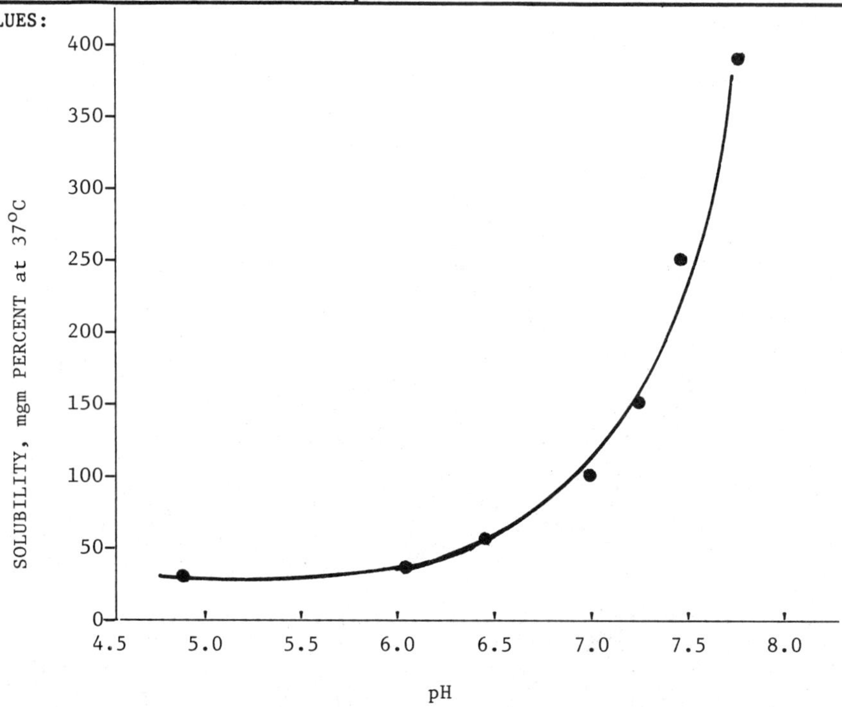

AUXILIARY INFORMATION

METHOD/APPARATUS/PROCEDURE:

An excess of acetyl sulfadiazine was stirred in boiling water , the soln was cooled to 37°C, the temp being maintained thermostatically, and 0.1N NaOH was added to increase pH. The pH was measured by means of a glass electrode-calomel half-cell system and was permitted to reach equilibrium before a reading was taken. The drug was then de-acetylated and the concn of sulfadiazine in soln was detd colorimetrically by withdrawing a sample through a filter-tip into a preheated micropipet.

SOURCE AND PURITY OF MATERIALS:

The source and purity of acetyl sulfadiazine were not specified.

Water was doubly distilled.

ESTIMATED ERROR:

Nothing specified.

REFERENCES:

COMPONENTS:

(1) Acetamide, N-[4-[(2-pyrimidinylamino)-sulfonyl]phenyl]- (acetyl sulfadiazine); $C_{12}H_{12}N_4O_3S$; [127-74-2]

(2) 1,2,3-Propanetricarboxylic acid, 2-hydroxy-, disodium salt (Na citrate); $C_6H_6Na_2O_7$; [144-33-2]

(3) Sodium hydroxide; NaOH; [1310-73-2]

(4) Water; H_2O; [7732-18-5]

ORIGINAL MEASUREMENTS:

Gilligan, D. R.; Plummer, M. N.

Proc. Soc. Exp. Biol. Med. 1943, *53*, 142-5.

VARIABLES:

pH

PREPARED BY:

R. Piekos

EXPERIMENTAL VALUES:

Solubility of acetyl sulfadiazine in M/10 Na citrate + NaOH solutions at 37°C

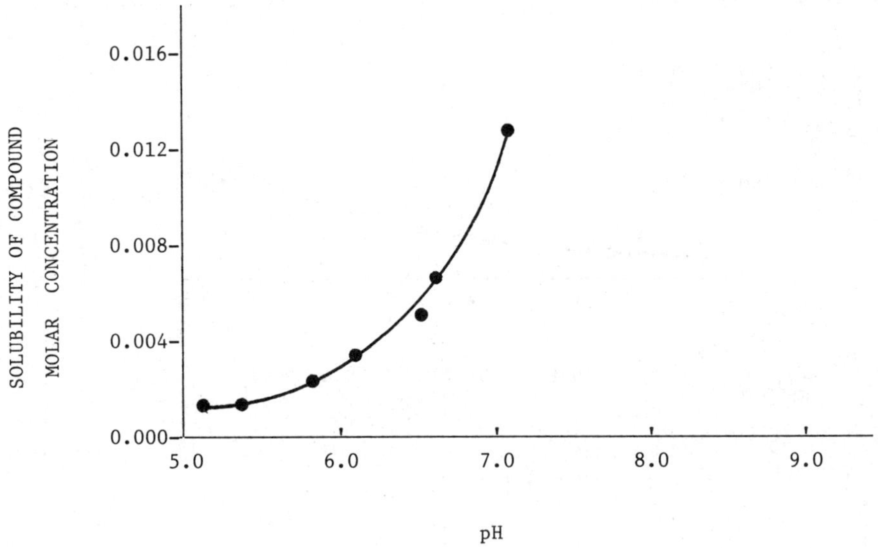

AUXILIARY INFORMATION

METHOD/APPARATUS/PROCEDURE:

An excess of acetyl sulfadiazine was shaken in M/10 Na citrate + NaOH solns of various pH values for 18 h in a water bath at 37°C, and filtered in an incubator room at this temp. The pH of the filtrate was measured immediately at room temp with a Beckmann glass electrode pH meter and appropriate corrections for the differences between room temp and 37°C were applied. The amt of dissolved compd was measured by the method of Bratton and Marshall (1).

SOURCE AND PURITY OF MATERIALS:

Acetyl sulfadiazine was supplied by Lederle Labs, Inc. The source and purity of the remaining materials were not specified.

ESTIMATED ERROR:

Nothing specified.

REFERENCES:

1. Bratton, A. C.; Marshall, E. K., Jr. *J. Biol. Chem.* 1939, *128*, 537.

COMPONENTS:

(1) Acetamide, N-[4-[(2-pyrimidinylamino)-sulfonyl]phenyl]- (acetyl sulfadiazine); $C_{12}H_{12}N_4O_3S$; [127-74-2]

(2) 1,2,3-Propanetricarboxylic acid, 2-hydroxy-, disodium salt (Na citrate); $C_6H_6Na_2O_7$; [144-33-2]

(3) Sodium hydroxide; NaOH; [1310-73-2]

(4) Water; H_2O; [7732-18-5]

ORIGINAL MEASUREMENTS:

Gilligan, D. R.; Garb, S.; Plummer, N.
Proc. Soc. Exp. Biol. Med. 1943, *52*, 248-50.

VARIABLES:

pH

PREPARED BY:

R. Piekos

EXPERIMENTAL VALUES:

The curve of sulobility of acetyl sulfadiazine in M/10 citrate plus NaOH buffer solutions at 37°C

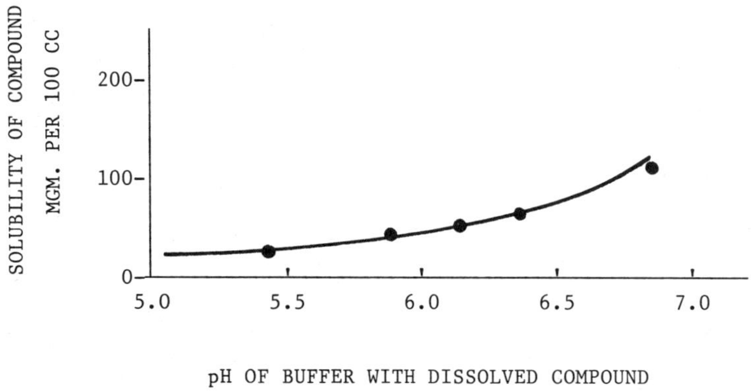

AUXILIARY INFORMATION

METHOD/APPARATUS/PROCEDURE:

An amt of acetyl sulfadiazine in large excess of that to be dissolved was added to M/10 citrate plus NaOH buffer solns shaken for 18 h in a water bath at 37°C and filtered at the same temp. The pH of the filtrate was measured immediately with a Beckmann pH meter and appropriate temp corrections were made. Acetyl sulfadiazine was assayed by the method of Bratton and Marshall (1).

SOURCE AND PURITY OF MATERIALS:

Acetyl sulfadiazine (purity not specified) was supplied by Lederle Labs, Inc. The source and purity of the remaining materials were not specified.

ESTIMATED ERROR:

NOthing specified.

REFERENCES:

1. Bratton, A. C.; Marshall, E. K., Jr. *J. Biol. Chem.* 1939, *128*, 537.

COMPONENTS:

(1) Acetamide, N-[4-[(2-pyrimidinylamino)-sulfonyl]phenyl]- (acetylsulfapyrimidine); $C_{12}H_{12}N_4O_3S$; [127-74-2]

(2) Phosphoric acid, monopotassium salt; KH_2PO_4; [7778-77-0]

(3) Water; H_2O; [7732-18-5]

ORIGINAL MEASUREMENTS:

Krüger-Thiemer, E.

Arch. Dermatol. Syphilis 1942, *183*, 90-116.

VARIABLES:

One temperature: ca 20°C; one pH: 4.37

PREPARED BY:

R. Piekos

EXPERIMENTAL VALUES:

Solubility of acetyl sulfapyrimidine in a 0.735M (10%) KH_2PO_4 solution of pH 4.37 at room temperature (about 20°C) is 0.0072 g% (2.5×10^{-4} mol dm^{-3} solution, compiler).

AUXILIARY INFORMATION

METHOD/APPARATUS/PROCEDURE:

Acetyl sulfapyrimidine (0.5 g) was dissolved in 10 cm^3 of the 0.735M (10%) KH_2PO_4 soln of pH 4.37, shaken for 2 h at room temp (about 20°C), and filtered. The filtrate was treated with equal vol of 2N HCl, and refluxed for 15 min. After proper diln, a 1-cm^3 aliquot was withdrawn, acidified, cooled, and the sulfonamide content was detd colorimetrically (as sulfapyrimidine) by the Marshall method modified by Kimmig (1) using an Authenrieth colorimeter. The pH was detd on an ultraionograph usisng a glass electrode.

SOURCE AND PURITY OF MATERIALS:

Acetyl sulfapyrimidine (source not specified) gave no coloration upon diazotization of its satd soln, thus showing absence of sulfapyrimidine. The source and purity of the remaining materials were not specified.

ESTIMATED ERROR:

Soly: precision ±5% (author).

Temp: not specified.

pH : ±0.05 pH unit (author).

REFERENCES:

1. Kimmig, J. *Arch. Dermatol.* 1938, *176*, 722; *Erg. Hyg.* 1941, *24*, 398.

COMPONENTS:

(1) Acetamide, N-[4-[(2-pyrimidinylamino)-sulfonyl]phenyl]- (acetyl sulfapyrimidine); $C_{12}H_{12}N_4O_3S$; [127-74-2]

(2) Phosphoric acid, disodium salt; Na_2HPO_4; [7558-94-4]

(3) Water; H_2O; [7732-18-5]

ORIGINAL MEASUREMENTS:

Krüger-Thiemer, E.

Arch. Dermatol. Syphilis 1942, *183*, 90-116.

VARIABLES:

One temperature: ca 20°C; one pH: 8.74

PREPARED BY:

R. Piekos

EXPERIMENTAL VALUES:

Solubility of acetyl sulfapyrimidine in a 0.705M (10%) Na_2HPO_4 solution of pH 8.74 at room temperature (about 20°C) is 0.548 g% (1.875×10^{-2} mol dm^{-3} solution, compiler).

AUXILIARY INFORMATION

METHOD/APPARATUS/PROCEDURE:

Acetyl sulfapyrimidine (0.5 g) was dissolved in 10 cm^3 of the 0.705M (10%) Na_2HPO_4 soln, shaken for 2 h at room temp (about 20°C), and filtered. The filtrate was treated with equal vol of 2N HCl and refluxed for 15 min. After proper diln, a 1-cm^3 aliquot was withdrawn, acidified, cooled, and the sulfonamide content was detd colorimetrically (as sulfapyrimidine) by the Marshall method modified by Kimmig (1) using an Authenrieth colorimeter. The pH was detd on an ultraionograph using a glass electrode.

SOURCE AND PURITY OF MATERIALS:

Acetyl sulfapyrimidine (source not specified) gave no coloration upon diazotization of its satd soln, thus showing absence of sulfapyrimidine. The source and purity of the remaining materials were not specified.

ESTIMATED ERROR:

Soly: precision ±5% (author).

Temp: not specified.

pH : ±0.05 pH unit (author).

REFERENCES:

1. Kimmig, J. *Arch. Dermatol.* *176*, 722; *Erg. Hyg.* 1941, *24*, 398.

COMPONENTS:

(1) Acetamide, N-[4-[(2-pyrimidinylamino)-sulfonyl]phenyl]- (acetyl sulfapyrimidine); $C_{12}H_{12}N_4O_3S$; [127-74-2]

(2) Phosphoric acid, disodium salt; Na_2HPO_4; [7558-94-4]

(3) Phosphoric acid, monopotassium salt; KH_2PO_4; [7778-77-0]

(4) Water; H_2O; [7732-18-5]

ORIGINAL MEASUREMENTS:

Krüger-Thiemer, E.

Arch. Dermatol. Syphilis 1942, *183*, 90-116.

VARIABLES:

Temperature, pH

PREPARED BY:

R. Piekos

EXPERIMENTAL VALUES:

Composition of 1/15M phosphate buffer solutions			pH	Solubility: Room temp (ca 20°C)		Solubility: 37°C	
Na_2HPO_4	KH_2PO_4	%Content		g%	10^3 mol dm^{-3} solution[a]	g%	10^3 mol dm^{-3} solution[a]
1.0	99.0	0.91	4.944	0.0082	0.28	-	-
10.0	90.0	0.91	5.906	0.0132	0.452	0.021	0.72
61.1	38.9	0.93	7.005	0.0840	2.87	0.166	5.68
9.5	0.5	0.733[b]	7.51	0.1810	6.19	-	-
94.7	5.3	0.95	8.018	0.2530	8.65	-	-

[a] Calculated by compiler.

b Molar content; 10% buffer solution.

AUXILIARY INFORMATION

METHOD/APPARATUS/PROCEDURE:

Acetyl sulfapyrimidine (0.5 g) was dissolved in 10 cm^3 of a buffer soln, shaken for 2 h at 20°C (or left for 48 h at 37°C), and filtered at respective temp. The filtrate was treated with equal vol of 2N HCl and refluxed for 15 min. After proper diln, a 1-cm^3 aliquot was withdrawn, acidified cooled, and the sulfonamide content was detd colorimetrically (as sulfapyrimidine) by the Marshall method modified by Kimmig (1) using the Authenrieth colorimeter. The pH was detd on an ultraionograph using a glass electrode.

SOURCE AND PURITY OF MATERIALS:

Acetyl sulfapyrimidine (source not specified) gave no coloration upon diazotization of its satd soln, thus showing absence of sulfapyrimidine. the source and purity of the remaining materials were not specified.

ESTIMATED ERROR:

Soly: precision ±5% (author).

Temp: not specified.

pH : ±0.05 pH unit (author).

REFERENCES:

1. Kimmig, J. *Arch. Dermatol.* 1938, *176* 722; *Erg. Hyg.* 1941, *24*, 398.

COMPONENTS:

(1) Acetamide, N-[4-[(2-pyrimidinylamino)-sulfonyl]phenyl]- (acetyl sulfadiazine); $C_{12}H_{12}N_4O_3S$; [127-74-2]

(2) Phosphoric acid, disodium salt; Na_2HPO_4; [7558-94-4]

(3) Phosphoric acid, monopotassium salt; KH_2PO_4; [7778-77-0]

(4) Water; H_2O; [7732-18-5]

VARIABLES:

pH

ORIGINAL MEASUREMENTS:

Gilligan, D. R.; Garb, S.; Plummer, N.
Proc. Soc.. Exp. Biol. Med. 1943, *52*, 248-50.

PREPARED BY:

R. Piekos

EXPERIMENTAL VALUES:

The curve of solubility of acetyl sulfadiazine in M/15 phosphate buffer solutions at 37°C.

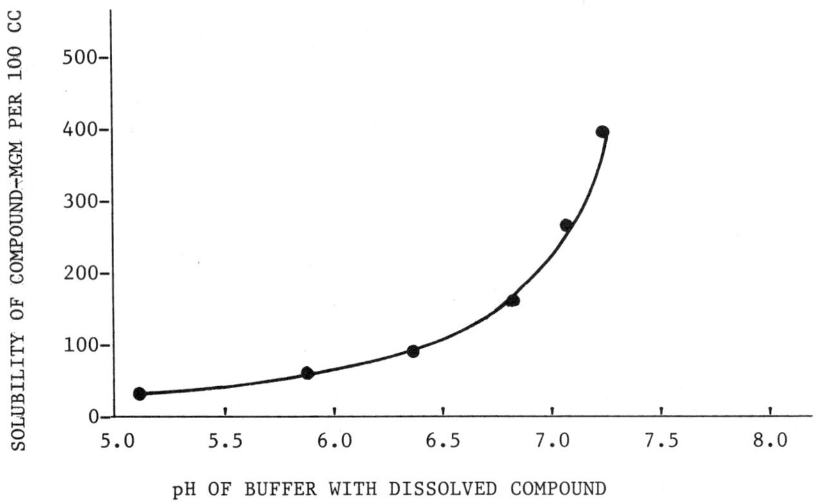

AUXILIARY INFORMATION

METHOD/APPARATUS/PROCEDURE:

An amt of acetyl sulfadiazine in large excess of that to be dissolved was added to M/15 phosphate buffer solns, shaken for 18 h in a water bath at 37°C and filtered at the same temp. The pH of the filtrate was measured immediately with a Beckmann pH meter and appropraite temp corrections were made. Acetyl sulfadiazine was assayed by the method of Bratton and Marshall (1).

SOURCE AND PURITY OF MATERIALS:

Acetyl sulfadiazine, (purity not specified) was supplied by Lederle Labs, Inc. The source and purity of the remaining materials were not specified.

ESTIMATED ERROR:

Nothing specified.

REFERENCES:

1. Bratton, A. C.; Marshall, E. K., Jr. *J. Biol. Chem.* 1939, *128*, 537.

COMPONENTS:

(1) Acetamide, N-[4-[(2-pyrimidinylamino)-sulfonyl]phenyl]- (acetyl sulfadiazine); $C_{12}H_{12}N_4O_3S$; [127-74-2]

(2) Phosphoric acid, disodium salt; Na_2HPO_4; [7558-94-4]

(3) Phosphoric acid, monopotassium salt; KH_2PO_4; [7778-77-0]

(4) Water; H_2O; [7732-18-5]

ORIGINAL MEASUREMENTS:

Gilligan, D. R.; Plummer, M. N.
Proc. Soc. Exp. Biol. Med. 1943, *53*, 142-5.

VARIABLES:

pH

PREPARED BY:

R. Piekos

EXPERIMENTAL VALUES:

Solubility of acetyl sulfadiazine in M/15 phosphate buffer solutions at 37°C

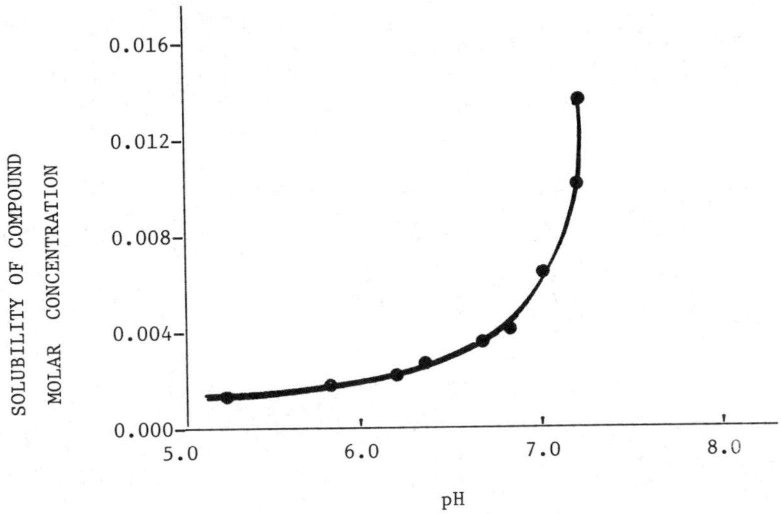

AUXILIARY INFORMATION

METHOD/APPARATUS/PROCEDURE:

An excess of acetyl sulfadiazine was shaken in M/15 phosphate buffer solns of various pH values for 18 h in a water bath at 37°C, and filtered in an incubator room at this temp. The pH of the filtrate was measured immediately at room temp with a Beckmann glass electrode pH meter and appropriate corrections for the differences between room temp and 37°C were applied. The amt of dissolved compd was measured by the method of Bratton and Marshall (1).

SOURCE AND PURITY OF MATERIALS:

Acetyl sulfadiazine was supplied by Lederle Labs, Inc. The source and purity of the remaining materials were not specified.

ESTIMATED ERROR:

Nothing specified.

REFERENCES:

1. Bratton, A. C.; Marshall, E. K., Jr. *J. Biol. Chem.* 1939, *128*, 537.

COMPONENTS:

(1) Acetamide, N-[4-[(2-pyrimidinylamino)-sulfonyl]phenyl]- (acetyl sulfadiazine); $C_{12}H_{12}N_4O_3S$; [127-74-2]

(2) Phosphoric acid, disodium salt; Na_2HPO_4 [7558-94-4]

(3) Phosphoric acid, monopotassium salt; KH_2PO_4; [7778-77-0]

(4) Water; H_2O; [7732-18-5]

ORIGINAL MEASUREMENTS:

Pulver, R.; Suter, R.
Schweiz. Med. Wochenschr. 1943, *73(13)*, 403-8.

PREPARED BY:

R. Piekos

EXPERIMENTAL VALUES:

pH	Solubility of acetyl sulfadiazine in M/15 phosphate buffers (according to Sørensen) at 20°C	
	mg%	10^3 mol dm^{-3} [a]
6.0	23	0.79
7.0	72	2.46
8.0	208	7.12

[a] Calculated by compiler.

AUXILIARY INFORMATION

METHOD/APPARATUS/PROCEDURE:

Nothing specified.

SOURCE AND PURITY OF MATERIALS:

Nothing specified.

ESTIMATED ERROR:

Nothing specified.

REFERENCES:

COMPONENTS:

(1) Acetamide, N-[4-[(2-pyrimidinylamino)-sulfonyl]phenyl]- (acetyl sulfapyrimidine); $C_{12}H_{12}N_4O_3S$; [127-74-2]

(2) Phosphoric acid, disodium salt; Na_2HPO_4; [7558-94-4]

(3) Phosphoric acid, monopotassium salt; KH_2PO_4; [7778-77-0]

(4) Water; H_2O; [7732-18-5]

ORIGINAL MEASUREMENTS:

Frisk, A. R.; Hagerman, G.; Helander, S.; Sjögren, B.

Hygiea 1946, *108(12)*, 639-51.

VARIABLES:

One temperature: 37°C; one pH: 6.1

PREPARED BY:

R. Piekos

EXPERIMENTAL VALUES:

Solubility of acetyl sulfapyrimidine in M/30 phosphate buffer of pH 6.1 at 37°C is 35 mg/100 ml solvent (1.2 x 10^{-3} mol dm^{-3}, compiler).

AUXILIARY INFORMATION

METHOD/APPARATUS/PROCEDURE:

An excess of acetyl sulfapyrimidine in the phosphate buffer was shaken at 37°C for 24 h. The concn of acetyl sulfapyrimidine was detd by the Bratton and Marshall method (1) using a photoelec colorimeter.

SOURCE AND PURITY OF MATERIALS:

Neither source nor purity of the materials was specified.

ESTIMATED ERROR:

Soly: precision ±4 mg/100 ml (authors).

Temp and pH: not specified.

REFERENCES:

1. Bratton, A. C.; Marshall, E. K., Jr. *J. Biol. Chem.* 1939, *128*, 537.

COMPONENTS:

(1) Acetamide, N-[4-[(2-pyrimidinylamino)-sulfonyl]phenyl]- (N^4-acetylsulfadiazine); $C_{12}H_{12}N_4O_3S$; [127-74-2]

(2) Phosphoric acid, disodium salt; Na_2HPO_4; [7558-94-4]

(3) Phosphoric acid, monopotassium salt; KH_2PO_4; [7778-77-0]

(4) Water; H_2O; [7732-18-5]

ORIGINAL MEASUREMENTS:

Hekster, Y. A.; Vree, T. B.; Damsma, J. E.; Friesen; W. T.
J. Antimicrob. Chemother. 1981, *8*, 133-44.

VARIABLES:

pH

PREPARED BY:

R. Piekos

EXPERIMENTAL VALUES:

pH	Solubility at 25°C	
	mg/l	10^3 mol dm^{-3} [a]
5.5	411	1.41
7.5	1620	5.54

[a] Calculated by compiler

AUXILIARY INFORMATION

METHOD/APPARATUS/PROCEDURE:

Satd solns of N^4-acetylsulfadiazine were prepd in phosphate buffers of pH 5.5 and 7.5 at room temp (25°C). The conc of the solute was measured by means of a Spectra Physics 3500B high-performance liquid chromatograph equipped with a column oven (Model 748) and a Pye-Unicam LC-UV spectrophotometric detector. The detector was connected to a 1-mV recorder. A stainless steel column (10 cm x 4.6 mm i.d.) was packed with Lichrosorb RPS, 5 μm, obtained from Chrompack. An injection loop of 100 μl was used. The oven temp was 40°C. Detection of the solute was performed at 260 nm.

SOURCE AND PURITY OF MATERIALS:

The source and purity of the materials were not specified.

ESTIMATED ERROR:

The detection limit of the solute by HPLC was 0.5 mg/l (authors). The error in temperature and pH was not specified.

REFERENCES:

COMPONENTS:	EVALUATOR:
(1) Benzenesulfonamide, 4-amino-N-(4-methyl-2-pyrimidinyl)- (sulfamerazine) $C_{11}H_{12}N_4O_2S$; [127-79-7] (2) Water	Anthony N. Paruta Department of Pharmaceutics University of Rhode Island Kingston, Rhode Island, USA and Ryszard Piekos Faculty of Pharmacy, University of Gdansk Gdansk, Poland 1986

CRITICAL EVALUATION:

Six reports (1-6_) on the solubility of sulfamerazine in water at 303K and 310K are in Table I.

Table I: Solubility of Sulfamerazine in water, 303K and 310K

	10^3 mol dm^{-3} (*indicates mol kg^{-1})	
Reference	303K	310K
1	-	1.20
2	-	2.50
3	-	0.965*
4	-	1.4
5	0.89	-
6	0.91	-

The two values given for 303K are very close and self consistant with the data at 310K (1-4), thus the recommended value is given as 9 x 10^{-4} mol dm^{-3}. At 310K, the values of Kikuth (2) and Sapoznikova et al. (3) were not considered further, because the equilibration time was insufficient or unreported. The results given by Roblin et al. (1) and Langecker (4) were derived for 24 hour equilibrium periods. The recommended value at 310K for sulfamerazine is 1.3 x 10^{-3} mol dm^{-3}.

REFERENCES:

(1) Roblin, R.O., Jr.; Williams, J.H.; Winnek, P.S.; English, J.P. *J. Am. Chem. Soc.* 1940, *62*, 2002-5

(2) Kikuth, W. *Med. Welt* 1943, *17(26/27)*, 483-6.

(3) Sapoznikova, N.V.; Postovakii, I. Ya. *Zh. Prikl. Khim.* 1944, *17*, 427-34.

(4) Langecker, H. *Arch. Exptl. Path. Pharmakol.* 1948, *205*, 291-301.

(5) Yamasaki, M.; Aoki, M.; Kamada, A.; Yata, N. *Yakusaigaku* 1967, *27(1)*, 37-40.

(6) Bhattacharyya, R.; Basu, U.P. *Indian Pharmacist* 1950, *6(3)* 77-8,86.

COMPONENTS:

(1) Benzenesulfonamide, 4-amino-N-(4-methyl-2-pyrimidinyl)- (sulfamerazine); $C_{11}H_{12}N_4O_2S$; [127-79-7]

(2) Water; H_2O; [7732-18-5]

ORIGINAL MEASUREMENTS:

Roblin, R. O., Jr.; Williams, J. H.; Winnek, P. S.; English, J. P.

J. Am. Chem. Soc. 1940, *62*, 2002-5.

VARIABLES:

One temperature: 37°C

PREPARED BY:

R. Piekos

EXPERIMENTAL VALUES:

Solubility of sulfamerazine in water at 37°C is 31.8 mg/100 cm^3 solution (1.20×10^{-3} mol dm^{-3}, compiler).

AUXILIARY INFORMATION

METHOD/APPARATUS/PROCEDURE:

Excess sulfonamide in water was heated and stirred on a steam bath for 30 min. The suspension was then agitated for 24 h in a thermostat at 37°C. A sample of the satd soln was withdrawn through a glass filter, dild, and analyzed by the Marshall method (1) using a General Electric recording spectrophotometer for comparing the colors developed with those of the standards.

SOURCE AND PURITY OF MATERIALS:

The sulfonamide, mp 235-6°C (dec, cor), was prepd by the authors. Anal: %C 49.6 (calcd 50.0); %H 4.4 (4.6); %N 21.1 (21.2). Purity of the water was not specified.

ESTIMATED ERROR:

Nothing specified.

REFERENCES:

1. Bratton, A. C.; Marshall, E. K., Jr. *J. Pharmacol.* 1939, *66*, 4.

COMPONENTS:

(1) Benzenesulfonamide, 4-amino-N-(4-methyl-2-pyrimidinyl)- (sulfamerazine); $C_{11}H_{12}N_4O_2S$; [127-79-7]

(2) Water; H_2O; [7732-18-5]

ORIGINAL MEASUREMENTS:

Kikuth, W.

Med. Welt. 1943, *17(26/27)*, 483-6.

VARIABLES:

One temperature: 37°C

PREPARED BY:

R. Piekos

EXPERIMENTAL VALUES:

Solubility of sulfamerazine in water at 37°C is 66.0 mg/100^3 solution (2.50×10^{-3} mol dm^{-3}).

AUXILIARY INFORMATION

METHOD/APPARATUS/PROCEDURE:

Nothing specified.

SOURCE AND PURITY OF MATERIALS:

Sulfamerazine: not specified.

The pH of the water was 7.0.

ESTIMATED ERROR:

Nothing specified.

REFERENCES:

COMPONENTS:	ORIGINAL MEASUREMENTS:
(1) Benzenesulfonamide, 4-amino-N-(4-methyl-2-pyrimidinyl)- (sulfamerazine); $C_{11}H_{12}N_4O_2S$; [127-79-7] (2) Water; H_2O; [7732-18-5]	Sapozhnikova, N. V; Postovskii, I. Ya. *Zh. Prikl. Khim.* 1944, *17*, 427-34.
VARIABLES: Temperature	PREPARED BY: R. Piekos

EXPERIMENTAL VALUES:

t/°C	Solubility	
	Weight%	10^3 mol kg^{-1} water[a]
37	0.0255[b]	0.965
50	0.050	1.89
75	0.172	6.52
99	0.319	12.1

[a] Calculated by compiler

[b] Calculated from the heat of dissolution (10,300 cal mol^{-1})

AUXILIARY INFORMATION

METHOD/APPARATUS/PROCEDURE:

Sulfamerazine was dissolved in water to form a satd soln which was occasionally agitated in a glass vessel immersed in a thermostat. The equilibrium was usually attained after 1 h. Five-to 100-cm^3 samples of the satd soln were placed in Pt crucibles or dishes and evapd to dryness at temps lower than 110-115°C. The residue was dried to const wt at 105-110°C and weighed.

SOURCE AND PURITY OF MATERIALS:

Pure, recrystd sulfamerazine was used. Its mp conformed to that reported in the literature.

Purity of the water was not specified.

ESTIMATED ERROR:

Soly: quite reliable results were obtained at 37, 50, and 75°C. At 99°C the accuracy was poor due to evapn of water during sampling (authors). Temp: ±0.05°C (authors).

REFERENCES:

COMPONENTS:

(1) Benzenesulfonamide, 4-amino-N-(4-methyl-2-pyrimidinyl)- (sulfamerazine); $C_{11}H_{12}N_4O_2S$; [127-79-7]

(2) Water; H_2O; [7732-18-5]

ORIGINAL MEASUREMENTS:

Langecker, H.

Arch. Exptl. Path. Pharmakol. 1948, *205*, 291-301.

VARIABLES:

Temperature

PREPARED BY:

R. Piekos

EXPERIMENTAL VALUES:

t/°C	Solubility	
	mg%	10^4 mol dm^{-3} [a]
25	21	7.9
37	37	14

[a] Calculated by compiler.

AUXILIARY INFORMATION

METHOD/APPARATUS/PROCEDURE:

An excess of sulfamerazine was boiled with water and left for 24 h in a vessel protected from access of CO_2. The sulfamerazine concn was detd colorimetrically by the method of Bratton and Marshall (1) using a Havemann colorimeter (2), as well as by microanal detn of the solid residue.

SOURCE AND PURITY OF MATERIALS:

Source and purity of the materials were not specified.

ESTIMATED ERROR:

Nothing specified.

REFERENCES:

1. Bratton, A. G.; Marshall, E. K., Jr. *J. Biol. Chem.* 1939, *128*, 537.
2. Havemann, R. *Klin. Wochenschr.* 1940, p. 503.

COMPONENTS:

(1) Benzenesulfonamide, 4-amino-N-(4-methyl-2-pyrimidinyl)- (sulfamerazine); $C_{11}H_{12}N_4O_2S$; [127-79-7]

(2) Water; H_2O; [7732-18-5]

ORIGINAL MEASUREMENTS:

Bhattacharyya, R.; Basu, U. P.

Indian Pharmacist 1950, *6(3)*, 77-8, 86.

VARIABLES:

One temperature: 30°C

PREPARED BY:

R. Piekos

EXPERIMENTAL VALUES:

Solubility of sulfamerazine in water at 30°C is 24 mg per 100 ml (9.1×10^{-4} mol dm^{-3}, compiler).

AUXILIARY INFORMATION

METHOD/APPARATUS/PROCEDURE:

A weighed sample of sulfamerazine was placed in a clean reagent bottle and a known vol of water was added. The mixt was shaken in a mech shaker at 80-100 strokes/min. After at least 24 h the mixt was filtered through a clean, dried and weighed sintered-glass crucible. At the end of the filtration the crucible was washed with about 1 ml of water, dried at 105°C for 2-3 h, cooled, and weighed to const wt.

SOURCE AND PURITY OF MATERIALS:

Neither source nor purity of the sulfamerazine was specified.

Doubly distd water was used.

ESTIMATED ERROR:

Soly: not specified.

Temp: ±0.2°C (authors).

REFERENCES:

COMPONENTS:

(1) Benzenesulfonamide, 4-amino-N-(4-methyl-2-pyrimidinyl)- (sulfamerazine); $C_{11}H_{12}N_4O_2S$; [127-79-7]

(2) Water; H_2O; [7732-18-5]

ORIGINAL MEASUREMENTS:

Yamazaki, M.; Aoki, M.; Kamada, A.; Yata, N. *Yakuzaigaku* 1967, *27(1)*, 37-40.

VARIABLES:

One temperature: 30°C

PREPARED BY:

R. Piekos

EXPERIMENTAL VALUES:

Solubility of sulfamerazine in water at 30°C is 0.89 mmol/L (0.24 g dm^{-3}, compiler).

AUXILIARY INFORMATION

METHOD/APPARATUS/PROCEDURE:

Sulfamerazine (0.5 g) was placed in an L-shaped tube together with 20 ml of water. The mixt was shaken in a thermostat until equilibrium was attained. The sulfamerazine was assayed in the supernatant spectrophotometrically at 545 nm on a Beckmann DU spectrophotometer. The results were taken from a calibration graph.

SOURCE AND PURITY OF MATERIALS:

Nothing specified.

ESTIMATED ERROR:

Soly: not specified.

Temp: ±1°C (authors).

REFERENCES:

COMPONENTS:

(1) Benzenesulfonamide, 4-amino-N-(4-methyl-2-pyrimidinyl)- (sulfamerazine); $C_{11}H_{12}N_4O_2S$; [127-79-7]

(2) Sodium hydroxide; NaOH; [1310-73-2]

(3) Water; H_2O; [7732-18-5]

ORIGINAL MEASUREMENTS:

Holz, E.; Garcia Onandia, T.; Holz, S. *Acta Cient. Venezolana* 1955, *6(2)*, 68-73.

VARIABLES:

Concentration of NaOH

PREPARED BY:

R. Piekos

EXPERIMENTAL VALUES:

Concentration of NaOH soln	Volume of the NaOH soln required to dissolve 1 g of sulfamerazine at 26°C	Solubility of sulfamerazine at 26°C
N	cm^3	mol dm^{-3} NaOH soln[a]
1/10	40.9	0.0925
1/4	15.8	0.239
1/2	8.0	0.473
1.00	4.0	0.946
1.25	3.2	1.18
1.30	3.3	1.15
1.40	8.6	0.440
1.50	13.2	0.287
2.00	36.4	0.104
2.50	156	0.0242

[a] Calculated by compiler

AUXILIARY INFORMATION

METHOD/APPARATUS/PROCEDURE:

Nothing specified.

SOURCE AND PURITY OF MATERIALS:

Nothing specified. Distd water was used.

ESTIMATED ERROR:

Nothing specified.

REFERENCES:

COMPONENTS:

(1) Benzenesulfonamide, 4-amino-N-(4-methyl-2-pyrimidinyl)- (sulfamerazine); $C_{11}H_{12}N_4O_2S$; [127-79-7]

(2) Phosphoric acid, disodium salt; Na_2HPO_4; [7558-94-4]

(3) Phosphoric acid, monopotassium salt; KH_2PO_4; [7778-77-0]

(4) Water; H_2O; [7732-18-5]

ORIGINAL MEASUREMENTS:

Gilligan, D. R.; Plummer, M. N.

Proc. Soc. Exp. Biol. Med. 1943, *53*, 142-5.

VARIABLES:

pH

PREPARED BY:

R. Piekos

EXPERIMENTAL VALUES:

Solubility of sulfamerazine in M/15 phosphate buffer solutions at 37°C

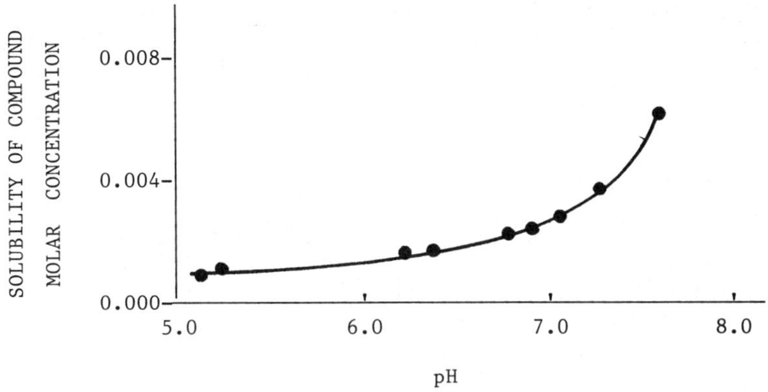

AUXILIARY INFORMATION

METHOD/APPARATUS/PROCEDURE:

An excess of sulfamerazine was shaken in M/15 phosphate buffer solns of various pH values for 18 h in a water bath at 37°C, and filtered in an incubator room at this temp. The pH of the filtrate was measured immediately at room temp with a Beckmann glass electrode pH meter and appropriate corrections for the differences between room temp and 37°C were applied. The amt of dissolved compd was measured by the method of Bratton and Marshall (1).

SOURCE AND PURITY OF MATERIALS:

Sulfamerazine was supplied by Lederle Labs, Inc. The source and purity of the remaining materials were not specified.

ESTIMATED ERROR:

Nothing specifed.

REFERENCES:

1. Bratton, A. C.; Marshall, E. K., Jr. *J. Biol. Chem.* 1939, *128*, 537.

COMPONENTS:	ORIGINAL MEASUREMENTS:
(1) Benzenesulfonamide, 4-amino-N-(4-methyl-2-pyrimidinyl)- (sulfamerazine); $C_{11}H_{12}N_4O_2S$; [127-79-7] (2) Phosphoric acid, disodium salt; Na_2HPO_4; [7558-94-4] (3) Phosphoric acid, monopotassium salt; KH_2PO_4; [7778-77-0] (4) Water; H_2O; [7732-18-5]	Frisk, A. R.; Hagerman, G.; Helander, S.; Sjögren, B. *Hygiea* <u>1946</u>, *108(12)*, 639-51.
VARIABLES:	PREPARED BY:
One temperature: 37°C; pH	R. Piekos

EXPERIMENTAL VALUES:

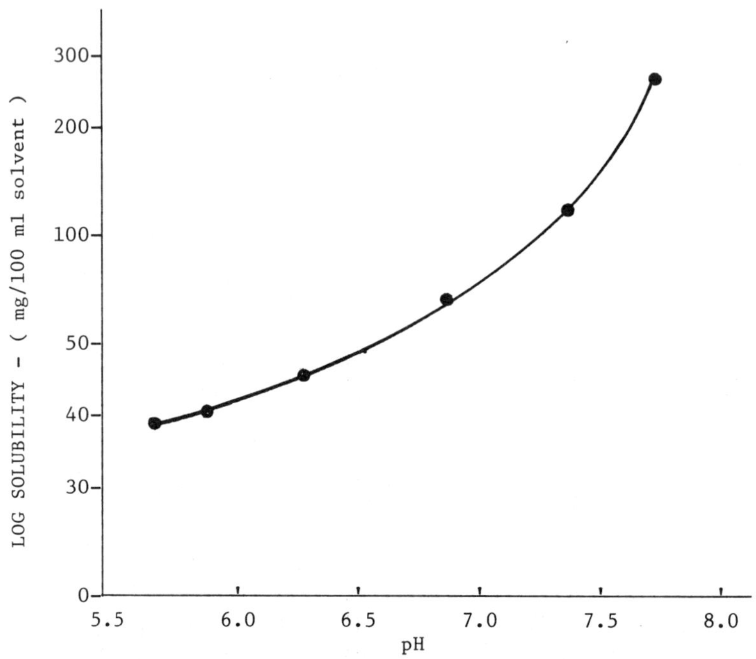

The solubility at pH 6.1 is 40 mg/100 ml solvent (1.5 x 10^{-3} mol dm^{-3}, compiler).

AUXILIARY INFORMATION

METHOD/APPARATUS/PROCEDURE:

An excess of sulfamethylpyrimidine in the phosphate buffer was shaken at 37°C for 24 h. The concn of the solute was detd by the Bratton and Marshall method (1) using a photoelec colorimeter.

SOURCE AND PURITY OF MATERIALS:

Neither source nor purity of the materials was specified.

ESTIMATED ERROR:

Soly: precision ±3 mg/100 ml (authors).
Temp and pH: not specified.

REFERENCES:

1. Bratton, A. C.; Marshall, E. K., Jr. *J. Biol. Chem.* <u>1939</u>, *128*, 537.

COMPONENTS:

(1) Benzenesulfonamide, 4-amino-N-(4-methyl-2-pyrimidinyl)- (sulfamerazine); $C_{11}H_{12}N_4O_2S$; [127-79-7]

(2) Phosphoric acid, disodium salt; Na_2HPO_4; [7558-94-4]

(3) Phosphoric acid, monopotassium salt; KH_2PO_4; [7778-77-0]

(4) Water; H_2O; [7732-18-5]

ORIGINAL MEASUREMENTS:

Langecker, H.

Arch. Exptl. Path. Pharmakol. 1948, *205*, 291-301.

VARIABLES:

pH

PREPARED BY:

R. Piekos

EXPERIMENTAL VALUES:

pH of the 1/15M phosphate buffer	Solubility at 37°C	
	mg%	10^3 mol dm^{-3} [a]
4.9	34	1.2
5.9	35	1.3
6.9	86	3.2
7.5	144	5.4

[a] Calculated by compiler

AUXILIARY INFORMATION

METHOD/APPARATUS/PROCEDURE:

An excess of sulfamerazine was added to a buffer soln and boiled for 1 h in a sealed ampul followed by keeping the ampul at 37°C. The concn of sulfamerazine was detd colorimetrically by the method of Bratton and Marshall (1) using a Havemann colorimeter (2), as well as by microanal detn of the solid residue.

SOURCE AND PURITY OF MATERIALS:

Source and purity of the materials were not specified.

ESTIMATED ERROR:

Nothing specified.

REFERENCES:

1. Bratton, A. C.; Marshall, E. K., Jr. *J. Biol. Chem.* 1939, *128*, 537.
2. Havemann, R. *Klin. Wochenschr.* 1940, p. 503.

COMPONENTS:	ORIGINAL MEASUREMENTS:
(1) Benzenesulfonamide, 4-amino-N-(4-methyl-2-pyrimidinyl)- (sulfamerazine)- $C_{11}H_{12}N_4O_2S$; [127-79-7] (2) Phosphoric acid, disodium salt; Na_2HPO_4; [7558-94-4] (3) Phosphoric acid, monopotassium salt; KH_2PO_4; [7778-77-0] (4) Water; H_2O; [7732-18-5]	Riess, W. *Intern. Congr. Chemotherapy, Proc., 3rd, Stuttgart* <u>1963,</u> *1,* 627-32.
VARIABLES: One temperature: 20°C; one pH: 7.4	PREPARED BY: R. Piekos

EXPERIMENTAL VALUES:

Solubility of sulfamerazine in M/15 Sörensen buffer solution (pH 7.4) at 20°C is 70 mg% (2.7×10^{-3} mol dm^{-3} solution, compiler).

AUXILIARY INFORMATION

METHOD/APPARATUS/PROCEDURE:	SOURCE AND PURITY OF MATERIALS:
Sörensen buffer solns of varying pH (7-8) were prepd, satd with sulfamerazine at 20°C, their pH was measured at equilibrium, and the sulfamerazine was assayed colorimetrically. The measured pH values were then plotted against concn, and the soly at pH 7.4 was detd by interpolation (personal communication).	Nothing specified. ESTIMATED ERROR: Nothing specified. REFERENCES:

COMPONENTS:	ORIGINAL MEASUREMENTS:
(1) Benzenesulfonamide, 4-amino-N-(4-methyl-2-pyrimidinyl)- (sulfamerazine); $C_{11}H_{12}N_4O_2S$; [127-79-7] (2) Phosphoric acid, disodium salt; Na_2HPO_4; [7558-94-4] (3) Phosphoric acid, monopotassium salt; KH_2PO_4; [7778-77-0] (4) Water; H_2O; [7732-18-5]	Yamazaki, M.; Aoki, M.; Kamada, A.; Yata, N. *Yakuzaigaku* 1967, *27(1)*, 37-40.
VARIABLES:	PREPARED BY:
One temperature: 30°C; one pH: 7.4	R. Piekos

EXPERIMENTAL VALUES:

Solubility of sulfamerazine in a phosphate buffer solution of pH 7.4 (μ= 0.17) at 30°C is 2.45 mmol/L (0.647 g dm^{-3}, compiler).

AUXILIARY INFORMATION

METHOD/APPARATUS/PROCEDURE:

Sulfamerazine (0,5 g) was placed in an L-shaped tube together with 20 ml of the buffer soln. The mixt was shaken in a thermostat until equilibrium was attained. The sulfamerazine was assayed in the supernatant spectrophotometrically at 545 nm on a Beckmann DU spectrophotometer. The results were taken from a calibration graph.

SOURCE AND PURITY OF MATERIALS:

Nothing specified.

ESTIMATED ERROR:

Soly and pH: not specified.

Temp: ±1°C (authors).

REFERENCES:

COMPONENTS:

(1) Benzenesulfonamide, 4-amino-N-(4-methyl-2-pyrimidinyl)- (sulfamerazine); $C_{11}H_{12}N_4O_2S$; [127-79-7]

(2) Phosphoric acid, disodium salt; Na_2HPO_4; [7558-94-4]

(3) Phosphoric acid, monopotassium salt; KH_2PO_4; [7778-77-0]

(4) Water; H_2O; [7732-18-5]

ORIGINAL MEASUREMENTS:

Hekster, Ch. A.; Vree, T. B.

Antibiotics Chemother. 1982, *31*, 22-118.

VARIABLES:

pH

PREPARED BY:

R. Piekos

EXPERIMENTAL VALUES:

pH	Solubility at 25°C	
	mg/l	10^4 mol dm^{-3} [a]
5.5	238	9.00
7.5[b]	840	31.8

[a] Calculated by compiler.

[b] Erroneous pH value of 7.0 is given in the article.

AUXILIARY INFORMATION

METHOD/APPARATUS/PROCEDURE:

The earlier developed method (1) was used (personal communication). Satd solns of sulfamerazine were prepd in phosphate buffers of pH 5.5 and 7.5 at 25°C. The concn of the solute was measured by means of a Spectra Physics 3500B high-performance liquid chromatograph equipped with a Model 748 column oven and a Pye-Unicam LC-UV spectrophotometric detector.

SOURCE AND PURITY OF MATERIALS:

Neither source nor the purity of the materials was specified.

ESTIMATED ERROR:

Soly: the detection limit of the solute by HPLC was 0.5 mg/l (authors).

The errors in temp and pH were not specified.

REFERENCES:

1. Hekster, Y. A.; Vree, T. B.; Damsma, J. E.; Friesen, W. T. *J. Antimicrob. Chemother* 1981, *8*, 133.

COMPONENTS:

(1) Benzenesulfonamide, 4-amino-N-(4-methyl-2-pyrimidinyl)- (sulfamerazine); $C_{11}H_{12}N_4O_2S$; [127-79-7]

(2) 1,2,3-Propanetricarboxylic acid, 2-hydroxy-, disodium salt (Na citrate); $C_6H_6Na_2O_7$; [144-33-2]

(3) Sodium hydroxide; NaOH; [1310-73-2]

(4) Water; H_2O; [7732-18-5]

ORIGINAL MEASUREMENTS:

Gilligan, D. R.; Plummer, M. N.

Proc. Soc. Exp. Biol. Med. 1943, *53*, 142-5.

VARIABLES:

pH

PREPARED BY:

R. Piekos

EXPERIMENTAL VALUES:

Solubility of sulfamerazine in M/10 Na citrate + NaOH solution at 37°C

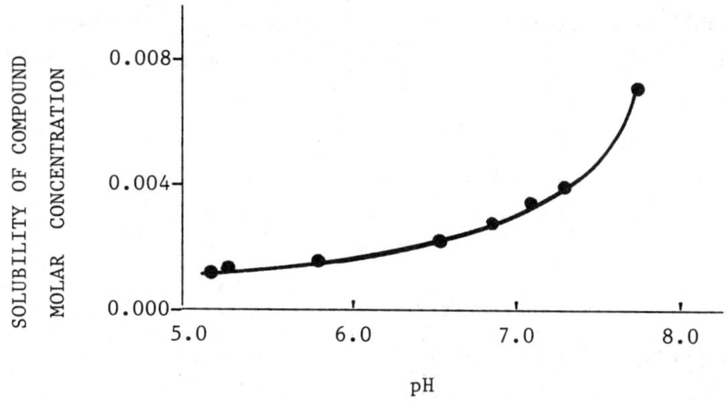

AUXILIARY INFORMATION

METHOD/APPARATUS/PROCEDURE:

An excess of sulfamerazine was shaken in M/10 Na citrate + NaOH solns of various pH values for 18 h in a water bath at 37°C, and filtered in an incubator at this temp. The pH of the filtrate was measured immediately at room temp with a Beckmann glass electrode pH meter and appropriate corrections for the differences between room temp and 37°C were applied. The amt of dissolved compd was measured by the method of Bratton and Marshall (1).

SOURCE AND PURITY OF MATERIALS:

Sulfamerazine was supplied by Lederle Labs, Inc. The source and purity of the remaining materials were not specified.

ESTIMATED ERROR:

Nothing specified.

REFERENCES:

1. Bratton, A. C.; Marshall, E. K., Jr. *J. Biol. Chem.* 1939, *128*, 537.

COMPONENTS:

(1) Benzenesulfonamide, 4-amino-N-(4-methyl-2-pyrimidinyl)- (sulfamerazine); $C_{11}H_{12}N_4O_2S$; [127-79-7]

(2) Benzene, methyl- (toluene); C_7H_8; [108-88-3]

(3) Mannitol; $C_6H_{14}O_6$; [87-78-5]

(4) Phosphoric acid, disodium salt; Na_2HPO_4; [7558-94-4]

(5) Phosphoric acid, monopotassium salt; KH_2PO_4; [7778-77-0]

(6) Sodium chloride; NaCl; [7647-14-5]

(7) Water; H_2O; [7732-18-5]

ORIGINAL MEASUREMENTS:

Sonnenberg, H.; Oelert, H.; Baumann, K. *Pflügers Arch. Ges. Physiol.* 1965, *286*, 171-80.

VARIABLES:

pH

PREPARED BY:

R. Piekos

EXPERIMENTAL VALUES:

Relative lipoid solubility determined on the basis of concentration measurements of sulfamerazine in perfusates[a,b] before (c_i) and after (c_e) equilibration with toluene

pH	$(100 - \frac{100\ c_e}{c_i})$
5[a]	8
8[b]	2

[a]Composition of perfusate: 110 mmol/l NaCl, 35 mmol/l mannitol in a phosphate buffer consisting of 98.8 ml of 0.022M KH_2PO_4 and 1.2 of 0.022M Na_2HPO_4.

[b]Composition of perfusate: 68 mmol/l NaCl, 100 mmol/l mannitol in a phosphate buffer consisting of 5.5 ml of 0.022M KH_2PO_4 and 94.5 ml of 0.022M Na_2HPO_4.

AUXILIARY INFORMATION

METHOD/APPARATUS/PROCEDURE:

Lipoid solubilities were detd by shaking equal volumes of the perfusate and toluene for 20 min and measuring the concn of sulfamerazine by the spectrophotometric method of Bratton and Marshall (1) in an aq phase before and after this procedure.

SOURCE AND PURITY OF MATERIALS:

None given.

ESTIMATED ERROR:

None given.

REFERENCES:

1. Bratton, A. C.; Marshall, E. K., Jr. *J. Biol. Chem.* 1939, *128*, 537.

COMPONENTS:

(1) Benzenesulfonamide, 4-amino-N-(4-methyl-2-pyrimidinyl)- (sulfamerazine); $C_{11}H_{12}N_4O_2S$; [127-79-7]

(2) Mannitol; $C_6H_{14}O_6$; [87-78-5]

(3) Methane, trichloro- (chloroform); $CHCl_3$; [67-66-3]

(4) Phosphoric acid, disodium salt; Na_2HPO_4; [7558-94-4]

(5) Phosphoric acid, monopotassium salt; KH_2PO_4; [7778-77-0]

(6) Sodium chloride; NaCl; [7647-14-5]

(7) Water; H_2O; [7732-18-5]

ORIGINAL MEASUREMENTS:

Sonnenberg, H.; Oelert, H.; Baumann, K. *Pflügers Arch. Ges. Physiol.* 1965, *286*, 171-80.

VARIABLES: pH

PREPARED BY: R. Piekos

EXPERIMENTAL VALUES:

Relative lipoid solubility determined on the basis of concentration measurements of sulfamerazine in perfusates[a,b] before (c_i) and after (c_e) equilibration with chloroform

pH	$(100 - \frac{100\ c_e}{c_i})$
5[a]	72
8[b]	34

[a]Composition of perfusate: 110 mmol/l NaCl, 35 mmol/l mannitol in a phosphate buffer consisting of 98.8 ml of 0.022M KH_2PO_4 and 1.2 ml of 0.022M Na_2HPO_4.

[b]Composition of perfusate: 68 mmol/l NaCl, 100 mmol/l mannitol in a phosphate buffer consisting of 5.5 ml of 0.022M KH_2PO_4 and 94.5 ml of 0.022M Na_2HPO_4.

AUXILIARY INFORMATION

METHOD/APPARATUS/PROCEDURE:

Lipoid solubilities were detd by shaking equal volumes of the perfusate and $CHCl_3$ for 20 min and measuring the concn of sulfamerazine by the spectrophotometric method of Bratton and Marshall (1) in the aq phase before and after this procedure.

SOURCE AND PURITY OF MATERIALS:

None given.

ESTIMATED ERROR:

None given.

REFERENCES:

1. Bratton, A. C.; Marshall, E. K., Jr. *J. Biol. Chem.* 1939, *128*, 537.

COMPONENTS:

(1) Benzenesulfonamide, 4-amino-N-(4-methyl-2-pyrimidinyl)- (sulfamerazine); $C_{11}H_{12}N_4O_2S$; [127-79-7]

(2) Ethanol; C_2H_6O; [64-17-5]

(3) 1,2,3-Propanetriol; $C_3H_8O_3$; [56-81-5]

(4) Water; H_2O; [7732-18-5]

ORIGINAL MEASUREMENTS:

Dolique, R.; Foucault, J.

Trav. soc. pharm. Montpellier 1952, *12*, 145-53.

VARIABLES:

One temperature: 26-28°C

PREPARED BY:

R. Piekos

EXPERIMENTAL VALUES:

Solubility of sulfamerazine in a mixture of 1,2,3-propanetriol and 95° ethanol (2:1 by wt) at 26-28°C is 0.45% (1.7 x 10^{-2} mol kg^{-1} solvent, compiler).

AUXILIARY INFORMATION

METHOD/APPARATUS/PROCEDURE:

The sulfamerazine content was detd by diazotization of the amine group in a cold acidified 0.1N KNO_2 soln. An excess of KNO_2 was detected by using iodinated starch.

SOURCE AND PURITY OF MATERIALS:

Nothing specified.

ESTIMATED ERROR:

Nothing specified.

REFERENCES:

COMPONENTS:	ORIGINAL MEASUREMENTS:
(1) Benzenesulfonamide, 4-amino-N-(4-methyl-2-pyrimidinyl)- (sulfamerazine); $C_{11}H_{12}N_4O_2S$; [127-79-7] (2) Ethanol; C_2H_6O; [64-17-5] (3) 1,2,3-Propanetriol; $C_3H_8O_3$; [56-81-5] (4) Urea; CH_4N_2O; [57-13-6] (5) Water; H_2O; [7732-18-5]	Dolique, R.; Foucault, J. *Trav. soc. pharm. Montpellier* <u>1952</u>, *12*, 145-53.
VARIABLES:	PREPARED BY:
One temperature: 26-28°C	R. Piekos

EXPERIMENTAL VALUES:

Solubility of sulfamerazine at 26-28°C in a saturated solution of urea in a mixture of 1,2,3-propanetriol and 95° ethanol (2:1 by wt), containing 54.5 g of urea per 100 g of the mixture, is 0.465% (1.76×10^{-2} mol kg^{-1}, compiler).

AUXILIARY INFORMATION

METHOD/APPARATUS/PROCEDURE:	SOURCE AND PURITY OF MATERIALS:
The sulfamerazine content was detd by diazotization of the amine group in a cold acidified 0.1N KNO_2 soln. An excess of KNO_2 was detected by using iodinated starch.	Nothing specified.
	ESTIMATED ERROR:
	Nothing specified.
	REFERENCES:

COMPONENTS:

(1) Benzenesulfonamide, 4-amino-N-(4-methyl-2-pyrimidinyl)- (sulfamerazine); $C_{11}H_{12}N_4O_2S$; [127-79-7]

(2) Ethanol; C_2H_6O; [64-17-5]

ORIGINAL MEASUREMENTS:

Sekikawa, H.; Nakano, M.; Arita, T.
Chem. Pharm. Bull. 1978, *26(1)*, 118-26.

VARIABLES:

Temperature

PREPARED BY:

R. Piekos

EXPERIMENTAL VALUES:

t/°C	Solubility[a] 10^3 mol dm^{-3} solution
10	3.57
20	4.46
30	6.32
40	8.99
50	12.8

[a]Original data are presented graphically.
The numerical values are given by the authors.

AUXILIARY INFORMATION

METHOD/APPARATUS/PROCEDURE:

After attaining equilibrium, sample solns were removed by a syringe and filtered quickly through a membrane filter (pore size 0.2 μ) and sulfamerazine was assayed spectrophotometrically at 270 nm using a Hitachi Type 200-20 spectrophotometer.

SOURCE AND PURITY OF MATERIALS:

Sulfamerazine, mp 235°C, was obtained from Na sulfamerazine (Tanabe Seiyaku Co.) by addn of HCl_{aq} and recrystn from EtOH. Abs EtOH was obtained by drying and distn of EtOH following the conventional procedures.

ESTIMATED ERROR:

Nothing specified.

REFERENCES:

COMPONENTS:	ORIGINAL MEASUREMENTS:
(1) Benzenesulfonamide, 4-amino-N-(4-methyl-2-pyrimidinyl)- (sulfamerazine); $C_{11}H_{12}N_4O_2S$; [127-79-7] (2) 2-Propanol; C_3H_8O; [67-63-0]	Burlage, H. M. *J. Am. Pharm. Assoc., Sci. Ed.* 1948, *37*, 345.

VARIABLES:	PREPARED BY:
One temperature: 25°C	R. Piekos

EXPERIMENTAL VALUES:

Solubility of sulfamerazine in 2-propanol at 25°C is 0.1740 g/100 cm^3 solution (6.583×10^{-3} mol dm^{-3}, compiler).

AUXILIARY INFORMATION

METHOD/APPARATUS/PROCEDURE:

Satd solns of sulfamerazine in 2-propanol were prepd at 25°C and definite vols of the solns were measured into tared dishes by means of standard pipets. The alcohol was allowed to evap at room temp and the residue was dried at 105°C. In the case of losses due to apparent decompn, the residue was dried in a desiccator (1).

SOURCE AND PURITY OF MATERIALS:

The sulfamerazine was manufd by Sharp and Dohme and was of the U.S.P. purity. The source and purity of 2-propanol were not specified.

ESTIMATED ERROR:

Nothing specified.

REFERENCES:

1. Burlage, H. M. *J. Pharm. Assoc., Sci. Ed.* 1947, *36(1)*, 16.

COMPONENTS:	ORIGINAL MEASUREMENTS:
(1) Benzenesulfonamide, 4-amino-N-(4-methyl-2-pyrimidinyl)- (sulfamerazine); $C_{11}H_{12}N_4O_2S$; [127-79-7] (2) Ethanol, 2-ethoxy-; $C_4H_{10}O_2$; [110-80-5]	Sunwoo, C.; Eisen, H. *J. Pharm. Sci.* 1971, *60*, 238-44.
VARIABLES:	PREPARED BY:
One temperature: 25°C	R. Piekos

EXPERIMENTAL VALUES:

The mole fraction solubility of sulfamerazine in 2-ethoxyethanol at 25°C is 0.0109 (3.29 g/100 g solution, compiler).

AUXILIARY INFORMATION

METHOD/APPARATUS/PROCEDURE:

Soly was detd by the method reported by Restaino and Martin (1). Sulfamerazine was assayed on a Coleman-Hitachi 124 double-beam spectrophotometer at 270 nm after diln of a sample with 95% ethanol or water.

SOURCE AND PURITY OF MATERIALS:

The USP sulfamerazine (American Cyanamid Co., Pearl River, N.Y.), recrystd from warm alcohol and an industrial grade 2-ethoxyethanol (Cellosolve solvent, Union Carbide, New York, N.Y.) were used.

ESTIMATED ERROR:

Temp: ±1.0° (authors).

Soly: the mean of 3 runs was given (authors).

REFERENCES:

1. Restaino, F. A.; Martin, A. N. *J. Pharm. Sci.* 1964, *53*, 636.

COMPONENTS:	ORIGINAL MEASUREMENTS:
(1) Benzenesulfonamide, 4-amino-N-(4-methyl-2-pyrimidinyl)- (sulfamerazine); $C_{11}H_{12}N_4O_2S$; [127-79-7] (2) Methane, trichloro- (chloroform); $CHCl_3$; [67-66-3]	Riess, W. *Intern. Congr. Chemotherapy, Proc., 3rd, Stuttgart* 1963, *1*, 627-32.
VARIABLES:	PREPARED BY:
One temperature	R. Piekos

EXPERIMENTAL VALUES:

Solubility of sulfamerazine in chloroform at 20°C is 37 mg% (1.4×10^{-3} mol dm^{-3} solution, compiler).

AUXILIARY INFORMATION

METHOD/APPARATUS/PROCEDURE:

Nothing specified.

SOURCE AND PURITY OF MATERIALS:

Nothing specified.

ESTIMATED ERROR:

Nothing specified.

REFERENCES:

COMPONENTS:

(1) Benzenesulfonamide, 4-amino-N-(4-methyl-2-pyrimidinyl)- (sulfamerazine); $C_{11}H_{12}N_4O_2S$; [127-79-7]

(2) Methane, trichloro- (chloroform); $CHCl_3$; [67-66-3]

ORIGINAL MEASUREMENTS:

Yamazaki, M.; Aoki, M.; Kamada, A.; Yata, N. *Yakuzaigaku* 1967, *27(1)*, 37-40.

VARIABLES:

One temperature: 30°C

PREPARED BY:

R. Piekos

EXPERIMENTAL VALUES:

Solubility of sulfamerazine in chloroform at 30°C is 3.08 mmol/L (0.814 g dm^{-3}, compiler).

AUXILIARY INFORMATION

METHOD/APPARATUS/PROCEDURE:

Sulfamerazine (0.5 g) was placed in an L-shaped tube together with 20 ml of chloroform. The mixt was shaken in a thermostat until equilibrium was attained. The sulfamerazine was assayed in the supernatant spectrophotometrically at 545 nm on a Beckmann DU spectrophotometer. The results were taken from a calibration graph.

SOURCE AND PURITY OF MATERIALS:

Nothing specified.

ESTIMATED ERROR:

Soly: not specified.

Temp: ±1°C (authors).

REFERENCES:

COMPONENTS:

(1) Benzenesulfonamide, 4-amino-N-(4-methyl-2-pyrimidinyl)- (sulfamerazine); $C_{11}H_{12}N_4O_2S$; [127-79-7]

(2) 2-Pyrrolidinone, 1-ethynyl-,polymers (poly(vinyl pyrrolidone)); $(C_6H_9NO)_x$; [9003-39-8] K-15

(3) Ethanol; C_2H_6O; [64-17-5]

ORIGINAL MEASUREMENTS:

Sekikawa, H.; Nakano, M.; Arita, T. *Chem. Pharm. Bull.* 1978, *26(1)*, 118-26.

VARIABLES:

Temperature

PREPARED BY:

R. Piekos

EXPERIMENTAL VALUES:

t/°C	M x 10^2 sulfamerazine solubilized by 1M vinyl pyrrolidone equivalent
10.0	0.631
20.0	0.770
30.0	0.950
40.0	1.16
50.0	1.39

AUXILIARY INFORMATION

METHOD/APPARATUS/PROCEDURE:

After attaining equilibrium, sample solns were removed by a syringe and filtered quickly through a membrane filter (pore size 0.2μ) and sulfamerazine was assayed spectrophotometrically at 270 nm using a Hitachi Type 200-20 spectrophotometer. No significant absorbance was found for poly(vinyl pyrrolidone).

SOURCE AND PURITY OF MATERIALS:

Sulfamerazine, mp 235°C, was obtained from Na sulfamerazine (Tanabe Seiyaku Co.) by addn of HCl_{aq} and recrystn from EtOH. Poly(vinyl pyrrolidone) K-15 was from Daiichi Pure Chemicals Co., Tokyo. Abs EtOH was obtained by drying and distn of EtOH following the conventional procedure.

ESTIMATED ERROR:

Nothing specified.

REFERENCES:

COMPONENTS:	ORIGINAL MEASUREMENTS:
(1) Zinc, bis[4-amino-*N*-(4-methyl-2-pyrimidinyl)benzenesulfonamidato-N^N,O]-(Zn(II) sulfamerazine); $C_{22}H_{20}N_8O_4S_2Zn$; [71496-63-4] (2) Water; H_2O; [7732-18-5]	Fox, Ch. L., Jr.; Modak, S.; Stanford, J. W.; Fox, P. L. *Scand. J. Plast. Reconstr. Surg.* 1979, *13(1)*, 89-94.
VARIABLES:	PREPARED BY:
One temperature: 28-30°C	R. Piekos

EXPERIMENTAL VALUES:

Solubility of Zn(II) sulfamerazine in water at room temperature (28-30°C)[a] is 25.2 mg% (4.24×10^{-4} mol dm^{-3} solution, compiler).

[a]Value given by one of the authors (S. M.) in personal communication.

AUXILIARY INFORMATION

METHOD/APPARATUS/PROCEDURE:

Satd soln of Zn(II) sulfamerazine was prepd in water and after 24 h aliquots from the clear supernatant were assayed for sulfamerazine content using the colorimetric method of Bratton and Marshall (1). The soly value was then calcd from the molecular formula.

SOURCE AND PURITY OF MATERIALS:

The Zn(II) sulfamerazine was prepd by the authors as follows: an inorg Zn salt was reacted with Na salt of sulfamerazine and the ppt was analyzed and characterized. No details were given, however.

Purity of the materials was not specified.

ESTIMATED ERROR:

Nothing specified.

REFERENCES:

1. Bratton, A. C.; Marshall, E. K, Jr. *J. Biol. Chem.* 1939, *120*, 537.

COMPONENTS:	EVALUATOR:
(1) Acetamide, N-[4-[[(4-methyl-2-pyrimidinyl)amino]sulfonyl]phenyl- (acetyl sulfamerazine) $C_{13}H_{14}N_4O_3S$; [127-73-1] (2) Water	Anthony N. Paruta Department of Pharmaceutics University of Rhode Island Kingston, Rhode Island, USA and Ryszard Piekos Faculty of Pharmacy, University of Gdansk Gdansk, Poland 1986

CRITICAL EVALUATION:

Table I: Solubility of Acetyl sulfamerazine in water, 310K

Reference	10^4 mol dm^{-3} (*indicates mol kg^{-1}) 310K
1	9.14
2	37.5
3	7.8*

There is a 15% difference in the two closer values (1,3). Roblin's (1) is based using a 24 hour equilibrium time, but that of Sapozhnikova and Postovskii (3) used only one hour, probably a presaturation condition, which is the probable reason for the lower value (1). The approximate solubility in water at 310K can be given as 8.5 x 10^{-4} mol dm^{-3}. This value is consistant with the parent compound, it being 14 times smaller.

REFERENCES:

(1) Roblin, R.O., Jr.; Williams, J.H.; Winnek, P.S.; English, J.P. *J. Am. Chem. Soc.* 1940, *62*, 2002-5.

(2) Kikuth, W. *Med. Welt* 1943, *17(26/27)*, 483-6.

(3) Sapozhnikova, N.V.; Postovskii, I. Ya. *Zh. Prikl. Khim.* 1944, *17*, 427-34.

COMPONENTS:

(1) Acetamide, N-[4-[[(4-methyl-2-pyrimidinyl)amino]sulfonyl]phenyl]- (acetyl sulfamerazine); $C_{13}H_{14}N_4O_3S$; [127-73-1]

(2) Water; H_2O; [7732-18-5]

ORIGINAL MEASUREMENTS:

Roblin, R. O., Jr.; Williams, J. H.; Winnek, P. S.; English, J. P.

J. Am. Chem. Soc. 1940, *62*, 2002-5.

VARIABLES:

One temperature: 37°C

PREPARED BY:

R. Piekos

EXPERIMENTAL VALUES:

Solubility of acetyl sulfamerazine in water at 37°C is 28.0 mg/100 cm^3 solution (9.14×10^{-4} mol dm^{-3}, compiler).

AUXILIARY INFORMATION

METHOD/APPARATUS/PROCEDURE:

Excess sulfonamide in water was heated and stirred on a steam bath for 30 min. The suspension was then agitated for 24 h in a thermostat at 37°C. A sample of the satd soln was withdrawn through a glass filter, dild, and analyzed by the Marshall method (1) using a General Electric recording spectrophotometer for comparing the colors developed with those of the standards.

SOURCE AND PURITY OF MATERIALS:

Acetyl sulfamerazine, mp 248-9°C (cor), was prepd by the authors. Anal: %C 51.0 (calcd 51.0); %H 4.8 (4.6); %N 17.9 (18.3). Purity of the water was not specified.

ESTIMATED ERROR:

Nothing specified.

REFERENCES:

1. Bratton, A. C.; Marshall, E. K., Jr. *J. Pharmacol.* 1939, *66*, 4.

COMPONENTS:	ORIGINAL MEASUREMENTS:
(1) Acetamide, N-[4-[[(4-methyl-2-pyrimidinyl)amino]sulfonyl]phenyl]- (acetyl sulfamerazine); $C_{13}H_{14}N_4O_3S$; [127-73-1] (2) Water; H_2O; [7732-18-5]	Kikuth, W. *Med. Welt* 1943, *17(26/27)*, 483-6.
VARIABLES:	**PREPARED BY:**
One temperature: 37°C	R. Piekos

EXPERIMENTAL VALUES:

Solubility of acetyl sulfamerazine in water at 37°C is 115 mg/100 cm^3 solution (3.75×10^{-3} mol dm^{-3} , compiler).

AUXILIARY INFORMATION

METHOD/APPARATUS/PROCEDURE:	SOURCE AND PURITY OF MATERIALS:
Nothing specified.	Acetyl sulfamerazine: not specified. The pH of the water was 7.8
	ESTIMATED ERROR:
	Nothing specified.
	REFERENCES:

COMPONENTS:

(1) Acetamide, N-[4-[[(4-methyl-2-pyrimidinyl)amino]sulfonyl]phenyl]- (acetyl sulfamerazine); $C_{13}H_{14}N_4O_3S$; [127-73-1]

(2) Water; H_2O; [7732-18-5]

ORIGINAL MEASUREMENTS:

Sapozhnikova, N. V.; Postovskii, I. Ya.
Zh. Prikl. Khim. 1944, *17*, 427-34.

VARIABLES:

Temperature

PREPARED BY:

R. Piekos

EXPERIMENTAL VALUES:

t/°C	Solubility	
	Weight%	10^3 mol kg^{-1} water[a]
37	0.024	0.78
75	0.170	5.56

[a]Calculated by compiler

AUXILIARY INFORMATION

METHOD/APPARATUS/PROCEDURE:

Acetyl sulfamerazine was dissolved in water to form a satd soln which was occasionally agitated in a glass vessel immersed in a thermostat. The equilibrium was usually attained after 1 h. Five - to 100-cm^3 samples of the satd soln were placed in Pt crucibles or dishes and evapd to dryness at temps lower than 110-115°C. The residue was dried to const wt at 105-110°C and weighed.

SOURCE AND PURITY OF MATERIALS:

Pure, recrystd acetyl sulfamerazine was used. Its mp conformed to that reported in the literature.

Purity of the water was not specified.

ESTIMATED ERROR:

Soly: quite reliable results were obtained (authors).

Temp: ±0.05°C (authors).

REFERENCES:

COMPONENTS:

(1) Acetamide, N-[4-[(4-methyl-2-pyrimidinyl)amino]sulfonyl]phenyl]- (acetyl sulfamerazine); $C_{13}H_{14}N_4C_3S$; [127-73-1]

(2) 1,2,3-Propanetricarboxylic acid, 2-hydroxy-, disodium salt (Na citrate); $C_6H_6Na_2O_7$; [144-33-2]

(3) Sodium hydroxide; NaOH; [1310-73-2]

(4) Water; H_2O; [7732-18-5]

ORIGINAL MEASUREMENTS:

Gilligan, D. R.; Plummer, M. N.

Proc. Soc. Exp. Biol. Med. 1943, *53*, 142-5.

VARIABLES:

pH

PREPARED BY:

R. Piekos

EXPERIMENTAL VALUES:

Solubility of acetyl sulfamerazine in M/10 Na citrate + NaOH solutions at 37°C

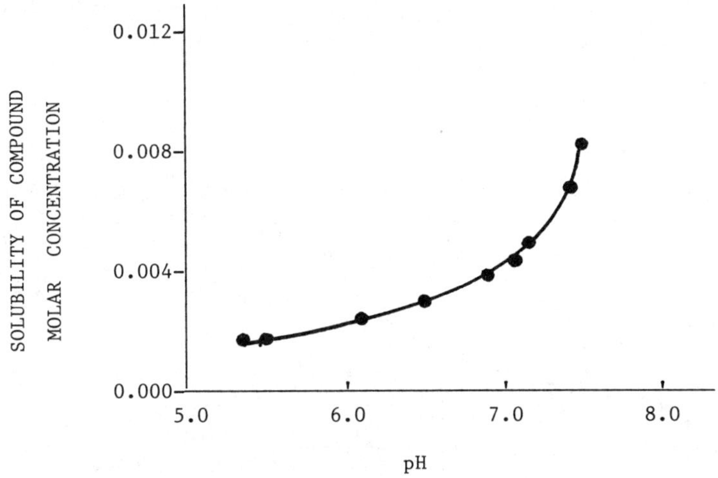

AUXILIARY INFORMATION

METHOD/APPARATUS/PROCEDURE:

An excess of acetyl sulfamerazine was shaken in M/10 Na citrate + NaOH solns of various pH values for 18 h in a water bath at 37°C, and filtered in an incubator room at this temp. The pH of the filtrate was measured immediately with a Beckmann glass electrode pH meter and appropriate corrections for the differences between room temp and 37°C were applied. The amt of dissolved compd was measured by the method of Bratton and Marshall (1).

SOURCE AND PURITY OF MATERIALS:

Acetyl sulfamerazine was supplied by Lederle Labs, Inc. The source and purity of the remaining materials were not specified.

ESTIMATED ERROR:

Nothing specified.

REFERENCES:

1. Bratton, A. C.; Marshall, E. K., Jr. *J. Biol. Chem.* 1939, *128*, 537.

COMPONENTS:

(1) Acetamide, N-[4-[[(4-methyl-2-pyrimidinyl)amino]sulfonyl]phenyl]- (acetyl sulfamerazine); $C_{13}H_{14}N_4C_3S$; [127-73-1]

(2) Phosphoric acid, disodium salt; Na_2HPO_4; [7558-94-4]

(3) Phosphoric acid, monopotassium salt; KH_2PO_4; [7778-77-0]

(4) Water; H_2O; [7732-18-5]

VARIABLES:

pH

ORIGINAL MEASUREMENTS:

Gilligan, D. R.; Plummer, M. N. *Proc. Soc. Exp. Biol. Med.* 1943, *53*, 142-5.

PREPARED BY:

R. Piekos

EXPERIMENTAL VALUES:

Solubility of acetyl sulfamerazine in M/15 phosphate buffer solutions at 37°C

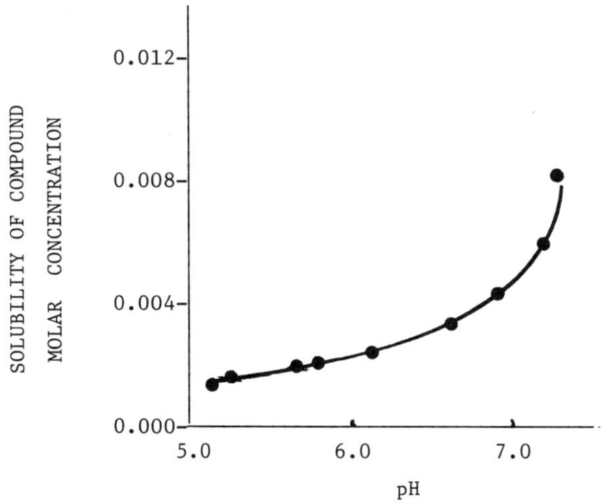

AUXILIARY INFORMATION

METHOD/APPARATUS/PROCEDURE:

An excess of acetyl sulfamerazine was shaken in M/15 phosphate buffer solns of various pH values for 18 h in a water bath at 37°C, and filtered in an incubator room at this temp. The pH of the filtrate was measured immediately at room temp with a Beckmann glass electrode pH meter and appropriate corrections for the differences between room temp and 37°C were applied. The amt of dissolved compd was measured by the method of Bratton and Marshall (1).

SOURCE AND PURITY OF MATERIALS:

Acetyl sulfamerazine was supplied by Lederle Labs, Inc. The source and purity of the remaining materials were not specified.

ESTIMATED ERROR:

Nothing specified.

REFERENCES:

1. Bratton, A. C.; Marshall, E. K., Jr. *J. Biol. Chem.* 1939, *128*, 537.

COMPONENTS:

(1) Acetamide, N-[4-[[(4-methyl-2-pyrimidinyl)amino]sulfonyl]phenyl]- (acetyl sulfamethylpyrimidine); $C_{13}H_{14}N_4O_3S$; [127-73-1]

(2) Phosphoric acid, disodium salt; Na_2HPO_4; [7558-94-4]

(3) Phosphoric acid, monopotassium salt; KH_2PO_4; [7778-77-0]

(4) Water; H_2O; [7732-18-5]

ORIGINAL MEASUREMENTS:

Frisk, A. R.; Hagerman, G.; Helander, S.; Sjögren, B.

Hygiea 1946, *108(12)*, 639-51.

VARIABLES:

One temperature: 37°C; one pH: 6.1

PREPARED BY:

R. Piekos

EXPERIMENTAL VALUES:

Solubility of acetyl sulfamethylpyrimidine in M/30 phosphate buffer of pH 6.1 at 37°C is 53 mg/100 ml solvent (1.7×10^{-3} mol dm^{-3}, compiler).

AUXILIARY INFORMATION

METHOD/APPARATUS/PROCEDURE:

An excess of acetyl sulfamethylpyrimidine in the phosphate buffer was shaken at 37°C for 24 h. The concn of the solute was detd by the Bratton and Marshall method (1) using a photoelec colorimeter.

SOURCE AND PURITY OF MATERIALS:

Neither source nor purity of the materials was specified.

ESTIMATED ERROR:

Soly: precision ±4 mg/100 ml (authors).

Temp and pH: not specified.

REFERENCES:

1. Bratton, A. C.; Marshall, E. K., Jr. *J. Biol. Chem.* 1939, *128*, 537.

COMPONENTS:

(1) Acetamide, N-[4-[[(4-methyl-2-pyrimidinyl)amino]sulfonyl]phenyl]- (N^4-acetylsulfamerazine); $C_{13}H_{14}N_4O_3S$; [127-73-1]

(2) Phosphoric acid, disodium salt; Na_2HPO_4; [7558-94-4]

(3) Phosphoric acid, monopotassium salt; KH_2PO_4; [7778-77-0]

(4) Water; H_2O; [7732-18-5]

ORIGINAL MEASUREMENTS:

Hekster, Ch. A.; Vree, T. B.

Antibiotics Chemother. 1982, *31*, 22-118.

VARIABLES: pH

PREPARED BY:

R. Piekos

EXPERIMENTAL VALUES:

pH	Solubility at 25°C	
	mg/l	10^3 mol dm^{-3} [a]
5.5	585	1.910
7.5[b]	2,300	7.508

[a]Calculated by compiler

[b]Erraneous pH value of 7.0 is given in the article.

AUXILIARY INFORMATION

METHOD/APPARATUS/PROCEDURE:

The earlier developed method (1) was used (personal communication). Satd solns of N^4-acetylsulfamerazine were prepd in phosphate buffers of pH 5.5 and 7.5 at 25°C. The concn of the solute was measured by means of a Spectra Physics 3500B high-performance liquid chromatograph equipped with a Model 748 column oven and a Pye-Unicam LC-UV spectrophotometric detector.

SOURCE AND PURITY OF MATERIALS:

Neither source nor the purity of the materials was specified.

ESTIMATED ERROR:

Soly: the detection limit of the solute by HPLC was 0.5 mg/l (authors).
The errors in temp and pH were not specified.

REFERENCES:

1. Hekster, Y. A.; Vree, T. B.; Damsma, J. E.; Friesen, W. T. *J. Antimicrob. Chemother.* 1981, *8*, 133.

COMPONENTS:

(1) Acetamide, N-[4-[[(4-methyl-2-pyrimidinyl)amino]sulfonyl]phenyl]- (acetylsulfamerazine); $C_{13}H_{14}N_4O_3S$; [127-73-1]

(2) Benzene,methyl- (toluene); C_7H_8; [108-88-3]

(3) Mannitol; $C_6H_{14}O_6$; [87-78-5]

(4) Phosphoric acid, disodium salt; Na_2HPO_4; [7558-94-4]

(5) Phosphoric acid, monopotassium salt; KH_2PO_4; [7778-77-0]

(6) Sodium chloride; NaCl; [7647-14-5]

(7) Water; H_2O; [7732-18-5]

ORIGINAL MEASUREMENTS:

Sonnenberg, H.; Oelert, H.; Baumann, K.

Pflügers Arch. Ges. Physiol. 1965, *286*, 171-80.

VARIABLES: pH

PREPARED BY:

R. Piekos

EXPERIMENTAL VALUES:

Relative lipoid solubility determined on the basis of concentration measurement of acetylsulfamerazine in perfusates [a,b] before (c_i) and after (c_e) equilibration with toluene

pH	$(100 - \frac{100\ c_e}{c_i})$
5[a]	8
8[b]	0

[a]Composition of perfusate: 110 mmol/l NaCl, 35 mmol/l mannitol in a phosphate buffer consisting of 98.8 ml of 0.022M KH_2PO_4 and 1.2 ml of 0.022M Na_2HPO_4.

[b]Composition of perfusate: 68 mmol/l NaCl, 100 mmol/l mannitol in a phospahte buffer consisting of 5.5 ml of 0.022M KH_2PO_4 and 94.5 ml of 0.022M Na_2HPO_4.

AUXILIARY INFORMATION

METHOD/APPARATUS/PROCEDURE:

Lipoid solubilities were detd by shaking equal volumes of the aperfusate and toluene with acetylsulfamerazine for 20 min and measuring the concn of acetylsulfamerazine by the spectrophotometric method of Bratton and Marshall (1) in an aq phase before and after this procedure.

SOURCE AND PURITY OF MATERIALS:

None given.

ESTIMATED ERROR:

None given.

REFERENCES:

1. Bratton, A. C.; Marshall, E. K., Jr. *J. Biol. Chem.* 1939, *128*, 537.

COMPONENTS:

(1) Acetamide, N-[4-[[(4-methyl-2-pyrimidinyl)amino]sulfonyl]phenyl]- (acetylsulfamerazine); $C_{13}H_{14}N_4O_3S$; [127-73-1]

(2) Mannitol; $C_6H_{14}O_6$; [87-78-5]

(3) Methane, trichloro- (chloroform); $CHCl_3$; [67-66-3]

(4) Phosphoric acid, disodium salt; Na_2HPO_4; [7558-94-4]

(5) Phosphoric acid, monopotassium salt; KH_2PO_4; [7778-77-0]

(6) Sodium chloride; NaCl; [7647-14-5]

(7) Water; H_2O; [7732-18-5]

ORIGINAL MEASUREMENTS:

Sonnenberg, H.; Oelert, H.; Baumann, K.

Pflügers Arch. Ges. Physiol. 1965, *286*, 171-80.

VARIABLES:

pH

PREPARED BY:

R. Piekos

EXPERIMENTAL VALUES:

Relative lipoid solubility determined on the basis of concentration measurements of acetylsulfamerazine in perfusates[a,b] before (c_1) and after (c_e) equilibration with chloroform

pH	$\left(100 - \frac{100\ c_e}{c_i}\right)$
5[a]	42
8[b]	0

[a]Composition of perfusate: 110 mmol/l NaCl, 35 mmol/l mannitol in a phosphate buffer conisting of 98.8 ml of 0.022M KH_2PO_4 and 1.2 ml of 0.022M Na_2HPO_4.

[b]Compostion of perfusate: 68 mmol/l NaCl, 100 mmpl/l mannitol in a phosphate buffer consisting of 5.5 ml of 0.022M KH_2PO_4 and 94.5 ml of 0.022M Na_2HPO_4.

AUXILIARY INFORMATION

METHOD/APPARATUS/PROCEDURE:

Lipoid solubilities were detd by shaking equal volumes of the perfusate and chloroform with acetylsulfamerazine for 20 min and measuring the concn of acetylsulfamerazine by the spectrophotometric method of Bratton and Marshall (1) in an aq phase before and after this procedure.

SOURCE AND PURITY OF MATERIALS:

None given.

ESTIMATED ERROR:

None given

REFERENCES:

1. Bratton, A. C.; Marshall, E. K., Jr. *J. Biol. Chem.* 1939, *128*, 537.

COMPONENTS:

(1) Benzenesulfonamide, 4-amino-N-(4-methoxy-2-pyrimidinyl)-; $C_{11}H_{12}N_4O_3S$; [3213-22-7]

(2) Water; H_2O; [7732-18-5]

ORIGINAL MEASUREMENTS:

Roblin, R. O., Jr.; Winnek, P. S.; English, J. P. *J. Am. Chem. Soc.* 1942, *64*, 567-70.

VARIABLES:

One temperature: 37°C

PREPARED BY:

R. Piekos

EXPERIMENTAL VALUES:

Solubility of 4-amino-N-(4-methoxy-2-pyrimidinyl)benzenesulfonamide in water at 37°C is 18.2 mg/100 cm^3 solution (6.49 10^{-4} mol dm^{-3}, compiler).

AUXILIARY INFORMATION

METHOD/APPARATUS/PROCEDURE:

Excess sulfonamide in water was heated and stirred on a steam bath for 30 min. The suspension was then agitated for 24 h in a thermostat at 37°C. A sample of the satd soln was withdrawn through a glass filter, dild, and analyzed by the Marshall method (1) using a General Electric recording spectrophotometer for comparing colors developed with those of the standards.

SOURCE AND PURITY OF MATERIALS:

The sulfonamide, mp 241-2°C, was prepd by the authors. Anal: %C 47.2 (calcd 47.1); %H 4.4 (4.3); %N 19.9 (20.0). Purity of the water was not specified.

ESTIMATED ERROR:

Nothing specified.

REFERENCES:

1. Bratton, A.C.; Marshall, E. K., Jr. *J. Pharmacol.* 1939, *66*, 4.

COMPONENTS:	ORIGINAL MEASUREMENTS:
(1) Benzenesulfonamide, 4-amino-N-(5-methoxy-2-pyrimidinyl)- (sulfamethoxydiazine); $C_{11}H_{12}N_4O_3S$; [651-06-9] (2) Poly(oxy-1,2,-ethanediyl), α-hydro-ω-hydroxy- (PEG 6000); $(C_2H_4O)_nH_2O$; [25322-68-3] 6000 (3) Water; H_2O; [7732-18-5]	Salib, N. N.; Ebian, A. R. *Pharm. Ind.* 1978, *40(3)*, 262-5.
VARIABLES:	PREPARED BY:
Concentration of PEG 6000	R. Piekos

EXPERIMENTAL VALUES:

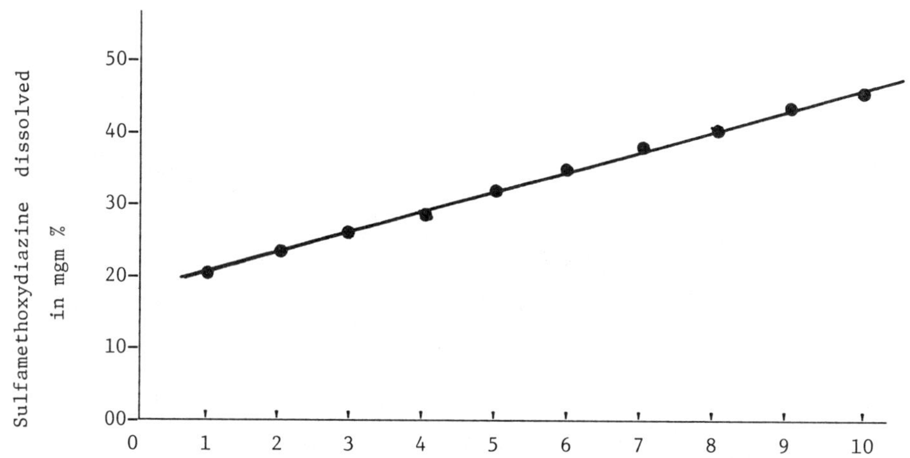

Effect of different concentrations of PEG 6000 on the solubility of sulfamethoxydiazine at 37±1°C .

AUXILIARY INFORMATION

METHOD/APPARATUS/PROCEDURE:	SOURCE AND PURITY OF MATERIALS:
An excess of sulfamethoxydiazine was added to bottles contg from 0 to 10% PEG 6000 solns. The bottles were rotated at 30 rpm for 24 h in a water bath at 37°C. After equilibrium had been attained, an aliquot was pipetted out, dild with 0.1N HCl, and the solute was assayed spectrophotometrically at 228 nm.	PEG 6000 was a product of Hoechst AG, West Germany. Sulfamethoxydiazine (Bayrena) was supplied by Bayer, Leverkusen. The purity of the materials was not specified.
	ESTIMATED ERROR: Soly: data points are means of 3 detns (authors). Temp: ±1°C (authors).
	REFERENCES:

COMPONENTS:

(1) Benzenesulfonamide, 4-amino-N-(5-methoxy-2-pyrimidinyl)- (sulfamethoxydiazine); $C_{11}H_{12}N_4O_3S$; [651-06-9]

(2) Poly(oxy-1,2-ethanediy) ,α -hydro- ω -hydroxy- PEG (20 000); $(C_2H_4O)_nH_2O$; [25322-68-3] 20 000

(3) Water; H_2O; [7732-18-5]

ORIGINAL MEASUREMENTS:

Salib, N. N.; Ebian, A. R.

Pharm. Ind. 1978, *40(3)*, 262-5.

VARIABLES:

Concentration of PEG 20 000

PREPARED BY:

R. Piekos

EXPERIMENTAL VALUES:

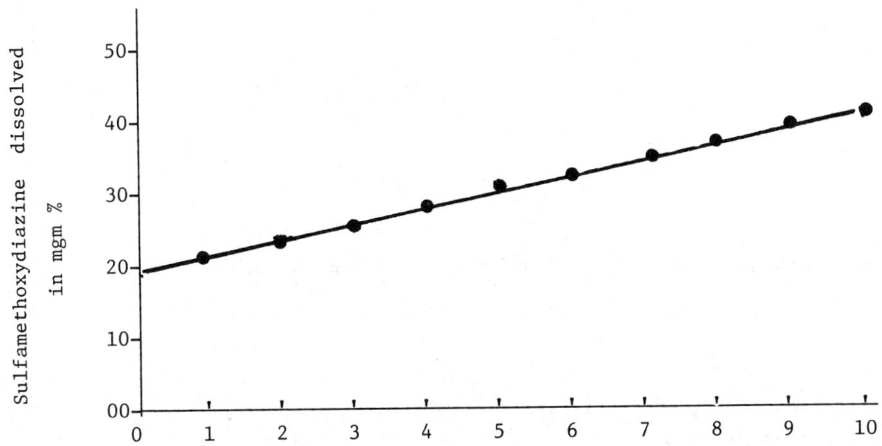

Effect of different concentrations of PEG 20 000 on the solubility of sulfamethoxydiazine at 37±1°C.

AUXILIARY INFORMATION

METHOD/APPARATUS/PROCEDURE:

An excess of sulfamethoxydiazine was added to bottles contg from 0 to 10% PEG 20 000 solns. The bottles were rotated at 30 rpm for 24 h in a water bath at 37°C. After equilibrium had been attained, an aliquot was pipetted out, dild with 0.1N HCl, and the solute was assayed spectrophotometrically at 228 nm.

SOURCE AND PURITY OF MATERIALS:

Sulfamethoxydiazine (Bayrena) was supplied by Bayer, Leverkusen. PEG 20 000 was a product of Hoechst AG, West Germany.

The purity of the materials was not specified.

ESTIMATED ERROR:

Soly: data points are means of 3 detns (authors).

Temp: ±1°C (authors).

REFERENCES:

COMPONENTS:	ORIGINAL MEASUREMENTS:
(1) Benzenesulfonamide, 4-amino-N-(5-methoxy-2-pyrimidinyl)- (sulfameter); $C_{11}H_{12}N_4O_3S$; [651-06-9] (2) Ethanol, 2-Ethoxy-; $C_4H_{10}O_2$; [110-80-5]	Sunwoo, C.; Eisen, H. *J. Pharm. Sci.* 1971, *60*, 238-44.
VARIABLES:	PREPARED BY:
One temperature: 25°C	R. Piekos

EXPERIMENTAL VALUES:

The mole fraction solubility of sulfameter in 2-ethoxyethanol at 25°C is 0.0119 (3.61 g/100 g solution, compiler).

AUXILIARY INFORMATION

METHOD/APPARATUS/PROCEDURE:

Soly was detd by the method reported by Restaino and Martin (1). Sulfameter was assayed on a Coleman-Hitachi 124 double-beam spectrophotometer at 271 nm after diln of a sample with 95% alcohol or water.

SOURCE AND PURITY OF MATERIALS:

Sulfameter (A. H. Robins Co., Richmond, Va) was recrystd from warm alcohol. 2-ethoxyethanol (Cellosolve solvent, Union Carbide, New York, N. Y.) was of industrial grade.

ESTIMATED ERROR:

Temp: ±1.0°C (authors).

Soly: the mean of 3 runs was given (authors).

REFERENCES:

1. Restaino, F. A.; Martin, A. N. *J. Pharm. Sci.* 1964, *53*, 636.

COMPONENTS:

(1) Benzenesulfonamide, 4-amino-N-(5-methoxy-2-pyrimidinyl)- (sulfamonomethoxine, SMM); $C_{11}H_{12}N_4O_3S$; [651-06-9]

(2) 1,4,7,10,13,16-Hexaoxacyclooctadecane (18-C-6); $C_{12}H_{24}O_6$; [17455-13-9]

(3) Hydrochloric acid; HCl; [7647-01-0]

(4) Water; H_2O; [7732-18-5]

ORIGINAL MEASUREMENTS:

Takayama, K.; Nambu, B.; Nagai, T. *Chem. Pharm. Bull.* 1978, *26(10)*, 2965-70.

VARIABLES:

Temperature

PREPARED BY:

R. Piekos

EXPERIMENTAL VALUES:

t/°C	Saturated concentration of sulfamonomethoxine after decomplexation of its 1:1 complex with 18-C-6 in 0.2N HCl 10^2M
30	0.390
35	0.473
40	0.571

AUXILIARY INFORMATION

METHOD/APPARATUS/PROCEDURE:

An excess of the complex was dissolved in 50 ml of 0.2N HCl. The sampling was done by a 1-ml pipet fitted with a G-4 glass filter. The concentratrion of the sulfonamide was detd by uv spectrophotometry after dilg with 0.2N HCl.

SOURCE AND PURITY OF MATERIALS:

SMM (Dai-ichi Pharmaceutical Co) was recrystd from a 30% (V/V) Me_2CO - H_2O system.
18-C-6 was of the reagent grade.
The complex was prepd by the authors.
Purity of the HCl soln was not specified.

ESTIMATED ERROR:

Nothing specified.

REFERENCES:

COMPONENTS:	ORIGINAL MEASUREMENTS:
(1) Benzenesulfonamide, 4-amino-N-(2-methoxy-5-pyrimidinyl)-; $C_{11}H_{12}N_4O_3S$; [71119-37-4] (2) Water; H_2O; [7732-18-5]	Roblin, R. O., Jr.; Winnek, P. S.; English, J. P. *J. Am. Chem. Soc.* 1942, *64*, 567-70.
VARIABLES:	PREPARED BY:
One temperature: 37°C	R. Piekos

EXPERIMENTAL VALUES:

Solubility of 4-amino-N-(2-methoxy-5-pyrimidinyl)benzenesulfonamide in water at 37°C is 9.2 mg/100 cm^3 solution (3.3 x 10^{-4} mol dm^{-3}, compiler).

AUXILIARY INFORMATION

METHOD/APPARATUS/PROCEDURE:

Excess sulfonamide in water was heated and stirred on a steam bath for 30 min. The suspension was then agitated for 24 h in a thermostat at 37°C. A sample of the satd soln was withdrawn through a glass filter, dild, and analyzed by the Marshall method (1) using a General Electric recording spectrophotometer for comparing the colors developed with those of the standards.

SOURCE AND PURITY OF MATERIALS:

The sulfonamide, mp 232-4°C (cor), was prepd by the authors. Anal: %C 47.3 (calcd 47.1); %H 4.0 (4.3); %N 20.1 (20.0). Purity of the water was not specified.

ESTIMATED ERROR:

Nothing specified.

REFERENCES:

1. Bratton, A. C.; Marshall, E. K., Jr. *J. Pharmacol.* 1939, *66*, 4.

COMPONENTS:	ORIGINAL MEASUREMENTS:
(1) Benzenesulfonamide, 4-amino-N-(6-methoxy-4-pyrimidinyl)- (sulfamonomethoxine); $C_{11}H_{12}N_4O_3S$; [1220-83-3] (2) Water; H_2O; [7732-18-5]	Yamazaki, M.; Aoki, M.; Kamada, A.; Yata, N. *Yakuzaigaku* 1967, *27(1)*, 37-40.
VARIABLES:	PREPARED BY:
One temperature: 30°C	R. Piekos

EXPERIMENTAL VALUES:

Solubility of sulfamonomethoxine in water at 30°C is 0.17 mmol/L (4.8×10^{-2} g dm^{-3}, compiler).

AUXILIARY INFORMATION

METHOD/APPARATUS/PROCEDURE:

Sulfamonomethoxine (0.5 g) was placed in an L-shaped tube together with 20 ml of water. The mixt was shaken in a thermostat until equilibrium was attained. The sulfamonomethoxine was assayed in the supernatant spectrophotometrically at 545 nm on a Beckmann DU spectrophotometer. The results were taken from a calibration graph.

SOURCE AND PURITY OF MATERIALS:

Nothing specified.

ESTIMATED ERROR:

Soly: not specified.

Temp: ±1°C (authors).

REFERENCES:

COMPONENTS:

(1) Benzenesulfonamide, 4-amino-N-(6-methoxy-4-pyrimidinyl)- (sulfamonomethoxine); $C_{11}H_{12}N_4O_3S$; [1220-83-3]

(2) Water; H_2O; [7732-18-5]

ORIGINAL MEASUREMENTS:

Ezerskii, M. L.; Per'kova, N. N.
Khim.-Farm. Zh. 1979, *13(11)*, 87-91.

VARIABLES:

Grinding regime

PREPARED BY:

R. Piekos

EXPERIMENTAL VALUES:

Specimen of sulfamonomethoxine	Solubility at room temperature	
	g/cm^3	10^4 mol dm^{-3} [a]
Commercial	0.000044	1.6
Commercial, ground in ball mill	0.000058	2.1
Commercial, ground in a jet mill	0.000054	1.9

[a] Calculated by compiler

AUXILIARY INFORMATION

METHOD/APPARATUS/PROCEDURE:

Satd solns were prepd by prolonged agitation of an excess of sulfamonomethoxine in water at room temp. The solns were then allowed to stand for 12 h and filtered. The concn of sulfamonomethoxine in the filtrate was detd nitritometrically by the method of the State Pharmacopeia X.

SOURCE AND PURITY OF MATERIALS:

Comm, pharmacopeial sulfamonomethoxine was used (source not specified). It was ground in a Pulverisette-5 lab ball mill or in a C-1266-00 jet mill.
Purity of the water was not specified.

ESTIMATED ERROR:

Nothing specified.

REFERENCES:

COMPONENTS:

(1) Benzenesulfonamide, 4-amino-N-(6-methoxy-4-pyrimidinyl)- (sulfamonomethoxine); $C_{11}H_{12}N_4O_3S$; [1220-83-3]

(2) Hydrochloric acid; HCl; [7647-01-0]

(3) Water; H_2O; [7732-18-5]

ORIGINAL MEASUREMENTS:

Ogata, H.; Shibazaki, T.; Inoue, T.; Ejima, A. *Chem. Pharm. Bull.* 1979, *27(6)*, 1281-6.

VARIABLES:

One temperature: 37°C

PREPARED BY:

R. Piekos

EXPERIMENTAL VALUES:

Solubility of sulfamonomethoxine in 0.1N HCl at 37°C is 1.336 mg/ml (4.766×10^{-3} mol dm^{-3}, compiler).

AUXILIARY INFORMATION

METHOD/APPARATUS/PROCEDURE:

A centrifuge tube contg 30 ml of 0.1N HCl and 0.5-3.0 g of the sulfamonomethoxine powder was tightly sealed and shaken at 37°C. The concn of the dissolved drug was detd spectrophotometrically following filtration through a Millipore filter (type EH, pore size 0.5 μm), and the procedure was repeated every 24 h until a const concn was obtained.

SOURCE AND PURITY OF MATERIALS:

Comm available 500-mg uncoated tablets of sulfamonomethoxine were used. Hydrochloric acid was of reagent grade.

ESTIMATED ERROR:

Nothing specified.

REFERENCES:

COMPONENTS:

(1) Benzenesulfonamide, 4-amino-N-(6-methoxy-4-pyrimidinyl)- (sulfamonomethoxine); $C_{11}H_{12}N_4O_3S$; [1220-83-3]

(2) Phosphoric acid, disodium salt; Na_2HPO_4; [7558-94-4]

(3) Phosphoric acid, monopotassium salt; KH_2PO_4; [7778-77-0]

(4) Water; H_2O; [7732-18-5]

ORIGINAL MEASUREMENTS:

Yamazaki, M.; Aoki, M.; Kamada, A.; Yata, N. *Yakuzaigaku* 1967, *27(1)*, 37-40.

VARIABLES:

One temperature: 30°C; one pH: 7.4

PREPARED BY:

R. Piekos

EXPERIMENTAL VALUES:

Solubility of sulfamonomethoxine in a phosphate buffer solution of pH 7.4 (μ = 0.17) at 30°C is 2.79 mmol/L (0.782 g dm^{-3}, compiler).

AUXILIARY INFORMATION

METHOD/APPARATUS/PROCEDURE:

Sulfamonomethoxine (0.5 g) was placed in an L-shaped tube together with 20 ml of the buffer soln. The mixt was shaken in a thermostat until equilibrium was attained. The sulfamonomethoxine was assayed in the supernatant spectrophotometrically at 545 nm on a Beckmann DU spectrophotometer. The results were taken from a calibration graph.

SOURCE AND PURITY OF MATERIALS:

Nothing specified.

ESTIMATED ERROR:

Soly and pH: not specified.

Temp: ±1°C (authors).

REFERENCES:

COMPONENTS:

(1) Benzenesulfonamide, 4-amino-N-(6-methoxy-4-pyrimidinyl)- (sulfamonomethoxine); $C_{11}H_{12}N_4O_3S$; [1220-83-3]

(2) Phosphoric acid, disodium salt; Na_2HPO_4; [7558-94-4]

(3) Phosphoric acid, monopotassium salt; KH_2PO_4; [7778-77-0]

(4) Water; H_2O; [7732-18-5]

ORIGINAL MEASUREMENTS:

Takayama, K.; Hasegawa, S.; Sasagawa, S.; Nambu, N.; Nagai, T.

Chem. Pharm. Bull. 1978, *26(1)*, 96-100.

VARIABLES:

One temperature: 30°C; one pH: 7.0

PREPARED BY:

R. Piekos

EXPERIMENTAL VALUES:

Saturated concentration of sulfamonomethoxine in a 1/15M phosphate buffer of pH 7.0 at 30°C is 3.47×10^{-3}M.

AUXILIARY INFORMATION

METHOD/APPARATUS/PROCEDURE:

The stationary disk method was employed (1). At appropriate time intervals, 3-ml aliquots of soln were withdrawn, the resultant amt of volume was compensated by adding the dissoln medium of the same temp. The concn was detd by the UV spectrophotometry.

SOURCE AND PURITY OF MATERIALS:

Sulfamonomethoxine (Dai-ichi Pharmaceutical Co.) mp 205°C, was prepd by heating a hydrate (recrystd from 30% (v/v) acetone-water) at 105°C for 4 h according to Japanese Pharmacopeia IX. The identity of the compd was confirmed by X-ray powder diffractometry, IR spectrophotometry and differential scanning calorimetry. The source and purity of the remaining materials were not specified.

ESTIMATED ERROR:

Nothing specified.

REFERENCES:

1. Hamada, Y.; Nambu, N.; Nagai, T. *Chem. Pharm. Bull.* 1975, *23*, 1205.

COMPONENTS:

(1) Benzenesulfonamide, 4-amino-N-(6-methoxy-4-pyrimidinyl)- (sulfamonomethoxine); $C_{11}H_{12}N_4O_3S$; [1220-83-3]

(2) Benzene; C_6H_6; [71-43-2]

ORIGINAL MEASUREMENTS:

Takayama, K.; Nambu, N.; Nagai, T. *Chem. Pharm. Bull.* 1977, *25*, 2608-12.

VARIABLES:

One temperature: 10°C

PREPARED BY:

R. Piekos

EXPERIMENTAL VALUES:

Solubility of sulfamonomethoxine in benzene at 10°C is 1.45 x 10^{-4} mol dm^{-3} [a].

[a] Numerical value supplied by the authors.

AUXILIARY INFORMATION

METHOD/APPARATUS/PROCEDURE:

The system was equilibrated in a sealed vial for 72 h at 10°. The satd soln was rapidly filtered through a Toyo filter paper No. 5B, 1 cm^3 of the filtrate was evapd at 40°C and the residue was dissolved in $CHCl_3$ to det the concn in the UV region using a Hitachi 124 spectrophotometer.

SOURCE AND PURITY OF MATERIALS:

Sulfamonomethoxine, m.p. 205°C, was a very pure compd supplied by Dai-ichi Pharmaceutical Co., Ltd.

Purity of the benzene was not specified.

ESTIMATED ERROR:

Nothing specified.

REFERENCES:

COMPONENTS:	ORIGINAL MEASUREMENTS:
(1) Benzenesulfonamide, 4-amino-N-(6-methoxy-4-pyrimidinyl)- (sulfamonomethoxine); $C_{11}H_{12}N_4O_3S$; [1220-83-3] (2) Methane, trichloro- (chloroform); $CHCl_3$; [67-66-3]	Yamazaki, M.; Aoki, M.; Kamada, A.; Yata, N. *Yakuzaigaku* 1967, *27(1)*, 37-40.
VARIABLES:	**PREPARED BY:**
One temperature: 30°C	R. Piekos

EXPERIMENTAL VALUES:

Solubility of sulfamonomethoxine in chloroform at 30°C is 2.63 mmol/L (0.737 g dm^{-3}, compiler).

AUXILIARY INFORMATION

METHOD/APPARATUS/PROCEDURE:

Sulfamonomethoxine (0.5 g) was placed in an L-shaped tube together with 20 ml of chloroform. The mixt was shaken in a thermostat until equilibrium was attained. The sulfamonomethoxine was assayed in the supernatant spectrophotometrically at 545 nm on a Beckmann DU spectrophotometer. The results were taken from a calibration graph.

SOURCE AND PURITY OF MATERIALS:

Nothing specified.

ESTIMATED ERROR:

Soly: not specified.

Temp: ±1°C (authors).

REFERENCES:

COMPONENTS:	ORIGINAL MEASUREMENTS:
(1) Benzenesulfonamide, 4-amino-N-(6-methoxy-4-pyrimidinyl)- (sulfamonomethoxine); $C_{11}H_{12}N_4O_3S$; [1220-83-3] (2) 1,4,7,10,13,16-Hexaoxacyclooctadecane (18-Crown-6); $C_{12}H_{24}O_6$; [17455-13-9] (3) Benzene; C_6H_6; [71-43-2]	Takayama, K.; Nambu, N.; Nagai, T. *Chem. Pharm. Bull.* 1977, *25*, 2608-12.
VARIABLES:	PREPARED BY:
Concentration of 18-Crown-6	R. Piekos

EXPERIMENTAL VALUES:

Concentration of 18-Crown-6 10^3 mol dm^{-3}	Solubility at 10°C [a] 10^4 mol dm^{-3}
1.0	1.65
2.0	1.99
3.0	2.02
4.0	2.04
5.0	2.05
6.0	1.90
7.0	1.65
8.0	1.68
9.0	1.64
10.0	1.48
12.0	1.55
14.0	1.50
16.0	1.55
18.0	1.56
20.0	1.63

[a] Numerical data supplied by the authors.

AUXILIARY INFORMATION

METHOD/APPARATUS/PROCEDURE:

The system was equilibrated in a sealed vial for 72 h at 10°C. The satd soln was rapidly filtered through a Toyo filter paper No. 5B, 1 cm^3 of the filtrate was evapd at 40°C and the residue was dissolved in $CHCl_3$ to det the concn in the UV region using a Hitachi 124 spectrophotometer.

SOURCE AND PURITY OF MATERIALS:

Sulfamonomethoxine, m. p. 205°C, was a very pure compd supplied by Dai-ichi Pharmaceutical Co., Ltd.
18-Crown-6 was of the reagent grade.
Purity of the benzene was not specified.

ESTIMATED ERROR:

None specified.

REFERENCES:

COMPONENTS:	ORIGINAL MEASUREMENTS:
Continued from previous page.	Takayama, K.; Nambu, N.; Nagai, T., *Chem. Pharm. Bull.* 1977, *25*, 2608-12.
VARIABLES:	**PREPARED BY:**
Concentration of 18-Crown-6	R. Piekos

EXPERIMENTAL VALUES:

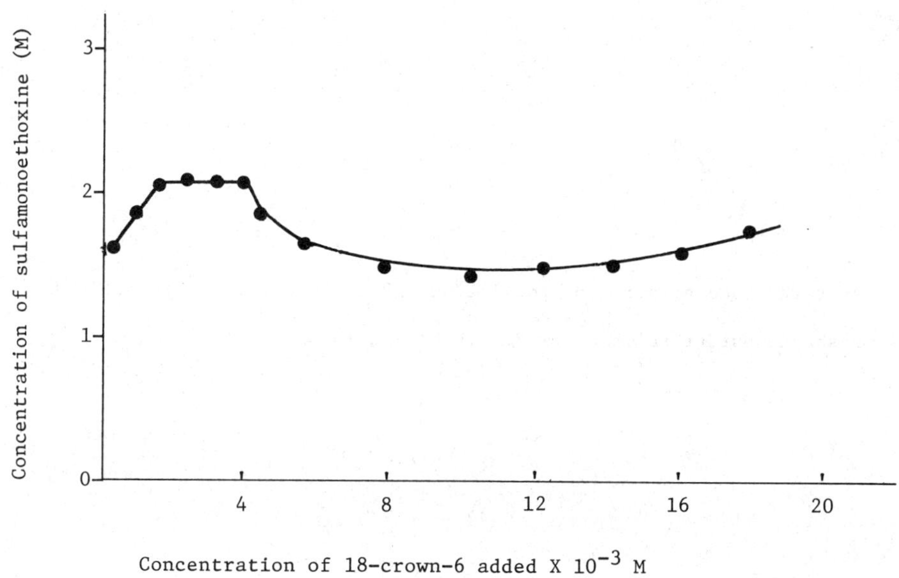

Solubility of Sulfamonomethoxine in Benzene as a Function of 18-Crown-6 Added

AUXILIARY INFORMATION

METHOD/APPARATUS/PROCEDURE:	SOURCE AND PURITY OF MATERIALS:
	ESTIMATED ERROR:
	REFERENCES:

COMPONENTS:

(1) Benzenesulfonamide, 4-amino-N-(6-methoxy-4-pyrimidinyl)- -1,4,7,10,13,16-Hexaoxacyclooctadecane complex 1:1 (SMM/18-Crown-6-complex); $C_{23}H_{36}N_4O_9S$; [65177-18-6]

(2) Phosphoric acid, disodium salt; Na_2HPO_4; [7558-94-4]

(3) Phosphoric acid, monopotassium salt; KH_2PO_4; [7778-77-0]

(4) Water; H_2O; [7732-18-5]

ORIGINAL MEASUREMENTS:

Takayama, K.; Hasegawa, S.; Sasagawa, S.; Nambu, N.; Nagai, T. *Chem. Pharm. Bull.* 1978, *26(1)*, 96-100.

VARIABLES:

One temperature: 30°C; one pH: 7.0

PREPARED BY:

R. Piekos

EXPERIMENTAL VALUES:

Saturated concentration of the 1:1 SMM/18-Crown-6-complex in a 1/15M phosphate buffer solution of 7.0 is 5.97×10^{-3}M.

AUXILIARY INFORMATION

METHOD/APPARATUS/PROCEDURE:

The stationary disk method was employed (1). At appropraite time intervals, 3-ml aliquots of soln were withdrawn, the resultant amt of vol was compensated by adding the dissoln medium at the same temp. The concn was detd by the UV spectrophotometry.

SOURCE AND PURITY OF MATERIALS:

The complex was prepd by the authors and contained a trace of the starting components. Its identity was confirmed by X-ray diffractometry, IR spectrophotometry, differenctial scanning calorimetry and other methods (2). The source and purity of the remaining materials were not specified.

ESTIMATED ERROR:

Nothing specified.

REFERENCES:

1. Hamada, Y.; Nambu, N.; Nagai, T. *Chem. Pharm. Bull.* 1975, *23*, 1205.
2. Takayama, K.; Nambu, N.; Nagai, T. *Chem. Pharm. Bull.* 1977, *25*, 2608.

COMPONENTS:

(1) Benzenesulfonamide, 4-amino-N-(5-methoxy-2-pyrimidinyl)- - 1,4,7,10,13,16-hexaoxycyclooctadecane complex (1:1) (SMM/18-C-6)); $C_{11}H_{12}N_4O_3S \cdot C_{12}H_{24}O_6$; [65177-18-6]

(2) Hydrochloric acid; HCl; [7647-01-0]

(3) Water; H_2O; [7732-18-5]

ORIGINAL MEASUREMENTS:

Takayama, K.; Nambu, N.; Nagai, T. *Chem. Pharm. Bull.* 1978, *26(10)*, 2965-70.

VARIABLES:

Temperature

PREPARED BY:

R. Piekos

EXPERIMENTAL VALUES:

t/°C	Saturated concentration of the complex in 0.2N HCl 10^2M
30	1.25
35	1.32
40	1.45

AUXILIARY INFORMATION

METHOD/APPARATUS/PROCEDURE:

An excess of sample powder was dissolved in 50 ml of 0.2N HCl. Sampling was done by a 1-ml pipet fitted with a G-4 glass filter. The concn of SMM in the filtrate was detd by uv spectrophotometry after dilg with 0.2N HCl.

SOURCE AND PURITY OF MATERIALS:

The complex was prepd by sealing 4 g of SMM anhydrate (Dai-ichi Pharmaceutical Co.) with 6 g of reagent-grade 18-C-6 and 100 ml of benzene in a flask and stiring well for 10 days at 10°C. Purity of the HCl soln was not specified.

ESTIMATED ERROR:

Nothing specified.

REFERENCES:

COMPONENTS:	ORIGINAL MEASUREMENTS:
(1) Benzenesulfonamide, 4-amino-N-(4-ethoxy-2-pyrimidinyl)-; $C_{12}H_{14}N_4O_3S$; [71138-72-2] (2) Water; H_2O; [7732-18-5]	Roblin, R. O., Jr.; Winnek, P. S. English, J. P. *J. Am. Chem. Soc.* 1942, *64*, 567-70.
VARIABLES:	PREPARED BY:
One temperature: 37°C	R. Piekos

EXPERIMENTAL VALUES:

Solubility of 4-amino-N-(4-ethoxy-2-pyrimidinyl)benzenesulfonamide in water at 37°C is 5.3 mg/100 cm^3 solution (1.8×10^{-4} mol dm^{-3}, compiler).

AUXILIARY INFORMATION

METHOD/APPARATUS/PROCEDURE:

Excess sulfonamide in water was heated and stirred on a steam bath for 30 min. The suspension was then agitated for 24 h in a thermostat at 37°C. A sample of the satd soln was withdrawn through a glass filter, dild, and analyzed by the Marshall method (1) using a General Electric recording spectrophotometer for comparing the colors developed with those of the standards.

SOURCE AND PURITY OF MATERIALS:

The sulfonamide, mp 255-6°C (cor), was prepd by the authors. Anal: %C 48.6 (calcd 49.0); %H 4.7 (4.8); %N 19.4 (19.0). Purity of the water was not specified.

ESTIMATED ERROR:

Nothing specified.

REFERENCES:

1. Bratton, A. C.; Marshall, E. K., Jr. *J. Pharmacol.* 1939, *66*, 4.

COMPONENTS:

(1) Benzenesulfonamide, 4-amino-(5-chloro-2-pyrimidinyl)-; $C_{10}H_9ClN_4O_2S$; [4482-46-6]

(2) Water; H_2O; [7732-18-5]

ORIGINAL MEASUREMENTS:

Roblin, R. O., Jr.; Winnek, P. S.; English, J. P. *J. Am. Chem. Soc.* 1942, *64*, 567-70.

VARIABLES:

One temperature: 37°C

PREPARED BY:

R. Piekos

EXPERIMENTAL VALUES:

Solubility of 4-amino-N-(5-chloro-2-pyrimidinyl)benzenesulfonamide in water at 37°C is 1.8 mg/100 cm^3 solution (6.3 x 10^{-5} mol dm^{-3}, compiler).

AUXILIARY INFORMATION

METHOD/APPARATUS/PROCEDURE:

Excess sulfonamide in water was heated and stirred on a steam bath for 30 min. The suspension was then agitated for 24 h in a thermostat at 37°C. A sample of the satd soln was withdrawn through a glass filter, dild, and analyzed by the Marshall method (1) using a General Electric recording spectrophotometer for comparing the colors developed with those of the standards.

SOURCE AND PURITY OF MATERIALS:

The sulfonamide, mp 246-7°C (cor), was prepd by the authors. Anal: %C 42.2 (calcd 42.2). Purity of the water was not specified.

ESTIMATED ERROR:

Nothing specified.

REFERENCES:

1. Bratton, A. C.; Marshall, E. K., Jr. *J. Pharmacol.* 1939, *66*, 4.

COMPONENTS:	ORIGINAL MEASUREMENTS:
(1) Benzenesulfonamide, 4-amino-N-(2-chloro-5-pyrimidinyl)-; $C_{10}H_9ClN_4O_2S$; [17103-49-0] (2) Water; H_2O; [7732-18-5]	Roblin, R. O., Jr.; Winnek, P. S.; English, J. P. *J. Am. Chem. Soc.* <u>1942,</u> *64,* 567-70.
VARIABLES:	PREPARED BY:
One temperature: 37°C	R. Piekos

EXPERIMENTAL VALUES:

Solubility of 4-amino-N-(2-chloro-5-pyrimidinyl)benzenesulfonamide in water at 37°C is 32.1 mg/100 cm^3 solution (1.13×10^{-3} mol dm^{-3}, compiler).

AUXILIARY INFORMATION

METHOD/APPARATUS/PROCEDURE:

Excess sulfonamide in water was heated and stirred on a steam bath for 30 min. The suspension was then agitated for 24 h in a thermostat at 37°C. A sample of the satd soln was withdrawn through a glass filter, dild, and analyzed by the Marshall method (1) using a General Electric recording spectrophotometer for comparing the colors developed with those of the standards.

SOURCE AND PURITY OF MATERIALS:

The sulfonamide, mp 206-7°C (cor), was prepd by the authors. Anal: %C 42.3 (calcd 42.2); %H 3.2 (3.2); %N 19.8 (19.7). Purity of the water was not specified.

ESTIMATED ERROR:

Nothing specified.

REFERENCES:

1. Bratton, A. C.; Marshall, E. K., Jr. *J. Pharmacol.* <u>1939</u>, *66*, 4.

COMPONENTS:	ORIGINAL MEASUREMENTS:
(1) Benzenesulfonamide, 4-amino-N-(4-amino-2-pyrimidinyl)-; $C_{10}H_{11}N_5O_2S$; [16806-00-1] (2) Water; H_2O; [7732-18-5]	Anderson, G. W.; Faith, H. E.; Marson, H.W.; Winnek, P. S.; Roblin, R. O., Jr. *J. Am. Chem. Soc.* 1942, *64*, 2902-5.
VARIABLES:	**PREPARED BY:**
One temperature: 37°C	R. Piekos

EXPERIMENTAL VALUES:

Solubility of 4-amino-N-(4-amino-2-pyrimidinyl)benzenesulfonamide in water at 37°C is 186 mg/100 cm^{-3} solution (7.01×10^{-3} mol dm^{-3}, compiler).

AUXILIARY INFORMATION

METHOD/APPARATUS/PROCEDURE:

Excess sulfonamide in water was heated and stirred on a steam bath for 30 min. The suspension was then agitated for 24 h in a thermostat. A sample of the satd soln was withdrawn through a glass filter, dild, and analyzed by the Marshall method (1) using a General Electric recording spectrophotometer for comparing the colors developed with those of the standards.

SOURCE AND PURITY OF MATERIALS:

The sulfonamide, mp 271-2°C, was prepd by the authors. Anal: %C 45.7 (calcd 54.3); %H 4.6 (4.2); %N 26.5 (26.4).
Purity of the water was not specified.

ESTIMATED ERROR:

Nothing specified.

REFERENCES:

1. Bratton, A. C.; Marshall, E. K., Jr. *J. Pharmacol.* 1939, *66*, 4.

COMPONENTS:	ORIGINAL MEASUREMENTS:
(1) Benzenesulfonamide, 4-amino-N-(2-amino-5-pyrimidinyl)-; $C_{10}H_{11}N_5O_2S$; [71119-38-5] (2) Water; H_2O; [7732-18-5]	Roblin, R. O., Jr.; Winnek, P. S.; English, J. P. *J. Am. Chem. Soc.* 1942, *64*, 567-70.
VARIABLES: One temperature: 37°C	PREPARED BY: R. Piekos

EXPERIMENTAL VALUES:

Solubility of 4-amino-N-(2-amino-5-pyrimidinyl)benzenesulfonamide in water at 37°C is 8.3 mg/100 cm^3 solution (3.1×10^{-4} mol dm^{-3}, compiler).

AUXILIARY INFORMATION

METHOD/APPARATUS/PROCEDURE:

Excess sulfonamide in water was heated and stirred on a steam bath for 30 min. The suspension was then agitated for 24 h in a thermostat at 37°C. A sample of the satd soln was withdrawn through a glass filter, dild, and analyzed by the Marshall method (1) using a General Electric recording spectrophotometer for comparing the colors developed with those of the standards.

SOURCE AND PURITY OF MATERIALS:

The sulfonamide, mp 293-8°C (cor), was prepd by the authors. Anal: %C 45.3 (calcd 45.3); %H 4.0 (4.1); %N 26.2 (26.4). Purity of the water was not specified.

ESTIMATED ERROR:

Nothing specified.

REFERENCES:

1. Bratton, A. C.; Marshall, E. K., Jr. *J. Pharmacol.* 1939, *66*, 4.

COMPONENTS:	ORIGINAL MEASUREMENTS:
(1) Benzenesulfonamide, 4-amino-N-[4-(diethylamino)-2-pyrimidinyl]-; $C_{14}H_{19}N_5O_2S$; [71119-24-9] (2) Water; H_2O; [7732-18-5]	Anderson, G. W.; Faith, H. E.; Marson, H.W.; Winnek, P. S.; Roblin, R. O., Jr. *J. Am. Chem. Soc.* 1942, *64*, 2902-5.
VARIABLES:	PREPARED BY:
One temperature: 37°C	R. Piekos

EXPERIMENTAL VALUES:

Solubility of 4-amino-N-[4-(diethylamino)-2-pyrimidinyl]benzenesulfonamide in water at 37°C is 4.2 mg/100 cm^3 solution (1.3 x 10^{-4} mol dm^{-3}, compiler).

AUXILIARY INFORMATION

METHOD/APPARATUS/PROCEDURE:

Excess sulfonamide in water was heated and stirred on a steam bath for 30 min. The suspension was then agitated for 24 h in a thermostat. A sample of the satd soln was withdrawn through a glass filter, dild, and analyzed by the Marshall method (1) using a General Electric recording spectrophotometer for comparing the colors developed with those of the standards.

SOURCE AND PURITY OF MATERIALS:

The sulfonamide, mp > 300°C (cor), was prepd by the authors. Anal: %C 52.5 (calcd 52.3); %H 5.8 (5.9); %N 21.7 (21.8).
Purity of the water was not specified.

ESTIMATED ERROR:

Nothing specified.

REFERENCES:

1. Bratton, A. C.; Marshall, E. K., Jr. *J. Pharmacol.* 1939, *66*, 4.

COMPONENTS:

(1) Benzenesulfonamide, 4-amino-N-(4, 5-dimethyl-2-pyrimidinyl)-; $C_{12}H_{14}N_4O_2S$; [4462-43-5]

(2) Water; H_2O; [7732-18-5]

ORIGINAL MEASUREMENTS:

Caldwell, W. T.; Kornfeld, E. C.; Donnell, C. K. *J. Am. Soc. Chem.* 1941, *63*, 2188-90.

VARIABLES:

One temperature: 29°C

PREPARED BY:

R. Piekos

EXPERIMENTAL VALUES:

Solubility of 4-amino-N-(4,5-dimethyl-2-pyrimidinyl)benzenesulfonamide in water at 29°C is 20 mg/100 ml solution (7×10^{-4} mol dm^{-3}, compiler).

AUXILIARY INFORMATION

METHOD/APPARATUS/PROCEDURE:

Soly was detd by weighing the residue obtained by evapg to dryness a known volume of soln satd at 29°C.

SOURCE AND PURITY OF MATERIALS:

The sulfonamide, mp 225-7-6.3°C (cor, recrystd from aq dioxane), was prepd by condensing 2-amino-4,5-dimethyl pyrimidine with acetylsulfanilyl chloride followed by hydrolysis with aq NaOH and pptn at pH 6. Anal: %N 20.09 (calcd 20.13). Purity of the water was not specified.

ESTIMATED ERROR:

Nothing specified.

REFERENCES:

COMPONENTS:	EVALUATOR:
(1) Benzenesulfonamide, 4-amino-N-(4,6-dimethyl-2-pyrimidinyl)- (sulfadimezine) $C_{12}H_{14}N_4O_2S$ [57-68-1] (2) Water	Anthony N. Paruta Department of Pharmaceutics University of Rhode Island Kingston, Rhode Island, USA and Ryszard Piekos Faculty of Pharmacy, University of Gdansk Gdansk, Poland 1986

CRITICAL EVALUATION:

Four reports (1-4) gave solubility values at 293K in water are given in Table I.

Table I: Solubility of Sulfadimezine in water at 293K

	10^3 mol dm^{-3} (*indicates mol kg^{-1})
Reference	293K
1	1.2*
2	1.2
3	3.3*
4	1.671*

The values given by Shkadova (3) and Gerencsér-Németh (4) are too high, the former (3) being 2.75 times higher and the latter (4) about 1.4 times higher. The recommended result is the value given by references (1,2) and can be stated as 1.2×10^{-3} mol dm^{-3} at 293K in water.

REFERENCES:

(1) Gusyakov, V.P.; Likhol'ot, N.M. *Farm. Zh. (kiev)* 1960, *15(3)*, 21-4.
(2) Likhol'ot, N.M. *Farm. Zh. (Kiev)* 1965, *20(5)*, 44-6.
(3) Shkadova, A.I. *Farm. Zh. (Kiev)* 1969, *24(3)*, 39-41.
(4) Gerencsér-Németh, M.; Harvath, M. *Gyógyszerészet* 1973, *17*, 417-21.

COMPONENTS:

(1) Benzenesulfonamide, 4-amino-N-(4,6-dimethyl-2-pyrimidinyl)- (sulfadimezine); $C_{12}H_{14}N_4O_2S$; [57-68-1]

(2) Water; H_2O; [7732-18-5]

ORIGINAL MEASUREMENTS:

Caldwell, W. T.; Kornfeld, E. C.; Donnell, C. K. *J. Am. Chem. Soc.* 1941, *63*, 2188-90.

VARIABLES:

One temperature: 29°C

PREPARED BY:

R. Piekos

EXPERIMENTAL VALUES:

Solubility of sulfadimezine in water at 29°C is 150 mg/100 ml solution (5.39×10^{-3} mol dm^{-3}, compiler).

AUXILIARY INFORMATION

METHOD/APPARATUS/PROCEDURE:

Soly was detd by weighing the residue obtained by evapg to dryness a known volume of soln satd at 29°C.

SOURCE AND PURITY OF MATERIALS:

Sulfadimezine, mp 178-80°C (cor, recrystd from water), was prepd by condensing 2-amino-4,6-dimethylpyrimidine with acetylsulfanilyl chloride followed by hydrolysis with aq NaOH and pptn at pH 6. Anal: %C 51.78 (calcd 51.78), %H 4.83 (5.07), %N 20.03, 20.17 (20.13). Purity of the water was not specified.

ESTIMATED ERROR:

Nothing specified.

REFERENCES:

COMPONENTS:

(1) Benzenesulfonamide, 4-amino-N-(4,6-dimethyl-2-pyrimidinyl)- (sulfadimezine); $C_{12}H_{14}N_4O_2S$; [57-68-1]

(2) Water; H_2O; [7732-18-5]

ORIGINAL MEASUREMENTS:

Roblin, R. O., Jr.; Winnek, P. S.; English, J. P. *J. Am. Chem. Soc.* 1942, *64*, 567-70.

VARIABLES:

One temperature: 37°C

PREPARED BY:

R. Piekos

EXPERIMENTAL VALUES:

Solubility of sulfadimezine in water at 37°C is 75 mg/100 cm^3 solution[a]

(2.7×10^{-3} mol dm^{-3}, compiler).

[a] The compound tends to remain supersaturated indefinitely in the presence of excess solid. The value represents approximate equilibrium when approached from low side (authors).

AUXILIARY INFORMATION

METHOD/APPARATUS/PROCEDURE:

Excess sulfonamide in water was heated and stirred on a steam bath for 30 min. The suspension was then agitated for 24 h in a thermostat at 37°C. A sample of the satd soln was withdrawn through a glass filter, dild, and analyzed by the Marshall method (1) using a General Electric recording spectrophotometer for comparing the colors developed with those of the standards.

SOURCE AND PURITY OF MATERIALS:

Sulfadiazine, mp 198-9°C (cor), was prepd by the authors. Anal: %C 52.0 (calcd 51.8); %H 5.1 (5.0); %N 20.0 (20.1). Purity of the water was not specified.

ESTIMATED ERROR:

Nothing specified.

REFERENCES:

1. Bratton, A. C.; Marshall, E. K., Jr. *J. Pharmacol.* 1939, *66*, 4.

COMPONENTS:

(1) Benzenesulfonamide, 4-amino-N-(4,6-dimethyl-2-pyrimidinyl)- (sulfadimezine); $C_{12}H_{14}N_4O_2S$; [57-68-1]

(2) Water; H_2O; [7732-18-5]

ORIGINAL MEASUREMENTS:

Gusyakov, V. P.; Likhol'ot, N. M.

Farm. Zh. (Kiev) 1960, *15(3)*, 21-4.

VARIABLES:

One temperature: 20°C

PREPARED BY:

R. Piekos

EXPERIMENTAL VALUES:

Solubility of sulfadimezine in water at 20°C is 0.033 g/100 g water (1.2 x 10^{-3} mol kg^{-1}, compiler).

AUXILIARY INFORMATION

METHOD/APPARATUS/PROCEDURE:

A small excess of sulfadimezine was equilibrated for 8 h in a 50-ml test tube with 20 ml of water. Aliquots were taken with a pipet fitted with a filter. Sulfadimezine was detd at 285 nm using a SF-4 spectrophotometer.

SOURCE AND PURITY OF MATERIALS:

Sulfadimezine was carefully washed with water, alcohol, and dried. Its purity conformed to the requirements of the State Pharmacopeia VIII.

Purity of the water was not specified.

ESTIMATED ERROR:

Soly: the accuracy corresponded to that of the colorimentric detns (authors:

Temp: not specified.

REFERENCES:

COMPONENTS:	ORIGINAL MEASUREMENTS:
(1) Benzenesulfonamide, 4-amino-N-(4,6-dimethyl-2-pyrimidinyl)- (sulfadimezine); $C_{12}H_{14}N_4O_2S$; [57-68-1] (2) Water; H_2O; [7732-18-5]	Likhol'ot, N. M.; *Farm. Zh. (Kiev)* 1965, *20(5)*, 44-6.
VARIABLES:	PREPARED BY:
One temperature: 20°C	R. Piekos

EXPERIMENTAL VALUES:

Solubility of sulfadimezine in water at 20°C is 0.033 g/100 ml (1. 2 x 10^{-3} mol dm^{-3}, compiler).

AUXILIARY INFORMATION

METHOD/APPARATUS/PROCEDURE:	SOURCE AND PURITY OF MATERIALS:
An earlier described method was employed (1) whereby a small excess of sulfadimezine was equilibrated with 20 ml of water for 8 h in a 50-ml test tube. Aliquots were withdrawn through a filter and sulfadimezine was assayed bromatometrically.	Nothing specified.
	ESTIMATED ERROR: Soly: not specified. Temp: ±0.1°C (authors).
	REFERENCES: 1. Gusyakov, V. P.; Likhol'ot, N. M. *Farm. Zh. (Kiev)* 1960, *15(8)*, 21.

COMPONENTS:

(1) Benzenesulfonamide, 4-amino-N-(4,6-dimethyl-2-pyrimidinyl)- (sulfadimezine); $C_{12}H_{14}N_4O_2S$; [57-68-1]

(2) Water; H_2O; [7732-18-5]

ORIGINAL MEASUREMENTS:

Shkadova, A. I.

Farm. Zh. (Kiev) 1969, *24(3)*, 39-41.

VARIABLES:

One temperature: 20°C

PREPARED BY:

R. Piekos

EXPERIMENTAL VALUES:

Solubility of sulfadimezine in water at 20°C is 0.33×10^{-2} mol/kg (9.2×10^{-2} g/100 g, compiler).

AUXILIARY INFORMATION

METHOD/APPARATUS/PROCEDURE:

A satd soln of sulfadimezine was equilibrated in a water thermostat at 20±0.1°C. The concn of sulfadimezine was detd by alkalimetric titrn.

SOURCE AND PURITY OF MATERIALS:

Purity of sulfadimezine conformed to the requirements of the State Pharmacopeia IX. Distd water was used.

ESTIMATED ERROR:

Soly: not specified.

Temp: ±0.1°C (author).

REFERENCES:

COMPONENTS:	ORIGINAL MEASUREMENTS:
(1) Benzenesulfonamide, 4-amino-N-(4,6-dimethyl-2-pyrimidinyl)- (sulfadimidine); $C_{12}H_{14}N_4O_2S$; [57-68-1] (2) Water; H_2O; [7732-18-5]	Gerencsér-Németh, M.; Horváth, M. *Gyógyszerészet* 1973, *17*, 417-21.
VARIABLES:	**PREPARED BY:**
One temperature: 20°C	R. Piekos

EXPERIMENTAL VALUES:

Solubility of sulfadimidine in water at 20°C is 0.0465 g/100 g solution (1.67 x 10^{-3} mol kg^{-1} water, compiler).

AUXILIARY INFORMATION

METHOD/APPARATUS/PROCEDURE:

A weighed excess of sulfadimidine was shaken with water in a shaker at 120 rpm for 6 h. The soln was then filtered, the residue was washed with the filtrate and finally with a small amt of distd water, dried and weighed.

SOURCE AND PURITY OF MATERIALS:

Sulfadimidine (source and purity not specified) was dried at 100°C for 3 h or over concn H_2SO_4 for 72 h. Its mp was 196.5-8°C. Distd water was used.

ESTIMATED ERROR:

Soly: precision ±0.0003 g/100 g (2 detns)-compiler.

Temp: not specified.

REFERENCES:

COMPONENTS:

(1) Benzenesulfonamide, 4-amino-N-(4,6-dimethyl-2-pyrimidinyl)- (sulfamethazine; sulfadimidine); $C_{12}H_{14}N_4O_2S$; [57-68-1]

(2) Water; H_2O; [7732-18-5]

ORIGINAL MEASUREMENTS:

Nasipuri, R. N.; Khalil, S. A. H.

J. Pharm. Sci. 1973, *62(3)*, 473-5.

VARIABLES:

One temperature: 24°C

PREPARED BY:

R. Piekos

EXPERIMENTAL VALUES:

Equilibrium solubility of sulfamethazine in water[a] at 24°C is 1.424 mmol/l (0.3963 g dm^{-3}, compiler).

[a]The pH of the saturated aqueous solution was 5.6

AUXILIARY INFORMATION

METHOD/APPARATUS/PROCEDURE:

Sulfamethazine was assayed spectrophotometrically at 230 nm in a 0.05M HCl soln using a Unicam SP 500 spectrophotometer.

SOURCE AND PURITY OF MATERIALS:

Sulfamethazine was a BP grade sulfadimidine (Imperial Chemical Industries, England). The purity of water was not specified.

ESTIMATED ERROR:

Soly: ±1.9% (not stated accuracy or reproducibility - authors).

Temp: ±0.2°C (authors).

REFERENCES:

COMPONENTS:

(1) Benzenesulfonamide, 4-amino-N-(4,6-dimethyl-2-pyrimidinyl)- (sulfadimidine); $C_{12}H_{14}N_4O_2S$; [57-68-1]

(2) Water; H_2O; [7732-18-5]

ORIGINAL MEASUREMENTS:

Abdel Hadi, I.; Mezősi, J.; Kedvessy, G.; Morvay, J. *Pharmazie* 1977, *32*, 791-3.

VARIABLES:

One temperature: 22°C

PREPARED BY:

R. Piekos

EXPERIMENTAL VALUES:

Solubility of cyrstalline forms I and II of sulfadimidine in water at 22°C is equal and amounts to 0.35 mg/ml water (1.3×10^{-3} mol dm^{-3}, compiler)

AUXILIARY INFORMATION

METHOD/APPARATUS/PROCEDURE:

Soly was detd as described by Martin (1). The system was equilibrated at 22°C for 48 h.

SOURCE AND PURITY OF MATERIALS:

Sulfadimidine was a comm product (Reanal, Hungary). Cryst. forms I and II, obtained by crystn from MeOH and EtOH resp, had both the mp of 197-8°C. Their identity was confirmed by IR spectroscopy, differential thermal analysis and X-ray diffraction patterns. Purity of the water was not specified.

ESTIMATED ERROR:

Nothing specified.

REFERENCES:

1. Martin, A. N. *Physical Pharmacy*, Lea and Febiger, Philadelphia 1962, p. 340.

COMPONENTS:

(1) Benzenesulfonamide, 4-amino-N-(4,6-dimethyl-2-pyrimidinyl)- (sulfadimezine); $C_{12}H_{14}N_4O_2S$; [57-68-1]

(2) Water; H_2O; [7732-18-5]

ORIGINAL MEASUREMENTS:

Ezerskii, M. L.; Per'kova, N. M.
Khim.-Farm. Zh. 1979, *13(11)*, 87-91.

VARIABLES:

Grinding regime

PREPARED BY:

R. Piekos

EXPERIMENTAL VALUES:

Specimen of sulfadiazine	Solubility at room temperature	
	g/cm^3	10^3 mol dm^{-3} [a]
Commercial	0.00036	1.3
Commercial, ground in a ball mill	0.00036	1.3
Commercial, ground in a jet mill	0.00038	1.4

[a] Calculated by compiler.

AUXILIARY INFORMATION

METHOD/APPARATUS/PROCEDURE:

Satd solns were prepd by prolonged agitation of an excess of sulfadimezine in water at room temp. The solns were then allowed to stand for 12 h and filtered. The concn of sulfadimezine in the filtrate was detd nitritometrically by the method of the State Pharmacopeia X.

SOURCE AND PURITY OF MATERIALS:

Comm, pharmacopeial sulfadimezine was used (source not specified). It was ground in a Pulverisette-5 lab ball mill or in a C-1266-00 jet mill.

Purity of the water was not specified.

ESTIMATED ERROR:

Nothing specified.

REFERENCES:

COMPONENTS:

(1) Benzenesulfonamide, 4-amino-N-(4,6-dimethyl-2-pyrimidinyl)- (sulfadimezine); $C_{12}H_{14}N_4O_2S$; [57-68-1]

(2) Lithium chloride; LiCl; [7447-41-8]

(3) Water; H_2O; [7732-18-5]

ORIGINAL MEASUREMENTS:

Likholet, N. M.; Gusyakov, V. P.

Med. Prom. SSSR 1963, *17(5)*, 28-31.

VARIABLES:

Concentration of LiCl

PREPARED BY:

R. Piekos

EXPERIMENTAL VALUES:

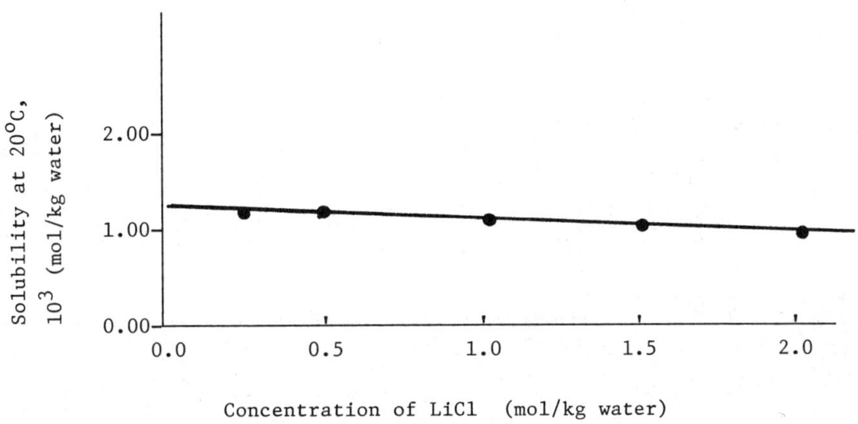

AUXILIARY INFORMATION

METHOD/APPARATUS/PROCEDURE:

Satd solns of sulfadimezine were prepd in 50-ml tightly closed ampuls in which 20 ml of a LiCl soln was placed and a small excess of sulfadimezine. The mixts were equilibrated for 18 h at 20°C. Aliquots were pipetted out through a filter, dild, and assayed spectrophotometrically at 285 nm on a SF-IV spectrophotometer.

SOURCE AND PURITY OF MATERIALS:

Sulfadimezine was purified by recrystn. Its purity was 98.5%. LiCl was purified by a recommended procedure (1). Purity of the water was not specified.

ESTIMATED ERROR:

Soly: measurements were repeated several times (authors).

Temp: ±0.1°C (authors).

REFERENCES:

1. Karyakin, Ya. V.; Angelov, I. I. *Chistye khimicheskye reaktivy,* Moscow, 1955.

COMPONENTS:	ORIGINAL MEASUREMENTS:
(1) Benzenesulfonamide, 4-amino-N-(4,6-dimethyl-2-pyrimidinyl)- (sulfadimezine); $C_{12}H_{14}N_4O_2S$; [57-68-1] (2) Sodium chloride; NaCl; [7647-14-5] (3) Water; H_2O; [7732-18-5]	Likholet, N. M.; Gusyakov, V. P. *Med. Prom. SSSR* 1963, *17(5)*, 28-31.

VARIABLES:	PREPARED BY:
Concentration of NaCl	R. Piekos

EXPERIMENTAL VALUES:

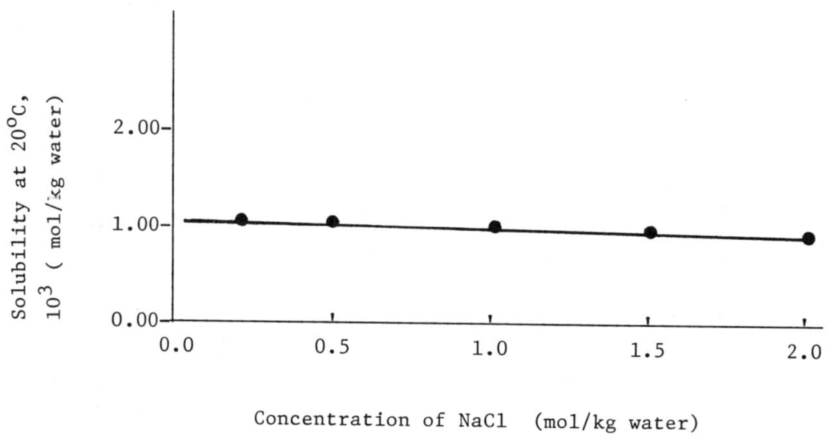

AUXILIARY INFORMATION

METHOD/APPARATUS/PROCEDURE:

Satd solns of sulfadimezine were prepd in 50 ml tightly closed ampuls in which 20 ml of a NaCl soln was placed and a small excess of sulfadimezine. The mixts were equilibrated for 18 h at 20°C. Aliquots were pipetted out through a filter, dild, and assayed spectrophotometrically at 285 nm on a SF-IV spectrophotometer.

SOURCE AND PURITY OF MATERIALS:

Sulfadimezine (source not specified) was purified by recrystn. Its purity was 98.5%. NaCl was purified by a recommended procedure.

Purity of the water was not specified.

ESTIMATED ERROR:

Soly: measurements were repreated several times (authors).

Temp: ±0.1°C (authors).

REFERENCES:

1. Karyakin, Ya. V.; Angelov, I. I. *Chistye khimicheskye reaktivy*, Moscow, 1955.

COMPONENTS:

(1) Benzenesulfonamide, 4-amino-N-(4 dimethyl-2-pyrimidinyl)- (sulfadimezine); $C_{12}H_{14}N_4O_2S$; [57-68-1]

(2) Potassium chloride; KCl; [7447-40-7]

(3) Water; H_2O; [7732-18-5]

ORIGINAL MEASUREMENTS:

Gusyakov, V. P.; Likhol'ot, N. M.

Farm. Zh. (Kiev) 1960, *15(3)*, 21-4.

VARIABLES:

Concentration of KCl

PREPARED BY:

R. Piekos

EXPERIMENTAL VALUES:

Concentration of KCl	Solubility at 20°C	
Weight %	g/100 g water	10^3 mol kg^{-1} [a]
0.74	0.031	1.11
1.82	0.030	1.01
3.59	0.033	1.19
6.93	0.032	1.15
12.97	0.025	0.90

[a] Calculated by compiler

AUXILIARY INFORMATION

METHOD/APPARATUS/PROCEDURE:

A small excess of sulfadimezine was equilibrated for 8 h in a 50-ml test tube with 20 ml of aqueous KCl soln. Aliquots were taken with a pipet fitted with a filter. Sulfadimezine was detd in the filtrate at 285 nm using a SF-4 spectrophotometer.

SOURCE AND PURITY OF MATERIALS:

Sulfadimezine was carefully washed with water, alcohol, and dried. Its purity conformed to the requirements of the State Pharmacopeia VIII.

KCl was doubly crystd. Purity of the water was not specified.

ESTIMATED ERROR:

Soly: the accuracy corresponded to that of colorimetric detns. (authors)

Temp: not specified.

REFERENCES:

COMPONENTS:	ORIGINAL MEASUREMENTS:
(1) Benzenesulfonamide, 4-amino-N-(4,6-dimethyl-2-pyrimidinyl)- (sulfadimezine); $C_{12}H_{14}N_4O_2S$; [57-68-1] (2) Potassium chloride: KCl; [7447-40-7] (3) Water; H_2O; [7732-18-5]	Likholet, N. M.; Gusyakov, V. P. *Med. Prom. SSSR* 1963, *17(5)*, 28-31.
VARIABLES:	PREPARED BY:
Concentration of KCl	R. Piekos

EXPERIMENTAL VALUES:

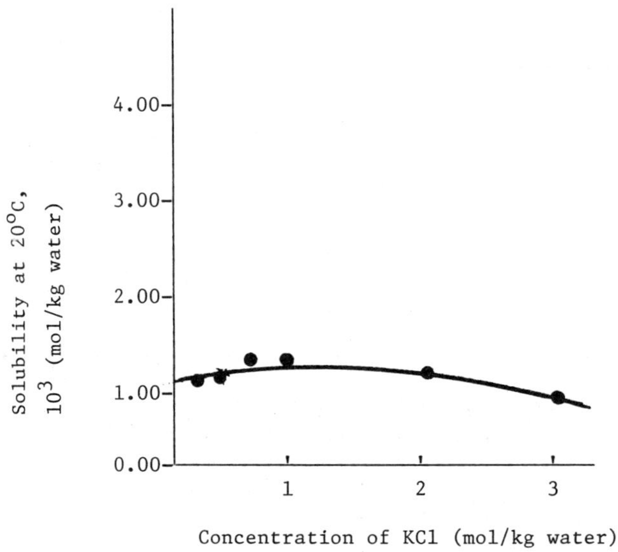

AUXILIARY INFORMATION

METHOD/APPARATUS/PROCEDURE:

Satd solns of sulfadimezine were prepd in 50-ml tightly closed ampuls in which 20 ml of a KCl soln was placed and a small excess of sulfadimezine. The mixts were equilibrated for 18 h at 20°C. Aliquots were pipetted out through a filter, dild, and assayed spectrophotometrically at 285 nm on a SF-IV spectrophotometer.

SOURCE AND PURITY OF MATERIALS:

Sulfadimezine (source not specified) was purified by recrystn. Its purity was 98.5%. KCl was purified by a recommended procedure (1). Purity of the water was not specified.

ESTIMATED ERROR:

Soly: measurements were repeated several times (authors).

Temp: ±0.1°C (authors).

REFERENCES:

1. Karyakin, Ya. V.; Angelov, I. I. *Chistye khimicheskye reaktivy*, Moscow, 1955.

COMPONENTS:

Continued from previous page

ORIGINAL MEASUREMENTS:

Likholet, N. M.; Gusyakov, V. P.
Med. Prom. SSSR 1963, *17(5)*, 28-31.

VARIABLES:

Concentration of KCl

PREPARED BY:

R. Piekos

EXPERIMENTAL VALUES:

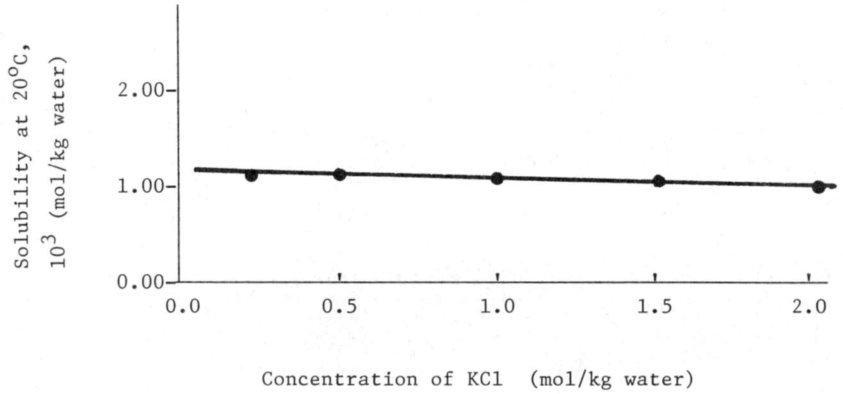

AUXILIARY INFORMATION

METHOD/APPARATUS/PROCEDURE:

SOURCE AND PURITY OF MATERIALS:

ESTIMATED ERROR:

REFERENCES:

COMPONENTS:

(1) Benzenesulfonamide, 4-amino-N-(4,6-dimethyl-2-pyrimidinyl)- (sulfadimezine); $C_{12}H_{14}N_4O_2S$; [57-68-1]

(2) Potassium bromide; KBr; [7758-02-3]

(3) Water; H_2O; [7732-18-5]

ORIGINAL MEASUREMENTS:

Gusyakov, V. P.; Likhol'ot, N. M.
Farm. Zh. (Kiev) 1960, *15(3)*, 21-4.

VARIABLES:

Concentration of KBr

PREPARED BY:

R. Piekos

EXPERIMENTAL VALUES:

Concentration of KBr	Solubility at 20°C	
Weight %	g/100 g water	10^3 mol kg^{-1} [a]
1.17	0.037	1.33
2.88	0.041	1.47
5.61	0.047	1.69
10.63	0.046	1.65
19.22	0.044	1.58

[a] Calculated by compiler

AUXILIARY INFORMATION

METHOD/APPARATUS/PROCEDURE:

A small excess of sulfadimezine was equilibrated for 8 h in a 50-ml test tube with 20 ml of aqueous KBr soln. Aliquots were taken with a pipet fitted with a filter. Sulfadimezine was detd in the filtrate at 285 nm using a SF-4 spectrophotometer.

SOURCE AND PURITY OF MATERIALS:

Sulfadimezine was carefully washed with water, alcohol, and dried. Its purity conformed to the requirements of the State Pharmacopeia VIII. KBr was doubly crystd. Purity of the water was not specified.

ESTIMATED ERROR:

Soly: the accuracy corresponded to that of colorimetric detns (authors).

Temp: not specified.

REFERENCES:

COMPONENTS:	ORIGINAL MEASUREMENTS:
(1) Benzenesulfonamide, 4-amino-N-(4,6-dimethyl-2-pyrimidinyl)- (sulfadimezine); $C_{12}H_{14}N_4O_2S$; [57-68-1] (2) Potassium bromide; KBr; [7758-02-3] (3) Water; H_2O; [7732-18-5]	Likholet, N. M.; Gusyakov, V. P. *Med. Prom. SSSR* 1963, *17(5)*, 28-31.
VARIABLES:	PREPARED BY:
Concentration of KBr	R. Piekos

EXPERIMENTAL VALUES:

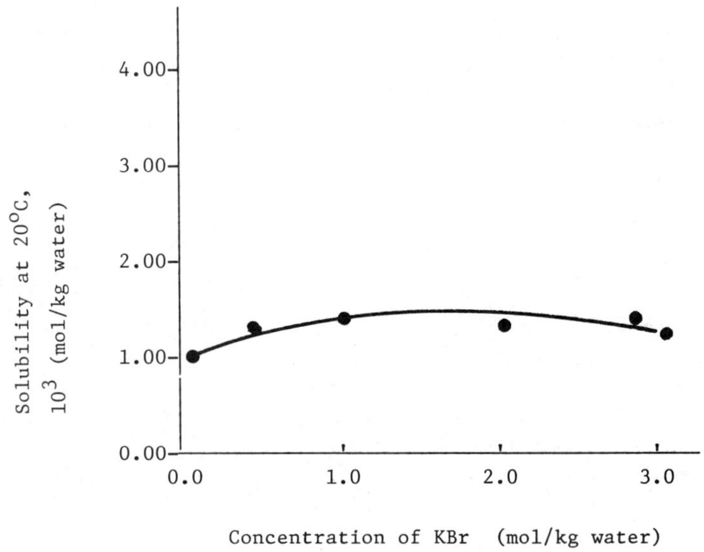

AUXILIARY INFORMATION

METHOD/APPARATUS/PROCEDURE:

Satd solns of sulfadimezine were prepd in 50-ml tightly closed ampuls in which 20 ml of a KBr soln was placed and a small excess of sulfadimezine. The mixts were equilibrated for 18 h at 20°C. Aliquots were pipetted out through a filter, dild, and assayed spectrophotometrically at 285 nm on a SP-IV spectrophotometer.

SOURCE AND PURITY OF MATERIALS:

Sulfadimezine (source not specified) was purified by recrystn. Its purity was 98.5%.
KBr was purified by a recommended procedure (1).
Purity of the water was not specified.

ESTIMATED ERROR:

Soly: measurements were repeated several times (authors).

Temp: ±0.1°C (authors).

REFERENCES:

1. Karyakin, Ya. V.; Angelov, I. I. *Chistye khimicheskye reaktivy*, Moscow, 1955.

COMPONENTS:

(1) Benzenesulfonamide, 4-amino-N-(4,6-dimethyl-2-pyrimidinyl)- (sulfadimezine); $C_{12}H_{14}N_4O_2S$; [57-68-1]

(2) Potassium iodide; KI; [7681-11-0]

(3) Water; H_2O; [7732-18-5]

ORIGINAL MEASUREMENTS:

Gusyakov, V. P.; Likhol'ot, N. M

Farm. Zh. (Kiev) 1960, *15(3)*, 21-4.

VARIABLES:

Concentration of KI

PREPARED BY:

R. Piekos

EXPERIMENTAL VALUES:

Concentration of KI	Solubility at 20°C	
Weight %	g/100 g water	10^3 mol kg^{-1} [a]
1.63	0.039	1.40
3.98	0.038	1.36
7.66	0.048	1.72
14.23	0.053	1.90

[a] Calculated by compiler

AUXILIARY INFORMATION

METHOD/APPARATUS/PROCEDURE:

A small excess of sulfadimezine was equilibrated for 8 h in a 50-ml test tube with 20 ml aqueous KI soln. Aliquots were taken with a pipet fitted with a filter. Sulfadimezine was detd in the filtrate at 285 nm using a SF-4 spectrophotometer.

SOURCE AND PURITY OF MATERIALS:

Sulfadimezine was carefully washed with water, alcohol, and dried. Its purity conformed to the requirements of the State Pharmacopeia VIII. KI was doubly crystd. Purity of the water was not specified.

ESTIMATED ERROR:

Soly: the accuracy corresponded to that of colorimetric detns (authors).

Temp: not specified.

REFERENCES:

COMPONENTS:	ORIGINAL MEASUREMENTS:
(1) Benzenesulfonamide, 4-amino-N-(4,6-dimethyl-2-pyrimidinyl)- (sulfadimezine); $C_{12}H_{14}N_4O_2S$; [57-68-1] (2) Potassium iodide; KI; [7681-11-0] (3) Water; H_2O; [7732-18-5]	Likholet, N. M.; Gusyakov, V. P. *Med. Prom. SSSR* <u>1963</u>, *17(5)*, 28-31.
VARIABLES:	PREPARED BY:
Concentration of KI	R. Piekos

EXPERIMENTAL VALUES:

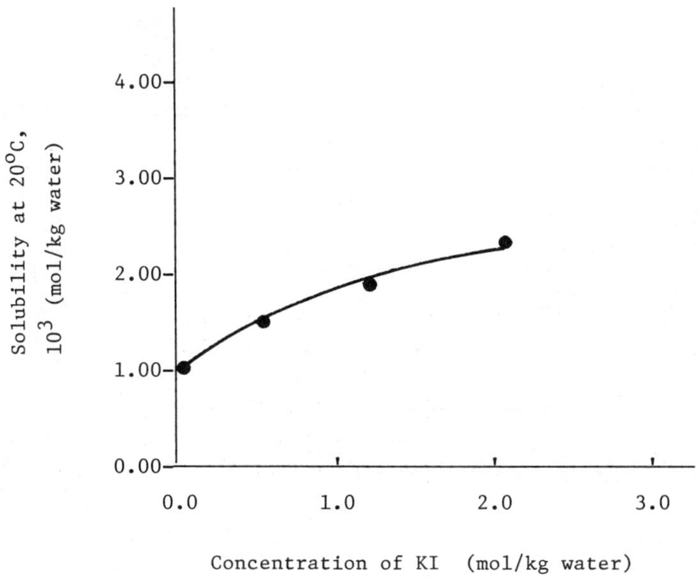

AUXILIARY INFORMATION

METHOD/APPARATUS/PROCEDURE:	SOURCE AND PURITY OF MATERIALS:
Satd solns of sulfadimezine were prepd in 50-ml tightly closed ampuls in which 20 ml of a KI soln was placed and a small excess of sulfadimezine. The mixts were equilibrated for 18 h at 20°C. Aliquots were pipetted out through a filter, dild, and assayed spectrophotometrically at 285 nm on a SF-IV spectrophotometer.	Sulfadimezine (source not specified) was purified by recrystn. Its purity was 98.5%. KI was purified by a recommended procedure (1). Purity of the water was not specified.
	ESTIMATED ERROR: Soly: measurements were repeated several times (authors). Temp: ±0.1°C (authors).
	REFERENCES: 1. Karyakin, Ya. V.; Angelov, I. I. *Chistye Khimicheskye reaktivy*, Moscow, <u>1955</u>.

COMPONENTS:

(1) Benzenesulfonamide, 4-amino-N-(4,6-dimethyl-2-pyrimidinyl)- (sulfadimezine); $C_{12}H_{14}N_4O_2S$; [57-68-1]

(2) Potassium thiocyanate; KSCN; [333-20-0]

(3) Water; H_2O; [7732-18-5]

ORIGINAL MEASUREMENTS:

Gusyakov, V. P.; Likhol'ot, N. M.

Farm. Zh. (Kiev) 1960, *15(3)*, 21-4.

VARIABLES:

Concentration of KSCN

PREPARED BY:

R. Piekos

EXPERIMENTAL VALUES:

Concentration of KSCN	Solubility at 20°C	
Weight %	g/100 g water	10^3 mol kg^{-1} [a]
0.96	0.036	1.2
2.37	0.039	1.4
4.63	0.047	1.6
8.85	0.064	2.3

[a] Calculated by compiler

AUXILIARY INFORMATION

METHOD/APPARATUS/PROCEDURE:

A samll excess of sulfadimezine was equilibrated for 8 h in a 50-ml test tube with 20 ml of aqueous KSCN soln. Aliquots were taken with a pipet fitted with a filter. Sulfadimezine was detd in the filtrate at 285 nm using a SF-4 spectrophotometer.

SOURCE AND PURITY OF MATERIALS:

Sulfadimezine was carefully washed with water, alcohol, and dried. Its purity conformed to the requirements of the State Pharmacopeia VIII. KSCN was doubly cyrstd. Purity of the water was not specified.

ESTIMATED ERROR:

Soly: the accuracy corresponded to that of colorimetric detns (authors).

Temp: not specified.

REFERENCES:

COMPONENTS:	ORIGINAL MEASUREMENTS:
(1) Benzenesulfonamide, 4-amino-N-(4,6-dimethyl-2-pyrimidinyl)- (sulfadimezine); $C_{12}H_{14}N_4O_2S$; [57-68-1] (2) Potassium thiocyanate; KCNS; [333-20-0] (3) Water; H_2O; [7732-18-5]	Likholet, N. M.; Gusyakov, V. P. *Med. Prom. SSSR.* 1963, *17(5)*, 28-31.
VARIABLES:	PREPARED BY:
Concentration of KCNS	R. Piekos

EXPERIMENTAL VALUES:

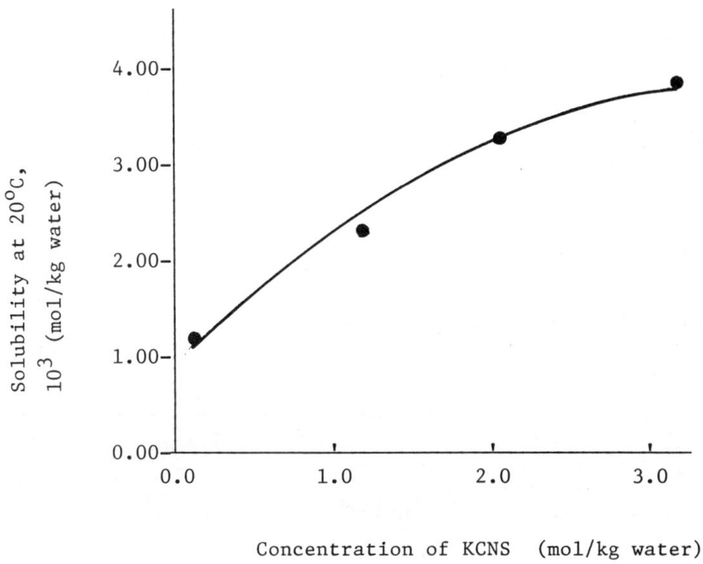

AUXILIARY INFORMATION

METHOD/APPARATUS/PROCEDURE:	SOURCE AND PURITY OF MATERIALS:
Satd solns of sulfadimezine were prepd in 50-ml tightly closed ampuls in which 20 ml of a KCNS soln was placed and a small excess of sulfadimezine. The mixts were equilibrated for 18 h at 20°C. Aliquots were pipetted out through a filter, dild, and assayed spectrophotometrically at 285 nm on a SF-IV spectrophotometer.	Sulfadimezine (source not specified) was purified by recrystn. Its purity was 98.5%. KCNS was purified by a recommended procedure (1). Purity of the water was not specified.
	ESTIMATED ERROR: Soly: measurements were repeated several times (authors). Temp: ±0.1°C (authors).
	REFERENCES: 1. Karyakin, Ya. V.; Angelov, I. I. *Chistye Khimicheskye reaktivy*, Moscow, 1955.

COMPONENTS:

(1) Benzenesulfonamide, 4-amino-N-(4,6-dimethyl-2-pyrimidinyl)- (sulfadimezine; $C_{12}H_{14}N_4O_2S$; [57-68-1]

(2) Magnesium chloride; $MgCl_2$; [7786-30-3]

(3) Water; H_2O; [7732-18-5]

ORIGINAL MEASUREMENTS:

Likholet, N. M.; Gusyakov, V. P.

Med. Prom. SSSR 1963, *17(5)*, 28-31.

VARIABLES:

Concentration of $MgCl_2$

PREPARED BY:

R. Piekos

EXPERIMENTAL VALUES:

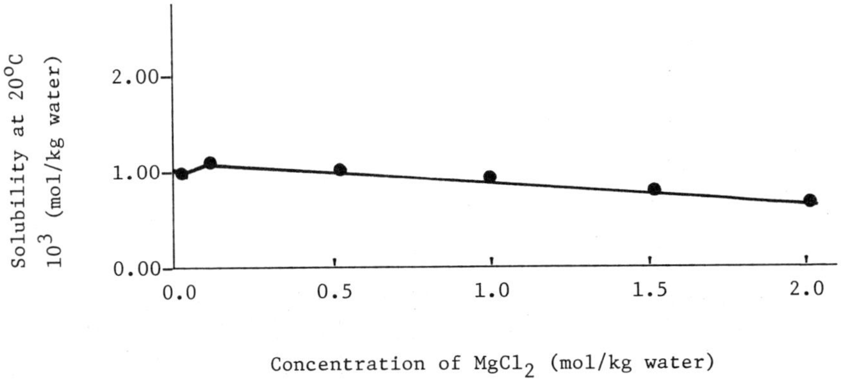

AUXILIARY INFORMATION

METHOD/APPARATUS/PROCEDURE:

Satd solns of sulfadimezine were prepd in 50-ml tightly closed ampuls in which 20 ml of a $MgCl_2$ soln was placed and a small excess of sulfadimezine. The mixts were equilibrated for 18 h at 20°C. Aliquots were pipetted out through a filter, dild, and assayed spectrophotometrically at 285 nm on a SF-IV spectrophotometer.

SOURCE AND PURITY OF MATERIALS:

Sulfadimezine (source not specified) was purified by recrystn. Its purity was 98.5%. $MgCl_2$ was purified by a recommended procedure.

Purity of the water was not specified.

ESTIMATED ERROR:

Soly: measurements were repeated several times (authors).

Temp: ±0.1°C (authors).

REFERENCES:

1. Karyakin, Ya. V.; Angelov, I. I. *Chistye Khimicheskye reaktivy*, Moscow, 1955.

COMPONENTS:

(1) Benzenesulfonamide, 4-amino-N-(4,6-dimethyl-2-pyrimidinyl)- (sulfadimezine); $C_{12}H_{14}N_4O_2S$; [57-68-1]

(2) Barium chloride; $BaCl_2$; [10361-37-2]

(3) Water; H_2O; [7732-18-5]

ORIGINAL MEASUREMENTS:

Likholet, N. M.; Gusyakov, V. P.

Med. Prom. SSSR. 1963, *17(5)*, 28-31.

VARIABLES:

Concentration of $BaCl_2$

PREPARED BY:

R. Piekos

EXPERIMENTAL VALUES:

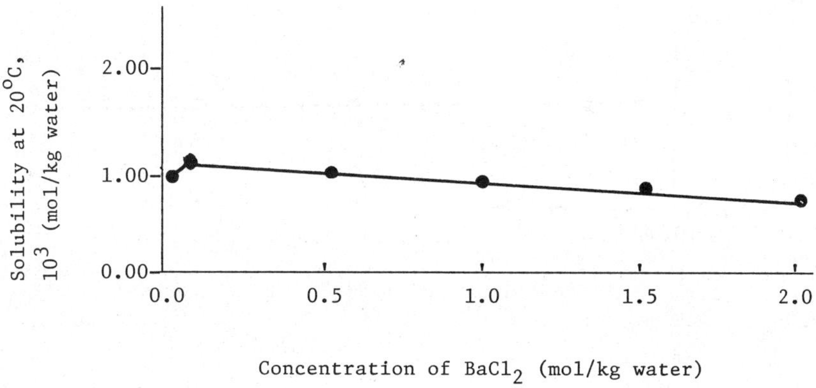

AUXILIARY INFORMATION

METHOD/APPARATUS/PROCEDURE:

Satd solns of sulfadimezine were prepd in 50-ml tightly closed ampuls in which 20 ml of a $BaCl_2$ soln was placed and a small excess of sulfadimezine. The mixts were equilibrated for 18 h at 20°C. Aliquots were pipetted out through a filter, dild, and assayed spectrophotometrically at 285 nm on a SF-IV spectrophotometer.

SOURCE AND PURITY OF MATERIALS:

Sulfadimezine (source not specified) was purified by recrystn. Its purity was 98.5%. $BaCl_2$ was purified by a recommended procedure (1).

Purity of the water was not specified.

ESTIMATED ERROR:

Soly: measurements were repeated several times (authors).

Temp: ±0.1°C (authors).

REFERENCES:

1. Karyakin, Ya. V.; Angelov, I. I. *Chistye khimicheskye reaktivy,* Moscow, 1955.

COMPONENTS:	ORIGINAL MEASUREMENTS:
(1) Benzenesulfonamide, 4-amino-N-(4,6-dimethyl-2-pyrimidinyl)- (sulfadimezine); $C_{12}H_{14}N_4O_2S$; [57-68-1] (2) Ammonium chloride; NH_4Cl; [12125-02-9] (3) Water; H_2O; [7732-18-5]	Likholet, N. M.; Gusyakov, V. P. *Med. Prom. SSSR* 1963, *17(5)*, 28-31.
VARIABLES:	**PREPARED BY:**
Concentration of NH_4Cl	R. Piekos

EXPERIMENTAL VALUES:

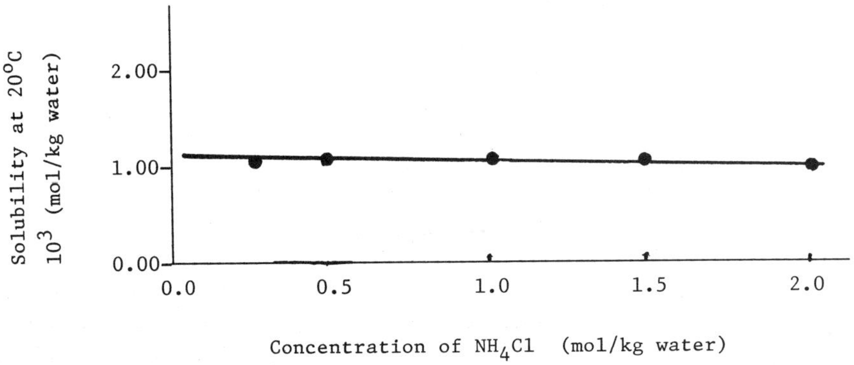

AUXILIARY INFORMATION

METHOD/APPARATUS/PROCEDURE:	SOURCE AND PURITY OF MATERIALS:
Satd solns of sulfadimezine were prepd in 50-ml tightly closed ampuls in which 20 ml of a NH_4Cl soln was placed and a small excess of sulfadimezine. The mixts were equilibrated for 18 h at 20°C. Aliquots were pipetted out through a filter, dild, and assayed spectrophotometrically at 285 nm on a SF-IV spectrophotometer.	Sulfadimezine (source not specified) was purified by crystn. Its purity was 98.5%. NH_4Cl was purified by a recommended procedure. Purity of the water was not specified.
	ESTIMATED ERROR: Soly: measurements were repeated several times (authors). Temp: ±0.1°C (authors).
	REFERENCES: 1. Karyakin, Ya. V.; Angelov, I. I. *Chistye Khimicheskye reaktivy*, Moscow, 1955.

COMPONENTS:

(1) Benzenesulfonamide, 4-amino-N-(4,6-dimeththyl-2-pyrimidinyl)- (sulfadimezine); $C_{12}H_{14}N_4O_2S$; [57-68-1]

(2) Calcium chloride; $CaCl_2$; [10043-52-4]

(3) Water; H_2O; [7732-18-5]

ORIGINAL MEASUREMENTS:

Likholet, N. M.; Gusyakov, V. P.

Med. Prom. SSSR 1963, *17(5)*, 28-31.

VARIABLES:

Concentration of $CaCl_2$

PREPARED BY:

R. Piekos

EXPERIMENTAL VALUES:

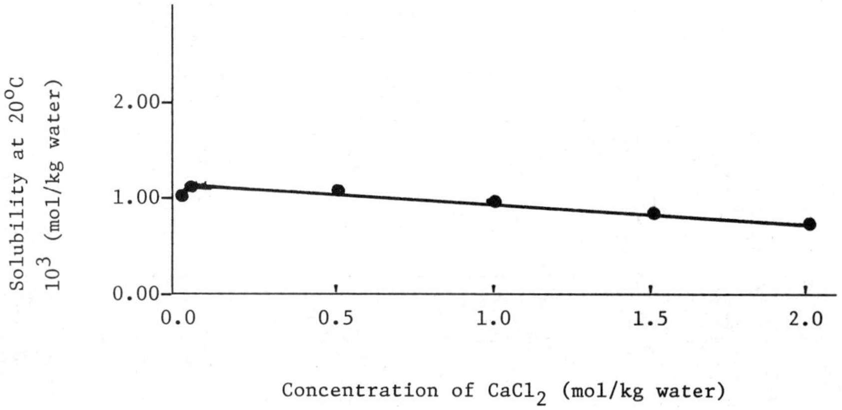

AUXILIARY INFORMATION

METHOD/APPARATUS/PROCEDURE:

Satd solns of sulfadimezine were prepd in 50-ml tightly closed ampuls in which 20 ml of a $CaCl_2$ soln was placed and a small excess of sulfadimezine. The mixts were equilibrated for 18 h at 20°C. Aliquots were pipetted out through a filter, dild, and assayed spectrophotometrically at 285 nm on a SF-IV spectrophotometer.

SOURCE AND PURITY OF MATERIALS:

Sulfadimezine (source not specified) was purified by recrystn. Its purity was 98.5%. $CaCl_2$ was purified by a recommended procedure (1).

Purity of the water was not specified.

ESTIMATED ERROR:

Soly: measurements were repeated several times (authors).

Temp: ±0.1°C (authors).

REFERENCES:

1. Karyakin, Ya. V.; Angelov, I. I. *Chistye Khimicheskye reaktivy*, Moscow, 1955.

COMPONENTS:

(1) Benzenesulfonamide, 4-amino-N-(4, 6-dimethyl-2-pyrimidinyl)- (sulfadimezine); $C_{12}H_{14}N_4O_2S$; [57-68-1]

(2) Hydrochloric acid; HCl; [7647-01-0]

(3) Sodium chloride; NaCl; [7647-14-5]

(4) Water; H_2O; [7732-18-5]

ORIGINAL MEASUREMENTS:

Portnov, M. A.; Zasosov, V. A.; Veselitskaya, T. A.; Mikhalev, V. A.

Med. Prom. SSSR 1964, *18(2)*, 36-9.

VARIABLES:

pH

PREPARED BY:

R. Piekos

EXPERIMENTAL VALUES:

pH	Solubility at 20°C	
	g/100 g solution	10^2 mol kg^{-1} [a]
1.1	0.65	2.33
1.5	0.15	0.54

[a] Calculated by compiler

AUXILIARY INFORMATION

METHOD/APPARATUS/PROCEDURE:

A test tube with an excess of sulfadimezine in a buffer soln was agitated for 6 h in a TS-15M thermostat and allowed to stand for 2 h. If, after 30 min of agitation, the pH differed from the required value, it was readjusted by the addn of a NaOH or HCl soln. A sample was withdrawn through a tube fitted with a filter and the sulfadimezine concn was detd by diazotization with potentiometric end point detection.

SOURCE AND PURITY OF MATERIALS:

Sulfadimezine (source not specified) conformed to the requirements of the State Pharmacopeia IX. The buffer solns were prepd from chemically pure reagents. Purity of the water was not specified.

ESTIMATED ERROR:

Soly and pH: not specified.

Temp: ±0.2°C (authors).

REFERENCES:

1. Portnov, M. A.; Veselitskaya, T. A.; Mikhalev, V. A. *Med. Prom. SSSR* 1962, *16(12)*, 27.

COMPONENTS:

(1) Benzenesulfonamide, 4-amino-N-(4,6-dimethyl-2-pyrimidinyl)- (sulfamethazine); $C_{12}H_{12}N_4O_2S$; [57-68-1]

(2) Phosphoric acid, disodium salt; Na_2HPO_4; [7558-94-4]

(3) Phosphoric acid, monopotassium salt; KH_2PO_4; [7778-77-0]

(4) Water; H_2O; [7732-18-5]

ORIGINAL MEASUREMENTS:

Gilligan, D. R.; Plummer, M. N.

Proc. Soc. Exp. Biol. Med. 1943, *53*, 142-5.

VARIABLES:

pH

PREPARED BY:

R. Piekos

EXPERIMENTAL VALUES:

Solubility of sulfamethazine in M/15 phosphate buffer solutions

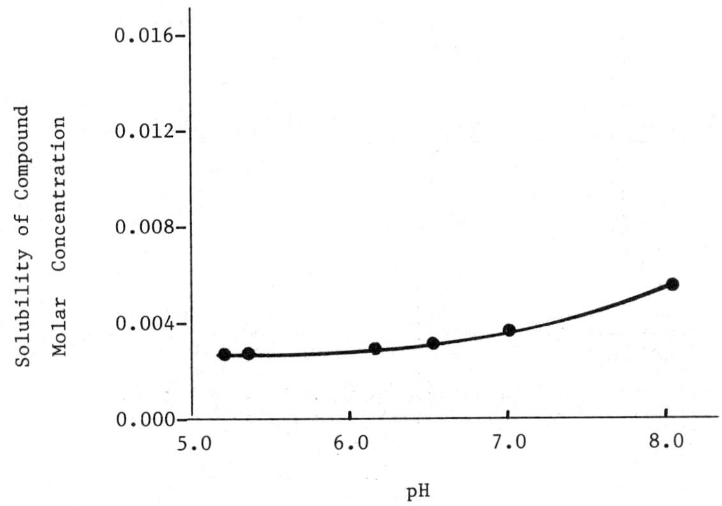

AUXILIARY INFORMATION

METHOD/APPARATUS/PROCEDURE:

An excess of sulfamethazine was shaken in M/15 phosphate buffer solns of various pH values for 18 h in a water bath at 37°C, and filtered in an incubator room at this temp. The pH of the filtrate was measured immediately at room temp with a Beckmann glass electrode pH meter and appropriate corrections for the differences between room temp and 37°C were applied. The amt of dissolved compd was measured by the method of Bratton and Marshall (1).

SOURCE AND PURITY OF MATERIALS:

Sulfamethazine was supplied by Lederle Labs, Inc. The source and purity of the remaining materials were not specified.

ESTIMATED ERROR:

Nothing specified.

REFERENCES:

1. Bratton, A. C.; Marshall, E. K., Jr. *J. Biol. Chem.* 1939, *128*, 537.

COMPONENTS:

(1) Benzenesulfonamide, 4-amino-N-(4,6-dimethyl-2-pyrimidinyl)- (Elkosin); $C_{12}H_{14}N_4O_2S$; [57-68-1]

(2) Phosphoric acid, disodium salt; Na_2HPO_4; [7558-94-4]

(3) Phosphoric acid, monopotassium salt; KH_2PO_4; [7778-77-0]

(4) Water; H_2O; [7732-18-5]

ORIGINAL MEASUREMENTS:

Meier, R.; Allemann, O.; von Meyenburg, H. *Schweiz. Med. Wochenschr.* 1944, *74(42)*, 1091-5.

VARIABLES:

pH

PREPARED BY:

R. Piekos

EXPERIMENTAL VALUES:

pH	Solubility of Elkosin in phosphate buffers at 37°C	
	mg%	10^3 mol dm^{-3} [a]
5.5	191.0	6.862
6.5	209.0	7.509
7.5	297.0	10.67

[a] Calculated by compiler

AUXILIARY INFORMATION

METHOD/APPARATUS/PROCEDURE:

Nothing specified.

SOURCE AND PURITY OF MATERIALS:

Nothing specified.

ESTIMATED ERROR:

Nothing specified.

REFERENCES:

COMPONENTS:	ORIGINAL MEASUREMENTS:
(1) Benzenesulfonamide, 4-amino-N-(4,6-dimethyl-2-pyrimidinyl)- (sulfadimidine); $C_{12}H_{14}N_4O_2S$; [57-68-1] (2) Phosphoric acid, disodium salt; Na_2HPO_4; [7558-94-4] (3) Phosphoric acid, monopotassium salt; KH_2PO_4; [7778-77-0] (4) Water; H_2O; [7732-18-5]	Riess, W. *Intern. Congr. Chemotherapy, Proc. 3rd, Stuttgart* 1963, *1*, 627-32.
VARIABLES: One temperature: 20°C; one pH: 7.4	PREPARED BY: R. Piekos

EXPERIMENTAL VALUES:

Solubility of sulfadimidine in a M/15 Sörensen buffer solution (pH 7.4) at 20°C is 60 mg %(2.2 x 10^{-3} mol dm^{-3}, compiler).

AUXILIARY INFORMATION

METHOD/APPARATUS/PROCEDURE:	SOURCE AND PURITY OF MATERIALS:
Sörensen buffer solns of pH varying between 7 and 8 were prepd, satd with sulfadimidine at 20°C, their pH was measured at equilibrium, and the sulfadimidine was assayed colorimetrically. The measured pH values were plotted against concn, and the soly at pH 7.4 was detd by interpolation (personal communication).	Nothing specified.
	ESTIMATED ERROR: Nothing specified.
	REFERENCES:

COMPONENTS:

(1) Benzenesulfonamide, 4-amino-N-(4,6-dimethyl-2-pyrimidinyl)- (sulfamethazine); $C_{12}H_{14}N_4O_2S$; [57-68-1]

(2) Phosphoric acid, disodium salt; Na_2HPO_4; [7558-94-4]

(3) 1,2,3-Propanetricarboxylic acid, 2-hydroxy- (citric acid); $C_6H_8O_7$; [77-92-9]

(4) Water; H_2O; [7732-18-5]

ORIGINAL MEASUREMENTS:

Bertazzoli, C.; Buogo, A.; Ciceri, C.; Ghione, M.; Turolla, E.; Zavaglio, V. *Minerva Med.* 1961, *52(40)*, 1789-96.

VARIABLES:

pH

PREPARED BY:

R. Piekos

EXPERIMENTAL VALUES:

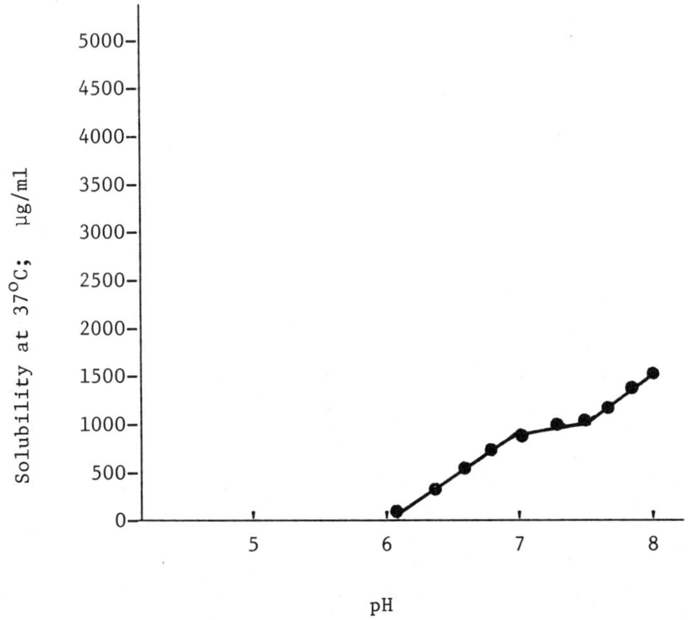

AUXILIARY INFORMATION

METHOD/APPARATUS/PROCEDURE:

The soly of sulfamethazine in McIlvaine's Na_2HPO_4 - citric acid buffer solns was detd under agitation at 37°C. No details were given.

SOURCE AND PURITY OF MATERIALS:

Nothing specified.

ESTIMATED ERROR:

Nothing specified.

REFERENCES:

COMPONENTS:

(1) Benzenesulfonamide, 4-amino-N-(4,6-dimethyl-2-pyrimidinyl)- (sulfadimezine); $C_{12}H_{14}N_4O_2S$; [57-68-1]

(2) Phosphoric acid, disodium salt; Na_2HPO_4; [7558-94-4]

(3) Phosphoric acid, monosodium salt; NaH_2PO_4; [7558-80-7]

(4) Water; H_2O; [7732-1805]

ORIGINAL MEASUREMENTS:

Portnov, M. A.; Zasosov, V. A.; Veselitskaya, T. A.; Mikhalev, V. A. *Med. Prom. SSSR* 1963, *18(2)*, 36-9.

VARIABLES:

pH

PREPARED BY:

R. Piekos

EXPERIMENTAL VALUES:

pH of phosphate buffer	Solubility at 20°C	
	g/100 g solution	10^2 mol kg^{-1} [a]
0.7	0.05	0.2
1.5	0.03	0.1
8.5	0.28	1.0

[a] Calculated by compiler

AUXILIARY INFORMATION

METHOD/APPARATUS/PROCEDURE:

A test tube with an excess of sulfadimezine in a buffer soln was agitated for 6 h in a TS-15M thermostat and allowed to stand for 2 h. If, after 30 min of agitation, the pH differed from the required value, it was readjusted by addn of a NaOH or HCl soln. A sample was withdrawn through a tube fitted with a filter and the sulfadimezine concn was detd by diazotization with potentiometric end point detection (1).

SOURCE AND PURITY OF MATERIALS:

Sulfadimezine (source not specified) conformed to the requirements of the State Pharmacopeia IX. The buffer solns were prepd from chemically pure H_3PO_4 and NaOH. Purity of the water was not specified.

ESTIMATED ERROR:

Soly and pH: not specified.

Temp: ±0.2°C (authors).

REFERENCES:

1. Portnov, M. A.; Veselitskaya, T. A.; Mikhalev, V. A. *Med. Prom. SSSR* 1962, *16(12)*, 27.

COMPONENTS:

(1) Benzenesulfonamide, 4-amino-N-(4,6-dimethyl-2-pyrimidinyl)- (sulfadimidine); $C_{12}H_{14}N_4O_2S$; [57-68-1]

(2) Phosphoric acid, disodium salt; Na_2HPO_4; [7558-94-4]

(3) Phosphoric acid, monopotassium salt; KH_2PO_4; [7778-77-0]

(4) Water; H_2O; [7732-18-5]

ORIGINAL MEASUREMENTS:

Hekster, Y. A.; Vree, T. B.; Damsma, J. E.; Friesen, W. T.

J. Antimicrob. Chemother. 1981, *8*, 133-44.

VARIABLES:

pH

PREPARED BY:

R. Piekos

EXPERIMENTAL VALUES:

pH	Solubility at 25°C mg/l	10^2 mol dm^{-3} [a]
5.5	438	0.157
7.5	7000	2.515

[a] Calculated by compiler

AUXILIARY INFORMATION

METHOD/APPARATUS/PROCEDURE:

Satd solns of sulfadimidine were prepd in phosphate buffers of pH 5.5 and 7.5 at room temp (25°C). The concn of sulfadimidine was measured by means of a Spectra Physics 3500B high-performance liquid chromatograph equipped with a column oven (Model 748) and a Pye-Unicam LC-UV spectrophotometric detector. The detector was connected to a 1-mV recorder. A stainless steel column (10 cm x 4.6 mm i.d.) was packed with Lichrosorb RPS, 5 μm, obtained from Chrompack. An injection loop of 100 μl was used. The oven temp was 40°C. Detection of sulfadimidine was performed at 260 nm.

SOURCE AND PURITY OF MATERIALS:

The source and purity of the materials were not specified.

ESTIMATED ERROR:

The detection limit of the solute by HPLC was 0.5 mg/l (authors). The error in temperature and pH were not specified.

REFERENCES:

COMPONENTS:

(1) Benzenesulfonamide, 4-amino-N-(4,6-dimethyl-2-pyrimidinyl)- (sulfamethazine); $C_{12}H_{14}N_4O_2S$; [57-68-1]

(2) 1,2,3-Propanetricarboxylic acid, 2-hydroxy-, disodium salt (Na citrate); $C_6H_6Na_2O_7$; [144-33-2]

(3) Sodium hydroxide; NaOH; [1310-73-2]

(4) Water; H_2O; [7732-18-5]

ORIGINAL MEASUREMENTS:

Gilligan, D. R.; Plummer, M. N.

Proc. Soc. Exp. Biol. Med. 1943, *53*, 142-5.

VARIABLES:

pH

PREPARED BY:

R. Piekos

EXPERIMENTAL VALUES:

Solubility of sulfamethazine in M/10 Na citrate + NaOH solution at 37°C

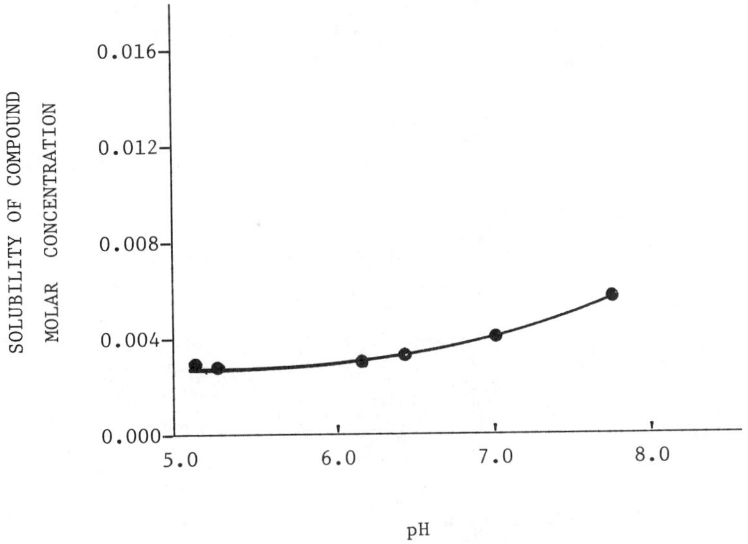

AUXILIARY INFORMATION

METHOD/APPARATUS/PROCEDURE:

An excess of sulfamethazine was shaken in M/10 Na citrate + NaOH solns of various pH values for 18 h in a water bath at 37°C, and filtered in an incubator at this temp. The pH of the filtrate was measured immediately at room temp with a Beckmann glass electrode pH meter and appropriate corrections for the differences between room temp and 37°C were applied. The amt of dissolved compd was measured by the method of Bratton and Marshall (1).

SOURCE AND PURITY OF MATERIALS:

Sulfamethazine was supplied by Lederle Labs, Inc. The source and purity of the remaining materials were not specified.

ESTIMATED ERROR:

Nothing specified.

REFERENCES:

1. Bratton, A. C.; Marshall, E. K., Jr. *J. Biol. Chem.* 1939, *128*, 537.

COMPONENTS:

(1) Benzenesulfonamide, 4-amino-N-(4,6-dimethyl-2-pyrimidinyl)- (sulfadimezine); $C_{12}H_{14}N_4O_2S$; [57-68-1]

(2) Phosphoric acid, disodium salt; Na_2HPO_4; [7558-94-4]

(3) 1,2,3-Propanetricarboxylic acid, 2-hydroxy- (citric acid); $C_6H_8O_7$; [77-92-9]

(4) Water; H_2O; [7732-18-5]

ORIGINAL MEASUREMENTS:

Likhol'ot, N. M. *Farm. Zh. (Kiev)* 1965, *20(5)*, 44-6.

VARIABLES:

pH

PREPARED BY:

R. Piekos

EXPERIMENTAL VALUES:

pH of McIllvaine's buffer solution	Solubility at 20°C g/100 ml	Solubility at 20°C 10^3 mol dm^{-3} [a]
4.1	0.034	1.2
5.1	0.037	1.3
5.9	0.039	1.4
6.5	0.045	1.6
6.9	0.054	1.9
7.5	0.093	3.3

[a] Calculated by compiler

AUXILIARY INFORMATION

METHOD/APPARATUS/PROCEDURE:

An earlier described method was employed (1) whereby a small excess of sulfadimezine was equilibrated with 20 ml of the McIlvaine's buffer soln for 8 h in a 50-ml test tube. Aliquots were withdrawn through a filter and sulfadimezine was assayed bromatometrically.

SOURCE AND PURITY OF MATERIALS:

Sulfadimezine: not specified.

McIlvaine's buffer solns were prepd from a 0.2M Na_2HPO_4 and a 0.1M citric acid solns. Source and purity of the buffer components were not specified.

ESTIMATED ERROR:

Soly: not specified.

Temp: ±0.1°C (authors).

pH : not specified.

REFERENCES:

1. Gusyakov, V. P.; Likhol'ot, N. M. *Farm. Zh. (Kiev)* 1960, *15(8)*, 21.

COMPONENTS:

(1) Benzenesulfonamide, 4-amino-N-(4,6-dimethyl-2-pyrimidinyl)- (sulfamethazine; sulfadimidine); $C_{12}H_{14}N_4O_2S$; [57-68-1]

(2) Benzoic acid; $C_7H_6O_2$; [65-85-0]

(3) Water; H_2O; [7732-18-5]

ORIGINAL MEASUREMENTS:

Nasipuri, R. N.; Khalil, S. A. H.

J. Pharm. Sci. 1973, *62(3)*; 473-5.

VARIABLES:

Sulfamethazine concentration, benzoic acid concentration.

PREPARED BY:

R. Piekos

EXPERIMENTAL VALUES:

Sulfamethazine concentration	Equilibrium solubility of sulfamethazine at fixed benzoic acid concentration of 8.18 mmoles/l, at 24°C	
mmoles/l	mmoles/l	g dm^{-3} [a]
2.16	0.076	0.0211
3.60	0.083	0.0231
7.20	0.173	0.0481
10.80	1.400	0.3897
14.40	1.442	0.4013
36.00	1.417	0.3944
72.00	1.457	0.4055
144.00	1.424	0.3963
360.00	1.421	0.3955

Initial benzoic acid concentration	Solubility of sulfamethazine[b]	
mg%	mg%	10^3mol dm^{-3}solution[a]
20	38.8	1.39
40	38.2	1.37
60	36.1	1.30
80	13.0	0.467
100	4.8	0.17

[a] Calculated by compiler.

[b] Initial sulfamethazine concentration = 0.2 g%

AUXILIARY INFORMATION

METHOD/APPARATUS/PROCEDURE:

The concn of both benzoic acid and sulfamerazine were detd in a filtrate spectrophotometrically by applying equations of Tinker and McBay (1) for two-component systems. Measurements were made using a Unicam SP 500 spectrophotometer at 230 and 300 nm (in 0.05N HCl). The validity of the eqns was tested by assaying known mixts of sulfamethazine and benzoic acid.

SOURCE AND PURITY OF MATERIALS:

Sulfamethazine was a BP grade sulfadimidine (Imperial Chemical Industries, England). Benzoic acid was of BP quality (British Drug Houses, England). Purity of the water was not specified.

ESTIMATED ERROR:

Soly: sulfamethazine and benzoic acid ±1.9 and ±2.3%, resp. (not stated accuracy or reproducibility - authors).

Temp: ±0.2°C (authors).

REFERENCES:

1. Tinker, R. B.; McBay, A. J. *J. Amer. Pharm. Assoc., Sci. Ed.* 1954, *43*, 315.

COMPONENTS:	ORIGINAL MEASUREMENTS:
(1) Benzenesulfonamide, 4-amino-N-(4,6-dimethyl-2-pyrimidinyl)- (sulfadimezine); $C_{12}H_{14}N_4O_2S$; [57-68-1] (2) Ethanol; C_2H_6O; [64-17-5] (3) Water; H_2O; [7732-18-5]	Shkadova, A. I. *Farm. Zh. (Kiev)* 1969, *24(3)*, 39-41.
VARIABLES:	PREPARED BY:
Concentration of ethanol	R. Piekos

EXPERIMENTAL VALUES:

Concentration of ethanol		Solubility at 20°C	
mole %	weight %	10^2 mol kg^{-1}	g/100 g [a]
0	0	0.33	0.092
10	22.14	0.48	0.13
20	39.01	1.07	0.298
30	52.31	2.18	0.607
40	63.04	2.85	0.793
50	71.90	3.12	0.868
60	79.33	3.07	0.854
70	85.65	2.96	0.824
80	91.10	2.38	0.662
90	95.83	0.98	0.27

[a] Calculated by compiler

AUXILIARY INFORMATION

METHOD/APPARATUS/PROCEDURE:	SOURCE AND PURITY OF MATERIALS:
Sulfadimezine was equilibrated with the solvent in a water thermostat at 20±0.1°C. The concn of sulfadimezine was detd by alkalimetric titration.	Purity of sulfadimezine conformed to the requirements of the State Pharmacopiea IX. The EtOH - water mixts were prepd from abs EtOH (purity and source not specified) and distd water.
	ESTIMATED ERROR: Soly: not specified. Temp: ±0.1°C (author).
	REFERENCES:

COMPONENTS:

(1) Benzenesulfonamide, 4-amino-N-(4,6-dimethyl-2-pyrimidinyl)- (sulfadimidine); $C_{12}H_{14}N_4O_2S$; [57-68-1]

(2) Sorbitan monooleate, polyoxyethylene derivatives (Tween 80); [9005-65-6]

(3) Water; H_2O; [7732-18-5]

ORIGINAL MEASUREMENTS:

Gerencsér-Németh, M.; Horbáth, M.

Gyógyszerészet 1973, *17*, 417-21.

VARIABLES:

Concentration of Tween 80

PREPARED BY:

R. Piekos

EXPERIMENTAL VALUES:

Concentration of Tween 80 Weight %	Solubility at 20°C	
	g/100 g soln[a]	10^3 mol kg^{-1} soln[b]
1	0.0530	1.90
	0.0518	1.86
3	0.0734	2.64
	0.0706	2.54
5	0.0922	3.31
	0.0914	3.28
8	0.1078	3.873
	0.1142	4.103

[a] Numerical values supplied by the authors.

[b] Calculated by compiler.

AUXILIARY INFORMATION

METHOD/APPARATUS/PROCEDURE:

An excess of sulfadimidine in an aq Tween 80 soln was shaken in a lab shaker at 120 rpm for 6 h. The soln was then filtered, the residue was washed first with the filtrate and finally with a small amt of water, dried and weighed.

SOURCE AND PURITY OF MATERIALS:

Sulfadimidine (source and purity not specified) was dried at 100°C for 3 h or over concd H_2SO_4 for 72 h. Its mp was 196.5-8°C. Distd water was used. Source and purity of Tween 80 were not specified.

ESTIMATED ERROR:

Nothing specified.

REFERENCES:

COMPONENTS:	ORIGINAL MEASUREMENTS:
(1) Benzenesulfonamide, 4-amino-N-(4,6-dimethyl-2-pyrimidinyl)- (sulfamethazine; sulfadimidine); $C_{12}H_{14}N_4O_2S$; [57-68-1] (2) 2-Pyrrolidinone, 1-ethenyl-, polymers (PVP); $(C_6H_9NO)_x$; [9003-39-8] (3) Water; H_2O; [7732-18-5]	Nasipuri, R. N.; Khalil, S. A. H. *J. Pharm. Sci.* 1973, *62(3)*, 473-5.
VARIABLES:	PREPARED BY:
One temperature: 24°C	R. Piekos

EXPERIMENTAL VALUES:

Solubility of sulfamethazine in an aqueous solution of PVP containing 5 mg% PVP, at 24°C, is 40.1 mg% (1.44×10^{-3} mol dm^{-3} solution, compiler).

AUXILIARY INFORMATION

METHOD/APPARATUS/PROCEDURE:	SOURCE AND PURITY OF MATERIALS:
Sulfamethazine was assayed spectrophotometrically at 230 nm in a 0.05N HCl soln using a Unicam SP 500 spectrophotometer.	Sulfamethazine was a BP grade sulfadimidine (Imperial Chemical Industries, England). PVP was a sample having an av mol wt of 40,000 (Plasdone k 29-32, GAF Corp., New York, N. Y.) Purity of the water was not specified.
	ESTIMATED ERROR:
	Soly: ±1.9% (not stated accuracy or reproducibility - authors). Temp: ±0.2°C (authors).
	REFERENCES:

COMPONENTS:

(1) Benzenesulfonamide, 4-amino-N-(4,6-dimethyl-2-pyrimidinyl)- (sulfamethazine; sulfadimidine); $C_{12}H_{14}N_4O_2S$; [57-68-1]

(2) Benzoic acid; $C_7H_6O_2$; [65-85-0]

(3) 2-Pyrrolidinone, 1-ethenyl-, polymers (PVP); $(C_6H_9NO)_x$; [9003-39-8]

(4) Water; H_2O; [7732-18-5]

ORIGINAL MEASUREMENTS:

Nasipuri, R. N.; Khalil, S. A. H.
J. Pharm. Sci. 1973, *62(3)*, 473-5.

VARIABLES:

Initial benzoic acid concentration

PREPARED BY:

R. Piekos

EXPERIMENTAL VALUES:

Initial benzoic acid concentration	Solubility of sulfamethazine at 24°C in a solution containing 5 mg% of PVP	
mg%	mg%	10^3 mol dm^{-3} solution[a]
20	39.9	1.43
40	40.4	1.45
60	40.2	1.44
80	39.8	1.43
100	39.9	1.43

[a] Calculated by compiler

AUXILIARY INFORMATION

METHOD/APPARATUS/PROCEDURE:

The concn of both benzoic acid and sulfamerazine were detd in a filtrate spectrophotometrically by applying equations of Tinker and McBay (1) for two-component systems. Measurements were made using a Unicam SP 500 spectrophotometer at 230 and 300 nm (in 0.05N HCl). The validity of the eqns was tested by assaying known mixts of sulfamethazine and benzoic acid.

SOURCE AND PURITY OF MATERIALS:

Sulfamethazine was a BP grade sulfadimidine (Imperial Chemical Industries, England). Benzoic acid was of BP quality (British Drug Houses, England). PVP (Plasdone k 29-32, GAF Corp., New York, N. Y.) was a sample having an av mol wt of 40,000. Purity of the water was not specified.

ESTIMATED ERROR:

Soly: sulfamethazine and benzoic acid ±1.9 and ±2.3%, resp. (not stated accuracy or reproducibility - authors).
Temp: ±0.2°C (authors).

REFERENCES:

1. Tinker, R. B.; McBay, A. J. *J. Amer. Pharm. Assoc., Sci. Ed.* 1954, *43*, 315.

COMPONENTS:

(1) Benzenesulfonamide, 4-amino-N-(4,6-dimethyl-2-pyrimidinyl)- (sulfamethazine); $C_{12}H_{14}N_4O_2S$; [57-68-1]

(2) Ethanol, 2-ethoxy-; $C_4H_{10}O_2$; [110-80-5]

ORIGINAL MEASUREMENTS:

Sunwoo, C.; Eisen, H.
J. Pharm. Sci. 1971, *60*, 238-44.

VARIABLES:

One temperature: 25°C

PREPARED BY:

R. Piekos

EXPERIMENTAL VALUES:

The mole fraction solubility of sulfamethazine in 2-ethoxyethanol at 25°C is 0.0184 mol/dm^3 (5.47 g/100 g solution, compiler).

AUXILIARY INFORMATION

METHOD/APPARATUS/PROCEDURE:

Soly was detd by the method reported by Restaino and Martin (1). Sulfamethazine was assayed on a Coleman-Hitachi 124 double-beam spectrophotometer at 270 nm after diln of a sample with 95% ethanol or water.

SOURCE AND PURITY OF MATERIALS:

The USP sulfamethazine (American Cyanamid Co., Pearl River, N.Y.), recrystd from warm alcohol and an industrial grade 2-ethoxyethanol (Cellosolve solvent, Union Carbide, New York, N. Y.) were used.

ESTIMATED ERROR:

Temp: ±1.0°C (authors).

Soly: the mean of 3 runs was given (authors).

REFERENCES:

1. Restaino, F. A.; Martin, A. N. *J. Pharm. Sci.* 1964, *53*, 636.

COMPONENTS:	ORIGINAL MEASUREMENTS:
(1) Benzenesulfonamide, 4-amino-N-(4,6-dimethyl-2-pyrimidinyl)- (sulfamethazine); $C_{12}H_{14}N_4O_2S$; [57-68-1] (2) 2-Propanone (acetone); C_3H_6O; [67-64-1]	Gutierrrez, F. H. *Anales fis. quim. (Madrid)* 1945, *41*, 537-60.
VARIABLES:	PREPARED BY:
Temperature	R. Piekos

EXPERIMENTAL VALUES:

t/°C	G[a]	E[b]	X_g/1[c]	mol/1[d] acetone	mmol/mol acetone	1:X_g^e	1 + X_{cc}^f
0	4.481	4.288	36.502	131.1	9.39	22.31	27.66
5	4.601	4.397	37.213	133.7	9.59	21.73	26.86
10	4.701	4.489	37.741	135.6	9.80	21.27	26.49
15	4.959	4.724	39.528	142.0	10.35	20.16	25.29
20	5.269	5.005	41.688	149.8	10.99	18.98	23.98
25	5.406	5.128	42.448	152.5	11.28	18.49	23.56
30	5.684	5.377	44.295	159.1	11.86	17.59	22.57
35	6.002	5.651	46.413	166.7	12.52	16.66	21.54
40	6.597	6.188	50.625	181.9	13.76	15.19	19.75
45	7.486	6.964	56.998	204.8	15.62	13.36	17.54
50	8.759	8.053	66.174	237.3	18.27	11.42	15.11

[a] $G = \frac{p\,100}{P - p}$, where p and P are the weights of solute and solution, resp.

[b] $E = \frac{G\,100}{G + 100}$, [c] g/1 acetone; [d] should be mmol/1 acetone (compiler);

[e] g of acetone required to dissolved 1 g of solute; [f] volume (cm^3) of acetone required to dissolve 1 g of solute.

AUXILIARY INFORMATION

METHOD/APPARATUS/PROCEDURE:

A special all-glass app was constructed enabling the prepn of satd solns, agitation by bubbling a stream of acetone-satd N, filtration, and distn off the solvent without contact with air. Two exchangeable dissoln vessels of 15 and 8 cm^3 working capacity were used depending on the soly of solute. The app was immersed in a thermostat. The vols of acetone used were 15 or 5 cm^3, and the equilibration time was 2-2.5 h. The satd solns were filtered, weighed, the solvent was distd off, the residues were dried at 105°C, weighed, and examd for the presence of solvated acetone.

SOURCE AND PURITY OF MATERIALS:

The source of the materials was not specified. Pure, anhyd acetone was used. The absence of impurities and water was confirmed by procedures of the German Pharmacopeia VI and Spanish Pharmacopeia VIII.

The purity of sulfamethazine was not specified.

ESTIMATED ERROR:

Soly: measurements were repeated until 2 values not differing in the second decimal were obtained (author).

Temp: ±0.1° C (author).

REFERENCES:

COMPONENTS:	ORIGINAL MEASUREMENTS:
(1) Benzenesulfonamide, 4-amino-N-(4,6-dimethyl-2-pyrimidinyl)- (sulfadimidine); $C_{12}H_{14}N_4O_2S$; [57-68-1] (2) Methane, trichloro- (chloroform); $CHCl_3$; [67-66-3]	Riess, W. *Intern. Congr. Chemotherapy, Proc. 3rd, Stuttgart* 1963, *1*, 627-32.
VARIABLES:	PREPARED BY:
One temperature: 20°C	R. Piekos

EXPERIMENTAL VALUES:

Solubility of sulfadimidine in chloroform at 20°C is 240 mg% (8.62 x 10^{-3} mol dm^{-3} solution, compiler).

AUXILIARY INFORMATION

METHOD/APPARATUS/PROCEDURE:	SOURCE AND PURITY OF MATERIALS:
Nothing specified.	Nothing specified.
	ESTIMATED ERROR:
	Nothing specified.
	REFERENCES:

COMPONENTS:	ORIGINAL MEASUREMENTS:
(1) Benzenesulfonamide, 4-amino-N-(4,6-dimethyl-2-pyrimidinyl)- (sulfadimezine); $C_{12}H_{14}N_4O_2S$; [57-68-1] (2) Poly(oxy-1,2-ethanediyl), α -hydro- ω -hydroxy- (PEG 400); $(C_2H_4O)_nH_2O$; [25322-68-3] 400	Gusyakov, V. P.; Likhol'ot, N. M.; Kutna, I. M. *Farm. Zh. (Kiev)* 1968, *23(6)*, 56-61.
VARIABLES:	**PREPARED BY:**
One temperature: 21-25°C	R. Piekos

EXPERIMENTAL VALUES:

Solubility of sulfadimezine in α-hydro-ω -hydroxypoly(oxy-1,2-ethanediyl) 400 at room temperature (21-25°C) is 10.4% by weight (0.417 mol kg^{-1} PEG 400, compiler).

AUXILIARY INFORMATION

METHOD/APPARATUS/PROCEDURE:	SOURCE AND PURITY OF MATERIALS:
Small quantities (2-4 mg) of sulfadimezine were added to a known quantity of PEG 400 under stirring until satn was attained.	Sulfadimezine: neither source nor purity was specified. PEG 400: source not specified; sp gr 1.127 g cm^{-3}; temp of solidification approx 6°C; refractive index 1.466 (temp not indicated).
	ESTIMATED ERROR:
	Nothing specified.
	REFERENCES:

COMPONENTS:	ORIGINAL MEASUREMENTS:
(1) Benzenesulfonamide, 4-amino-N-(4,6-dimethyl-2-pyrimidinyl)- hemihydrate (sulfamethazine hemihydrate); $C_{12}H_{14}N_4O_2S\cdot1/2H_2O$; [82537-68-6] (2) Sodium hydroxide; NaOH; [1310-73-2] (3) Water; H_2O; [7732-18-5]	Rose, F. L.; Martin, A. R.; Bevan, H.G.L. *J. Pharm. Exp. Therap.* 1943, *77*, 127-42.
VARIABLES:	PREPARED BY:
pH	R. Piekos

EXPERIMENTAL VALUES:

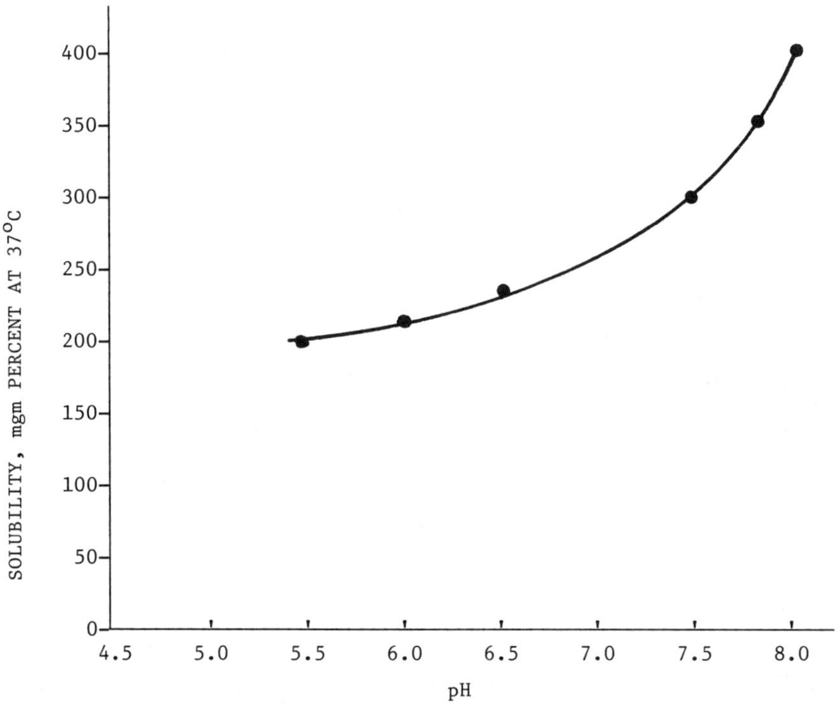

AUXILIARY INFORMATION

METHOD/APPARATUS/PROCEDURE:

An excess of sulfamethazine hemihydrate was stirred in boiling water, the soln was cooled to 37°C, the temp being maintained thermostatically, and 0.1N NaOH was added to increase the pH. The pH was measured by means of a glass electrode - calomel half-cell system and was permitted to reach equilibrium before a reading was taken. The concn of sulfamethazine hemihydrate in soln was detd colorimetrically by withdrawing a sample through a filter-tip into a preheated micropipet.

SOURCE AND PURITY OF MATERIALS:

The source and purity of sulfamethazine hemihydrate and of NaOH were not specified. Water was doubly distilled.

ESTIMATED ERROR:

Nothing specified.

REFERENCES:

COMPONENTS:

(1) Cobalt, bis[4-amino-N-(4,6-dimethyl-2-pyrimidinyl)benzenesulfonamidato-N^N,O]-, hydrate; $C_{24}H_{26}CoN_8O_4S_2 \cdot nH_2O$; [86729-24-0]

(2) Hydrochloric acid; HCl; [7647-01-0]

(3) Water; H_2O; [7732-18-5]

VARIABLES:

pH

ORIGINAL MEASUREMENTS:

Tskitishvili, M. G.; Mikadze, I. I. *Soobshch. Akad. Nauk Gruz. SSR* 1978, *89(3)*, 589-92.

PREPARED BY:

R. Piekos

EXPERIMENTAL VALUES:

K_{so} over the HCl concentration range 2.5×10^{-2} - 2.5×10^{-5} mol dm^{-3}, at 25°C, is 3.35×10^{-13}.

AUXILIARY INFORMATION

METHOD/APPARATUS/PROCEDURE:

In a glass vessel, a mixt of 100 ml of HCl of appropriate concn and the solute was placed and shaken for 6 h in a water thermostat at 25°C. After attaining equilibrium, the pH of the soln was measured and the Co^{2+} and S content was detd to calculate K_{so}.

SOURCE AND PURITY OF MATERIALS:

Nothing specified.

ESTIMATED ERROR:

Nothing specified.

REFERENCES:

COMPONENTS:	ORIGINAL MEASUREMENTS:
(1) Benzenesulfonamide, 4-amino-N-(5-methoxy-2-pyrimidinyl)- (sulfamethoxydiazine); $C_{11}H_{12}N_4O_3S$; [651-06-9] (2) Poly(oxy-1,2-ethanediy), α-hydro- ω-hydroxy- PEG (20 000); $(C_2H_4O)_nH_2O$; [25322-68-3] 20 000 (3) Water; H_2O; [7732-18-5]	Salib, N. N.; Ebian, A. R. *Pharm. Ind.* 1978, *40(3)*, 262-5.
VARIABLES:	PREPARED BY:
Concentration of PEG 20 000	R. Piekos

EXPERIMENTAL VALUES:

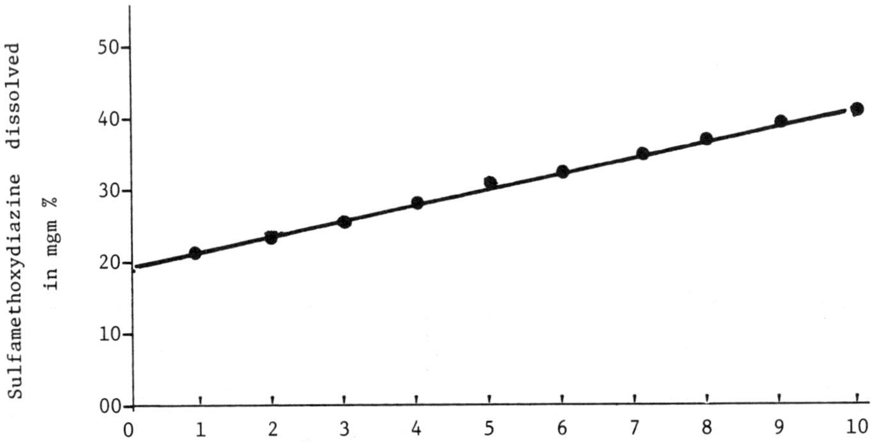

Effect of different concentrations of PEG 20 000 on the solubility of sulfamethoxydiazine at 37±1°C.

AUXILIARY INFORMATION

METHOD/APPARATUS/PROCEDURE:	SOURCE AND PURITY OF MATERIALS:
An excess of sulfamethoxydiazine was added to bottles contg from 0 to 10% PEG 20 000 solns. The bottles were rotated at 30 rpm for 24 h in a water bath at 37°C. After equilibrium had been attained, an aliquot was pipetted out, dild with 0.1N HCl, and the solute was assayed spectrophotometrically at 228 nm.	Sulfamethoxydiazine (Bayrena) was supplied by Bayer, Leverkusen. PEG 20 000 was a product of Hoechst AG, West Germany. The purity of the materials was not specified.
	ESTIMATED ERROR: Soly: data points are means of 3 detns (authors). Temp: ±1°C (authors).
	REFERENCES:

COMPONENTS:

(1) Copper, bis[4-amino-N-(4,6-dimethyl-2-pyrimidinyl)benzenesulfonamidato-N^N,O]-, hydrate; $C_{24}H_{26}CuN_8O_4S_2 \cdot nH_2O$; [86729-23-9]

(2) Hydrochloric acid; HCl; [7647-01-0]

(3) Water; H_2O; [7732-18-5]

ORIGINAL MEASUREMENTS:

Tskitishvili, M. G.; Mikadze, I. I.
Soobshch. Akad. Nauk Gruz. SSR 1978, *89(3)*, 589-92.

VARIABLES:

pH

PREPARED BY:

R. Piekos

EXPERIMENTAL VALUES:

K_{so} over the HCl concentration range 2.5×10^{-2} - 2.5×10^{-5} mol dm^{-3}, at 25°C, is 1.77×10^{-16}.

AUXILIARY INFORMATION

METHOD/APPARATUS/PROCEDURE:

In a glass vessel, a mixt of 100 ml of HCl of appropriate concn and the solute was placed and shaken for 6 h in a water thermostat at 25°C. After attaining equilibrium, the pH of the soln was measured and the Cu^{2+} and S content was detd to calculate K_{so}.

SOURCE AND PURITY OF MATERIALS:

Nothing specified.

ESTIMATED ERROR:

Nothing specified.

REFERENCES:

COMPONENTS:

(1) Manganese, bis[4-amino-N-(4,6-dimethyl-2-pyrimidinyl)benzenesulfonamidato-N^N,O]-hydrate; $C_{24}H_{26}MnN_8O_4S_2 \cdot nH_2O$; [84812-75-9]

(2) Hydrochloric acid; HCl; [7647-01-0]

(3) Water; H_2O; [7732-18-5]

ORIGINAL MEASUREMENTS:

Tskitishvili, M. G.; Shvelashvili, A. E.; Mikadze, I. I.; Zhorzholiani, N. B.; Chrelashvili, M. V. *Izv. Akad. Nauk Gruz. SSR, Ser. Khim.* 1981, *7(4)*, 300-4.

VARIABLES:

pH

PREPARED BY:

R. Piekos

EXPERIMENTAL VALUES:

Concentration of HCl (mol/l)	pH	10^9 K_{so} at 25°C
1.0×10^{-2}	6.40	7.91
5.0×10^{-3}	6.77	7.94
2.5×10^{-3}	6.96	7.91
1.0×10^{-3}	7.54	7.92
5.0×10^{-4}	7.82	7.90
2.5×10^{-4}	7.90	7.92
1.0×10^{-4}	8.10	7.94
5.0×10^{-5}	8.12	7.97
1.5×10^{-5}	8.15	8.00
	Mean	7.94

To be continued on the next page.

AUXILIARY INFORMATION

METHOD/APPARATUS/PROCEDURE:

The earlier described apparatus and method was used (1): in a glass vessel, a mixt of 100 ml of HCl of appropriate concn and the solute were placed and shaken for 6 h in a water thermostat at 25°C. After attaining equilibrium , the pH of the soln was measured and the Mn^{2+} and S content was determined to calculate K_{so}. The pH was measured on a pH-673 pH meter.

SOURCE AND PURITY OF MATERIALS:

0.1M solutions of chem pure $Mn(OAc)_2$, monosodium salt of sulfadimezine, HCl as well as doubly distd water were used. The source of the materials was no specified.

ESTIMATED ERROR:

K_{so} : std deviation 3×10^{-11} (compiler).

Temp and pH: not specified.

REFERENCES:

1. Tskitishvili, M. G.; Mikadze, I. I. *Soobshch. Akad. Nauk Gruz. SSR* 1978, *89(3)*, 589.

COMPONENTS:	ORIGINAL MEASUREMENTS:
(1) Nickel,bis[4-amino-N-(4,6-dimethyl-2-pyrimidinyl)benzenesulfonamidato-N^N,O]-dihydrate; $C_{24}H_{26}N_8NiO_4S_2 \cdot 2H_2O$; [84812-74-8] (2) Hydrochloric acid; HCl; [7647-01-0] (3) Water; H_2O; [7732-18-5]	Tskitishvili, M. G.; Shvelashvili, A. E.; Mikadze, I. I.; Zhorzholiani, N. B.; Chrelashvili, M. V. *Izv. Akad. Nauk Gruz. SSR. Ser. Khim.* 1981, *7(4)*, 300-4.
VARIABLES:	PREPARED BY:
pH	R. Piekos

EXPERIMENTAL VALUES:

Concentration of HCl (mol/l)	pH	10^{14} K_{so} at 25°C
2.5×10^{-2}	6.08	1.61
1.0×10^{-2}	6.50	1.63
5.0×10^{-3}	7.37	1.66
2.5×10^{-3}	8.29	1.60
1.0×10^{-3}	8.85	1.60
5.0×10^{-4}	9.10	1.66
2.5×10^{-4}	9.25	1.64
1.0×10^{-4}	9.34	1.67
5.0×10^{-5}	9.35	1.67
2.5×10^{-5}	9.37	1.63
	Mean	1.64

AUXILIARY INFORMATION

METHOD/APPARATUS/PROCEDURE:

The earlier described apparatus and method was used (1): in a glass vessel, a mixt of 100 ml of HCl of appropriate concn and the solute were placed and shaken for 6 h in a water thermostat at 25°C. After attaining equilibrium, the pH of the soln was measured and the Ni^{2+} and S content was determined to calculate K_{so}. The pH was measured on a pH-673 pH meter.

SOURCE AND PURITY OF MATERIALS:

0.1M solns of chem. pure $Ni(OAc)_2$, monosodium salt of sulfadimezine, and HCl as well as doubly distd water were used. The source of the materials was not specified.

ESTIMATED ERROR:

K_{so} : std deviation 3×10^{-16} (compiler)/
Temp and pH: not specified.

REFERENCES:

1. Tskitishvili, M. G.; Mikadze, I. I. *Soobshch. Akad. Nauk Gruz. SSR* 1978, *89(3)*, 589.

COMPONENTS:

(1) Benzenesulfonamide, 4-amino-N-(4,6-dimethyl-2-pyrimidinyl)-, monosilver salt (Ag sulfamethazine); $C_{12}H_{13}AgN_4O_2S$; [53081-02-0]

(2) Nitric acid; HNO_3; [7697-37-2]

(3) Potassium nitrate; KNO_3; [7757-79-1]

(4) Water; H_2O; [7732-18-5]

ORIGINAL MEASUREMENTS:

Nesbitt, R. U., Jr.; Sandmann, B. J.
J. Pharm. Sci. 1978, *67(7)*, 1012-17.

VARIABLES:

pH

PREPARED BY:

R. Piekos

EXPERIMENTAL VALUES:

Comparison of Total Silver Sulfamethazine Molar Solubility, S. Determined by the Method of Known Subtraction with the Molar Concentration of the Silver Ion Determined by Direct Potentiometry on Identical Samples at 25±0.1°, 0.1M Ionic Strength, in Nitric Acid Buffer

	pH 2.186		pH 3.970	
	$S \times 10^3$	$[Ag^+] \times 10^3$	$S \times 10^4$	$[Ag^+] \times 10^4$
	1.194	1.185	0.9920	1.028
	1.191	1.185	0.9689	1.003
	1.191	1.171	0.9813	1.041
	1.218	1.194	1.1010	1.170
	1.194	1.157	0.9945	1.053
	1.204	1.166	-	-
Mean	1.198	1.176	1.0100	1.059

AUXILIARY INFORMATION

METHOD/APPARATUS/PROCEDURE:

Mixts of 100 mg Ag sulfamethazine and 25 or 27 ml of the nitric acid buffer were placed in paraffin-coated vials, adjusted to an ionic strength 0.1M with KNO_3 and rotated end over end in a themostated bath until equilibrium soly was obtained (3-7 days). After filtration through 20M glass filtering crucibles, the solns were analyzed at 25±0.1°C in paraffin-coated beakers for Ag^+ ions with a silver-ion selective electrode (No. 94-16, Orion Res., Cambridge, Mass) standardized at the temp indicated and 0.1M ionic strength . The pH was measured with a triple-purpose pH electrode (Corning Sci. Instruments, Medfield, Mass) standardized using buffers meeting NBS requirements. The nitric acid buffers were prepd by diln of 0.1M HNO_3 and were adjusted to an ionic strength of 0.1M with KNO_3.

SOURCE AND PURITY OF MATERIALS:

All reagents used were anal or USP grade.

Ag sulfamethazine was prepd by the method of Rosenzweig and Fuchs (1) and recrystd from ammonia (2). Water had a sp cond of (1-10) $\times 10^{-7}$ ohm^{-1} cm^{-1}.

The source of the reagents was not specified.

ESTIMATED ERROR:

Soly: when tested by one-way analysis of variance, the means displayed in the 1st Table were found not to be statistically different at the 1% confidence level (authors).

REFERENCES:

1. Rosenzweig, S.; Fuchs, W. *U. S. pat.* 2,536,095 (1951).

(2) Sandmann, B. J.; Nesbitt, R. U., Jr.; Sandmann, R. A. *J. Pharm. Sci.* 1974, *63*, 948.

COMPONENTS:

(1) Benzenesulfonamide, 4-amino-N-(4,6-dimethyl-2-pyrimidinyl)-, monosilver salt (Ag sulfamethazine); $C_{12}H_{13}AgN_4O_2S$; [53081-02-0]

(2) Nitric acid; HNO_3; [7697-37-2]

(3) Potassium nitrate; KNO_3; [7757-79-1]

(4) Water; H_2O; [7732-18-5]

ORIGINAL MEASUREMENTS:

Nesbitt, R. U., Jr.; Sandmann, B. J.

J. Pharm. Sci. 1978, *67(7)*, 1012-17.

VARIABLES:

pH

PREPARED BY:

R. Piekos

EXPERIMENTAL VALUES:

Continued from previous page.

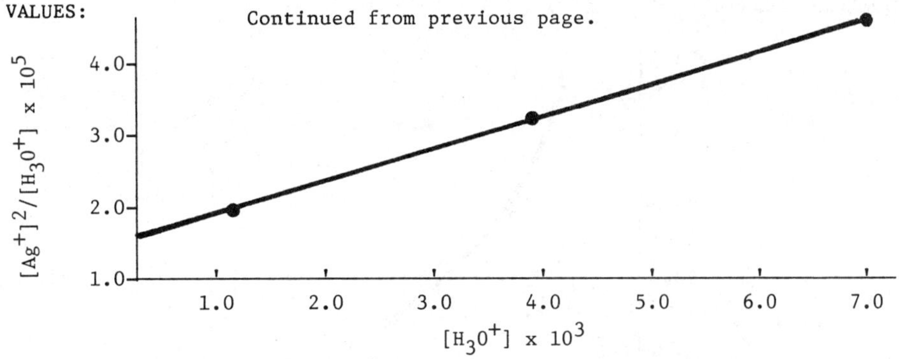

Equilibrium values of $[Ag^+]^2/[H_3O^+]$ versus $[H_3O^+]$ for silver sulfamethazine in nitric acid buffer at 0.1M ionic strength and 25 ± 0.1°

Calculation of the solubility product of Ag sulfamethazine[a], K_s, at 25 ± 0.1°C and 0.1M ionic strength.

[a]from eq. $K_s = f_o S^2$, where $f_o = \left(1 + \frac{[H_3O^+]}{K_2} + \frac{[H_3O^+]^2}{K_1 K_2}\right)^{-1}$

S is the total molar solubility, and K_1 and K_2 are the apparent dissociation constants of the N^4- (amino) and N^1- (amido) hydrogens of sulfamethazine, respectively.

AUXILIARY INFORMATION

METHOD/APPARATUS/PROCEDURE:

SOURCE AND PURITY OF MATERIALS:

ESTIMATED ERROR:

REFERENCES:

COMPONENTS:	ORIGINAL MEASUREMENTS:
Continued from previous page.	Nesbitt, R. U., Jr.; Sandmann, B. J. *J. Pharm. Sci.* 1977, *66(4)*, 519-22.
VARIABLES: pH	PREPARED BY: R. Piekos

EXPERIMENTAL VALUES:

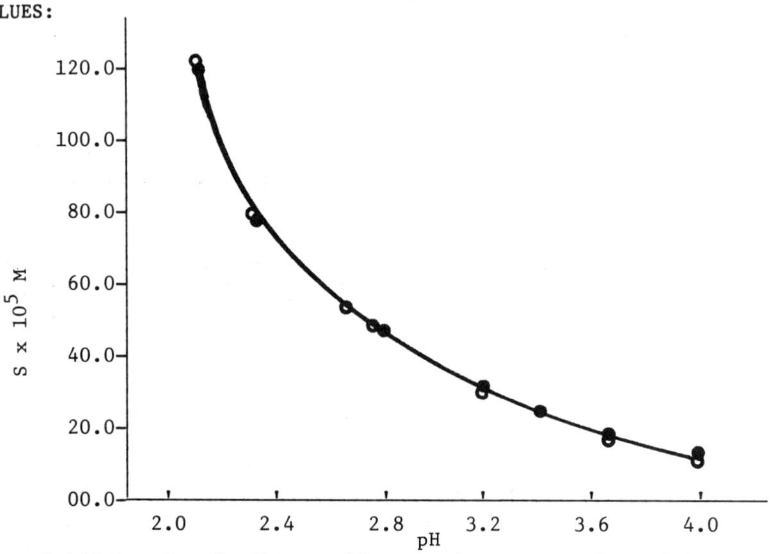

Molar solubility, S, of silver sulfamethazine versus pH at 0.1M ionic strength and 25 ± 0.1°. Key: o, calculated from Eq. 7, and ●, experimental values.

Eq. 7: $$S^2 = [Ag^+]^2 = \frac{[H_3O^+]^2K_s}{K_1\ \ K_2} + \frac{[H_3O^+]K_s}{K_2}$$

AUXILIARY INFORMATION

METHOD/APPARATUS/PROCEDURE:

SOURCE AND PURITY OF MATERIALS:

ESTIMATED ERROR:

Temp: ±0.1° (authors).

pH : accuracy ±0.001 pH unit (authors).

REFERENCES:

COMPONENTS:

Continued from previous page.

ORIGINAL MEASUREMENTS:

Nesbitt, R. U., Jr.; Sandmann, B. J.
J. Pharm. Sci. 1978, *67(7)*, 1012-17.

VARIABLES:

pH

PREPARED BY:

R. Piekos

EXPERIMENTAL VALUES:

pH	f_o	S^2	K_s[a]
2.174	2.551×10^{-6}	1.477×10^{-6}	3.77×10^{-12}
2.421	6.323×10^{-6}	5.897×10^{-7}	3.72×10^{-12}
2.668	1.455×10^{-5}	2.606×10^{-7}	3.79×10^{-12}
2.934	3.265×10^{-5}	1.151×10^{-7}	3.76×10^{-12}
		Mean	$(3.76 \pm 0.03)10^{-12}$

[a] K_s reported as mean ± SD.

AUXILIARY INFORMATION

METHOD/APPARATUS/PROCEDURE:

SOURCE AND PURITY OF MATERIALS:

ESTIMATED ERROR:

REFERENCES:

COMPONENTS:

(1) Benzenesulfonamide, 4-amino-N-(4,6-dimethyl-2-pyrimidinyl)-, monosilver salt (Ag sulfamethazine); $C_{12}H_{13}AgN_4O_2S$; [53081-02-0]

(2) 4-Morpholinepropanesulfonic acid; $C_7H_{15}NO_4S$; [1132-61-2]

(3) 4-Morpholinepropanesulfonic acid, sodium salt; $C_7H_{14}NNaO_4S$; [71119-22-7]

(4) Potassium nitrate; KNO_3; [7757-79-1]

(5) Water; H_2O; [7732-18-5]

ORIGINAL MEASUREMENTS:

Nesbitt, R. U., Jr.; Sandmann, B. J.

J. Pharm. Sci. 1978, *67(7)*, 1012-17.

VARIABLES:

Hydronium-ion concentration

PREPARED BY:

R. Piekos

EXPERIMENTAL VALUES:

Equilibrium values of S^2 (S = total molar solubility) versus $[H_3O^+]$ for Ag sulfamethazine in 0.05M 4-morpholinepropanesulfonic acid buffer at 0.1M ionic strength (KNO_3) and 25±0.1°C.

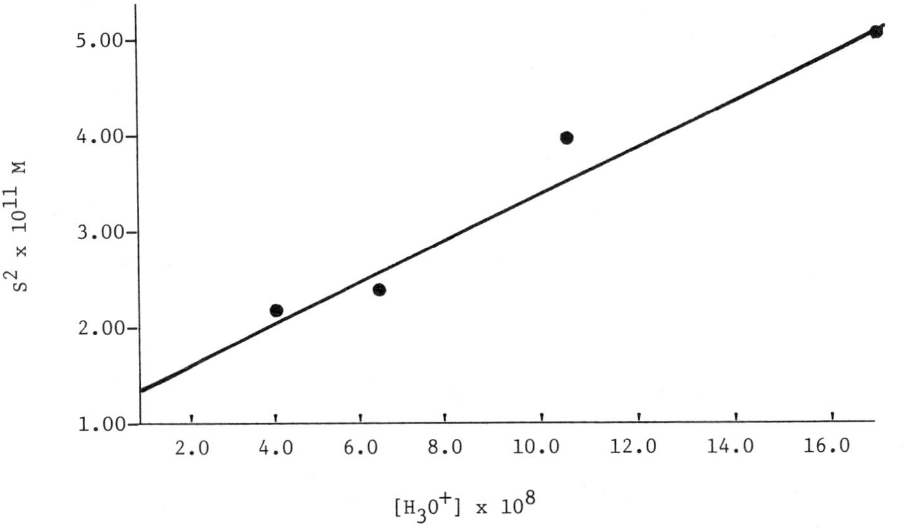

AUXILIARY INFORMATION

METHOD/APPARATUS/PROCEDURE:

Mixts of 100 mg Ag sulfamethazine and 25 or 27 ml of the 4-morpholinepropanesulfonic acid buffer were placed in paraffin-coated vials, adjusted to an ionic strength 0.1M with KNO_3, and rotated end over end in a thermostated bath until equilibrium soly was obtained (3-7 days). After filtration through 20M glass filtering crucibles, the solns were analyzed at 25±0.1°C in paraffin-coated beakers for Ag+ ions with a silver-ion selective electrode (No. 94-16, Orion Res., Cambridge, Mass) standardized at the temp indicated and 0.1M ionic strength. The pH was measured with a triple-purpose pH electrode (Corning Sci. Instruments, Medfield, Mass) standardized using buffers meeting NBS requirements. The buffers were prepd with a total molar concn of 0.05M and adjusted to an ionic strength of 0.1M with KNO_3.

SOURCE AND PURITY OF MATERIALS:

All reagents used were anal or USP grade.

Ag sulfamethazine was prepd by the method of Rosenzweig and Fuchs (1) and recrystd from ammonia (2). Water had a sp cond of (1-10) x 10^{-7} ohm^{-1} cm^{-1}.

The buffer soln was from US Biochem. Corp., Cleveland, Ohio (purity not specified).

ESTIMATED ERROR:

Soly: not specified.

Temp: ±0.1° (authors).

pH : accuracy ±0.001 pH unit (authors).

REFERENCES:

1. Rosenzweig, S.; Fuchs, W. *U.S. pat.* 2,536,095 (1951)
2. Sandmann, B. J.; Nesbitt, R. U., Jr.; Sandmann, R. A. *J. Pharm. Sci.* 1974, *63*, 948.

COMPONENTS:	ORIGINAL MEASUREMENTS:
(1) Zinc, (*T*-4)-bis[4-amino-*N*-(4,6-dimethyl-2-pyrimidinyl)benzenesulfonamidato-N^N,O]-(Zn(II) sulfamethazine); $C_{24}H_{24}N_8O_4S_2Zn$; [71261-83-1] (2) Water; H_2O; [7732-18-5]	Fox, Ch. L., Jr.; Modak, S.; Stanford, J. W.; Fox, P. L. *Scand. J. Plast. Reconstr. Surg.* 1979, *13(1)*, 89-94.
VARIABLES:	PREPARED BY:
One temperature: 28-30°C	R. Piekos

EXPERIMENTAL VALUES:

Solubility of Zn(II) sulfamethazine in water at room temperature (28-30°C)[a] is 8.6 mg% (1.4×10^{-4} mol dm^{-3} solution, compiler).

[a]Value given by one of the authors (S.M.) in personal communication.

AUXILIARY INFORMATION

METHOD/APPARATUS/PROCEDURE:

Satd soln of Zn(II) sulfamethazine was prepd in water and after 24 h aliquots from the clear supernatant were assayed for sulfamethazine content using the colorimetric method of Bratton and Marshall (1). The soly value was then calcd from the molecular formula.

SOURCE AND PURITY OF MATERIALS:

The Zn(II) sulfamethazine was prepd by the authors as follows: an inorg Zn salt was reacted with Na salt of sulfamethazine and the ppt was analyzed and characterized. No details were given, however. Purity of the materials was not specified.

ESTIMATED ERROR:

Nothing specified.

REFERENCES:

1. Bratton, A. C.; Marshall, E. K., Jr. *J. Biol. Chem.* 1939, *120*, 537.

COMPONENTS:	ORIGINAL MEASUREMENTS:
(1) Acetamide, N-[4-[[(4,6-dimethyl-2-pyrimidinyl)amino]sulfonyl]phenyl]- (acetyl sulfamethazine); $C_{14}H_{16}N_4O_3S$; [100-90-3] (2) Sodium hydroxide; NaOH; [1310-73-2] (3) Water; H_2O; [7732-18-5]	Rose, F. L.; Martin, A. R.; Bevan, H.G.L. *J. Pharm. Exp. Therap.* 1943, *77*, 127-42.
VARIABLES:	PREPARED BY:
pH	R. Piekos

EXPERIMENTAL VALUES:

Solubility at pH 6.5, at 37°C, was 117 mg/100 cm^3 (3.65×10^{-3} mol dm^{-3}, compiler).

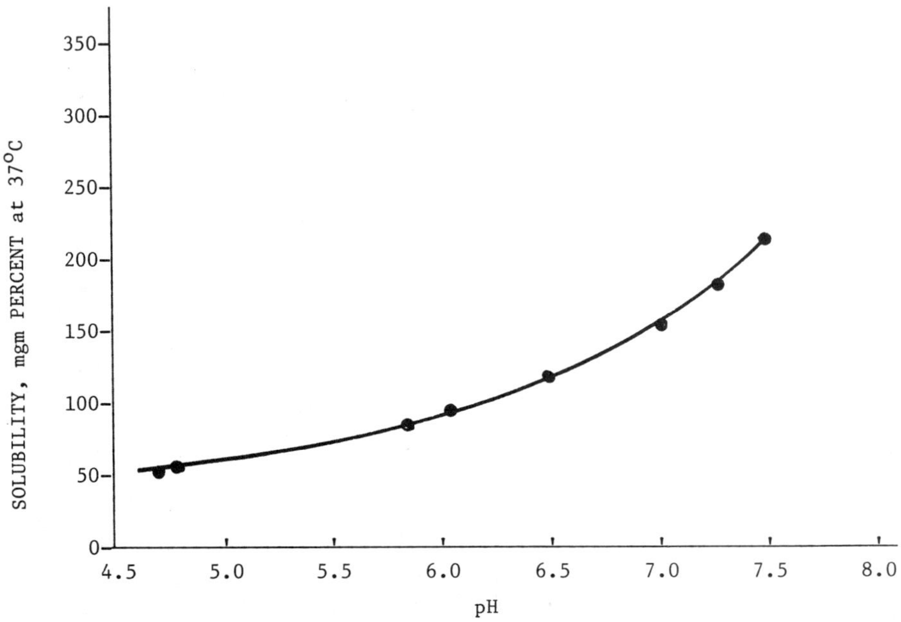

AUXILIARY INFORMATION

METHOD/APPARATUS/PROCEDURE:	SOURCE AND PURITY OF MATERIALS:
An excess of acetyl sulfamethazine was stirred in boiling water, the soln was cooled to 37°C, the temp being maintained thermostatically, and 0.1N NaOH was added to increase the pH. The pH was measured by means of a glass electrode - calomel half-cell system and was permitted to reach equilibrium before a reading was taken. The drug was then de-acetylated and the concn of sulfamethazine in soln was detd colorimetrically by withdrawing a sample through a filter-tip into a preheated micropipet.	The source and purity of acetyl sulfamethazine and of NaOH was not specified. ESTIMATED ERROR: Nothing specified. REFERENCES:

COMPONENTS:

(1) Acetamide, N-[4-[[(4,6-dimethyl-2-pyrimidinyl)amino]sulfonyl]phenyl]- (acetyl sulfamethazine); $C_{14}H_{16}N_4O_3S$; [100-90-3]

(2) Phosphoric acid, disodium salt; Na_2HPO_4; [7558-94-4]

(3) Phosphoric acid, monopotassium salt; KH_2PO_4; [7778-77-0]

(4) Water; H_2O; [7732-18-5]

ORIGINAL MEASUREMENTS:

Gilligan, D. R.; Plummer, M. N.

Proc. Soc. Exp. Biol. Med. <u>1943</u>, *53*, 142-5.

VARIABLES:

pH

PREPARED BY:

R. Piekos

EXPERIMENTAL VALUES:

Solubility of acetyl sulfamethazine in M/15 phosphate buffer solutions

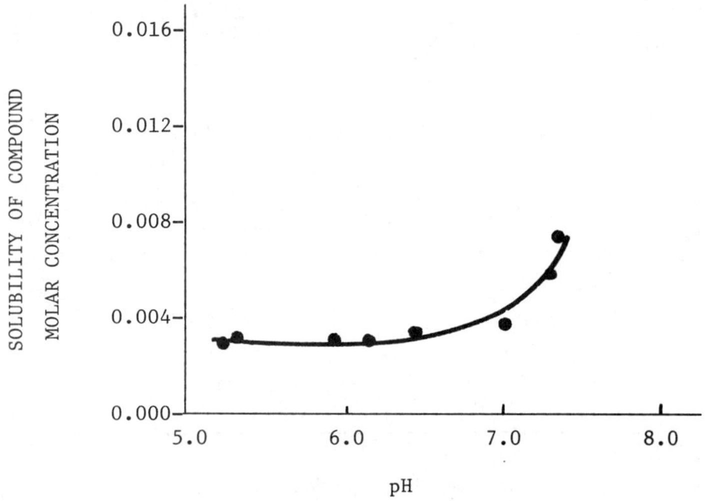

AUXILIARY INFORMATION

METHOD/APPARATUS/PROCEDURE:

An excess of acetyl sulfamethazine was shaken in M/15 phosphate buffer solns of various pH values for 18 h in a water bath at 37°C, and filtered in an incubator room at this temp. The pH of the filtrate was measured immediately at room temp with a Beckmann glass electrode pH meter and appropriate corrections for the differences between room temp and 37°C were applied. The amt of dissolved compd was measured by the method of Bratton and Marshall (1).

SOURCE AND PURITY OF MATERIALS:

Acetyl sulfamethazine was supplied by Lederle Labs, Inc. The source and purity of the remaining materials were not specified.

ESTIMATED ERROR:

Nothing specified.

REFERENCES:

1. Bratton, A. C.; Marshall, E. K., Jr. *J. Biol. Chem.* <u>1939</u>, *128*, 537.

COMPONENTS:

(1) Acetamide, N-[4-[[(4,6-dimethyl-2-pyrimidinyl)amino]sulfonyl]phenyl]- (acetyl Elkosin); $C_{14}H_{16}N_4O_3S$; [100-90-3]

(2) Phosphoric acid, disodium salt; Na_2HPO_4; [7558-94-4]

(3) Phosphoric acid, monopotassium salt; KH_2PO_4; [7778-77-0]

(4) Water; H_2O; [7732-18-5]

ORIGINAL MEASUREMENTS:

Meier, R; Allemann, O., von Meyenburg, H. *Schweiz. Med. Wochenschr.* 1944, *74(42)*, 1091-5.

VARIABLES:

pH

PREPARED BY:

R. Piekos

EXPERIMENTAL VALUES:

pH	Solubility of acetyl Elkosin in phosphate buffers at 37°C	
	mg%	10^3 mol dm^{-3} [a]
5.5	115.0	3.590
6.5	117.0	3.652
7.5	176.0	5.494

[a] Calculated by compiler.

AUXILIARY INFORMATION

METHOD/APPARATUS/PROCEDURE:

Nothing specified.

SOURCE AND PURITY OF MATERIALS:

Nothing specified.

ESTIMATED ERROR:

Nothing specified.

REFERENCES:

COMPONENTS:

(1) Acetamide, N-[4-[[(4,6-dimethyl-2-pyrimidinyl)amino]sulfonyl]phenyl]- (N^4-acetylsulfadimidine); $C_{14}H_{16}N_4O_3S$; [100-90-3]

(2) Phosphoric acid, disodium salt; Na_2HPO_4; [7558-94-4]

(3) Phosphoric acid, monopotassium salt; KH_2PO_4; [7778-77-0]

(4) Water; H_2O; [7732-18-5]

ORIGINAL MEASUREMENTS:

Hekster, Y. A.; Vree, T. B. Damsma, J. E.; Friesen, W. T.

J. Antimicrob, Chemother. 1981, *8*, 133-44.

VARIABLES:

pH

PREPARED BY:

R. Piekos

EXPERIMENTAL VALUES:

pH	Solubility at 25°C mg/l	10^3 mol dm^{-3} [a]
5.5	752	2.35
7.5	1340	4.18

[a] Calculated by compiler.

AUXILIARY INFORMATION

METHOD/APPARATUS/PROCEDURE:

Satd solns of N^4- acetylsulfadimidine were prepd in phosphate buffers of pH 5.5 and 7.5 at room temp (25°C). The concn of the solute was measured by means of a Spectra Physics 3500B high-performance liquid chromatograph equipped with a column oven (Model 748) and a Pye-Unicam LC-UV spectrophotometric detector. The detector was connected to a 1-mV recorder. A stainless steel column (10 cm x 4.6 mm i.d.) was packed with Lichrosorb RPS, 5 μm, obtained from Chrompack. An injection loop of 100 μl was used. The oven temp was 40°C. Detection of the solute was performed at 260 nm.

SOURCE AND PURITY OF MATERIALS:

The source and purity of the materials were not specified.

ESTIMATED ERROR:

The detection limit of the solute by HPLC was 0.5 mg/l (authors.

The error in temperature and pH was not specified.

REFERENCES:

COMPONENTS:

(1) Acetamide, N-[4-[[(4,6-dimethyl-2-pyrimidinyl)amino]sulfonyl]phenyl]-
$C_{14}H_{16}N_4O_3S$; [100-90-3]

(2) 1,2,3-Propanetricarboxylic acid, 2-hydroxy-, disodium salt (Na citrate)
$C_6H_6Na_2O_7$; [144-33-2]

(3) Sodium hydroxide; NaOH; [1310-73-2]

(4) Water; H_2O; [7732-18-5]

ORIGINAL MEASUREMENTS:

Gilligan, D. R.; Plummer, M. N.
Proc. Soc. Exp. Biol. Med. 1943, *53*, 142-5.

VARIABLES:

pH

PREPARED BY:

R. Piekos

EXPERIMENTAL VALUES:

Solubility of acetyl sulfamethazine in M/10 Na citrate + NaOH solutions at 37°C

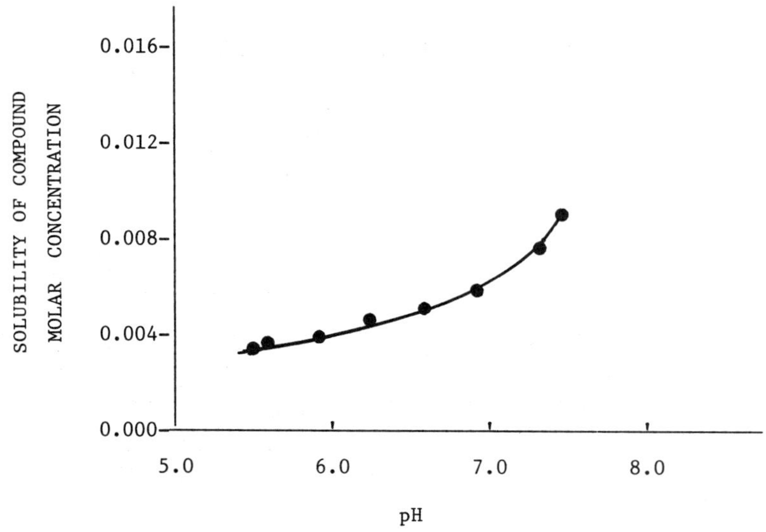

AUXILIARY INFORMATION

METHOD/APPARATUS/PROCEDURE:

An excess of acetyl sulfamethazine was shaken in M/10 Na citrate + NaOH solns of various pH values for 18 h in a water bath at 37°C, and filtered in an incubator room at this temp. The pH of the filtrate was measured immediately with a Beckmann glass electrode pH meter and appropriate corrections for the differences between room temp and 37°C were applied. the amt of dissolved compd was measured by the method of Bratton and Marshall (1).

SOURCE AND PURITY OF MATERIALS:

Acetyl sulfamethazine was supplied by Lederle Labs, Inc. The source and purity of the remaining materials were not specified.

ESTIMATED ERROR:

Nothing specified.

REFERENCES:

1. Bratton, A. C.; Marshall, E. K., Jr. *J. Biol. Chem.* 1939, *128*, 537.

COMPONENTS:

(1) Acetamide, N-[[4-(acetylamino)phenyl]-sulfonyl]-N-(4,6-dimethyl-2-pyrimidinyl)- (N^1, N^4-diacetylsulfamidine); $C_{16}H_{18}N_4O_4S$; [59224-69-0]

(2) Phosphoric acdi, disodium salt; Na_2HPO_4; [7558-94-4]

(3) Phosphoric acid, monopotassium salt; KH_2PO_4; [7778-77-0]

(4) Water; H_2O; [7732-18-5]

ORIGINAL MEASUREMENTS:

Hekster, Ch. A.; Vree, T. B. *Antibiotics Chemother.* 1982, *31*, 22-118.

VARIABLES:

pH

PREPARED BY:

R. Piekos

EXPERIMENTAL VALUES:

pH	Solubility at 25°C	
	mg/l	10^5 mol dm^{-3} [a]
5.5	11.8	3.25
7.5[b]	9.3	2.6

[a] Calculated by compiler

[b] Erroneous pH value of 7.0 is given in the article.

AUXILIARY INFORMATION

METHOD/APPARATUS/PROCEDURE:

The earlier developed method (1) was used (personal communication). Satd solns of N^1,N^4-diacetylsulfadimidine were prepd in phosphate buffers of pH 5.5 and 7.5 at 25°C. The concn of the solute was measured by means of a Spectra Physics 3500B high-performance liquid chromatograph equipped with a Model 748 column oven and a Pye-Unicam LC-UV spectrophotometric detector.

SOURCE AND PURITY OF MATERIALS:

Neither source nor the purity of the materials was specified.

ESTIMATED ERROR:

Soly: the detection limit of the solute by HPLC was 0.5 mg/l (authors).

The errors in temp and pH were not specified.

REFERENCES:

1. Hekster, Y. A; Vree, T. B.; Damsma, J. E.; Friesen, W. T. *J. Antimicrob. Chemother.* 1981, *8*, 133.

COMPONENTS:

(1) Benzenesulfonamide, 4-amino-N-(4-methyl-5-n-pentyl-2-pyrimidinyl)-; $C_{16}H_{22}N_4O_2S$; [71119-35-2]

(2) Water; H_2O; [7732-18-5]

ORIGINAL MEASUREMENTS:

Caldwell, W. T.; Kornfeld, E. C.; Donnell, C. K. *J. Am. Chem. Soc.* 1941, *63*, 2188-90.

VARIABLES:

One temperature: 29°C

PREPARED BY:

R. Piekos

EXPERIMENTAL VALUES:

Solubility of 4-amino-N-(4-methyl-5-n-pentyl-2-pyrimidinyl)benzenesulfonamide in water at 29°C is 2.8 mg/100 ml solution (8.4×10^{-5} mol dm^{-3}- compiler).

AUXILIARY INFORMATION

METHOD/APPARATUS/PROCEDURE:

Soly was detd by weighing the residue obtained by evapg to dryness a known volume of soln satd at 29°C.

SOURCE AND PURITY OF MATERIALS:

The sulfonamide, mp 188-90°C (cor, recrystd from aq EtOH), was prepd by condensing 2-amino-4-methyl-5-n-pentylpyrimidine with acetylsulfanilyl chloride followed by hydrolysis with aq NaOH and pptn at pH 6. Anal: %N 16.67 (calcd 16.75). Purity of the water was not specified.

ESTIMATED ERROR:

Nothing specified.

REFERENCES:

COMPONENTS:	ORIGINAL MEASUREMENTS:
(1) Benzenesulfonamide, 4-amino-N-(4,6-dimethoxy-2-pyrimidinyl)- (sulfadimethoxypyrimidine); $C_{12}H_{14}N_4O_4S$; [155-91-9] (2) Phosphoric acid, disodium salt; Na_2HPO_4; [7558-94-4] (3) 1,2,3-Propanetricaboxylic acid, 2-hydroxy- (citric acid); $C_6H_8O_7$; [77-92-9] (4) Water; H_2O; [7732-18-5]	Bertazzoli, C.; Buogo, A.; Ciceri, C; Ghione, M.; Turolla, E.; Zavaglio, V. *Minerva Med.* 1961, *52(40)*, 1789-96.
VARIABLES: pH	PREPARED BY: R. Piekos

EXPERIMENTAL VALUES:

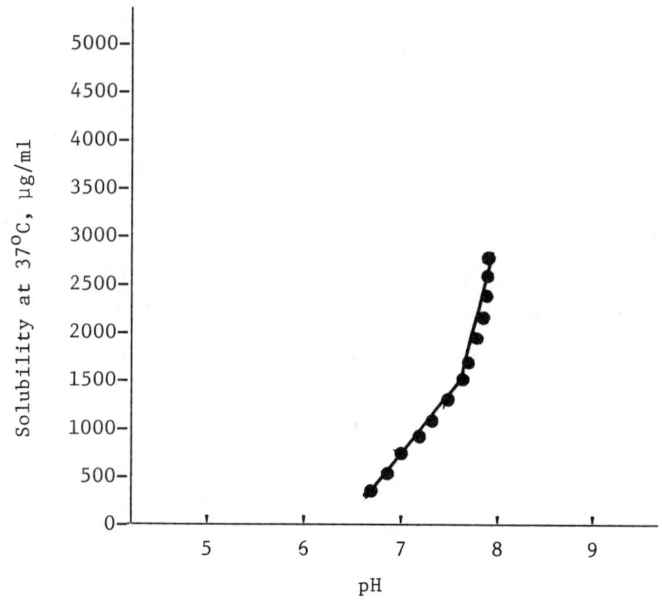

AUXILIARY INFORMATION

METHOD/APPARATUS/PROCEDURE:	SOURCE AND PURITY OF MATERIALS:
The soly of sulfadimethoxypyrimidine in McIlvaine's Na_2HPO_4 - citric acid buffer solns was detd under agitation at 37°C. No details were given.	Nothing specified.
	ESTIMATED ERROR: Nothing specified.
	REFERENCES:

COMPONENTS:

(1) Benzenesulfonamide, 4-amino-N-(2-methyl-4-pyrimidinyl)-; $C_{11}H_{12}N_4O_2S$; [599-84-8]

(2) Water; H_2O; [7732-18-5]

ORIGINAL MEASUREMENTS:

Roblin, R. O., Jr.; Winnek, P. S.; English, J. P.; *J. Am. Chem. Soc.* 1942, *64*, 567-70.

VARIABLES:

One temperature: 37°C

PREPARED BY:

R. Piekos

EXPERIMENTAL VALUES:

Solubility of 4-amino-N-(2-methyl-4-pyrimidinyl)benzenesulfonamide in water at 37°C is 623 mg/100 cm^3 solution (2.36×10^{-2} mol dm^{-3}, compiler).

AUXILIARY INFORMATION

METHOD/APPARATUS/PROCEDURE:

Excess sulfonamide in water was heated and stirred on a steam bath for 30 min. The suspension was then agitated for 24 h in a thermostat at 37°C. A sample of the satd soln was withdrawn through a glass filter, dild, and analyzed by the Marshall method (1) using a General electric recording spectrophotometer for comparing the colors developed with those of the standards.

SOURCE AND PURITY OF MATERIALS:

The sulfonamide, mp 207-8°C (cor), was prepd by the authors. Anal: %C 50.2 (calcd 50.0); %H 4.6 (4.6); %N 21.4 (21.2).

Purity of the water was not specified.

ESTIMATED ERROR:

Nothing specified.

REFERENCES:

1. Bratton, A. C.; Marshall, E. K., Jr. *J. Pharmacol.* 1939, *66*, 4.

COMPONENTS:	EVALUATOR:
(1) Benzenesulfonamide, 4-amino-N-(2-6-dimethyl-4-pyrimidinyl)- (sulfasomidine); $C_{12}H_{14}N_4O_2S$; [515-64-0] (2) Water	Anthony N. Paruta Department of Pharmaceutics University of Rhode Island Kingston, Rhode Island, USA and Ryszard Piekos Faculty of Pharmacy, University of Gdansk Gdansk, Poland 1986

CRITICAL EVALUATION:

The value of 6.86×10^{-3} mol dm^{-3} at 310K as given by Kaneniwa et al. (1) is based on 3-5 days equilibration and accurate spectrophotometric analytical method. The temperature variation was well controlled, ±0.05K. Five years later, Goto et al. (2), also using several days of equilibration and a spectrophotometric analysis provided a value of 6.75×10^{-3} mol dm^{-3}, in good agreement with the earlier one. The recommended value is given as 6.80×10^{-3} mol dm^{-3} in water at 310K.

The two solubility values (3,4) in ethanol are also in close agreement. In 1977, Mauger et al. (3) using a 24 hour equilibrium period and spectrophotometric analysis to report a value of 9.45×10^{-3} mol dm^{-3} in ethanol at 298K. Martin and Miralles (4) determined a value of 9.52×10^{-3} mol dm^{-3} at 298K, in excellent agreement with the earlier value. The solubilities in water and ethanol at 310K are 6.8 and 13.9×10^{-3} mol dm^{-3}, roughly a two fold increase in ethanol.

REFERENCES:

(1) Kaneniwa, N.; Watari, N.; Iijima, H. *Chem. Pharm. Bull.* 1978, *26(9)*, 2603-14.

(2) Goto, S.; Komatsu, M.; Tagawa, K.; Kawata, M. *Chem. Pharm. Bull.* 1983, *31(1)*, 256-61.

(3) Mauger, J.W.; Paruta, A.N.; Gerraughty, R.J.; *J. Pharm. Sci.* 1972, *61(1)*, 94-7.

(4) Martin, A.; Miralles, M.J. *J. Pharm. Sci.* 1982, *71(4)*, 439-42.

COMPONENTS:	ORIGINAL MEASUREMENTS:
(1) Benzenesulfonamide, 4-amino-N-(2,6-dimethyl-4-pyrimidinyl)- (sulfisomidine); $C_{12}H_{14}N_4O_2S$; [515-64-0] (2) Water; H_2O; [7732-18-5]	Yamazaki, M.; Aoki, M.; Kamada, A.; Yata, N. *Yakuzaigaku* 1967, *27(1)*, 37-40.
VARIABLES:	PREPARED BY:
One temperature: 30°C	R. Piekos

EXPERIMENTAL VALUES:

Solubility of sulfisomidine in water at 30°C is 5.65 mmol/L (1.57 g dm^{-3}, compiler).

AUXILIARY INFORMATION

METHOD/APPARATUS/PROCEDURE:	SOURCE AND PURITY OF MATERIALS:
Sulfisomidine (0.5 g) was placed in an L-shaped tube together with 20 ml of water. The mixt was shaken in a thermostat until equilibrium was attained. The sulfisomidine was assayed in the supernatant spectrophotometrically at 545 nm on a Beckmann DU spectrophotometer. The results were taken from a calibration graph.	Nothing specified.
	ESTIMATED ERROR: Soly: not specified. Temp: ±1°C (authors)
	REFERENCES:

COMPONENTS:

(1) Benzenesulfonamide, 4-amino-N-(2,6-dimethyl -4-pyrimidinyl)- (sulfisomidine); $C_{12}H_{14}N_4O_2S$; [515-64-0]

(2) Water; H_2O; [7732-18-5]

ORIGINAL MEASUREMENTS:

Kaneniwa, N.; Watari, N.; Iijima, H. *Chem. Pharm. Bull.* 1978, *26(9)*, 2603-14.

VARIABLES:

One temperature: 37°C

PREPARED BY:

R. Piekos

EXPERIMENTAL VALUES:

Solubility of sulfisomidine in water at 37°C is 1.91 mg/ml solution (6.86×10^{-3} mol dm^{-3}, compiler).

AUXILIARY INFORMATION

METHOD/APPARATUS/PROCEDURE:

An excess of sulfisomidine was placed in a flask contg 25 ml of water. The flask was shaken (2 strokes/s at the amplitude of 3 cm) in a thermostatically controlled water bath at 37°C. One-ml sample was withdrawn every 6 h (total equilibration period was 3-5 days) using a warmed Millipore filter syringe with a filter pore size of 0.45 μ (Millipore HAWP 01300) and the filtrate was dild with water and assayed spectrophotometrically (1).

SOURCE AND PURITY OF MATERIALS:

Commercial sulfisomidine of the Japanese Pharmacopeia grade and distd water were used.

ESTIMATED ERROR:

Soly: not specified.

Temp: ±0.05°C (authors).

REFERENCES:

1. Kaneniwa, N.; Watari, N. *Chem. Pharm. Bull.* 1974, *22*, 1699.

COMPONENTS:

(1) Benzenesulfonamide, 4-amino-N-(2,6-dimethyl-4-pyrimidinyl)- (sulfisomidine); $C_{12}H_{14}N_4O_2S$; [515-64-0]

(2) Water; H_2O; [7732-18-5]

ORIGINAL MEASUREMENTS:

Goto, S.; Komatsu, M.; Tagawa, K.; Kawata, M. *Chem. Pharm. Bull.* 1983, *31(1)*, 256-61.

VARIABLES:

Temperature

PREPARED BY:

R. Piekos

EXPERIMENTAL VALUES:

t/°C	Solubility g/l	Solubility 10^3 mol dm^{-3} [a]
37	1.88	6.75
55	3.35	12.0

[a] Calculated by compiler

AUXILIARY INFORMATION

METHOD/APPARATUS/PROCEDURE:

A 3 g sample of sulfisomidine powder was accurately weighed into a 20-ml ampul and 10 ml of water was added. The ampul was sealed, placed in a const temp (37° or 55°C) bath and allowed to stand for several days. The equilibrium concn of the solute was measured spectrophotometrically at 538 nm after diazotization with the 0.1% Tsuda reagent (1).

SOURCE AND PURITY OF MATERIALS:

Sulfisomidine had mp 245-9°C (decomp). The purity of water was not specified.

ESTIMATED ERROR:

Nothing specified.

REFERENCES:

1. Tsuda, K.; Matsunaga, S. *Yakugaku Zasshi* 1942, *62*, 362.

COMPONENTS:

(1) Benzenesulfonamide, 4-amino-N-(2,6-dimethyl-4-pyrimidinyl)- (sulfisomidine); $C_{12}H_{14}N_4O_2S$; [515-64-0]

(2) Hydrochloric acid; HCl; [7647-01-0]

(3) Water; H_2O; [7732-18-5]

ORIGINAL MEASUREMENTS:

Ogata, H.; Shibazaki, T.; Inoue, T.; Ejima, A. *Chem. Pharm. Bull.* 1979, *27(6)*, 1281-6.

VARIABLES:

One temperature: 37°C

PREPARED BY:

R. Piekos

EXPERIMENTAL VALUES:

Solubility of sulfisomidine in 0.1N HCl at 37°C is 49.418 mg/ml (0.17755 mol dm^{-3}, compiler).

AUXILIARY INFORMATION

METHOD/APPARATUS/PROCEDURE:

A centrifuge tube contg 30 ml of 0.1N HCl and 0.5-3.0 g of the sulfisomidine powder was tightly sealed and shaken at 37°C. The concn of the dissolved drug was detd spectrophotometrically following filtration through a Millipore filter (type EH, pore size 0.5 μm), and the procedure was repeated every 24 h until a const concn was obtained.

SOURCE AND PURITY OF MATERIALS:

Comm available 500-mg uncoated tablets of sulfisomidine were used.
Hydrochloric acid was of reagent grade.

ESTIMATED ERROR:

Nothing specified.

REFERENCES:

COMPONENTS:	ORIGINAL MEASUREMENTS:
(1) Benzenesulfonamide, 4-amino-N-(2,6-dimethyl-4-pyrimidinyl)- (sulfisomidine); $C_{12}H_{14}N_4O_2S$; [515-64-0] (2) Carbonic acid, monosodium salt; $NaHCO_3$; [144-55-8] (3) Water; H_2O; [7732-18-5]	Takubo, T.; Matsumaru, H. Tsuchiya, S.; Hiura, M. *Chem. Pharm. Bull.* 1973, *21(7)*, 1440-5.
VARIABLES:	PREPARED BY:
One temperature: 37°C; one pH: 8.4	R. Piekos

EXPERIMENTAL VALUES:

Solubility of sulfisomidine in a $NaHCO_3$ solution (1.680 g $NaHCO_3$/100 ml water) of pH 8.4 at 37°C is 9.32 mg/ml solution[a] (3.35×10^{-2} mol dm^{-3} solution, compiler).

[a]Numerical value to the graphical data given by one of the authors (S. T.) in personal communication.

AUXILIARY INFORMATION

METHOD/APPARATUS/PROCEDURE:

Aliquots of the $NaHCO_3$ solution were placed in glass-stoppered flask with excess of sulfisomidine. The flasks were allowed to stand at 37±1°C and shaken vigorously for 4 h until equilibrium was established. One ml of the supernatant was removed by means of a filter pipet and sulfisomidine was assayed by the previously reported method (1).

SOURCE AND PURITY OF MATERIALS:

The sulfisomidine was of pharmaceutical grade. The source and purity of $NaHCO_3$ were not specified.
Distd water was used.

ESTIMATED ERROR:

Soly and pH: not specified.
Temp: ±1°C (authors).

REFERENCES:

1. Takubo, T.; Tsuchiya, S.; Hiura, M. *Yakuzaigaku* 1971, *31*, 298.

COMPONENTS:

(1) Benzenesulfonamide, 4-amino-N-(2,6-dimethyl-4-pyrimidinyl)- (sulfisomidine); $C_{12}H_{14}N_4O_2S$; [515-64-0]

(2) Carbonic acid, disodium salt; Na_2CO_3; [497-19-8]

(3) Water; H_2O; [7732-18-5]

ORIGINAL MEASUREMENTS:

Takubo, T.; Matsumaru, H.; Tsuchiya, S.; Hiura, M.

Chem. Pharm. Bull. 1973, *21(7)*, 1440-5.

VARIABLES:

One temperature: 37°C; one pH: 11.3

PREPARED BY:

R. Piekos

EXPERIMENTAL VALUES:

Solubility of sulfisomidine in a Na_2CO_3 solution (2.210 g Na_2CO_3/100 ml water) of pH 11.3 at 37°C is 37.38 mg/ml solution[a] (0.1343 mol dm^{-3} solution, compiler).

[a]Numerical value to the graphical data was given by one of the author (S. T.) in personal communication.

AUXILIARY INFORMATION

METHOD/APPARATUS/PROCEDURE:

Aliquots of the Na_2CO_3 soln were placed in glass-stoppered flasks with excess of sulfisomidine. The flasks were allowed to stand at 37±1°C and shaken vigorously for 4 h until equilibrium was established. One ml of the supenatant was removed by means of a filter pipet and sulfisomidine was assayed by the previously reported method (1).

SOURCE AND PURITY OF MATERIALS:

The sulfisomidine was of pharmaceutical grade. The source and purity of Na_2CO_3 were not specified.

Distd water was used.

ESTIMATED ERROR:

Soly and pH: not specified.

Temp: ±1°C (authors).

REFERENCES:

1. Takubo, T.; Tsuchiya, S.; Hiura, M. *Yakuzaigaku* 1971, *31*, 298.

COMPONENTS:

(1) Benzenesulfonamide, 4-amino-N-(2,6-dimethyl-4-pyrimidinyl)- (sulfisomidine); $C_{12}H_{14}N_4O_2S$; [515-64-0]

(2) Carbonic acid, disodium salt; Na_2CO_3; [497-19-8]

(3) Carbonic acid, monosodium salt; $NaHCO_3$; [144-55-8]

(4) Water; H_2O; [7732-18-5]

ORIGINAL MEASUREMENTS:

Takubo, T.; Matsumaru, H.; Tsuchiya, S.; Hiura, M.

Chem. Pharm. Bull. 1973, *21(7)*, 1440-5.

VARIABLES:

pH

PREPARED BY:

R. Piekos

EXPERIMENTAL VALUES:

Na_2CO_3	$NaHCO_3$	pH	Solubility at 37°C	
g/100 ml water	g/100 ml water		mg/ml soln[a]	$10\ mol\ dm^{-3}$ soln[b]
0.212	1.512	9.1	11.82	0.4247
0.848	1.008	9.8	25.70	0.9234
1.908	0.168	10.7	34.36	1.2345

[a] Numerical values to the graphical data were given by one of the authors (S. T.) in personal communication.

[b] Calculated by compiler.

AUXILIARY INFORMATION

METHOD/APPARATUS/PROCEDURE:

Aliquots of the carbonate buffer solns were placed in glass-stoppered flasks with excess of sulfisomidine. The flasks were allowed to stand at 37±1°C and shaken vigorously for 4 h until equilibrium was established. One ml of the supernatant was removed by means of a filter pipet and sulfisomidine was assayed by the previously reported method (1).

SOURCE AND PURITY OF MATERIALS:

The sulfisomidine was of pharmaceutical grade. The source and purity of Na_2CO_3 and $NaHCO_3$ were not specified.
Distd water was used.

ESTIMATED ERROR:

Soly and pH: not specified.
Temp: ±1°C (authors).

REFERENCES:

1. Takubo, T.; Tsuchiya, S.; Hiura, M. *Yakuzaigaku* 1971, *31*, 298.

COMPONENTS:	ORIGINAL MEASUREMENTS:
(1) Benzenesulfonamide, 4-amino-N-(2,6-dimethyl-4-pyrimidinyl)- (sulfaisodimidine); $C_{12}H_{14}N_4O_2S$; [515-64-0] (2) Phosphoric acid, disodium salt; Na_2HPO_4; [7558-94-4] (3) Phosphoric acid, monopotassium salt; KH_2PO_4; [7778-77-0] (4) Water; H_2O; [7732-18-5]	Riess, W. *Intern. Congr. Chemotherapy, Proc. 3rd, Stuttgart* 1963, *1*, 627-32.
VARIABLES:	**PREPARED BY:**
One temperature: 20°C; one pH: 7.4	R. Piekos

EXPERIMENTAL VALUES:

Solubility of sulfaisodimidine in a M/15 Sörensen buffer solution (pH 7.4) at 20°C is 190 mg% (6.83 x 10^{-3} mol dm^{-3} solution, compiler).

AUXILIARY INFORMATION

METHOD/APPARATUS/PROCEDURE:

Sörensen buffer solns of pH varying between 7 and 8 were prepd, satd with sulfaisodimidine at 20°C, their pH was measured at equilibrium, and the sulfaisodimidine was assayed colorimetrically. The measured pH values were then plotted against concn, and the soly at pH 7.4 was detd by interpolation (personal communication).

SOURCE AND PURITY OF MATERIALS:

Nothing specified.

ESTIMATED ERROR:

Nothing specified.

REFERENCES:

COMPONENTS:

(1) Benzenesulfonamide, 4-amino-N-(2,6-dimethyl-4-pyrimidinyl)- (sulfisomidine); $C_{12}H_{14}N_4O_2S$; [515-64-0]

(2) Phosphoric acid, disodium salt; Na_2HPO_4; [7558-94-4]

(3) Phosphoric acid, monopotassium salt; KH_2PO_4; [7778-77-0]

(4) Water; H_2O; [7732-18-5]

ORIGINAL MEASUREMENTS:

Yamazaki, M.; Aoki, M.; Kamada, A.; Yata, N.

Yakuzaigaku 1967, *27(1)*, 37-40.

VARIABLES:

One temperature: 30°C; one pH: 7.4

PREPARED BY:

R. Piekos

EXPERIMENTAL VALUES:

Solubility of sulfisomidine in a phosphate buffer solution of pH 7.4 (μ = 0.17) at 30°C is 8.53 mmol/L (2.37 g dm^{-3}, compiler).

AUXILIARY INFORMATION

METHOD/APPARATUS/PROCEDURE:

Sulfisomidine (0.5 g) was placed in an L-shaped tube together with 20 ml of the buffer soln. The mixt was shaken in a thermostat until equilibrium was attained. The sulfisomidine was assayed in the supernatant spectrophotometrically at 545 nm on a Beckmann DU spectrophotometer. The results were taken from a calibration graph.

SOURCE AND PURITY OF MATERIALS:

Nothing specified.

ESTIMATED ERROR:

Soly and pH: not specified.

Temp: ±1°C (authors).

REFERENCES:

COMPONENTS:

(1) Benzenesulfonamide, 4-amino-*N*-(2,6-dimethyl-4-pyrimidinyl)- (sulfisomidine); $C_{12}H_{14}N_4O_2S$; [515-64-0]

(2) Phosphoric acid, disodium salt; Na_2HPO_4; [7558-94-4]

(3) Phosphoric acid, monopotassium salt; KH_2PO_4; [7778-77-0]

(4) Water; H_2O; [7732-18-5]

ORIGINAL MEASUREMENTS:

Hekster, Ch. A.; Vree, T. B.

Antibiotics Chemother. 1982, *31*, 22-118.

VARIABLES:

pH

PREPARED BY:

R. Piekos

EXPERIMENTAL VALUES:

pH	Solubility at 25°C	
	mg/l	10^3 mol dm^{-3} [a]
5.5	1,580	5.677
7.5[b]	2,480	8.910

[a] Calculated by compiler

[b] Erroneous pH value of 7.0 is given in the article.

AUXILIARY INFORMATION

METHOD/APPARATUS/PROCEDURE:

The earlier developed method (1) was used (personal communication). Satd solns of sulfisomidine were prepd in phosphate buffers of pH 5.5 and 7.5 at 25°C. The concn of the solute was measured by means of a Spectra Physics 3500B high-performance liquid chromatograph equipped with a Model 748 column oven and a Pye-Unicam LC-UV spectrophotometric detector.

SOURCE AND PURITY OF MATERIALS:

Neither source nor the purity of the materials was specified.

ESTIMATED ERROR:

Soly: the detection limit of the solute by HPLC was 0.5 mg/l (authors).

The errors in temp and pH were not specified.

REFERENCES:

1. Hekster, Y. A.; Vree, T. B.; Damsma, J. E.; Friesen, W. T. *J. Antimicrob. Chemother.* 1981, *8*, 133.

COMPONENTS:

(1) Benzenesulfonamide, 4-amino-N-(2,6-dimethyl-4-pyrimidinyl)- (sulfisomidine); $C_{12}H_{14}N_4O_2S$; [515-64-0]

(2) Phosphoric acid, disodium salt; Na_2HPO_4; [7558-94-4]

(3) 1,2,3-Propanetricarboxylic acid, 2-hydroxy- (citric acid); $C_6H_8O_7$; [77-92-9]

(4) Water; H_2O; [7732-18-5]

ORIGINAL MEASUREMENTS:

Takubo, T.; Matsumaru, H.; Tsuchiya, S.; Hiura, M.

Chem. Pharm. Bull. 1973, *21(7)*, 1440-5.

VARIABLES:

pH

PREPARED BY:

R. Piekos

EXPERIMENTAL VALUES:

Citric acid	Na_2HPO_4	pH	Solubility at 37°C	
g/100 ml water	g/100 ml water		mg/ml soln[a]	10^2 mol dm^{-3} soln[b]
1.680	0.572	3.1	2.98	1.071
1.260	1.144	4.2	1.92	0.690
0.840	1.716	5.8	1.48	0.532
0.420	2.228	6.8	1.81	0.650

[a] Numerical values to the graphical data were given by one of the authors (S. T.) in personal communication.

[b] Calculated by compiler.

AUXILIARY INFORMATION

METHOD/APPARATUS/PROCEDURE:

Aliquots of the buffer soln were placed in glass-stoppered flasks with excess of sulfisomidine. The flasks were allowed to stand at 37±1°C and shaken vigorously for 4 h until equilibrium was established. One ml of the supernatant was removed by means of a filter pipet and sulfisomidine was assayed by the previously reported method (1).

SOURCE AND PURITY OF MATERIALS:

The sulfisomidine was of pharmaceutical grade. The source and purity of the citric acid and Na_2HPO_4 were not specified. Distd water was used.

ESTIMATED ERROR:

Soly and pH: not specified.

Temp: ±1°C (authors).

REFERENCES:

1. Takubo, T.; Tsuchiya, S.; Hiura, M. *Yakuzaigaku* 1971, *31*, 298.

COMPONENTS:

(1) Benzenesulfonamide, 4-amino-N-(2,6-dimethyl-4-pyrimidinyl)- (sulfisomidine); $C_{12}H_{14}N_4O_2S$; [515-64-0]

(2) 1,2,3-Propanetricarboxylic acid, 2-hydroxy- (citric acid); $C_6H_8O_7$ [77-92-9]

(3) Water; H_2O; [7732-18-5]

ORIGINAL MEASUREMENTS:

Takubo, T.; Matsumaru, H.; Tsuchiya, S.; Hiura, M.

Chem. Pharm. Bull. 1973, *21(7)*, 1440-5.

VARIABLES:

One temperature: 37°C; one pH: 2.1

PREPARED BY:

R. Piekos

EXPERIMENTAL VALUES:

Solubility of sulfisomidine in a citric acid solution (2.100 g citric acid per 100 ml water) of pH 2.1 at 37°C is 6.21 mg/ml solution[a] (2.23×10^{-2} mol dm^{-3} solution, compiler).

[a]Numerical value to the graphical data given by one of the authors (S. T.) in personal communication.

AUXILIARY INFORMATION

METHOD/APPARATUS/PROCEDURE:

Aliquots of the citric acid soln were placed in glass-stoppered flasks with excess of sulfisomidine. The flasks were allowed to stand at 37±1°C and shaken vigorously for 4 h until equilibrium was established. One ml of the supernatant was removed by means of a filter pipet and sulfisomidine was assayed by the previously reported procedure (1).

SOURCE AND PURITY OF MATERIALS:

The sulfisomidine was of pharmaceutical grade. The source and purity of the citric acid were not specified.

Distd water was used.

ESTIMATED ERROR:

Soly and pH: not specified.

Temp: ±1°C (authors).

REFERENCES:

1. Takubo, M; Tsuchiya, S; Hiura, M. *Yakuzaigaku* 1971, *31*, 298.

COMPONENTS:

(1) Benzenesulfonamide, 4-amino-N-(2,6-dimethyl-4-pyrimidinyl)- (sulfisomidine); $C_{12}H_{14}N_4O_2S$; [515-64-0]

(2) Ethanol; C_2H_6O; [64-17-5]

(3) Water; H_2O; [7732-18-5]

ORIGINAL MEASUREMENTS:

Martin, A,; Miralles, M. J.

J. Pharm. Sci. 1982, *71(4)*, 439-42.

VARIABLES:

One temperature: 25°C

PREPARED BY:

R. Piekos

EXPERIMENTAL VALUES:

Mole fraction solubility of sulfisomidine in 94.94% (v/v) ethanol at 25°C is 0.001110 (numerical value given in personal communication to compiler).

AUXILIARY INFORMATION

METHOD/APPARATUS/PROCEDURE:

About 20 ml of 95% (label) ethanol was introduced into screw-capped vials contg an excess of sulfisomidine. The vials were agitated for 72 h in a shaker bath maintained at 25±0.2°C. After equilibrium was obtained, a filtered aliquot was pipetted, using an automatic micropipet, into a volumetric flask and appropriately dild with MeOH. The solns were analyzed in a Beckmann Model 25 spectrophotometer at 281.5 nm.

SOURCE AND PURITY OF MATERIALS:

Sulfisomidine, mp 515.6 K, was from Sigma Chemical, St. Louis, Mo. The 94.94% (v/v) ethanol (labeled 95%), d = 0.8139 g cm^{-3} at 25°C, was from Fisher Scientific, Fair Lawn, N. J. Its purity was not specified.

ESTIMATED ERROR:

Soly: detns were run in triplicate (authors).

Temp: ±0.2°C (authors).

REFERENCES:

COMPONENTS:

(1) Benzenesulfonamide, 4-amino-N-(2,6-dimethyl-4-pyrimidinyl)- (sulfisomidine); $C_{12}H_{14}N_4O_2S$; [515-64-0]

(2) Ethanol; C_2H_6O; [64-17-5]

(3) Formic acid; H_2CO_2; [64-18-6]

(4) Sodium formate; $HCNaO_2$; [141-53-7]

(5) Water; H_2O; [7732-18-5]

ORIGINAL MEASUREMENTS:

Martin, A.; Miralles, M. J.

J. Pharm. Sci. 1982, *71(4)*, 439-42.

VARIABLES:

Concentration of ethanol

PREPARED BY:

R. Piekos

EXPERIMENTAL VALUES:

% Ethanol in aqueous buffer[a]	Mole fraction solubility[b] at 25°C
95	0.001320
90	0.001460
80	0.001660
50	0.000858

[a]The buffer solution was prepared by mixing together 29.9107 g HCO_2H, 2.0500 g NaOH, and 470.0393 g distilled water. Its pH was 2.59 (personal communication to compiler).

[b]Numerical values given in personal communication to compiler.

AUXILIARY INFORMATION

METHOD/APPARATUS/PROCEDURE:

About 20 ml of the solvent mixt was introduced into screw-capped vials contg an excess of sulfisomidine. The vials were agitated for 72 h in a shaker bath maintained at 25±0.2°C. After equilibrium was obtained, a filtered aliquot was pipetted, using an automatic micropipette, into a volumetric flask and appropriately dild with MeOH. The solns were analyzed in a Beckmann Model 25 spectrophotometer at 281.5 nm.

SOURCE AND PURITY OF MATERIALS:

Sulfisomidine, mp 515,6 K, was from Sigma Chemicals, St. Louis, Mo. Ethanol (94.94% v/v) and NaOH were from Fisher Scientific, Fair Lawn, N. J.

The source and purity of formic acid were not specified.

Distilled water was used.

ESTIMATED ERROR:

Soly: detns were run in triplicate (authors).

TEmp: ±0.2°C (authors).

REFERENCES:

COMPONENTS:

(1) Benzenesulfonamide, 4-amino-N-(2,6-dimethyl-4-pyrimidinyl)- (sulfisomidine); $C_{12}H_{14}N_4O_2S$; [515-64-0]

(2) Methanol; CH_4O; [67-56-1]

ORIGINAL MEASUREMENTS:

Mauger, J. W.; Paruta, A.N.; Gerraughty, R. J. *J. Pharm. Sci.* 1972, *61(1)*, 94-7.

VARIABLES:

Temperature

PREPARED BY:

R. Piekos

EXPERIMENTAL VALUES:

t/°C	Mole fraction solubility (x 10^4)
25	11.2
30	12.7
37	16.5

AUXILIARY INFORMATION

METHOD/APPARATUS/PROCEDURE:

A const temp bath contg screw-capped bottles with sulfisomidine in excess and methanol was rotated for 24 h. Samples were withdrawn through a pledget of glass wool into a pipet, which was wiped clean and allowed to drain into a volumetric flask. Solute concns were detd by spectrophotometric assay at predetd wavelengths using a Cary model 16 spectrophotometer (1).

SOURCE AND PURITY OF MATERIALS:

Sulfisomidine: lot E2498 from Ciba Pharmaceutical Co.

Methanol: spectrophotometric grade solvent, Mallinckdrodt Chem Works.

ESTIMATED ERROR:

Soly: av values of 3 detns are given (authors).

Temp: ±0.1°C (authors).

REFERENCES:

1. Paruta, A. N.; Mauger, J. W. *J. Pharm. Sci.* 1971, *60*, 432.

COMPONENTS:

(1) Benzenesulfonamide, 4-amino-N-(2,6-dimethyl-4-pyrimidinyl)- (sulfisomidine); $C_{12}H_{14}N_4O_2S$; [515-64-0]

(2) Methanol; CH_4O; [67-56-1]

ORIGINAL MEASUREMENTS:

Mauger, J. W.; Petersen, H., Jr.; Alexander, K. S.; Paruta, A. N. *Drug. Dev. Ind. Pharm.* 1977, *3(2)*, 163-83.

VARIABLES:

Temperature

PREPARED BY:

R. Piekos

EXPERIMENTAL VALUES:

$t/^{\circ}C$	Solubility		
	mg/ml	10^3 X [a]	10^2 mol dm^{-3} [b]
25	7.64	1.12	2.74
30	8.67	1.27	3.15
37	11.2	1.65	4.02

[a] X = mole fraction

[b] Calculated by compiler

AUXILIARY INFORMATION

METHOD/APPARATUS/PROCEDURE:

A const temp bath contg screw-capped bottles with sulfisomidine in excess and methanol was rotated for 24 h. Samples were withdrawn through a pledget of glass wool into a pipet, which was wiped clean and allowed to drain into a volumetric flask. Soly was detd from absorbance and previously ascertained Beer's law plots detd on a Cary model 16 spectrophotometer (1).

SOURCE AND PURITY OF MATERIALS:

Sulfisomidine: lot E2498, Ciba Pharmaceutical Co. Its mp agreed with the literature value. Methanol was a spectrograde solvent from the Mallinckrodt Chemical Works.

ESTIMATED ERROR:

Soly: av of at least 3 detn is reported (authors).

Temp: $\pm 0.1^{\circ}C$ (authors).

REFERENCES:

1. Mauger, J. W.; Paruta, A. N.; Gerraughty, R. J. *J. Pharm. Sci.* 1972, *61(1)*, 94.

COMPONENTS:

(1) Benzenesulfonamide, 4-amino-N-(2,6-dimethyl-4-pyrimidinyl)- (sulfisomidine); $C_{12}H_{14}N_4O_2S$; [515-64-0]

(2) Ethanol; C_2H_6O; [64-17-5]

ORIGINAL MEASUREMENTS:

Mauger, J. W.; Paruta, A. N.; Gerraughty, R. J. *J. Pharm. Sci.* 1972, *61(1)*, 94-7.

VARIABLES:

Temperature

PREPARED BY:

R. Piekos

EXPERIMENTAL VALUES:

t/°C	Mole fraction solubility (x 10^4)
25	5.53
30	6.38
37	8.20

AUXILIARY INFORMATION

METHOD/APPARATUS/PROCEDURE:

A const temp bath contg screw-capped bottles with sulfisomidine in excess and ethanol was rotated for 24 h. Samples were withdrawn through a pledget of glass wool into a pipet, which was wiped clean and allowed to drain into a volumetric flask. Solute concns were detd by spectrophotometric assay at predetd wavelengths using a Cary model 16 spectrophotometer (1).

SOURCE AND PURITY OF MATERIALS:

Sulfisomidine: lot E2498 from Ciba Pharmaceutical Co.

Ethanol was from the U. S. Industrial Chemicals Co. (purity not specified).

ESTIMATED ERROR:

Soly: av values of 3 detns are given (authors).

Temp: ±0.1°C (authors).

REFERENCES:

1. Paruta, A. N.; Mauger, J. W. *J. Pharm. Sci.* 1971, *60*, 432.

COMPONENTS:

(1) Benzenesulfonamide, 4-amino-N-(2,6-dimethyl-4-pyrimidinyl)- (sulfisomidine); $C_{12}H_{14}N_4O_2S$; [515-64-0]

(2) Ethanol; C_2H_6O; [64-17-5]

ORIGINAL MEASUREMENTS:

Mauger, J. W.; Petersen, H., Jr.; Alexander, K. S.; Paruta, A. N.

Drug Dev. Ind. Pharm. 1977, *3(2)*, 163-83.

VARIABLES:

Temperature

PREPARED BY:

R. Piekos

EXPERIMENTAL VALUES:

t/°C	Solubility		
	mg/ml	10^4 X [a]	10^2 mol dm^{-3} [b]
25	2.63	5.53	0.945
30	3.02	6.38	1.08
37	3.86	8.20	1.39

[a] X = mole fraction

[b] Calculated by compiler.

AUXILIARY INFORMATION

METHOD/APPARATUS/PROCEDURE:

A const temp bath contg screw-capped bottles with sulfisomidine in excess and ethanol was rotated for 24 h. Samples were withdrawn through a pledget of glass wool into a pipet, which was wiped clean and allowed to drain into a volumetric flask. Soly was detd from absorbance and previously ascertained Beer's law plots detd on a Cary model 16 spectrophotometer (1).

SOURCE AND PURITY OF MATERIALS:

Sulfisomidine: lot E2498, Ciba Pharmaceutical Co. Its mp agreed with the literature value. Ethanol was from the U.S. Industrial Chemicals Co. Its refractive index value and density agreed with literature values.

ESTIMATED ERROR:

Soly: av of at least 3 detns is reported (authors).

Temp: ±0.1°C (authors).

REFERENCES:

1. Mauger, J. W.; Paruta, A. N.; Gerraughty, R. J. *J. Pharm. Sci.* 1972, *61(1)*, 94.

COMPONENTS:	ORIGINAL MEASUREMENTS:
(1) Benzenesulfonamide, 4-amino-N-(2,6-dimethyl-4-pyrimidinyl)- (sulfisomidine); $C_{12}H_{14}N_4O_2S$; [515-64-0] (2) Ethanol; C_2H_6O; [64-17-5]	Martin, A.; Miralles, M. J. *J. Pharm. Sci.* 1982, *71(4)*, 439-42.
VARIABLES:	PREPARED BY:
One temperature: 25°C	R. Piekos

EXPERIMENTAL VALUES:

Mole fraction solubility of sulfisomidine in absolute ethanol at 25°C is 0.000557 (numerical value given in personal communication to compiler).

AUXILIARY INFORMATION

METHOD/APPARATUS/PROCEDURE:

About 20 ml of abs ethanol was introduced into screw-capped vials contg an excess of sulfisomidine. The vials were agitated for 72 h in a shaker bath maintained at 25±0.2°C. After equilibrium was obtained, a filtered aliquot was pipetted, using an automatic micropipette, into a volumetric flask and appropriately dild with MeOH. The solns were analyzed in a Beckmann Model 25 spectrophotometer at 281.5 nm.

SOURCE AND PURITY OF MATERIALS:

Sulfisomidine, mp 515.6 K, was from Sigma Chemicals, St. Louis, Mo., and abs ethanol was from Commercial Solvent, Terre Haute, Ind., and contained 0.57% (v/v) water. The reagents were tested for identity and purity.

ESTIMATED ERROR:

Soly: detns were run in triplicate (authors).

Temp: ±0.2°C (authors).

REFERENCES:

COMPONENTS:	ORIGINAL MEASUREMENTS:
(1) Benzenesulfonamide, 4-amino-N-(2,6-dimethyl-4-pyrimidinyl)- (sulfisomidine); $C_{12}H_{14}N_4O_2S$; [515-64-0] (2) Ethanol, 2-ethoxy-, $C_4H_{10}O_2$; [110-80-5]	Sunwoo, C.; Eisen, H. *J. Pharm. Sci.* 1971, *60*, 238-44.
VARIABLES:	PREPARED BY:
One temperature: 25°C	R. Piekos

EXPERIMENTAL VALUES:

The mole fraction solubility of sulfisomidine in 2-ethoxyethanol at 25°C is 0.0111 (3.35 g/100 g solution, compiler).

AUXILIARY INFORMATION

METHOD/APPARATUS/PROCEDURE:

Soly was detd by the method reported by Restaino and Martin (1). Sulfisomidine was assayed on a Coleman-Hitachi 124 double-beam spectrophotometer at 270 nm after diln of a sample with 95% alcohol or water.

SOURCE AND PURITY OF MATERIALS:

Sulfisomidine (Ciba Pharmaceutical Co., Summit, N. J.) was recrystd from warm alcohol. 2-Ethoxyethanol was of industrial grade (Cellosolve solvent, Union Carbide, New York, N. Y.).

ESTIMATED ERROR:

Temp: ±1.0°C (authors).

Soly: the mean of 3 runs was given (authors).

REFERENCES:

1. Restaino, F. A.; Martin, A. N. *J. Pharm. Sci.* 1964, *53*, 636.

COMPONENTS:

(1) Benzenesulfonamide, 4-amino-N-(2,6-dimethyl-4-pyrimidinyl)- (sulfisomidine); $C_{12}H_{14}N_4O_2S$; [515-64-0]

(2) 1-Propanol; C_3H_8O; [71-23-8]

ORIGINAL MEASUREMENTS:

Mauger, J. W.; Paruta, A. N.; Gerraughty, R. J. *J. Pharm. Sci.* 1972, *61(1)*, 94-7.

VARIABLES:

Temperature

PREPARED BY:

R. Piekos

EXPERIMENTAL VALUES:

t/°C	Mole fraction solubility (x 10^4)
25	4.23
30	4.89
37	6.48

AUXILIARY INFORMATION

METHOD/APPARATUS/PROCEDURE:

A const temp bath contg screw-capped bottles with sulfisomidine in excess and 1-propanol was rotated for 24 h. Samples were withdrawn through a pledget of glass wool into a pipet, which was wiped clean and allowed to drain into a volumetric flask. Solute concns were detd by spectrophotometric assay at predetd wavelengths using a Cary model 16 spectrophotometer (1).

SOURCE AND PURITY OF MATERIALS:

Sulfisomidine: lot E2498 from Ciba Pharmaceutical Co.

1-Propanol was a Baker Analyzed Reagent, J. T. Baker Chemical Co.

ESTIMATED ERROR:

Soly: av values of 3 detns are given (authors).

Temp: ±0.1°C (authors).

REFERENCES:

1. Paruta, A. N.; Mauger, J. W. *J. Pharm. Sci.* 1971, *60*, 432.

COMPONENTS:	ORIGINAL MEASUREMENTS:
(1) Benzenesulfonamide, 4-amino-N-(2,6-dimethyl-4-pyrimidinyl)- (sulfisomidine); $C_{12}H_{14}N_4O_2S$; [515-64-0] (2) 1-Propanol; C_3H_8O; [71-23-8]	Mauger, J. W.; Petersen, H., Jr. Alexander, K. S.; Paruta, A. N. *Drug. Dev. Ind. Pharm.* 1977, *3(2)*, 163-83.
VARIABLES:	PREPARED BY:
Temperature	R. Piekos

EXPERIMENTAL VALUES:

t/°C	Solubility		
	mg/ml	10^4X^a	10^3 mol dm^{-3} [b]
25	1.57	4.23	5.64
30	1.81	4.89	6.50
37	2.39	6.48	8.59

[a] X = mole fraction

[b] Calculated by compiler

AUXILIARY INFORMATION

METHOD/APPARATUS/PROCEDURE:

A const temp bath contg screw-capped bottles with sulfisomidine in excess and 1-propanol was rotated for 24 h. Samples were withdrawn through a pledget of glass wool into a pipet, which was wiped clean and allowed to drain into a volumetric flask. Soly was detd from absorbance and previously ascertained Beer's law plots detd on a Cary model 16 spectrophotometer (1).

SOURCE AND PURITY OF MATERIALS:

Sulfisomidine: lot E2498, Ciba Pharmaceutical Co. Its mp agreed with the literature value. 1-Propanol was Baker Analyzed Reagent (J.T. Baker Chemical Co.). Its refractive index value and density agreed with literature values.

ESTIMATED ERROR:

Soly: av of at least 3 detns is reported (authors).

Temp: ±0.1°C (authors).

REFERENCES:

1. Mauger, J. W.; Paruta, A. N.; Gerraughty, R. J. *J. Pharm. Sci.* 1972, *61(1)*, 94.

COMPONENTS:

(1) Benzenesulfonamide, 4-amino-N-(2,6-dimethyl-4-pyrimidinyl)- (sulfisomidine); $C_{12}H_{14}N_4O_2S$; [515-64-0]

(2) 1-Butanol; $C_4H_{10}O$; [71-36-3]

ORIGINAL MEASUREMENTS:

Mauger, J. W.; Paruta, A. N.; Gerraughty, R. J. *J. Pharm. Sci.* 1972, *61(1)*, 94-7.

VARIABLES:

Temperature

PREPARED BY:

R. Piekos

EXPERIMENTAL VALUES:

$t/^{\circ}C$	Mole fraction solubility ($\times 10^4$)
25	3.44
30	4.17
37	5.56

AUXILIARY INFORMATION

METHOD/APPARATUS/PROCEDURE:

A const temp bath contg screw-capped bottles with sulfisomidine in excess and 1-butanol was rotated for 24 h. Samples were withdrawn through a pledget of glass wool into a pipet, which was wiped clean and allowed to drain into a volumetric flask. Solute concns were detd by spectrophotometric assay at predtd wavelengths using a Cary model 16 spectrophotometer (1).

SOURCE AND PURITY OF MATERIALS:

Sulfisomidine: lot E2498 from Ciba Pharmaceutical Co.

1-Butanol was from the Mallinckrodt Chemical Works.

ESTIMATED ERROR:

Soly: av values of 3 runs are given (authors).

Temp: $\pm 0.1^{\circ}C$ (authors).

REFERENCES:

1. Paruta, A. N.; Mauger, J. W. *J. Pharm. Sci.* 1971, *60*, 432.

COMPONENTS:	ORIGINAL MEASUREMENTS:
(1) Benzenesulfonamide, 4-amino-N-(2,6-dimethyl-4-pyrimidinyl)- (sulfisomidine); $C_{12}H_{14}N_4O_2S$; [515-64-0] (2) 1-Butanol; $C_4H_{10}O$; [71-36-3]	Mauger, J. W.; Petersen, H., Jr.; Alexander, K. S.; Paruta, A. N. *Drug Dev. Ind. Pharm.* 1977, *3(2)*, 163-83.
VARIABLES:	PREPARED BY:
Temperature	R. Piekos

EXPERIMENTAL VALUES:

t/°C	Solubility		
	mg/ml	10^4 X^a	10^3 mol dm^{-3} [b]
25	1.04	3.44	3.74
30	1.26	4.17	4.53
37	1.67	5.56	6.00

[a] X = mole fraction

[b] Calculated by compiler

AUXILIARY INFORMATION

METHOD/APPARATUS/PROCEDURE:

A const temp bath contg screw-capped bottles with sulfisomidine in excess and 1-butanol was rotated for 24 h. Samples were withdrawn through a pledget of glass wool into a pipet, which was wiped clean and allowed to drain into a volumetric flask. Soly was detd from absorbance and previously ascertained Beer's law plots detd on a Cary model 16 spectrophotometer (1).

SOURCE AND PURITY OF MATERIALS:

Sulfisomidine: lot 2498, Ciba Pharmaceutical Co. Its mp agreed with the literature value. 1-Butanol was from the Mallinckrodt Chemical Works. Its refractive index value and density agreed with literature values.

ESTIMATED ERROR:

Soly: av of at least 3 detns is reported (authors).

Temp: ±0.1°C (authors).

REFERENCES:

1. Mauger, J. W.; Paruta, A. N.; Gerraughty, R. J. *J. Pharm. Sci.* 1972, *61(1)*, 94.

COMPONENTS:

(1) Benzenesulfonamide, 4-amino-N-(2,6-dimethyl-4-pyrimidinyl)- (sulfisomidine); $C_{12}H_{14}N_4O_2S$; [515-64-0]

(2) 1-Pentanol; $C_5H_{12}O$; [71-41-0]

ORIGINAL MEASUREMENTS:

Mauger, J. W.; Paruta, A. N.; Gerraughty, R. J. *J. Pharm. Sci.* 1972, *61(1)*, 94-7.

VARIABLES:

Temperature

PREPARED BY:

R. Piekos

EXPERIMENTAL VALUES:

t/°C	Mole fraction solubility (x 10^4)
25	2.84
30	3.43
37	4.54

AUXILIARY INFORMATION

METHOD/APPARATUS/PROCEDURE:

A const temp bath contg screw-capped bottles with sulfisomidine in excess and 1-pentanol was rotated for 24 h. Samples were withdrawn through a pledget of glass wool into a pipet, which was wiped clean and allowed to drain into a volumetric flask. Solute concns were detd by spectrophotometric assay at predetd wavelengths using a Cary model 16 spectrophotometer (1).

SOURCE AND PURITY OF MATERIALS:

Sulfisomidine: lot E2498 from Ciba Pharmaceutical Co.

1-Pentanol was from Fisher Scientific Co.

ESTIMATED ERROR:

Soly: av values of 3 runs are given (authors).

Temp: ±0.1°C (authors).

REFERENCES:

1. Paruta, A. N.; Mauger, J. W. *J. Pharm. Sci.* 1971, *60*, 432.

COMPONENTS:

(1) Benzenesulfonamide, 4-amino-N-(2,6-dimethyl-4-pyrimidinyl)- (sulfisomidine); $C_{12}H_{14}N_4O_2S$; [515-64-0]

(2) 1-Pentanol; $C_5H_{12}O$; [71-41-0]

ORIGINAL MEASUREMENTS:

Mauger, J. W.; Petersen, H., Jr.; Alexander, K. S.; Paruta, A. N. *Drug. Dev. Ind. Pharm.* 1977, *3(2)*, 163-83.

VARIABLES:

Temperature

PREPARED BY:

R. Piekos

EXPERIMENTAL VALUES:

t/°C	Solubility		
	mg/ml	10^4 X[a]	10^3 mol dm^{-3} [b]
25	0.73	2.84	2.62
30	0.88	3.43	3.16
37	1.16	4.54	4.17

[a] X = mole fraction

[b] Calculated by compiler

AUXILIARY INFORMATION

METHOD/APPARATUS/PROCEDURE:

A const temp bath contg screw-capped bottles with sulfisomidine in excess and 1-pentanol was rotated for 24 h. Samples were withdrawn through a pledget of glass wool into a pipet, which was wiped clean and allowed to drain into a volumetric flask. Soly was detd from absorbance and previously ascertained Beer's law plots detd on a Cary model 16 spectrophotometer (1).

SOURCE AND PURITY OF MATERIALS:

Sulfisomidine: lot E2498, Ciba Pharmaceutical Co. Its mp agreed with the literature value. 1-Pentanol was from Fisher Scientific Co. Its refractive index value and density agreed with literature values.

ESTIMATED ERROR:

Soly: av of at least 3 detns is reported (authors).

Temp: ±0.1°C (authors).

REFERENCES:

1. Mauger, J. W.; Paruta, A. N.; Gerraughty, R. J. *J. Pharm. Sci.* 1972, *61(1)*, 94.

COMPONENTS:	ORIGINAL MEASUREMENTS:
(1) Benzenesulfonamide, 4-amino-N-(2,6-dimethyl-4-pyrimidinyl)- (sulfisomidine); $C_{12}H_{14}N_4O_2S$; [515-64-0] (2) 1-Octanol; $C_8H_{18}O$; [111-87-5]	Mauger, J. W.; Paruta, A. N.; Gerraughty, R. J. *J. Pharm. Sci.* 1972, *61(1)*, 94-7.
VARIABLES:	PREPARED BY:
Temperature	R. Piekos

EXPERIMENTAL VALUES:

t/°C	Mole fraction solubility (x 10^4)
25	1.36
30	1.83
37	2.44

AUXILIARY INFORMATION

METHOD/APPARATUS/PROCEDURE:

A const temp bath contg screw-capped bottles with sulfisomidine in excess and 1-octanol was rotated for 24 h. Samples were withdrawn through a pledget of glass wool into a pipet, which was wiped clean and allowed to drain into a volumetric flask. Solute concns were detd by spectrophotometric assay at predetd wavelengths using a Cary model 16 spectrophotometer (1).

SOURCE AND PURITY OF MATERIALS:

Sulfisomidine: lot E2498 from Ciba Pharmaceutical Co.

1-Pentanol was from Fisher Scientific Cp.

ESTIMATED ERROR:

Soly: av values of 3 runs are given (authors).

Temp: ±0.1°C (authors).

REFERENCES:

1. Paruta, A. N.; Mauger, J. W. *J. Pharm. Sci.* 1971, *60*, 432.

COMPONENTS:

(1) Benzenesulfonamide, 4-amino-N-(2,6-dimethyl-4-pyrimidinyl)- (sulfisomidine); $C_{12}H_{14}N_4O_2S$; [515-64-0]

(2) 1-Octanol; $C_8H_{18}O$; [111-87-5]

ORIGINAL MEASUREMENTS:

Mauger, J. W.; Petersen, H., Jr.; Alexander, K. S.; Paruta, A. N. *Drug Dev. Ind. Pharm.* 1977, *3(2)*, 163-83.

VARIABLES:

Temperature

PREPARED BY:

R. Piekos

EXPERIMENTAL VALUES:

t/°C	Solubility		
	mg/ml	10^4 X^a	10^3 mol dm^{-3} [b]
25	0.24	1.36	0.86
30	0.32	1.83	1.15
37	0.43	2.44	1.55

[a] X = mole fraction

[b] Calculated by compiler

AUXILIARY INFORMATION

METHOD/APPARATUS/PROCEDURE:

A const temp bath contg screw-capped bottles with sulfisomidine in excess and 1-octanol was rotated for 24 h. Samples were withdrawn through a pledget of glass wool into a pipet, which was wiped clean and allowed to drain into a volumetric flask. Soly was detd from absorbance and previously ascertained Beer's law plots detd on a Cary model 16 spectrophotometer (1).

SOURCE AND PURITY OF MATERIALS:

Sulfisomidine: lot E2498, Ciba Pharmaceutical Co. Its mp agreed with the literature value. 1-Octanol was from Fisher Scientific Co. Its refractive index value and density agreed with literature values.

ESTIMATED ERROR:

Soly: av of a least 3 detns is reported (authors).

Temp: ±0.1°C (authors).

REFERENCES:

1. Mauger, J. W.; Paruta, A. N.; Gerraughty, R. J. *J. Pharm. Sci.* 1972, *61(1)*, 94.

COMPONENTS:

(1) Benzenesulfonamide, 4-amino-N-(2,6-dimethyl-4-pyrimidinyl)- (sulfisomidine); $C_{12}H_{14}N_4O_4S$; [515-64-0]

(2) 1-Decanol; $C_{10}H_{22}O$ [112-30-1]

ORIGINAL MEASUREMENTS:

Mauger, J. W.; Paruta, A. N.; Gerraughty, R. J.; *J. Pharm. Sci.* 1972, *61(1)*, 94-7.

VARIABLES:

Temperature

PREPARED BY:

R. Piekos

EXPERIMENTAL VALUES:

t/°C	Mole fraction solubility (x 10^4)
25	1.80
30	2.04
37	2.53

AUXILIARY INFORMATION

METHOD/APPARATUS/PROCEDURE:

A const temp bath contg screw-capped bottles with sulfisomidine in excess and 1-decanol was rotated for 24 h. Samples were withdrawn through a pledget of glass wool into a pipet, which was wiped clean and allowed to drain into a volumetric flask. Solute concns were detd by spectrophotometric assay at predetd wavelengths using a Cary model 16 spectrophotometer (1).

SOURCE AND PURITY OF MATERIALS:

Sulfisomidine: lot E2498 from Ciba Pharmaceutical Co.

1-Decanol was from Matheson, Coleman and Bell.

ESTIMATED ERROR:

Soly: av values of 3 runs are given (authors).

Temp: ±0.1°C (authors).

REFERENCES:

1. Paruta, A. N.; Mauger, J. W. *J. Pharm. Sci.* 1971, *60*, 432.

COMPONENTS:	ORIGINAL MEASUREMENTS:
(1) Benzenesulfonamide, 4-amino-N-(2,6-dimethyl-4-pyrimidinyl)- (sulfisomidine); $C_{12}H_{14}N_4O_2S$; [515-64-0] (2) 1-Decanol; $C_{10}H_{22}O$; [112-30-1]	Mauger, J. W.; Petersen, H., Jr.; Alexander, K. S.; Paruta, A. N. *Drug Dev. Ind. Pharm.* 1977, *3(2)*, 163-83.
VARIABLES:	PREPARED BY:
Temperature	R. Piekos

EXPERIMENTAL VALUES:

t/°C	Solubility		
	mg/ml	10^4 X^a	10^3 mol dm^{-3} a
25	0.26	1.80	0.93
30	0.30	2.04	1.08
37	0.37	2.53	1.33

[a] X = mole fraction

[b] Calculated by compiler

AUXILIARY INFORMATION

METHOD/APPARATUS/PROCEDURE:

A const temp bath contg screw-capped bottles with sulfisomidine in excess and 1-decanol was rotated for 24 h. Samples were withdrawn through a pledget of glass wool into a pipet, which was wiped clean and allowed to drain into a volumetric flask. Soly was detd from absorbance and previously ascertained Beer's law plots detd on a Cary model 16 spectrophotometer (1).

SOURCE AND PURITY OF MATERIALS:

Sulfisomidine: lot E2498, Ciba Pharmaceutical Co. Its mp agreed with the literature value. 1-Decanol was from Matheson, Coleman and Bell. Its refractive index value and density agreed with literature values.

ESTIMATED ERROR:

Soly: av of at least 3 detns is reported (authors).

Temp: ±0.1°C (authors).

REFERENCES:

1. Mauger, J. W.; Paruta, A. N.; Gerraughty, R. J. *J. Pharm. Sci.* 1972, *61(1)*, 94.

COMPONENTS:	ORIGINAL MEASUREMENTS:
(1) Benzenesulfonamide, 4-amino-N-(2,6-dimethyl-4-pyrimidinyl)- (sulfisomidine); $C_{12}H_{14}N_4O_2S$; [515-64-0] (2) n-Hexane; C_6H_{14}; [110-54-3]	Martin, A.; Miralles, M. J. *J. Pharm. Sci.* 1982, *71(4)*, 439-42.
VARIABLES:	PREPARED BY:
One temperature: 25°C	R. Piekos

EXPERIMENTAL VALUES:

Mole fraction solubility of sulfisomidine in n-hexane at 25°C is 0.000038 (numerical value given in personal communication - compiler).

AUXILIARY INFORMATION

METHOD/APPARATUS/PROCEDURE:

About 20 ml of n-hexane was introduced into screw-capped vials contg an excess of sulfisomidine. The vials were agitated for 72 h in a shaker bath maintained at 25±0.2°C. After equilibrium was obtained, a filtered aliquot was pipetted, using an automatic micropipette, into a volumetric flask and appropriately dild with MeOH. The solns were analyzed in a Beckmann Model 25 spectrophotometer at 281.5 nm.

SOURCE AND PURITY OF MATERIALS:

Sulfisomidine, mp 515.6 K, was from Sigma Chemicals, St. Louis, Mo., and n-hexane was from Aldrich Chemical, Milwaukee, Wis. They were tested for identity and purity.

ESTIMATED ERROR:

Soly: the detn was made in triplicate (authors).

Temp: ±0.2°C (authors).

REFERENCES:

COMPONENTS:	ORIGINAL MEASUREMENTS:
(1) Benzenesulfonamide, 4-amino-N-(2,6-dimethyl-4-pyrimidinyl)- (sulfaisodimidine); $C_{12}H_{14}N_4O_2S$; [515-64-0] (2) Methane, trichloro- (chloroform); $CHCl_3$; [67-66-3]	Riess, W. *Intern. Congr. Chemotherapy, Proc., 3rd, Stuttgart* 1963, *1*, 627-32.
VARIABLES:	PREPARED BY:
One temperature: 20°C	R. Piekos

EXPERIMENTAL VALUES:

Solubility of sulfaisodimidine in chloroform at 20°C is 38 mg%

(1.4 x 10^{-3} mol dm^{-3} solution, compiler).

AUXILIARY INFORMATION

METHOD/APPARATUS/PROCEDURE:	SOURCE AND PURITY OF MATERIALS:
Nothing specified.	Nothing specified.
	ESTIMATED ERROR:
	Nothing specified.
	REFERENCES:

COMPONENTS:	ORIGINAL MEASUREMENTS:
(1) Benzenesulfonamide, 4-amino-N-(2,6-dimethyl-4-pyrimidinyl)- (sulfisomidine); $C_{12}H_{14}N_4O_2S$; [515-64-0] (2) Methane, trichloro- (chloroform); $CHCl_3$; [67-66-3]	Yamazaki, M.; Aoki, M.; Kamada, A.; Yata, N. *Yakuzaigaku* 1967, *27(1)*, 37-40.
VARIABLES:	PREPARED BY:
One temperature: 30°C	R. Piekos

EXPERIMENTAL VALUES:

Solubility of sulfisomidine in chloroform at 30°C is 1.63 mmol/L (0.454 g dm^{-3}, compiler).

AUXILIARY INFORMATION

METHOD/APPARATUS/PROCEDURE:	SOURCE AND PURITY OF MATERIALS:
Sulfisomidine (0.5 g) was placed in an L-shaped tube together with 20 ml of chloroform. The mixt was shaken in a thermostat until equilibrium was attained. The sulfisomidine was assayed in the supernatant spectrophotometrically at 545 nm on a Beckmann DU spectrophotometer. The results were taken from a calibration graph.	Nothing specified. ESTIMATED ERROR: Soly: not specified. Temp: ±1°C (authors). REFERENCES:

COMPONENTS:	ORIGINAL MEASUREMENTS:
(1) Benzenesulfonamide, 4-amino-N-(2,6-dimethyl-4-pyrimidinyl)- (sulfisomidine); $C_{12}H_{14}N_4O_2S$; [515-64-0] (2) Ethanol; C_2H_6O; [64-17-5] (3) n-Hexane; C_6H_{14}; [110-54-3]	Martin, A.; Miralles, M. J. *J. Pharm. Sci.* 1982, *71(4)*, 439-42.
VARIABLES:	**PREPARED BY:**
Concentration of n-hexane	R. Piekos

EXPERIMENTAL VALUES:

% n-Hexane	Mole fraction solubility[a] at 25°C
75	0.000037
60	0.000088
40	0.000207
20	0.000392
10	0.000473

[a]Numerical values given in personal communication to compiler.

AUXILIARY INFORMATION

METHOD/APPARATUS/PROCEDURE:

About 20 ml of the solvent mixt was introduced into screw-capped vials contg an excess of sulfisomidine. The vials were agitated for 72 h in a shaker bath maintained at 25±0.2°C. After equilibrium was obtained, a filtered aliquot was pipetted, using an automatic micropipette, into a volumetric flask and appropriately dild with MeOH. The solns were analyzed in a Beckmann Model 25 spectrophotometer at 281.5 nm.

SOURCE AND PURITY OF MATERIALS:

Sulfisomidine, mp 515.6 K, was from Sigma Chemicals, St. Louis, Mo., and abs ethanol was from Commercial Solvent, Terre Haute, Ind., and contained 0.57% (v/v) water, n-Hexane was from Aldrich Chemical, Milwaukee, Wis. The reagents were tested for identity and purity.

ESTIMATED ERROR:

Soly: detns were made in triplicate (authors).

Temp: ±0.2°C (authors).

REFERENCES:

COMPONENTS:	ORIGINAL MEASUREMENTS:
(1) Zinc, (T-4)-bis[4-amino-N-(2,6-dimethyl-4-pyrimidinyl)benzenesulfonamidato-N^N,O]- (Zn(II) sulfisomidine); $C_{24}H_{24}N_8O_4S_2Zn$; [71261-88-6] (2) Water; H_2O; [7732-18-5]	Fox, Ch. L., Jr.; Modak, S.; Stanford, J. W.; Fox, P. L. *Scand. J. Plast. Reconstr. Surg.* 1979, *13(1)*, 89-94.
VARIABLES:	PREPARED BY:
One temperature: 28-30°C	R. Piekos

EXPERIMENTAL VALUES:

Solubility of Zn(II) sulfisomidine in water at room temperature (28-30°C)[a] is 171.0 mg% (2.758×10^{-3} mol dm^{-3} solution, compiler).

[a]Value given by one of the authors (S. M.) in personal communication.

AUXILIARY INFORMATION

METHOD/APPARATUS/PROCEDURE:

Satd soln of Zn(II) sulfisomidine was prepd in water and after 24 h aliquots from the clear supernatant were assayed for sulfisomidine content using the colorimetric method of Bratton and Marshall (1). The soly was then calculated from the molecular formula.

SOURCE AND PURITY OF MATERIALS:

The Zn(II) sulfisomidine was prepd by the authors as follows: an inorg Zn salt was reacted with Na salt of sulfisomidine and the ppt was analyzed and characterized. No details were given, however.
Purity of the materials was not specified.

ESTIMATED ERROR:

Nothing specified.

REFERENCES:

1. Bratton, A. C.; Marshall, E. K., Jr. *J. Biol. Chem.* 1939, *120*, 537.

COMPONENTS:

(1) Acetamide, N-[4-[[(2,6-dimethyl-4-pyrimidinyl)amino]sulfonyl]phenyl]- (N^4-acetylsulfisomidine); $C_{14}H_{16}N_4O_3S$; [3163-31-3]

(2) Phosphoric acid, disodium salt; Na_2HPO_4; [7558-94-4]

(3) Phosphoric acid, monopotassium salt; KH_2PO_4; [7778-77-0]

(4) Water; H_2O; [7732-18-5]

ORIGINAL MEASUREMENTS:

Hekster, Ch. A.; Vree, T. B. *Antibiotics Chemother.* 1982, *31*, 22-118.

VARIABLES:

pH

PREPARED BY:

R. Piekos

EXPERIMENTAL VALUES:

pH	Solubility at 25°C mg/l	10^4 mol dm^{-3} [a]
5.5	53	1.65
7.5[b]	128	3.99

[a] Calculated by compiler.

[b] Erraneous pH vaue of 7.0 is given in the article.

AUXILIARY INFORMATION

METHOD/APPARATUS/PROCEDURE:

The earlier developed method (1) was used (personal communication). Satd solns of N^4-acetylsulfisomidine were prepd in phosphate buffers of pH 5.5 and 7.5 at 25°C. The concn of the solute was measured by means of a Spectra Physics 3500B high-performance liquid chromatograph equipped with a Model 748 column oven and a Pye-Unicam LC-UV spectrophotometric detector.

SOURCE AND PURITY OF MATERIALS:

Neither source nor the purity of the materials was specified.

ESTIMATED ERROR:

Soly: the detection limit of the solute by HPLC was 0.5 mg/l (authors).
The errors in temp and pH were not specified.

REFERENCES:

1. Hekster, Y. A.; Vree, T. B.; Damsma, J. E.; Friesen, W. T. *J. Anitmicrob. Chemother.* 1981, *8*, 133.

COMPONENTS:	EVALUATOR:
(1) Benzenesulfonamide, 4-amino-N-(2-6-dimethoxy-4-pyrimidinyl) (sulfadimethoxine); $C_{12}H_{14}N_4O_4S$; [122-11-2] (2) Water or Aqueous HCl	Anthony N. Paruta Department of Pharmaceutics University of Rhode Island Kingston, Rhode Island, USA and Ryszard Piekos Faculty of Pharmacy, University of Gdansk Gdansk, Poland 1986

CRITICAL EVALUATION:

The solubility of sulfadimethoxine in water at 310K was reported by Watari, Hanano and Kaneniwa in two different journals (1,2); the determined value of the solubility for sulfadimethoxine given as 1.49×10^{-4} mol dm^{-3}. Except for the source of the material all experimental conditions were identical. With the assumption that the commercial grade (1), purified by crystallization, was equivalent to the pharmacopeial grade (2), the recommended value of solubility in water at 310K is 1.49×10^{-4} mol dm^{-3}.

The reported solubilities at 310K in 0.1 mol L^{-1} HCl solution are divergent (1,3). Watari (1) gave a value of 0.793×10^{-3} mol dm^{-3} which is some four tenths of that given by Ogata et al. (3). The latter (3) value refers to commercial tablets of sulfadimethoxine which may contain water-soluble non-active components that could interfere with the analysis. Though both methods and procedures were quite similar, Watari et al. (1) used a different pore sized millipore filter and the recommended technique of withdrawal to prevent thermal shock. It would seem that both reported equilibrium data, a 3-5 day period by Watari et al. (1) and the invarient asymptope of a dissolution profile by Ogata et al. (3). While it would be difficult to ascertain a recommended value with this limited data, an approximate range of $1\text{-}2 \times 10^{-3}$ mol dm^{-3} in aqueous 0.1 mol L^{-1} HCl at 310K can be given. It is instructive to note that the solubility value of the protonated compound is about 10 times greater than that of pure water. Although $1\text{-}2 \times 10^{-3}$ mol dm^{-3} is a low level of solubility, it is 10 times greater than the solubility in the absence of a high concentration of protons.

REFERENCES:

(1) Watari, N.; Hanano, M.; Kaneniwa, N. *Chem. Pharm. Bull.* 1980, *28(7)*, 2221-5.
(2) Watari, N.; Kaneniwa, N.; Hanano, M. *Int. J. Pharm.* 1980, *6(2)*, 155-66.
(3) Ogata, H.; Shibazaki, T.; Inoue, T.; Ejima, A. *Chem. Pharm. Bull.* 1979, *27(6)*, 1281-6.

COMPONENTS:

(1) Benzenesulfonamide, 4-amino-N-(2,6-dimethoxy-4-pyrimidinyl)- (sulfadimethoxine); $C_{12}H_{14}N_4O_4S$; [122-11-2]

(2) Water; H_2O; [7732-18-5]

ORIGINAL MEASUREMENTS:

Yamazaki, M.; Aoki, M.; Kamada, A.; Yata, N. *Yakuzaigaku* 1967, *27(1)*, 37-40.

VARIABLES:

One temperature: 30°C

PREPARED BY:

R. Piekos

EXPERIMENTAL VALUES:

Solubility of sulfadimethoxine in water at 30°C is 0.10 mmol/L (3.1×10^{-2} g dm^{-3}, compiler).

AUXILIARY INFORMATION

METHOD/APPARATUS/PROCEDURE:

Sulfadimethoxine (0.5 g) was placed in an L-shaped tube together with 20 ml of water. The mixt was shaken in a thermostat until equilibrium was attained. The sulfadimethoxine was assayed in the supernatant spectrophotometrically at 545 nm on a Beckmann DU spectrophotometer. The results were taken from a calibration graph.

SOURCE AND PURITY OF MATERIALS:

Nothing specified.

ESTIMATED ERROR:

Soly: not specified.

Temp: ±1°C (authors).

REFERENCES:

COMPONENTS:

(1) Benzenesulfonamide, 4-amino-N-(2,6-dimethoxy-4-pyrimidinyl)- (sulfadimethoxine); $C_{12}H_{14}N_4O_4S$; [122-11-2]

(2) Water; H_2O; [7732-18-5]

ORIGINAL MEASUREMENTS:

Watari, N.; Hanano, M.; Kaneniwa, N. *Chem. Pharm. Bull.* 1980, *28(7)*, 2221-5.

VARIABLES:

One temperature: 37°C

PREPARED BY:

R. Piekos

EXPERIMENTAL VALUES:

Solubility of sulfadimethoxine in water at 37°C is 4.63 mg/100 ml (1.49×10^{-4} mol dm^{-3}, compiler).

AUXILIARY INFORMATION

METHOD/APPARATUS/PROCEDURE:

The earlier reported app and procedure was employed (1): an excess of sulfadimethoxine, required to saturate the medium, was placed in a flask contg 25 ml of water. The flask was shaken (2 strokes/s) at the amplitude of 3 cm in a thermostatically controlled water bath at 37±0.05°C. One-ml sample was removed every 6 h (total equilibration period was 3-5 days) using a warmed Millipore filter syringe with a filter pore size of 0.45 μ (Millipore HAWP 01300), and the filtrate was dild with water and assayed spectrophotometrically.

SOURCE AND PURITY OF MATERIALS:

Comm sulfadimethoxine was recrystd from EtOAc and sieved through a Ro-Tap testing sieve shaker.
Distd water was used.

ESTIMATED ERROR:

Soly: not specified.
Temp: ±0.05°C (authors).

REFERENCES:

1. Kaneniwa, N.; Watari, N. *Chem. Pharm. Bull.* 1974, *22*, 1699.

COMPONENTS:

(1) Benzenesulfonamide, 4-amino-N-(2,6-dimethoxy-4-pyrimidinyl)- (sulfadimethoxine); $C_{12}H_{14}N_4O_4S$; [122-11-2]

(2) Water; H_2O; [7732-18-5]

ORIGINAL MEASUREMENTS:

Watari, N.; Kaneniwa, N.; Hanano, M.
Int. J. Pharm. 1980, *6(2)*, 155-66.

VARIABLES:

One temperature: 37°C

PREPARED BY:

R. Piekos

EXPERIMENTAL VALUES:

Solubility of sulfadimethoxine in water at 37°C is 4.63 mg/100 ml (1.49×10^{-4} mol dm^{-3}, compiler).

AUXILIARY INFORMATION

METHOD/APPARATUS/PROCEDURE:

The earlier developed method was employed (1), whereby an excess of sulfadimethoxine, required to saturate medium, was placed in a flask contg 25 ml of water. The flask was stroken (2 strokes/s) at an amplitude of 3 cm, in a thermostatically controlled bath. One-ml sample was removed every 6 h (total equilibration time was 3-5 days) using a warmed Millipore filter syringe with a filter pore size of 0.45 μ (Millipore HAWP 01300) and the filtrate was dild with water and assayed spectrophotometrically.

SOURCE AND PURITY OF MATERIALS:

Sulfadimethoxine was of the Japanese Pharmacopeia grade.
Distilled water was used.

ESTIMATED ERROR:

Soly: not specified.
Temp: ±0.05°C (authors).

REFERENCES:

1. Kaneniwa, N.; Watari, N.
Chem. Pharm. Bull. 1974, *22*, 1699.

COMPONENTS:

(1) Benzenesulfonamide, 4-amino-N-(2,6-dimethoxy-4-pyrimidinyl)- (sulfadimethoxime); $C_{12}H_{14}N_4O_4S$; [122-11-2]

(2) Hydrochloric acid; HCl; [7647-01-0]

(3) Water; H_2O; [7732-18-5]

ORIGINAL MEASUREMENTS:

Ogata, H.; Shibazaki, T.; Inoue, T.; Ejima, A.

Chem. Pharm. Bull. 1979, *27(6)*, 1281-6.

VARIABLES:

One temperature: 37°C

PREPARED BY:

R. Piekos

EXPERIMENTAL VALUES:

Solubility of sulfadimethoxine in 0.1N HCl at 37°C is 0.606 mg/ml (1.95×10^{-3} mol dm^{-3}, compiler).

AUXILIARY INFORMATION

METHOD/APPARATUS/PROCEDURE:

A centrifuge tube contg 30 ml of 0.1N HCl and 0.5-3.0 g of the sulfadimethoxine powder was tightly sealed and shaken at 37°C. The concn of the dissolved drug was detd spectrophotometrically following filtration through a Millipore filter (type EH, pore size 0.5 μm), and the procedure was repeated until a const concn was obtained.

SOURCE AND PURITY OF MATERIALS:

Comm available 250-mg uncoated tablets of sulfadimethoxine were used. Hydrochloric acid was of reagent grade.

ESTIMATED ERROR:

Nothing specified.

REFERENCES:

COMPONENTS:

(1) Benzenesulfonamide, 4-amino-N-(2,6-dimethoxy-4-pyrimidinyl)- (sulfadimethoxine); $C_{12}H_{14}N_4O_4S$; [122-11-2]

(2) Hydrochloric acid; HCl; [7647-01-0]

(3) Water; H_2O; [7732-18-5]

ORIGINAL MEASUREMENTS:

Watari, N.; Hanano, M.; Kaneniwa, N.; *Chem. Pharm. Bull.* 1980, *28(7)*, 2221-5.

VARIABLES:

One temperature: 37°C

PREPARED BY:

R. Piekos

EXPERIMENTAL VALUES:

Solubility of sulfadimethoxine in 0.1N HCl solution at 37°C is 24.6 mg/100 ml (7.93×10^{-4} mol dm^{-3} - compiler).

AUXILIARY INFORMATION

METHOD/APPARATUS/PROCEDURE:

The earlier reported app and procedure was employed (1): an excess of sulfadimethoxine, required to saturate the medium, was placed in a flask contg 25 ml of 0.1N HCl soln. The flask was shaken (2 strokes/s) at the amplitude of 3 cm in a thermostatically controlled water bath at 37±0.05°C. One-ml sample was removed every 6 h (total equilibration period was 3-5 days) using a warmed Millipore filter syringe with a filter pore size of 0.45 μ (Millipore HAWP 01300), and the filtrate was dild with water and assayed spectrophotometrically.

SOURCE AND PURITY OF MATERIALS:

Comm sulfadimethoxine was recrystd from EtOAc and sieved through a Ro-Tap testing sieve shaker.
Source and purity of the 0.1N HCl soln were not specified.

ESTIMATED ERROR:

Soly: not specified.
Temp: ±0.05°C (authors).

REFERENCES:

1. Kaneniwa, N.; Watari, N. *Chem. Pharm. Bull.* 1974, *22*, 1699.

COMPONENTS:	ORIGINAL MEASUREMENTS:
(1) Benzenesulfonamide, 4-amino-N-(2,6-dimethoxy-4-pyrimidinyl)- (sulfadimethoxine); $C_{12}H_{14}N_4O_4S$; [122-11-2] (2) Phosphoric acid, disodium salt; Na_2HPO_4; [7558-94-4] (3) Phosphoric acid, monopotassium salt; KH_2PO_4; [7778-77-0] (4) Water; H_2O; [7732-18-5]	Riess, W. *Intern. Congr. Chemotherapy, Proc. 3rd, Stuttgart* <u>1963</u>, *1*, 627-32.
VARIABLES:	PREPARED BY:
One temperature: 20°C; one pH 7.4	R. Piekos

EXPERIMENTAL VALUES:

Solubility of sulfadimethoxine in a M/15 Sörensen buffer solution (pH 7.4) at 20°C is 28 mg% (9.0×10^{-4} mol dm^{-3} solution, compiler).

AUXILIARY INFORMATION

METHOD/APPARATUS/PROCEDURE:	SOURCE AND PURITY OF MATERIALS:
Sörensen buffer solns of pH varying between 7 and 8 were prepd, satd with sulfadimethoxine at 20°C, their pH was measured at equilibrium, and the sulfadimethoxine was assayed colorimetrically. The measured pH values were plotted against concn, and the soly at pH 7.4 was detd by interpolation (personal communication).	Nothing specified.
	ESTIMATED ERROR: Nothing specified.
	REFERENCES:

COMPONENTS:	ORIGINAL MEASUREMENTS:
(1) Benzenesulfonamide, 4-amino-N-(2,6-dimethoxy-4-pyrimidinyl)- (sulfadimethoxine); $C_{12}H_{14}N_4O_4S$; [122-11-2] (2) Phosphoric acid, disodium salt; Na_2HPO_4; [7558-94-4] (3) Phosphoric acid, monopotassium salt; KH_2PO_4; [7778-77-0] (4) Water; H_2O; [7732-18-5]	Yamazaki, M.; Aoki, M.; Kamada, A.; Yata, N. *Yakuzaigaku* 1967, *27(1)*, 37-40.
VARIABLES:	PREPARED BY:
One temperature: 30°C; one pH: 7.4	R. Piekos

EXPERIMENTAL VALUES:

Solubility of sulfadimethoxine in a phosphate buffer solution of pH 7.4 (μ = 0.17) at 30°C is 1.73 mmol/L (0.537 g dm^{-3}, compiler).

AUXILIARY INFORMATION

METHOD/APPARATUS/PROCEDURE:

Sulfadimethoxine (0.5 g) was placed in an L-shaped tube together with 20 ml of the buffer soln. The mixt was shaken in a thermostat until equilibrium was attained. The sulfadimethoxine was assayed in the supernatant spectrophotometrically at 545 nm on a Beckmann DU spectrophotometer. The results were taken from a calibration graph.

SOURCE AND PURITY OF MATERIALS:

Nothing specified.

ESTIMATED ERROR:

Soly and pH: not specified.

Temp: ±1°C (authors).

REFERENCES:

COMPONENTS:

(1) Benzenesulfonamide, 4-amino-N-(2,6-dimethoxy-4-pyrimidinyl)- (sulfadimethoxine); $C_{12}H_{14}N_4O_4S$; [122-11-2]

(2) Methanol; CH_4O; [67-56-1]

ORIGINAL MEASUREMENTS:

Mauger, J. W.; Paruta, A. N.; Gerraughty, R. J. *J. Pharm. Sci.* 1972, *61(1)*,

VARIABLES:

Temperature

PREPARED BY:

R. Piekos

EXPERIMENTAL VALUES:

t/°C	Mole fraction solubility (x 10^4)
25	11.6
30	13.9
37	17.7

AUXILIARY INFORMATION

METHOD/APPARATUS/PROCEDURE:

A const temp bath contg screw-capped bottles with sulfadimethoxine and methanol was rotated for 24 h. Samples were withdrawn through a pledget of glass wool into a pipet, which was wiped clean and allowed to drain into a volumetric flask. Solute concns were detd by spectrophotometric assay at predetd wavelengths using a Cary model 16 spectrophotometer (1).

SOURCE AND PURITY OF MATERIALS:

Sulfadimethoxine: lot 203057, Hoffmann-La Roche, Inc.

Methanol: spectrophotometric grade solvent, Mallinckrodt Chem Works.

ESTIMATED ERROR:

Soly: av values of 3 detns are given (authors).

Temp: ±0.1°C (authors).

REFERENCES:

1. Paruta, A. N.; Mauger, J. W. *J. Pharm. Sci.* 1971, *60*, 432.

COMPONENTS:

(1) Benzenesulfonamide, 4-amino-N-(2,6-dimethoxy-4-pyrimidinyl)- (sulfadimethoxine); $C_{12}H_{14}N_4O_4S$; [122-11-2]

(2) Methanol; CH_4O; [67-56-1]

ORIGINAL MEASUREMENTS:

Mauger, J. W.; Petersen, H., Jr.; Alexander, K. S.; Paruta, A. N. *Drug Dev. Ind. Pharm.* 1977, *3(2)*, 163-83.

VARIABLES:

Temperature

PREPARED BY:

R. Piekos

EXPERIMENTAL VALUES:

t/°C	Solubility mg/ml	10^3 X [a]	10^2 mol dm^{-3} [b]
25	8.84	1.16	2.85
30	10.5	1.39	3.38
37	13.4	1.77	4.32

[a] X = mole fraction

[b] Calculated by compiler.

AUXILIARY INFORMATION

METHOD/APPARATUS/PROCEDURE:

A const temp bath contg screw-capped bottles with sulfadimethoxine in excess and methanol was rotated for 24 h. Samples were withdrawn through a pledget of glass wool into a pipet, which was wiped clean and allowed to drain into a volumetric flask. Soly was detd from absorbance and previously ascertained Beer's law plots detd on a Cary model 16 spectrophotometer (1).

SOURCE AND PURITY OF MATERIALS:

Sulfadimethoxine: lot 203027, Hoffman-La Roche, Inc. Its mp agreed with the literature value. Methanol was a spectrograde solvent from the Mallinckrodt Chemical Works.

ESTIMATED ERROR:

Soly: av of at least 3 detns is reported (authors).

Temp: ±0.1°C (authors).

REFERENCES:

1. Mauger, J. W.; Paruta, A. N.; Gerraughty, R. J. *J. Pharm. Sci.* 1972, *61(1)*, 94.

COMPONENTS:

(1) Benzenesulfonamide, 4-amino-N-(2,6-dimethoxy-4-pyrimidinyl)- (sulfadimethoxine); $C_{12}H_{14}N_4O_4S$; [122-11-2]

(2) Ethanol; C_2H_6O; [64-17-5]

ORIGINAL MEASUREMENTS:

Mauger, J. W.; Paruta, A. N.; Gerraughty, R. J. *J. Pharm. Sci.* 1972, *61(1)*, 94-7.

VARIABLES:

Temperature

PREPARED BY:

R. Piekos

EXPERIMENTAL VALUES:

t/°C	Mole fraction solubility (x 10^4)
25	7.14
30	8.58
37	11.00

AUXILIARY INFORMATION

METHOD/APPARATUS/PROCEDURE:

A const temp bath contg screw-capped bottles with sulfadimethoxine in excess and ethanol was rotated for 24 h. Samples were withdrawn through a pledget of glass wool into a pipet, which was wiped clean and allowed to drain into a volumetric flask. Solute concns were detd by spectrophotometric assay at predetd wavelengths using a Cary model 16 spectrophotometer (1).

SOURCE AND PURITY OF MATERIALS:

Sulfadimethoxine: lot 203057, Hoffmann - La Roche, Inc.

Ethanol (purity not specified) was from U.S. Industrial Chemicals Co.

ESTIMATED ERROR:

Soly: av values of 3 detns are given (authors).

Temp: ±0.1°C (authors).

REFERENCES:

1. Paruta, A. N.; Mauger, J. W. *J. Pharm. Sci.* 1971, *60*, 432.

COMPONENTS:

(1) Benzenesulfonamide, 4-amino-N-(2,6-dimethoxy-4-pyrimidinyl)- (sulfadimethoxine); $C_{12}H_{14}N_4O_4S$; [122-11-2]

(2) Ethanol; C_2H_6O; [64-17-5]

ORIGINAL MEASUREMENTS:

Mauger, J. W.; Petersen, H., Jr.; Alexander, K. S.; Paruta, A. N.
Drug Dev. Ind. Pharm. 1977, *3(2)*, 163-83.

VARIABLES:

Temperature

PREPARED BY:

R. Piekos

EXPERIMENTAL VALUES:

t/°C	Solubility		
	mg/ml	10^4 X [a]	10^2 mol dm^{-3} [b]
25	3.78	7.14	1.22
30	4.52	8.58	1.46
37	5.74	11.00	1.85

[a] X = mole fraction

[b] Calculated by compiler.

AUXILIARY INFORMATION

METHOD/APPARATUS/PROCEDURE:

A const temp bath contg screw-capped bottles with sulfadimethoxine in excess and ethanol was rotated for 24 h. Samples were withdrawn through a pledget of glass wool into a pipet, which was wiped clean and allowed to drain into a volumetric flask. Soly was detd from absorbance and previously ascertained Beer's law plots detd on a Cary model 16 spectrophotometer (1).

SOURCE AND PURITY OF MATERIALS:

Sulfadimethoxine: lot 203027, Hoffmann La Roche, Inc. Its mp agreed with the literature value. Ethanol was from the U. S. Industrial Chemicals Co. Its refractive index value and density agreed with literature values.

ESTIMATED ERROR:

Soly: av of at least 3 detns is reported (authors).

Temp: ±0.1°C (authors).

REFERENCES:

1. Mauger, J. W.; Paruta, A. N.; Gerraughty, R. J. *J. Pharm. Sci.* 1972, *61(1)*, 94.

COMPONENTS:

(1) Benzenesulfonamide, 4-amino-N-(2,6-dimethoxy-4-pyrimidinyl)- (sulfadimethoxine); $C_{12}H_{14}N_4O_4S$; [122-11-2]

(2) 1-Propanol; C_3H_8O; [71-23-8]

ORIGINAL MEASUREMENTS:

Mauger, J. W.; Paruta, A. N.; Gerraughty, R. J. *J. Pharm. Sci.* 1972, *61(1)*, 94-7.

VARIABLES:

Temperature

PREPARED BY:

R. Piekos

EXPERIMENTAL VALUES:

t/°C	Mole fraction solubility (x 10^4)
25	4.71
30	5.63
37	7.79

AUXILIARY INFORMATION

METHOD/APPARATUS/PROCEDURE:

A const temp bath contg screw-capped bottles with sulfadimethoxine in excess and 1-propanol was rotated for 24 h. Samples were withdrawn through a pledget of glass wool into a pipet, which was wiped clean and allowed to drain into a volumetric flask. Solute concns were detd by spectrophotometric assay at predetd wavelengths using a Cary model 16 spectrophotometer (1).

SOURCE AND PURITY OF MATERIALS:

Sulfadimethoxine: lot 203057, Hoffmann - La Roche, Inc.

1-Propanol: Baker Analyzed Reagent, J. T. Baker Chemical Co.

ESTIMATED ERROR:

Soly: av values of 3 detns are given (authors).

Temp: ±0.1°C (authors).

REFERENCES:

1. Paruta, A. N.; Mauger, J. W. *J. Pharm. Sci.* 1971, *60*, 432.

COMPONENTS:

(1) Benzenesulfonamide, 4-amino-N-(2,6-dimethoxy-4-pyrimidinyl)- (sulfadimethoxine); $C_{12}H_{14}N_4O_4S$; [122-11-2]

(2) 1-Propanol; C_3H_8O; [71-23-8]

ORIGINAL MEASUREMENTS:

Mauger, J. W.; Petersen, H., Jr.; Alexander, K. S.; Paruta, A. N. *Drug Dev. Ind. Pharm.* 1977, *3(2)*, 163-83.

VARIABLES:

Temperature

PREPARED BY:

R. Piekos

EXPERIMENTAL VALUES:

t/°C	Solubility		
	mg/ml	10^4 X [a]	10^2 mol dm^{-3} [b]
25	1.95	4.71	0.63
30	2.32	5.63	0.75
37	3.20	7.79	1.03

[a] X = mole fraction

[b] Calculated by compiler

AUXILIARY INFORMATION

METHOD/APPARATUS/PROCEDURE:

A const temp bath contg screw-capped bottles with sulfadimethoxine in excess and 1-propanol was rotated for 24 h. Samples were withdrawn through a pledget of glass wool into a pipet, which was wiped clean and allowed to drain into a volumetric flask. Soly was detd from absorbance and previously ascertained Beer's law plots detd on a Cary model 16 spectrophotometer (1).

SOURCE AND PURITY OF MATERIALS:

Sulfadimethoxine: lot 203027 Hoffmann-La Roche, Inc. Its mp agreed with the literature value. 1-Propanol was a Baker Analyzed Reagent (J. T. Baker Chemical Co.) Its refractive index value and density agreed with literature values.

ESTIMATED ERROR:

Soly: av of at least 3 detns is reported (authors).

Temp: ±0.1°C (authors).

REFERENCES:

1. Mauger, J. W.; Paruta, A. N. Gerraughty, R. J. *J. Pharm. Sci.* 1972, *61(1)*, 94.

COMPONENTS:

(1) Benzenesulfonamide, 4-amino-N-(2,6-dimethoxy-4-pyrimidinyl)- (sulfadimethoxine); $C_{12}H_{14}N_4O_4S$; [122-11-2]

(2) 1-Butanol; $C_4H_{10}O$; [71-36-3]

ORIGINAL MEASUREMENTS:

Mauger, J. W.; Paruta, A. N.; Gerraughty, R. J. *J. Pharm. Sci.* 1972, *61(1)*, 94-7.

VARIABLES:

Temperature

PREPARED BY:

R. Piekos

EXPERIMENTAL VALUES:

t/°C	Mole fraction solubility (x 10^4)
25	3.89
30	5.26
37	6.70

AUXILIARY INFORMATION

METHOD/APPARATUS/PROCEDURE:

A const temp bath contg screw-capped bottles with sulfadimethoxine in excess and 1-butanol was rotated for 24 h. Samples were withdrawn through a pledget of glass wool into a pipet, which was wiped clean and allowed to drain into a volumetric flask. Solute concns were detd by spectrophotometric assay at predetd wavelengths using a Cary model 16 spectrophotometer (1).

SOURCE AND PURITY OF MATERIALS:

Sulfadimethoxine: lot 203057, Hoffmann-La Roche, Inc.

1-Butanol was from Mallinckrodt Chem Works.

ESTIMATED ERROR:

Soly: av values of 3 detns are given (authors).

Temp: ±0.1°C (authors).

REFERENCES:

1. Paruta, A. N.; Mauger, J. W. *J. Pharm. Sci.* 1971, *60*, 432.

COMPONENTS:

(1) Benzenesulfonamide, 4-amino-N-(2,6-dimethoxy-4-pyrimidinyl)- (sulfadimethoxine); $C_{12}H_{14}N_4O_4S$; [122-11-2]

(2) 1-Butanol; $C_4H_{10}O$; [71-36-3]

ORIGINAL MEASUREMENTS:

Mauger, J. W.; Petersen, H., Jr.; Alexander, K. S.; Paruta, A. N. *Drug Dev. Ind. Pharm.* 1977, *3(2)*, 163-83.

VARIABLES:

Temperature

PREPARED BY:

R. Piekos

EXPERIMENTAL VALUES:

t/°C	Solubility		
	mg/ml	10^4 X [a]	10^3 mol dm^{-3} [b]
25	1.31	3.89	4.22
30	1.77	5.26	5.70
37	2.25	6.70	7.25

[a] X = mole fraction

[b] Calculated by compiler

AUXILIARY INFORMATION

METHOD/APPARATUS/PROCEDURE:

A const temp bath contg screw-capped bottles with sulfadimethoxine in excess and 1-butanol was rotated for 24 h. Samples were withdrawn through a pledget of glass wool into a pipet, which was wiped clean and allowed to drain into a volumetric flask. Soly was detd from absorbance and previously ascertained Beer's law plots detd on a Cary model 16 spectrophotometer (1).

SOURCE AND PURITY OF MATERIALS:

Sulfadimethoxine: lot 203027, Hoffmann-La Roche Inc. Its mp agreed with the literature value. 1-Butanol was from the Mallinckrodt Chemical Works. Its refractive index value and density agreed with literature values.

ESTIMATED ERROR:

Soly: av of at least 3 detns is reported (authors).

Temp: ±0.1°C (authors).

REFERENCES:

1. Mauger, J. W.; Paruta, A. N.; Gerraughty, R. J. *J. Pharm. Sci.* 1972, *61(1)*, 94.

COMPONENTS:

(1) Benzenesulfonamide, 4-amino-N-(2,6-dimethoxy-4-pyrimidinyl)- (sulfadimethoxine); $C_{12}H_{14}N_4O_4S$; [122-11-2]

(2) 1-Pentanol; $C_5H_{12}O$; [71-41-0]

ORIGINAL MEASUREMENTS:

Mauger, J. W.; Paruta, A. N.; Gerraughty, R. J. *J. Pharm. Sci.* 1972, *61(1)*, 94-7.

VARIABLES:

Temperature

PREPARED BY:

R. Piekos

EXPERIMENTAL VALUES:

t/°C	Mole fraction solubility (x 10^4)
25	3.41
30	4.41
37	5.65

AUXILIARY INFORMATION

METHOD/APPARATUS/PROCEDURE:

A const temp bath contg screw-capped bottles with sulfadimethoxine in excess and 1-pentanol was rotated for 24 h. Samples were withdrawn through a pledget of glass wool into a pipet, which was wiped clean and allowed to drain into a volumetric flask. Solute concns were detd by spectrophotometric assay at predetd wavelengths using a Cary model 16 spectrophotometer.

SOURCE AND PURITY OF MATERIALS:

Sulfadimethoxine: lot 203057, Hoffmann-La Roche, Inc.

1-Pentanol was from Fisher Scientific Co.

ESTIMATED ERROR:

Soly: av values of 3 detns are given (authors).

Temp: ±0.1°C (authors).

REFERENCES:

1. Paruta, A. N.; Mauger, J. W. *J. Pharm. Sci.* 1971, *60*, 432.

COMPONENTS:

(1) Benzenesulfonamide, 4-amino-N-(2,6-dimethoxy-4-pyrimidinyl)- (sulfadimethoxine); $C_{12}H_{14}N_4O_4S$; [122-11-2]

(2) 1-Pentanol; $C_5H_{12}O$; [71-41-0]

ORIGINAL MEASUREMENTS:

Mauger, J. W.; Petersen, H., Jr.; Alexander, K. S.; Paruta, A. N. *Drug Dev. Ind. Pharm.* 1977 *3(2)*, 163-83.

VARIABLES:

Temperature

PREPARED BY:

R. Piekos

EXPERIMENTAL VALUES:

t/°C	Solubility		
	mg/ml	10^4 X [a]	10^3 mol dm^{-3} [b]
25	0.98	3.41	3.16
30	1.26	4.41	4.06
37	1.60	5.65	5.16

[a] X = mole fraction

[b] Calculated by compiler

AUXILIARY INFORMATION

METHOD/APPARATUS/PROCEDURE:

A const temp bath contg screw-capped bottles with sulfadimethoxine in excess and 1-pentanol was rotated for 24 h. Samples were withdrawn through a pledget of glass wool into a pipet, which was wiped clean and allowed to drain into a volumetric flask. Soly was detd from absorbance and previously ascertained Beer's law plots detd on a Cary model 16 spectrophotometer (1).

SOURCE AND PURITY OF MATERIALS:

Sulfadimethoxine: lot 203027, Hoffmann-La Roche, Inc. Its mp agreed with the literature value. 1-Pentanol was from Fisher Scientific Co. Its refractive index value and density agreed with literature values.

ESTIMATED ERROR:

Soly: av of at least 3 detns is reported (authors).

Temp: ±0.1°C (authors).

REFERENCES:

1. Mauger, J. W.; Paruta, A. N.; Gerraughty, R. J. *J. Pharm. Sci.* 1972, *61(1)*, 94.

COMPONENTS:

(1) Benzenesulfonamide, 4-amino-N-(2,6-dimethoxy-4-pyrimidinyl)- (sulfadimethoxine); $C_{12}H_{14}N_4O_4S$; [122-11-2]

(2) 1-Octanol; $C_8H_{18}O$; [111-87-5]

ORIGINAL MEASUREMENTS:

Mauger, J. W.; Paruta, A. N.; Gerraughty, R. J. *J. Pharm. Sci.* 1972, *61(1)*, 94-7.

VARIABLES:

Temperature

PREPARED BY:

R. Piekos

EXPERIMENTAL VALUES:

t/°C	Mole fraction solubility (x 10^4)
25	2.04
30	2.78
37	3.59

AUXILIARY INFORMATION

METHOD/APPARATUS/PROCEDURE:

A const temp bath contg screw-capped bottles with sulfadimethoxine in excess and 1-octanol was rotated for 24 h. Samples were withdrawn through a pledget of glass wool into a pipet, which was wiped clean and allowed to drain into a volumetric flask. Solute concns were detd by spectrophotometric assay at predetd wavelengths using a Cary model 16 spectrophotometer (1).

SOURCE AND PURITY OF MATERIALS:

Sulfadimethoxine: lot 203057; Hoffmann-La Roche, Inc.

1-Octanol was from Mallinckrodt Chem Works.

ESTIMATED ERROR:

Soly: av of 3 detns are given (authors).

Temp: ±0.1°C (authors).

REFERENCES:

1. Paruta, A. N.; Mauger, J. W. *J. Pharm. Sci.* 1971, *60*, 432.

COMPONENTS:

(1) Benzenesulfonamide, 4-amino-N-(2,6-dimethoxy-4-pyrimidinyl)- (sulfadimethoxine); $C_{12}H_{14}N_4O_4S$; [122-11-2]

(2) 1-Octanol; $C_8H_{18}O$; [111-87-5]

ORIGINAL MEASUREMENTS:

Mauger, J. W.; Petersen, H., Jr.; Alexander, K. S.; Paruta, A. N.
Drug Dev. Ind. Pharm. 1977, *3(2)*, 163-83.

VARIABLES:

Temperature

PREPARED BY:

R. Piekos

EXPERIMENTAL VALUES:

$t/^{\circ}C$	Solubility		
	mg/ml	10^4 X [a]	10^3 mol dm^{-3} [b]
25	0.40	2.04	1.29
30	0.54	2.78	1.74
37	0.70	3.59	2.26

[a] X = mole fraction

[b] Calculated by compiler

AUXILIARY INFORMATION

METHOD/APPARATUS/PROCEDURE:

A const temp bath contg screw-capped bottles with sulfadimethoxine in excess and 1-octanol was rotated for 24 h. Samples were withdrawn through a pledget of glass wool into a pipet, which was wiped clean and allowed to drain into a volumetric flask. Soly was detd from absorbance and previously ascertained Beer's law plots detd on a Cary model 16 spectrophotometer (1).

SOURCE AND PURITY OF MATERIALS:

Sulfadimethoxine: lot 203027; Hoffmann-La Roche, Inc. Its mp agreed with the literature value. 1-Octanol was from the Fisher Scientific Co. Its refractive index value and density agreed with literature values.

ESTIMATED ERROR:

Soly: av of at least 3 detns is reported (authors).

Temp: $\pm 0.1^{\circ}C$ (authors).

REFERENCES:

1. Mauger, J. W.; Paruta, A. N.; Gerraughty, R. J. *J. Pharm. Sci.* 1972, *61(1)*, 94.

COMPONENTS:

(1) Benzenesulfonamide, 4-amino-N-(2,6-dimethoxy-4-pyrimidinyl)- (sulfadimethoxine); $C_{12}H_{14}N_4O_4S$; [122-11-2]

(2) 1-Decanol; $C_{10}H_{22}O$; [112-30-1]

ORIGINAL MEASUREMENTS:

Mauger, J. W.; Paruta, A. N.; Gerraughty, R. J. *J. Pharm. Sci.* 1972, *61(1)*, 94-7.

VARIABLES:

Temperature

PREPARED BY:

R. Piekos

EXPERIMENTAL VALUES:

t/°C	Solubility mg/ml	10^4 X [a]	10^3 mol dm^{-3} [b]
25	0.36	2.24	1.16
30	0.44	2.69	1.42
37	0.54	3.37	1.74

[a] X = mole fraction

[b] Calculated by compiler

AUXILIARY INFORMATION

METHOD/APPARATUS/PROCEDURE:

A const temp bath contg screw-capped bottles with sulfadimethoxine in excess and 1-decanol was rotated for 24 h. Samples were withdrawn through a pledget of glass wool into a pipet, which was wiped clean and allowed to drain into a volumetric flask. Solute concns were detd by spectrophotometric assay at predetermined wavelengths using a Cary model 16 spectrophotometer (1).

SOURCE AND PURITY OF MATERIALS:

Sulfadimethoxine: lot 203027, Hoffmann-La Roche, Inc. M.p. agreed with the literature value. 1-Decanol was purchased from Matheson, Coleman and Bell. Refractive index value and density agreed with those reported in the literature.

ESTIMATED ERROR:

Temp: 0.1°C (authors).

Soly: not specified.

REFERENCES:

1. Paruta, A. N.; Mauger, J. W. *J. Pharm. Sci.* 1971, *60*, 432.

COMPONENTS:

(1) Benzenesulfonamide, 4-amino-N-(2,6 -dimethoxy-4-pyrimidinyl)- (sulfadimethoxine); $C_{12}H_{14}N_4O_4S$; [122-11-2]

(2) 1-Decanol; $C_{10}H_{22}O$; [112-30-1]

ORIGINAL MEASUREMENTS:

Mauger, J. W.; Petersen, H., Jr.; Alexander, K. S.; Paruta, A. N. *Drug Dev. Ind. Pharm.* 1977, *3(2)*, 163-83.

VARIABLES:

Temperature

PREPARED BY:

R. Piekos

EXPERIMENTAL VALUES:

t/°C	Solubility		
	mg/ml	10^4 X [a]	10^3 mol dm^{-3} [b]
25	0.36	2.24	1.16
30	0.44	2.69	1.42
37	0.54	3.37	1.74

[a] X = mole fraction

[b] Calucalted by compiler

AUXILIARY INFORMATION

METHOD/APPARATUS/PROCEDURE:

The soly was detd by the method of Paruta et al. (1): Screw-capped bottles with sulfadimethoxine in excess and 1-decanol were rotated in a const temp bath for 24 h. Samples were withdrawn through a pledget of glass wool into a pipet, which was wiped clean and allowed to drain into a volumetric flask. Soly was detd from absorbance and previously ascertained Beer's law plots detd on a Cary model 16 spectrophotometer.

SOURCE AND PURITY OF MATERIALS:

Sulfadimethoxine: lot 203027, Hoffman-La Roche, Inc. M.p agreed with that of literature. 1-Decanol was purchased from Matheson, Coleman and Bell. Refractive index value and density agreed with those reported in the literature.

ESTIMATED ERROR:

Temp: ±0.1°C (authors).

Soly: not specified.

REFERENCES:

1. Paruta, A. N.; Mauger, J. W.; Gerraughty, R. J. *J. Pharm. Sci.* 1972, *61*, 94.

COMPONENTS:	ORIGINAL MEASUREMENTS:
(1) Benzenesulfonamide, 4-amino-N-(2,6-dimethoxy-4-pyrimidinyl)- (sulfadimethoxine); $C_{12}H_{14}N_4O_4S$; [122-11-2] (2) Methane, trichloro- (chloroform); $CHCl_3$; [67-66-3]	Riess, W. *Intern. Congr. Chemotherapy, Proc., 3rd, Stuttgart* 1963, *1*, 627-32.
VARIABLES:	PREPARED BY:
One temperature: 20°C	R. Piekos

EXPERIMENTAL VALUES:

Solubility of sulfadimethoxine in chloroform at 20°C is 134 mg% (4.32×10^{-3} mol dm^{-3} solution, compiler).

AUXILIARY INFORMATION

METHOD/APPARATUS/PROCEDURE:	SOURCE AND PURITY OF MATERIALS:
Nothing specified.	Nothing specified.
	ESTIMATED ERROR:
	Nothing specified.
	REFERENCES:

COMPONENTS:	ORIGINAL MEASUREMENTS:
(1) Benzenesulfonamide, 4-amino-N-(2,6-dimethoxy-4-pyrimidinyl)- (sulfadimethoxine); $C_{12}H_{14}N_4O_4S$; [122-11-2] (2) Methane, trichloro- (chloroform); $CHCl_3$; [67-66-3]	Yamazaki, M.; Aoki, M; Kamada, A.; Yata, N. *Yakuzaigaku* 1967, *27(1)*, 37-40.
VARIABLES:	PREPARED BY:
One temperature: 30°C	R. Piekos

EXPERIMENTAL VALUES:

Solubility of sulfadimethoxine in chloroform at 30°C is 7.30 mmol/L (2.26 g dm^{-3}, compiler).

AUXILIARY INFORMATION

METHOD/APPARATUS/PROCEDURE:

Sulfadimethoxine (0.5 g) was placed in an L-shaped tube together with 20 ml of chloroform. The mixt was shaken in a thermostat until equilibrium was attained. The sulfadimethoxine was assayed in the supernatant spectrophotometrically at 545 nm on a Beckmann DU spectrophotometer. The results were taken from a calibration graph.

SOURCE AND PURITY OF MATERIALS:

Nothing specified

ESTIMATED ERROR:

Soly: not specified.

Temp: ±1°C (authors).

REFERENCES:

COMPONENTS:	ORIGINAL MEASUREMENTS:
(1) Copper, bis[4-amino-N-(2,6-dimethoxy-4-pyrimidinyl)benzenesulfonamidato]-, hydrate; $C_{24}H_{26}CuN_8O_8S_2 \cdot nH_2O$; [86729-19-3] (2) Hydrochloric acid; HCl; [7647-01-0] (3) Water; H_2O; [7732-18-5]	Tskitishvili, M. G.; Mikadze, I. I. *Soobshch. Akad. Nauk Gruz. SSR* 1978, *89(3)*, 589-92.
VARIABLES:	PREPARED BY:
pH	R. Piekos

EXPERIMENTAL VALUES:

K_{so} over the HCl concentration range 2.5×10^{-2} - 2.5×10^{-5} mol dm^{-3}, at 25°C, is 1.63×10^{-14}.

AUXILIARY INFORMATION

METHOD/APPARATUS/PROCEDURE:	SOURCE AND PURITY OF MATERIALS:
In a glass vessel, a mixt of 100 ml of HCl of appropriate concn and the solute was placed and shaken for 6 h in a water thermostat at 25°C. After attaining equilibrium, the pH of the soln was measured and the Cu^{2+} and S content was detd to calculate K_{so}.	Nothing specified.
	ESTIMATED ERROR:
	Nothing specified.
	REFERENCES:

COMPONENTS:

(1) Cobalt; bis[4-amino-N-(2,6-dimethoxy-4-pyrimidinyl)benzenesulfonamidato]- hydrate; $C_{24}H_{26}CoN_8O_8S_2 \cdot nH_2O$; [86729-20-6]

(2) Hydrochloric acid; HCl; [7647-01-0]

(3) Water; H_2O; [7732-18-5]

ORIGINAL MEASUREMENTS:

Tskitishvili, M. G.; Mikadze, I. I.

Soobshch. Akad. Nauk Gruz. SSR 1978, *89(3)*, 589-92.

VARIABLES:

pH

PREPARED BY:

R. Piekos

EXPERIMENTAL VALUES:

K_{so} over the HCl concentration range 2.5×10^{-2} - 2.5×10^{-5} mol dm^{-3}, at 25°C, is 2.28×10^{-11}.

AUXILIARY INFORMATION

METHOD/APPARATUS/PROCEDURE:

In a glass vessel, a mixt of 100 ml of HCl of appropriate concn and the solute was placed and shaken for 6 h in a water thermostat at 25°C. After attaining equilibrium, the pH of the soln was measured and the Co^{2+} and S content was detd to calculate K_{so}.

SOURCE AND PURITY OF MATERIALS:

Nothing specified.

ESTIMATED ERROR:

Nothing specified.

REFERENCES:

COMPONENTS:

(1) Copper, bis[4-amino-N-(2,6-dimethoxy-4-pyrimidinyl)benzenesulfonamidato]-, hydrate; $C_{24}H_{26}CuN_8O_8S_2 \cdot nH_2O$; [86729-19-3]

(2) Hydrochloric acid; HCl; [7647-01-0]

(3) Water; H_2O; [7732-18-5]

ORIGINAL MEASUREMENTS:

Tskitishvili, M.G.; Mikadze, I.I. *Soobshch. Akad. Nauk Gruz. SSR* 1978, *89(3)*, 589-92.

VARIABLES:

pH

PREPARED BY:

R. Piekos

EXPERIMENTAL VALUES:

K_{so} over the HCl concentration range 2.5×10^{-2} - 2.5×10^{-5} mol dm^{-3}, at 25°C, is 1.63×10^{-14}.

AUXILIARY INFORMATION

METHOD/APPARATUS/PROCEDURE:

In a glass vessel, a mixt of 100 ml of HCl of appropriate concn and the solute was placed and shaken for 6 h in a water thermostat at 25°C. After attaining equilibrium, the pH of the soln was measured and the Cu^{2+} and S content was detd to calculate K_{s0}.

SOURCE AND PURITY OF MATERIALS:

Nothing specified.

ESTIMATED ERROR:

Nothing specified.

REFERENCES:

COMPONENTS:	ORIGINAL MEASUREMENTS:
(1) Magnesium, (T-4)-bis[4-amino-N-(2,6-dimethoxy-4-pyrimidinyl)benzenesulfonamidato]-hydrate; $C_{24}H_{26}MgN_8O_8S_2 \cdot nH_2O$; [84812-81-7] (2) Hydrochloric acid; HCl; [7647-01-0] (3) Water; H_2O; [7732-18-5]	Tskitishvili, M. G. Shvelashvili, A. E.; Mikadze, I. I.; Zhorzholiani, N. B.; Chrelashvili, M. V. *Izv. Akad. Nauk Gruz. SSR, Ser. Khim.* 1981, *7(4)*, 300-4.
VARIABLES:	PREPARED BY:
pH	R. Piekos

EXPERIMENTAL VALUES:

K_{so} over the HCl concentration range 5.0×10^{-3} to 1.5×10^{-5} mol dm^{-3}, at 25^oC, is 6.33×10^{-5}.

AUXILIARY INFORMATION

METHOD/APPARATUS/PROCEDURE:

The earlier described apparatus and method was used (1): in a glass vessel, a mixt of 100 ml of HCl of appropriate concn and the solute were placed and shaken for 6 h in a water thermostat at 25^oC. After attaining equilibrium, the pH of the soln was measured and the Mg^{2+} and S content was determined to calculate K_{so}. The pH was measured on a pH-673 pH meter.

SOURCE AND PURITY OF MATERIALS:

0.1M solns of chem pure $Mg(OAc)_2$, monosodium salt of sulfadimethoxine and HCl as well as doubly distd water were used. The source of the materials was not specified.

ESTIMATED ERROR:

Nothing specified.

REFERENCES:

1. Tskitishvili, M. G,; Mikadze, I. I. *Soobshch. Akad. Nauk Gruz. SSR* 1978, *89(3)*, 589.

COMPONENTS:	ORIGINAL MEASUREMENTS:
(1) Manganese, bis[4-amino-N-(2,6-dimethoxy-4-pyrimidinyl)benzenesulfonamidato]-hydrate; $C_{24}H_{26}MnN_8O_8S_2 \cdot nH_2O$; [84812-80-6] (2) Hydrochloric acid; HCl; [7647-01-0] (3) Water; H_2O; [7732-18-5]	Tskitishvili, M. G.; Shvelashvili, A. E.; Mikadze, I. I.; Zhorzholiani, N. B.; Chrelashvili, M. V. *Izv. Akad. Nauk Gruz. SSR, Ser. Khim.* 1981 *7(4)*, 300-4.
VARIABLES:	PREPARED BY:
pH	R. Piekos

EXPERIMENTAL VALUES:

Concentration of HCl (mol/l)	pH	10^9 K_{s0} at 25^oC
5.0×10^{-3}	6.54	7.45
2.5×10^{-3}	7.02	7.40
1.0×10^{-3}	7.09	7.41
5.0×10^{-4}	7.16	7.39
2.5×10^{-4}	7.21	7.43
1.0×10^{-4}	7.26	7.39
5.0×10^{-5}	7.27	7.43
1.5×10^{-5}	7.30	7.38
	Mean	7.41

AUXILIARY INFORMATION

METHOD/APPARATUS/PROCEDURE:

The earlier described apparatus and method was used (1): in a glass vessel, a mixt of 100 ml of HCl of appropriate concn and the solute were placed and shaken for 6 h in a water thermostat at 25^oC. After attaining equilibrium, the pH of the soln was measured and the Mn^{2+} and S content was determined to calculate K_{s0} The pH was measured on a pH-673 pH meter.

SOURCE AND PURITY OF MATERIALS:

0.1M solns of chem. pure $Mn(OAc)_2$, monosodium salt of sulfadimethoxine, and HCl as well as doubly distd water were used. The source of the materials was not specified.

ESTIMATED ERROR:

K_{s0}: std deviation 2×10^{-11} (compiler).

Temp and pH: not specified.

REFERENCES:

1. Tskitishvili, M. G.; Mikadze, I. I. *Soobshch. Akad. Nauk Gruz. SSR* 1978, *89(3)*, 589.

COMPONENTS:	ORIGINAL MEASUREMENTS:
(1) Nickel, bis[4-amino-N-(2,6,-dimethoxy-4-pyrimidinyl)benzenesulfonamidato]-hydrate; $C_{24}H_{26}N_8NiO_8S_2 \cdot nH_2O$; [84812-79-3] (2) Hydrochloric acid; HCl; [7647-01-0] (3) Water; H_2O; [7732-18-5]	Tskitishvili, M. G.; Shvelashvili, A. E.; Mikadze, I. I.; Zhorzholiani, N. B.; Chrelashvili, M. V. *Izv. Akad. Nauk Gruz. SSR. Ser. Khim.* 1981, *7(4)*, 300-4.
VARIABLES:	PREPARED BY:
pH	R. Piekos

EXPERIMENTAL VALUES:

Concentration of HCl (mol/l)	pH	10^9 K_{so} at 25°C
2.5×10^{-2}	6.37	3.17
1.0×10^{-2}	6.66	3.20
5.0×10^{-3}	6.85	3.16
2.5×10^{-3}	6.96	3.17
1.0×10^{-3}	7.05	3.13
5.0×10^{-4}	7.10	3.18
2.5×10^{-4}	7.16	3.16
1.0×10^{-4}	7.63	3.16
5.0×10^{-5}	7.80	3.11
2.5×10^{-5}	7.82	3.15
	Mean	3.16

AUXILIARY INFORMATION

METHOD/APPARATUS/PROCEDURE:

The earlier described apparatus and method was used (1): in a glass vessel, a mixt of 100 ml of HCl of appropriate concn and the solute were placed and shaken for 6 h in a water thermostat at 25°C. After attaining equilibrium, the pH of the soln was measured and the Ni^{2+} and S content was determined to calculate K_{so} The pH was measured on a pH-673 pH meter.

SOURCE AND PURITY OF MATERIALS:

0.1M solns of chem pure $Ni(OAc)_2$, monosodium salt of sulfadimethoxine, HCl as well as doubly distd water were used. The source of the materials was not specified.

ESTIMATED ERROR:

K_{so}: std deviation 2.5×10^{-11} (compiler).

Temp and pH: not specified.

REFERENCES:

1. Tskitishvili, M. G.; Mikadze, I. I. *Soobshch. Akad. Nauk Gruz. SSR* 1978, *89(3)*, 589.

COMPONENTS:

(1) Benzenesulfonamide, 4-amino-N-(5,6-dimethoxy-4-pyrimidinyl)- (sulfadoxine); $C_{12}H_{14}N_4O_4S$; [2447-57-6]

(2) Phosphoric acid, disodium salt; Na_2HPO_4; [7558-94-4]

(3) Phosphoric acid, monopotassium salt; KH_2PO_4; [7778-77-0]

(4) Water; H_2O; [7732-18-5]

ORIGINAL MEASUREMENTS:

Hekster, Ch. A.; Vree, T. B.
Antibiotics Chemother. 1982, *31*, 22-118.

VARIABLES:

PH

PREPARED BY:

R. Piekos

EXPERIMENTAL VALUES:

pH	Solubility at 25°C	
	mg/l	10^4 mol dm^{-3} [a]
5.5	186	5.99
7.5[b]	2,387	76.92

[a] Calculated by compiler.

AUXILIARY INFORMATION

METHOD/APPARATUS/PROCEDURE:

The earlier developed method (1) was used (personal communication). Satd solns of sulfadoxine were prepd in phosphate buffers of pH 5.5 and 7.5 at 25°C. The concn of the solute was measured by means of a Spectra Physics 3500B high-performance liquid chromatograph equipped with a Model 748 column oven and a Pye-Unicam LC-UV spectrophotometric detector.

SOURCE AND PURITY OF MATERIALS:

Neither source nor the purity of the materials was specified.

ESTIMATED ERROR:

Soly: the detection limit of the solute by HPLC was 0.5 mg/l (authors).
The errors in temp and pH were not specified.

REFERENCES:

1. Hekster, Y. A.; Vree, T. B.; Damsma, J. E.; Friesen, W. T. *J. Antimicrob. Chemother.* 1981, *8*, 133.

COMPONENTS:

(1) Acetamide, N-[4-[[(5,6-dimethoxy-4-pyrimidinyl)amino]sulfonyl]phenyl]- (N^4-acetylsulfadoxine); $C_{14}H_{16}N_4O_5S$; [5018-54-2]

(2) Phosphoric acid, disodium salt; Na_2HPO_4; [7558-94-4]

(3) Phosphoric acid, monopotassium salt; KH_2PO_4; [7778-77-0]

(4) Water; H_2O; [7732-18-5]

VARIABLES:

pH

ORIGINAL MEASUREMENTS:

Hekster, Ch. A.; Vree, T. B.

Antibiotics Chemother. 1982, *31*, 22-118.

PREPARED BY:

R. Piekos

EXPERIMENTAL VALUES:

pH	Solubility at 25°C	
	mg/l	10^4 mol dm^{-3} [a]
5.5	221	6.27
7.5[b]	3,420	97.06

[a] Calculated by compiler

[b] Erroneous pH value of 7.0 is given in the article.

AUXILIARY INFORMATION

METHOD/APPARATUS/PROCEDURE:

The earlier developed method (1) was used (personal communication). Satd solns of N^4-acetylsulfadoxine were prepd in phosphate buffers of pH 5.5 and 7.5 at 25°C. The concn of the solute was measured by means of a Spectra Physics 3500B high-performance liquid chromatograph equipped with a Model 748 column oven and a Pye-Unicam LC-UV spectrophotometric detector.

SOURCE AND PURITY OF MATERIALS:

Neither source nor the purity of the materials was specified.

ESTIMATED ERROR:

Soly: the detection limit of the solute by HPLC was 0.5 mg/l (authors).

The errors in temp and pH were not specified.

REFERENCES:

1. Hekster, Y. A.; Vree, T. B.; Damsma, J. E.; Friesen, W. T. *J. Antimicrob. Chemother.* 1981, *8*, 133.

COMPONENTS:

(1) Benzoic acid, 5-[[4-[[(2,4-dimethoxy-6-pyrimidinyl)amino]sulfonyl] phenyl]azo]-2-hydroxy- (salazodimethoxine); $C_{19}H_{17}N_5O_7S$; [40016-88-4]

(2) Water; H_2O; [7732-18-5]

ORIGINAL MEASUREMENTS:

Ezerskii, M. L; Per'kova, N. N.
Khim.-Farm. Zh. 1979, *13(11)*, 87-91.

VARIABLES:

Grinding regime

PREPARED BY:

R. Piekos

EXPERIMENTAL VALUES:

Specimen of salazodimethoxine	Solubility at room temperature	
	g/cm^3	10^4 mol dm^{-3} [a]
Commercial	0.000075	1.6
Commercial, ground in a ball mill	0.000080	1.7
Commercial, ground in a jet mill	0.000100	2.2

[a] Calculated by compiler

AUXILIARY INFORMATION

METHOD/APPARATUS/PROCEDURE:

Satd solns were prepd by prolonged agitation of an excess of salazodimethoxine in water at room temp. The solns were then allowed to stand for 12 and h and filtered. The concn of the solute in the filtrate was detd spectrophotometrically at 462 nm in a 1-cm cuvet.

SOURCE AND PURITY OF MATERIALS:

Comm, pharmacopeial salazodimethoxine was used (source not specified). It was ground in a Pulverisette-5 lab mill or in a C-1266-00 jet mill. Purity of the water was not specified.

ESTIMATED ERROR:

Nothing specified.

REFERENCES:

COMPONENTS:

(1) Pyrimidine, 2,5-bis[[(4-aminophenyl)sulfonyl]amino]-; $C_{16}H_{16}N_6O_4S_2$; [71119-39-6]

(2) Water; H_2O; [7732-18-5]

ORIGINAL MEASUREMENTS:

Roblin, R. O., Jr.; Winnek, P. S.; English, J. P. *J. Am. Chem. Soc.* 1942, *64*, 567-70.

VARIABLES:

One temperature: 37°C

PREPARED BY:

R. Piekos

EXPERIMENTAL VALUES:

Solubility of 2,5-bis[[(4-aminophenyl)sulfonyl]amino]pyrimidine in water at 37°C is 2.2 mg/100 cm^3 solution (5.2×10^{-5} mol dm^{-3}, compiler).

AUXILIARY INFORMATION

METHOD/APPARATUS/PROCEDURE:

Excess sulfonamide in water was heated and stirred on a steam bath for 30 min. The suspension was then agitated for 24 h in a thermostat at 37°C. A sample of the satd soln was withdrawn through a glass filter, dild, and analyzed by the Marshall method (1) using a General Electric recording spectrophotometer for comparing the colors developed with those of the standards.

SOURCE AND PURITY OF MATERIALS:

The sulfonamide, mp 231-2°C (cor), was prepd by the authors. Anal: %C 45.4 (calcd 45.6); %H 4.0 (3.8); %N 20.1 (20.1)
Purity of the water was not specified.

ESTIMATED ERROR:

Nothing specified.

REFERENCES:

1. Bratton, A. C.; Marshall, E. K., Jr. *J. Pharmacol.* 1939, *66*, 4.

COMPONENTS:	ORIGINAL MEASUREMENTS:
(1) Benzenesulfonamide, 4-amino-N-5-[2,4(1H,3H)-pyrimidinedionyl]-; $C_{10}H_{10}N_4O_4S$; [6912-98-7] (2) Water; H_2O; [7732-18-5]	Roblin, R. O., Jr.; Williams, J. H.; Winnek, P. S.; English, J. P. *J. Am. Chem. Soc.* <u>1940</u>, *62*, 2002-5.
VARIABLES:	PREPARED BY:
One temperature: 37°C	R. Piekos

EXPERIMENTAL VALUES:

Solubility of 4-amino-N-5-[2,4(1H,3H)-pyrimidinedionyl]benzenesulfonamide in water at 37°C is 48.6 mg/100 cm^3 solution (1.72 x 10^{-3} mol dm^{-3}, compiler).

AUXILIARY INFORMATION

METHOD/APPARATUS/PROCEDURE:

Excess sulfonamide in water was heated and stirred on a steam bath for 30 min. The suspension was then agitated for 24 h in a thermostat at 37°C. A sample of the satd soln was withdrawn through a glass filter, dild, and analyzed by the Marshall method (1) using a General Electric recording spectrophotometer for comparing the colors developed with those of the standards.

SOURCE AND PURITY OF MATERIALS:

The sulfonamide, mp 277-9°C (dec, cor), was prepd by the authors. Anal: %C 42.8 calcd (42.6); %H 3.4 (3.5); %N 20.2 (19.9). Purity of the water was not specified.

ESTIMATED ERROR:

Nothing specified.

REFERENCES:

1. Bratton, A. C.; Marshall, E. K., Jr. *J. Pharmacol.* <u>1939</u>, *66*, 4.

COMPONENTS:

(1) Benzenesulfonamide, 4-amino-N-pyrazinyl- (sulfapyrazine); $C_{10}H_{10}N_4O_2S$; [116-44-9]

(2) 2-Propanol; C_3H_8O; [67-63-0]

ORIGINAL MEASUREMENTS:

Burlage, H. M.

J. Am. Pharm. Assoc., Sci. Ed. 1948, *37*, 345.

VARIABLES:

One temperature: 25°C

PREPARED BY:

R. Piekos

EXPERIMENTAL VALUES:

Solubility of sulfapyrazine in 2-propanol at 25°C is 0.0290 g/100 cm^3 solution (1.16 x 10^{-3} mol dm^{-3}, compiler).

AUXILIARY INFORMATION

METHOD/APPARATUS/PROCEDURE:

Satd solns of sulfapyrazine in 2-propanol were prepd at 25°C and definite vols of the solns were measured into tared dishes by means of standard pipets. The alcohol was allowed to evap at room temp and the residue was dried at 105°C. In the case of losses due to apparent decompn, the residue was dried in a dessicator (1).

SOURCE AND PURITY OF MATERIALS:

The sulfapyrazine N.N.R. was manufd by Mead Johnson. The source and purity of 2-propanol were not reported.

ESTIMATED ERROR:

Nothing specified.

REFERENCES:

1. Burlage, H. M. *J. Am. Pharm. Assoc., Sci. Ed.* 1947, *36(1)*, 16.

COMPONENTS:	ORIGINAL MEASUREMENTS:
(1) Benzenesulfonamide, 4-amino-N-pyrazinyl-, monosodium salt (sodium sulfapyrazine); $C_{10}H_{10}N_4O_2S{\cdot}Na$; [547-31-9] (2) 2-Propanol; C_3H_8O; [67-63-0]	Burlage, H. M. *J. Am. Pharm. Assoc., Sci. Ed.* 1948, *37*, 345.
VARIABLES: One temperature: 25°C	PREPARED BY: R. Piekos

EXPERIMENTAL VALUES:

Solubility of sodium sulfapyrazine in 2-propanol at 25°C is 0.7080 g/100 cm^3 solution (2.600×10^{-2} mol dm^{-3}, compiler).

AUXILIARY INFORMATION

METHOD/APPARATUS/PROCEDURE:

Satd solns of sodium sulfapyrazine in 2-propanol were prepd at 25°C and definite vols of the solns were measured into tared dishes by means of standard pipets. The alcohol was allowed to evap at room temp and the residue was dried at 105°C. In the case of losses due to apparent decompn, the residue was dried in a desiccator (1).

SOURCE AND PURITY OF MATERIALS:

The sodium pyrazine N.N.R. was manufd by Mead Johnson. The source and purity of 2-propanol were not specified.

ESTIMATED ERROR:

Nothing specified.

REFERENCES:

1. Burlage, H. M.
J. Am. Pharm. Assoc., Sci. Ed. 1947, *36(1)*, 16.

COMPONENTS:	ORIGINAL MEASUREMENTS:
(1) Benzenesulfonamide, 4-amino-N-(3-methoxypyrazinyl)- (sulfamethoxypyrazine); $C_{11}H_{12}N_4O_3S$; [152-47-6] (2) Phosphoric acid, disodium salt; Na_2HPO_4; [7558-94-4] (3) 1,2,3-Propanetricarboxylic acid, 2-hydroxy- (citric acid); $C_6H_8O_7$; [77-92-9] (4) Water; H_2O; [7732-18-5]	Bertazzoli, C.; Buogo, A.; Ciceri, C.; Ghione, M.; Turolla, E.; Zavaglio, V. *Minerva Med.* 1961, *52(40)*, 1789-96.
VARIABLES: pH	PREPARED BY: R. Piekos

EXPERIMENTAL VALUES:

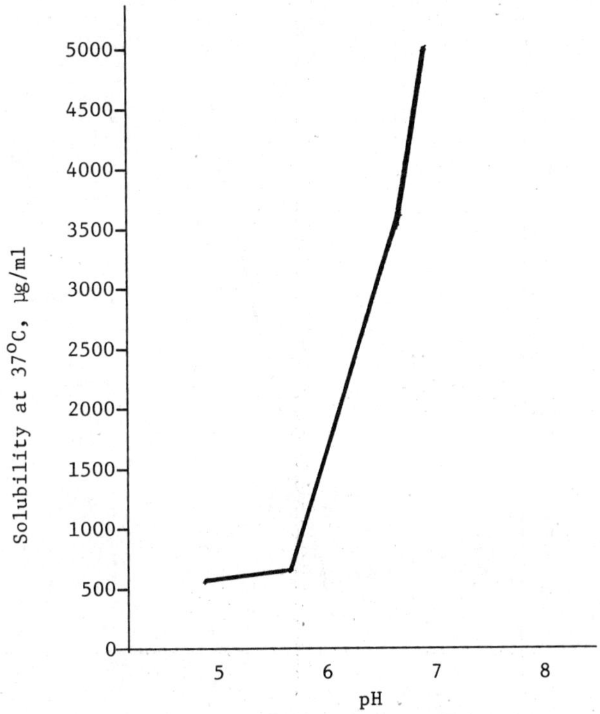

AUXILIARY INFORMATION

METHOD/APPARATUS/PROCEDURE:	SOURCE AND PURITY OF MATERIALS:
The soly of sulfamethoxypyrazine in McIlvaine's Na_2HPO_4 - citric acid buffer solns was detd under agitation at 37°C. No details were given.	Nothing specified.
	ESTIMATED ERROR: Nothing specified.
	REFERENCES:

COMPONENTS:	ORIGINAL MEASUREMENTS:
(1) Acetamide, N-[4-[[(3-methoxypyrazinyl)-amino]sulfonyl]phenyl]- (acetyl sulfamethoxypyrazine); $C_{13}H_{14}N_4O_4S$; [655-78-7] (2) Phosphoric acid, disodium salt; Na_2HPO_4; [7558-94-4] (3) 1,2,3-Propanetricarboxylic acid, 2-hydroxy- (citric acid); $C_6H_8O_7$; [77-92-9] (4) Water; H_2O; [7732-18-5]	Bertazzoli, C.; Buogo, A.; Ciceri, C.; Ghione, M.; Turolla, E.; Zavaglio, V. *Minerva Med.* 1961, *52(40)*, 1789-96.
VARIABLES: pH	PREPARED BY: R. Piekos

EXPERIMENTAL VALUES:

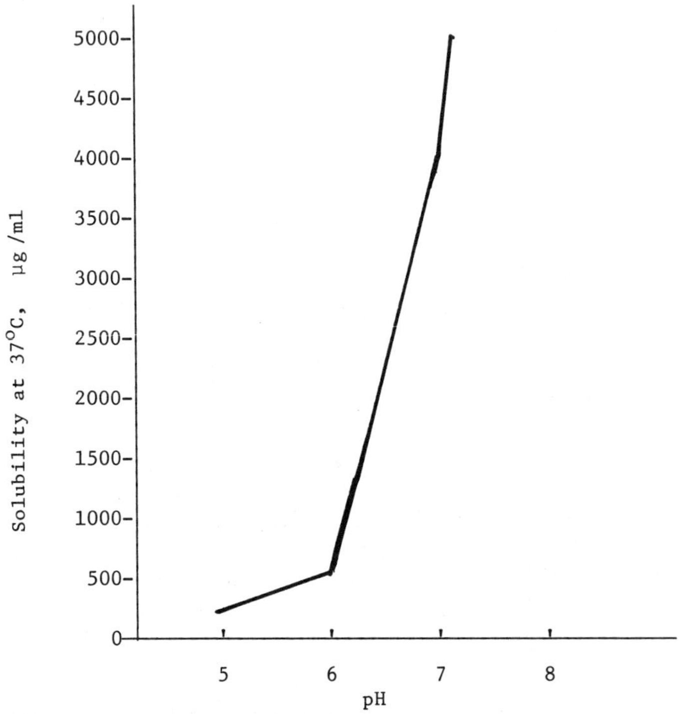

AUXILIARY INFORMATION

METHOD/APPARATUS/PROCEDURE:	SOURCE AND PURITY OF MATERIALS:
The solubility of acetyl sulfamethoxypyrazine in McIlvaine's Na_2HPO_4 - citric acid buffer solutions was determined under agitation at 37°C. No details were given.	Nothing specified.
	ESTIMATED ERROR: Nothing specified.
	REFERENCES:

COMPONENTS:

(1) Benzenesulfonamide, 4-amino-N-(4,6-diamino-1,3,5-triazin-2-yl)-; $C_9H_{11}N_7O_2S$; [51249-11-7]

(2) Water; H_2O; [7732-18-5]

ORIGINAL MEASUREMENTS:

Anderson, G. W.; Faith, H. E.; Marson, H. W.; Winnek, P. S.; Roblin, R. O., Jr. *J. Am. Chem. Soc.* 1942, *64*, 2902-5.

VARIABLES:

One temperature: 37°C

PREPARED BY:

R. Piekos

EXPERIMENTAL VALUES:

Solubility of 4-amino-N-(4,6-diamino-1,3,5-triazin-2-yl)benzenesulfonamide in water at 37°C is 728 mg/100 cm^3 solution (2.59×10^{-2} mol dm^{-3}, compiler).

AUXILIARY INFORMATION

METHOD/APPARATUS/PROCEDURE:

Excess sulfonamide in water was heated and stirred on a steam bath for 30 min. The suspension was then agitated for 24 h in a thermostat. A sample of the satd soln was withdrawn through a glass filter, dild, and analyzed by the Marshall method (1) using a General Electric recording spectrophotometer for comparing the colors developed with those of the standards.

SOURCE AND PURITY OF MATERIALS:

The sulfonamide, mp 290-5°C (cor), was prepd by the authors. Anal: %C 38.9 (calcd 38.4); %H 4.3 (3.9); %N 34.7 (34.9).

Purity of the water was not specified.

ESTIMATED ERROR:

Nothing specified.

REFERENCES:

1. Bratton, A. C.; Marshall, E. K., Jr. *J. Pharmacol.* 1939, *66*, 4.

COMPONENTS:	ORIGINAL MEASUREMENTS:
(1) Benzenesulfonamide, 4-amino-N-2-quinoxalinyl- (sulfaquinoxaline); $C_{14}H_{12}N_3O_2S$; [59-40-5] (2) Hydrochloric acid; HCl; [7647-01-0] (3) Water; H_2O; [7732-18-5]	Paál, T.; Regősz, P. *Gyógyszerészet* 1973, *17*, 59-63.

VARIABLES:	PREPARED BY:
Concentration of HCl	R. Piekos

EXPERIMENTAL VALUES:

Concentration of HCl N	Concentration of the most concentrated real solution of sulfaquinoxaline at 26°C mol dm^{-3} solvent
5	3×10^{-3} (9×10^{-2})[a]
1	$< 2 \times 10^{-3}$
0.1	$< 2 \times 10^{-3}$

[a]Concentration of the most concentrated metastable solution that could be prepared without precipitation of the solute

AUXILIARY INFORMATION

METHOD/APPARATUS/PROCEDURE:

Satd solns were prepd by addn of increasing quantities of aq HCl to weighed quantities of sulfaquinoxaline. After the dissoln had been completed, the soln was stirred with a magnetic stirrer and allowed to stand for 24 h. The soln was considered stable, if it remained clear during a 24-h period. If the solute pptd out from the clear soln, the soln was considered metastable.

SOURCE AND PURITY OF MATERIALS:

Sulfaquinoxaline was a product of Chinoin Pharm and Chem Works. Its purity was 99.6% as detd by diazotization. The source and purity of the remaining materials were not specified.

ESTIMATED ERROR:

Soly: accuracy ±10% (authors).

Temp: ±3°C (authors).

REFERENCES:

COMPONENTS:

(1) Benzenesulfonamide, 4-amino-N-2-quinoxalinyl- (sulfaquinoxaline); $C_{14}H_{12}N_3O_2S$; [59-40-5]

(2) Perchloric acid; $HClO_4$; [7601-90-3]

(4) Water; H_2O; [7732-18-5]

ORIGINAL MEASUREMENTS:

Paál, T.; Regősz, P.

Gyógyszerészet 1973, *17*, 59-63.

VARIABLES:

Concentration of $HClO_4$

PREPARED BY:

R. Piekos

EXPERIMENTAL VALUES:

Concentration of $HClO_4$ N	Concentration of the most concentrated real solution of sulfaquinoxaline at 26°C mol dm^{-3} solvent
5	2×10^{-2} (o.67)[a]
1	$< 2 \times 10^{-3}$
0.1	$< 2 \times 10^{-3}$

[a]Concentration of the most concentrated metastable solution that could be prepared without precipitation of the solute

AUXILIARY INFORMATION

METHOD/APPARATUS/PROCEDURE:

Satd solns were prepd by addn of increasing amts of aq $HClO_4$ to weighed quantities of sulfaquinoxaline. After the dissoln had been completed, the soln was stirred with a magnetic stirrer and allowed to stand for 24 h. The soln was considered stable, if it remained clear during a 24-h period. If the solute pptd out from the clear soln, the soln was considered metastable.

SOURCE AND PURITY OF MATERIALS:

Sulfaquinoxaline was a product of Chinoin Pharm and Chem Works. Its purity was 99.6% as detd by diazotization. The source and purity of the remaining materials were not specified.

ESTIMATED ERROR:

Soly: accuracy ±10% (authors).

Temp: ±3°C (authors).

REFERENCES:

COMPONENTS:	ORIGINAL MEASUREMENTS:
(1) Benzenesulfonamide, 4-amino-N-2-quinoxalinyl- (sulfaquinoxaline); $C_{14}H_{12}N_3O_2S$; [59-40-5] (2) Nitric acid; HNO_3; [7697-37-2] (3) Water; H_2O; [7732-18-5]	Paál, T.; Regősz, P. *Gyógyszerészet* 1973, *17*, 59-63.
VARIABLES:	PREPARED BY:
Concentration of HNO_3	R. Piekos

EXPERIMENTAL VALUES:

Concentration of HNO_3 N	Concetration of the most concentrated real solution of sulfaquinoxaline at 26°C mol dm^{-3} solvent
5	2×10^{-2} (0.2)[a]
1	$< 2 \times 10^{-3}$
0.1	$< 2 \times 10^{-3}$

[a]Concentration of the most concentrated metastable solution that could be prepared without precipitation of the solute

AUXILIARY INFORMATION

METHOD/APPARATUS/PROCEDURE:

Satd solns were prepd by addn of increasing amts of aq HNO_3 to weighed quantities of sulfaquinoxaline. After the dissoln had been completed, the soln was stirred with a magnetic stirrer and allowed to stand for 24 h. The soln was considered stable, if it remained clear during a 24-h period. If the solute pptd out from the clear soln, the soln was considered metastable.

SOURCE AND PURITY OF MATERIALS:

Sulfaquinoxaline was a product of Chinoin Pharm and Chem Works. Its purity was 99.6% as detd by diazotization.
The source and purity of the remaining materials were not specified.

ESTIMATED ERROR:

Soly: accuracy ±10% (authors).
Temp: ±3°C (authors).

REFERENCES:

COMPONENTS:	ORIGINAL MEASUREMENTS:
(1) Benzenesulfonamide, 4-amino-N-[5,6,7,8-tetrahydro-5-methyl-8-(2-propyl)-2-quinozalinyl]-; $C_{18}H_{24}N_4O_2S$; [71119-36-3] (2) Water; H_2O; [7732-18-5]	Caldwell, W. T.; Kornfeld, E. C.; Donnell, E. K. *J. Am. Chem. Soc.* 1941, *63*, 2188-90.
VARIABLES:	PREPARED BY:
One temperature: 29°C	R. Piekos

EXPERIMENTAL VALUES:

Solubility of 4-amino-N-[5,6,7,8-tetrahydro-5-methyl-8-(2-propyl)-2-quinozalinyl]benzenesulfonamide in water at 29°C is 2.4 mg/100 ml solution (6.6×10^{-5} mol dm^{-3}, compiler).

AUXILIARY INFORMATION

METHOD/APPARATUS/PROCEDURE:	SOURCE AND PURITY OF MATERIALS:
Soly was detd by weighing the residue obtained by evapg to dryness a known volume of soln satd at 29°C.	The sulfonamide, mp 185-7°C (cor, recrystd from aq EtOH), was prepd by condensing 2-amino-5,6,7,8-tetrahydro-5-methyl-8-(2-propyl)-quinazoline with acetylsulfanilyl chloride followed by hydrolysis with aq NaOH and pptn at pH 6. Anal: %N 15.58 (calcd 15.55). Purity of the water was not specified. ESTIMATED ERROR: Nothing specified. REFERENCES:

COMPONENTS:	ORIGINAL MEASUREMENTS:
(1) Benzenesulfonamide, 4-amino-N-(5,6,7,8-tetrahydro-2-quinazolinyl)-; $C_{14}H_{16}N_4O_2S$; [71119-34-1] (2) Water; H_2O; [7732-18-5]	Caldwell, W. T.; Kornfeld, E. C.; Donnell, C. K. *J. Am. Chem. Soc.* 1941, *63*, 2188-90.

VARIABLES:	PREPARED BY:
One temperature: 29°C	R. Piekos

EXPERIMENTAL VALUES:

Solubility of 4-amino-N-(5,6,7,8-tetrahydro-2-quinazolinyl)benzenesulfonamide in water at 29°C is 6.8 mg/100 ml solution (2.2×10^{-4} mol dm^{-3} - compiler).

AUXILIARY INFORMATION

METHOD/APPARATUS/PROCEDURE:

Soly was detd by weighing the residue obtained by evapg to dryness a known volume of soln satd at 29°C.

SOURCE AND PURITY OF MATERIALS:

The sulfonamide, mp 255-6°C (cor, recrystd from aq dioxane) was pred by condensing 2-amino-5,6,7,8-tetrahydroquinazoline with acetylsulfanilyl chloride followed by hydrolysis with aq NaOH and pptn at pH 6. Anal: %N 18.40 (calcd 18.41). Purity of the water was not specified.

ESTIMATED ERROR:

Nothing specified.

REFERENCES:

COMPONENTS:	ORIGINAL MEASUREMENTS:
(1) Benzenesulfonamide, 4-amino-N-(5,6,7,8-tetrahydro-8,9,9-trimethyl-5,8-methano-quinazolin-2-yl)-; $C_{18}H_{22}N_4O_2S$; [71720-66-6] (2) Water; H_2O; [7732-18-5]	Caldwell, W. T.; Kornfeld, E. C.; Donnell, C. K. *J. Am. Chem. Soc.* 1941, *63*, 2188-90.
VARIABLES:	PREPARED BY:
One temperature: 29°C	R. Piekos

EXPERIMENTAL VALUES:

Solubility of 4-amino-N-(5,6,7,8-tetrahydro-8,9,9-trimethyl-5,8-methanoquinazolin-2-yl)benzenesulfonamide in water at 29°C is 3.0 mg/100 ml solution (8.4×10^{-5} mol dm^{-3} - compiler).

AUXILIARY INFORMATION

METHOD/APPARATUS/PROCEDURE:	SOURCE AND PURITY OF MATERIALS:
Soly was detd by weighing the residue obtained by evapg to dryness a known volume of soln satd at 29°C.	The sulfonamide, mp 276-7°C (cor, recrystd from aq dioxane), was prepd by condensing 2-amino-5,6,7,8-tetrahydro-8,9,9-trimethyl-5,8-methanoquinazoline with acetylsulfanilyl chloride followed by hydrolysis with aq NaOH and pptn at pH 6. Anal: %N 15.39 (calcd 15.63). Purity of the water was not specified.
	ESTIMATED ERROR:
	Nothing specified.
	REFERENCES:

COMPONENTS:

(1) Benzenesulfonamide, 4-amino- (sulfanilamide); $C_6H_8N_2O_2S$; [63-74-1]

(2) Acetamide, N-[(4-aminosulfonyl)phenyl]- (acetyl sulfanilamide); [121-61-9]

(3) Water; H_2O; [7732-18-5]

ORIGINAL MEASUREMENTS:

Langecker, H.

Arch. Exptl. Path. Pharmakol. 1948, *205*, 291-301.

VARIABLES:

One temperature: 37°C

PREPARED BY:

R. Piekos

EXPERIMENTAL VALUES:

Solubility of sulfanilamide and acetyl sulfanilamide in a saturated solution of both compounds in water at 37°C is 1620 mg% (9.4 x 10^{-2} mol dm^{-3}, compiler) and 375 mg% (1.75 x 10^{-2} mol dm^{-3}, compiler), respectively.

AUXILIARY INFORMATION

METHOD/APPARATUS/PROCEDURE:

A mixt of sulfanilamide and acetyl sulfanilamide was boiled with water and the components were detd colorimetrically by the method of Bratton and Marshall (1) using a Havemann colorimeter (2)

SOURCE AND PURITY OF MATERIALS:

Nothing specified.

ESTIMATED ERROR:

Nothing specified.

REFERENCES:

1. Bratton, A. C.; Marshall, E. K., Jr. *J. Biol. Chem.* 1939, *128*, 537.
2. Havemann, R. *Klin. Wochenschr.* 1940, p. 503.

COMPONENTS:	ORIGINAL MEASUREMENTS:
(1) Benzenesulfonamide, 4-amino-N-2-thiazolyl- (sulfathiazole); $C_9H_9N_3O_2S_2$; [72-14-0] (2) Acetamide, N-[4-[(2-thiazolylamino)sulfonyl]phenyl]- (acetyl sulfathiazole); $C_{11}H_{11}N_3O_3S_2$; [127-76-4] (3) Water; H_2O; [7732-18-5]	Langecker, H. *Arch. Exptl. Path. Pharmakol.* 1948, *205*, 291-301.
VARIABLES:	PREPARED BY:
One temperature: 37°C	R. Piekos

EXPERIMENTAL VALUES:

Solubility of sulfathiazole and acetyl sulfathiazole in a saturated solution of both compounds in water at 37°C is 110 mg% (4.3×10^{-3} mol dm^{-3}, compiler) and 14 mg% (5.5×10^{-4} mol dm^{-3}, compiler), respectively.

AUXILIARY INFORMATION

METHOD/APPARATUS/PROCEDURE:

A mixt of sulfathiazole and acetyl sulfathiazole was boiled with water and the components were detd colorimetrically by the method of Bratton and Marshall (1) using a Havemann colorimeter (2).

SOURCE AND PURITY OF MATERIALS:

Nothing specified.

ESTIMATED ERROR:

Nothing specified.

REFERENCES:

1. Bratton, A. C.; Marshall, E. K., Jr. *J. Biol. Chem.* 1939, *128*, 537.
2. Havemann, R. *Klin. Wochenschr.* 1940, p. 503.

COMPONENTS:

(1) Benzenesulfonamide, 4-amino-N-(5-ethyl-1,3,4-thiadiazol-2-yl)- (sulfaethylthiadiazole); $C_{10}H_{12}N_4O_2S_2$; [94-19-9]

(2) Acetamide, N-[4-[[(5-ethyl-1,3,4-thiadiazol-2-yl)amino]sulfonyl]phenyl]- (acetyl sulfaethylthiadiazole); $C_{12}H_{14}N_4O_3S_2$; [1037-51-0]

(3) Water; H_2O; [7732-18-5]

ORIGINAL MEASUREMENTS:

Langecker, H.
Arch. Exptl. Path. Pharmakol. 1948, *205*, 291-301.

VARIABLES:

One temperature: 37°C

PREPARED BY:

R. Piekos

EXPERIMENTAL VALUES:

Solubility of sulfaethylthiadiazole and acetyl sulfaethylthiadiazole in a saturated solution of both compounds in water at 37°C is 117 mg% (4.1 x 10^{-3} mol dm^{-3}, compiler) and 40 mg% (1.2 x 10^{-3} mol dm^{-3}, compiler), respectively.

AUXILIARY INFORMATION

METHOD/APPARATUS/PROCEDURE:

A mixt of sulfaethylthiadiazole and acetyl sulfaethylthiadiazole was boiled with water and the components were detd colorimetrically by the method of Bratton and Marshall (1) using a Havemann colorimeter (2).

SOURCE AND PURITY OF MATERIALS:

Nothing specified.

ESTIMATED ERROR:

Nothing specified.

REFERENCES:

1. Bratton, A. C.; Marshall, E. K., Jr. *J. Biol. Chem.* 1939, *128*, 537.
2. Havemann, R. *Klin. Wochenschr.* 1940, p. 503.

COMPONENTS:

(1) Benzenesulfonamide, 4-amino-N-2-pyrimidinyl- (sulfadiazine, SD); $C_{10}H_{10}N_4O_2S$; [68-35-9]

(2) Benzenesulfonamide, 4-amino-N-2-thiazolyl- (sulfathiazole, ST); $C_9H_9N_3O_2S_2$; [72-14-0]

(3) Sodium hydroxide; NaOH; [1310-73-2]

(4) Water; H_2O; [7732-18-5]

ORIGINAL MEASUREMENTS:

Garcia Onandia, A.; Holz, E.; Holz, S. *Acta Cient. Venezolana* 1955, *6(4)*, 157-63.

VARIABLES:

SD/ST ratio

PREPARED BY:

R. Piekos

EXPERIMENTAL VALUES:

Composition of sulfonamide mixt SD/ST	Volume of 1.5N NaOH solution required to dissolve the mixture at 25°C
g/g	cm^3
0.1/0.1	0.55
0.1/0.2	0.85
0.1/0.3	1.125
0.1/0.4	1.375
0.1/0.5	1.675
0.2/0.1	0.85
0.3/0.1	1.125
0.4/0.1	1.35
0.5/0.1	1.675

AUXILIARY INFORMATION

METHOD/APPARATUS/PROCEDURE:

Nothing specified.

SOURCE AND PURITY OF MATERIALS:

Nothing specified.

ESTIMATED ERROR:

Nothing specified.

REFERENCES:

COMPONENTS:

(1) Benzenesulfonamide, 4-amino-N-2-pyrimidinyl- (sulfadiazine, SD); $C_{10}H_{10}N_4O_2S$; [68-35-9]

(2) Benzenesulfonamide, 4-amino-N-2-thiazolyl- (sulfathiazole, ST); $C_9H_9N_3O_2S_2$; [72-14-0]

(3) Sodium hydroxide; NaOH; [1310-73-2]

(4) Water; H_2O; [7732-18-5]

ORIGINAL MEASUREMENTS:

Holz, E.; Garcia Onandia, A.; Holz, S. *Acta Cient. Venezolana* 1955, *6(2)*, 68-73.

VARIABLES:

SD/ST ratio

PREPARED BY:

R. Piekos

EXPERIMENTAL VALUES:

Composition of sulfonamide mixt SD/ST	Volume of 1N NaOH soln required to dissolve the mixture at 26°C
g/g	cm^3
0.1/0.1	0.825
0.1/0.2	1.225
0.1/0.3	1.625
0.1/0.4	2.025
0.1/0.5	2.425
0.2/0.1	1.25
0.3/0.1	1.675
0.4/0.1	2.075
0.5/0.1	2.5

AUXILIARY INFORMATION

METHOD/APPARATUS/PROCEDURE:

Nothing specified.

SOURCE AND PURITY OF MATERIALS:

Not specified. Distd water was used.

ESTIMATED ERROR:

Not specified.

REFERENCES:

COMPONENTS:	ORIGINAL MEASUREMENTS:
(1) Benzenesulfonamide, 4-amino-N-2-thiazolyl- (sulfathiazole); $C_9H_9N_3O_2S_2$; [72-14-0] (2) Benzenesulfonamide, 4-amino-N-2-pyrimidinyl- (sulfadiazine); $C_{10}H_{10}N_4O_2S$; [68-35-9] (3) Phosphoric acid, disodium salt; Na_2HPO_4; [7558-94-4] (4) Phosphoric acid, monopotassium salt; KH_2PO_4; [7778-77-0] (5) Water; H_2O; [7732-18-5]	Frisk, A. R.; Hagerman, G.; Helander, S.; Sjögren, B. *Hygiea* <u>1946</u>, *108(12)*, 639-51. PREPARED BY: R. Piekos

VARIABLES:
One temperature: 37°C; one pH: 6.1

EXPERIMENTAL VALUES:

Solubility of a mixture of sulfathiazole and sulfadiazine in M/30 phosphate buffer of pH 6.1 at 37°C is 113 mg/100 ml solvent.

AUXILIARY INFORMATION

METHOD/APPARATUS/PROCEDURE:

An excess of sulfathiazole and sulfadiazine in the phosphate buffer was shaken at 37°C for 24 h. The concn of the sulfonamides was detd by the Bratton and Marshall method (1) using a photoelec colorimeter.

SOURCE AND PURITY OF MATERIALS:

Neither source nor purity of the materials was specified.

ESTIMATED ERROR:

Soly: precision ±11 mg/100 ml (authors).
Temp and pH: not specified.

REFERENCES:

1. Bratton, A. C.; Marshall, E. K., Jr. *J. Biol. Chem.* <u>1939</u>, *128*, 537.

COMPONENTS:

(1) Benzenesulfonamide, 4-amino-N-(4-methyl-2-pyrimidinyl)- (sulfamerazine, SM); $C_{11}H_{12}N_4O_2S$; [127-79-7]

(2) Benzenesulfonamide, 4-amino-N-2-thiazolyl- (sulfathiazole, ST); $C_9H_9N_3O_2S_2$; [72-14-0]

(3) Sodium Hydroxide; NaOH; [1310-73-2]

(4) Water; H_2O; [7732-18-5]

ORIGINAL MEASUREMENTS:

Garcia Onandia, A.; Holz, E.; Holz, S. *Acta Cient. Venezolana* 1955, *6(4)*, 157-63.

VARIABLES:

SM/ST ratio

PREPARED BY:

R. Piekos

EXPERIMENTAL VALUES:

Composition of sulfonamide mixt SM/ST	Volume of 1.5N NaOH soultion required to dissolve the mixture at 26°C
g/g	cm^3
0.1/0.1	0.54
0.1/0.2	0.8
0.1/0.3	1.1
0.1/0.4	1.375
0.1/0.5	1.625
0.2/0.1	0.8
0.3/0.1	1.1
0.4/0.1	1.325
0.5/0.1	1.6

AUXILIARY INFORMATION

METHOD/APPARATUS/PROCEDURE:

Nothing specified.

SOURCE AND PURITY OF MATERIALS:

Nothing specified.

ESTIMATED ERROR:

Nothing specified.

REFERENCES:

COMPONENTS:

(1) Benzenesulfonamide, 4-amino-N-(4-methyl-2-pyrimidinyl)- (sulfamerazine, SM); $C_{11}H_{12}N_4O_2S$; [127-79-7]

(2) Benzenesulfonamide, 4-amino-N-2-thiazolyl- (sulfathiazole, ST); $C_9H_9N_3O_2S_2$; [72-14-0]

(3) Sodium hydroxide; NaOH;]1310-73-2]

(4) Water; H_2O; [7732-18-5]

ORIGINAL MEASUREMENTS:

Holz, E.; Garcia Onandia, A.; Holz, S. *Acta Cient. Venezolana* 1955, *6(2)*, 68-73.

VARIABLES:

SM/ST ratio

PREPARED BY:

R. Piekos

EXPERIMENTAL VALUES:

Composition of sulfonamide mixture SM/ST	Volume of 1N NaOH soln required to dissolve the mixture at 26°C
g/g	cm^3
0.1/0.1	0.8
0.1/0.2	1.2
0.1/0.3	1.6
0.1/0.4	2.0
0.1/0.5	2.5
0.2/0.1	1.2
0.3/0.1	1.6
0.4/0.1	2.0
0.5/0.1	2.4

AUXILIARY INFORMATION

METHOD/APPARATUS/PROCEDURE:

Nothing specified.

SOURCE AND PURITY OF MATERIALS:

Nothing specified. Distd water was used.

ESTIMATED ERROR:

Nothing specified.

REFERENCES:

COMPONENTS:

(1) Benzenesulfonamide, 4-amino-N-2-thiazolyl- (sulfathiazole); $C_9H_9N_3O_2S_2$; [72-14-0]

(2) Benzenesulfonamide, 4-amino-N-(4-methyl-2-pyrimidinyl)- (sulfamethylpyrimidine); $C_{11}H_{12}N_4O_2S$; [127-79-7]

(3) Phosphoric acid, disodium salt; Na_2HPO_4; [7558-94-4]

(4) Phosphoric acid, monopotassium salt; KH_2PO_4; [7778-77-0]

(5) Water; H_2O; [7732-18-5]

ORIGINAL MEASUREMENTS:

Frisk, A. R.; Hagerman, G.; Helander, S.; Sjögren, B. *Hygiea* 1946, *108(12)*, 639-51.

VARIABLES:

One temperature: 37°C; one pH: 6.1

PREPARED BY:

R. Piekos

EXPERIMENTAL VALUES:

Solubility of a mixture of sulfathiazole and sulfamethylpyrimidine in M/30 phosphate buffer of pH 6.1 at 37°C is 135 mg/100 ml solvent.

AUXILIARY INFORMATION

METHOD/APPARATUS/PROCEDURE:

An excess of sulfathiazole and sulfamethylpyrimidine in the phosphate buffer was shaken at 37°C for 24 h. The concn of the sulfonamides was detd by the Bratton and Marshall method (1) using a photoelec colorimeter.

SOURCE AND PURITY OF MATERIALS:

Neither source nor purity of the materials was specified.

ESTIMATED ERROR:

Soly: precision ±14 mg/100 ml (authors).

Temp and pH: not specified.

REFERENCES:

1. Bratton, A. C.; Marshall, E. K., Jr. *J. Biol. Chem.* 1939, *128*, 537.

COMPONENTS:

(1) Benzenesulfonamide, 4-amino-N-2-pyrimidinyl- (sulfadiazine, SD); $C_{10}H_{10}N_4O_2S$; [68-35-9]

(2) Benzenesulfonamide, 4-amino-N-(4-methyl-2-pyrimidinyl)- (sulfamerazine, SM); $C_{11}H_{12}N_4O_2S$; [127-79-7]

(3) Sodium hydroxide; NaOH; [1310-73-2]

(4) Water; H_2O; [7732-18-5]

ORIGINAL MEASUREMENTS:

Garcia Onandia, A.; Holz, E.; Holz, S. *Acta Cient. Venezolana* 1955, *6(4)*, 157-63.

VARIABLES:

SD/SM ratio

PREPARED BY:

R. Piekos

EXPERIMENTAL VALUES:

Composition of sulfonamide mixt SD/SM	Volume of 1.5N NaOH solution required to dissolve the mixture at 26°C
g/g	cm^3
0.1/0.1	0.55
0.1/0.2	0.825
0.1/0.3	1.075
0.1/0.4	1.35
0.1/0.5	1.6
0.2/0.1	0.825
0.3/0.1	1.1
0.4/0.1	1.375
0.5/0.1	1.675

AUXILIARY INFORMATION

METHOD/APPARATUS/PROCEDURE:

Nothing specified.

SOURCE AND PURITY OF MATERIALS:

Nothing specified.

ESTIMATED ERROR:

Nothing specified.

REFERENCES:

COMPONENTS:

(1) Benzenesulfonamide, 4-amino-N-2-pyrimidinyl- (sulfadiazine, SD); $C_{10}H_{10}N_4O_2S$; [68-35-9]

(2) Benzenesulfonamide, 4-amino-N-(4-methyl-2-pyrimidinyl)- (sulfamerazine, SM); $C_{11}H_{12}N_4O_2S$; [127-79-7]

(3) Sodium hydroxide; NaOH; [1310-73-2]

(4) Water; H_2O; [7732-18-5]

ORIGINAL MEASUREMENTS:

Holz, E.; Garcia Onandia, A.; Holz, S. *Acta Cient. Venezolana* 1955, *6(2)*, 68-73.

VARIABLES:

SD/SM ratio

PREPARED BY:

R. Piekos

EXPERIMENTAL VALUES:

Composition of sulfonamide mixt SD/SM	Volume of 1N NaOH soln required to dissolve the mixture at 26°C
0.1/0.1	0.825
0.1/0.2	1.2
0.1/0.3	1.6
0.1/0.4	2.0
0.1/0.5	2.4
0.2/0.1	1.225
0.3/0.1	1.65
0.4/0.1	2.075
0.5/0.1	2.5

AUXILIARY INFORMATION

METHOD/APPARATUS/PROCEDURE:

Nothing specified.

SOURCE AND PURITY OF MATERIALS:

Nothing specified. Distd water was used.

ESTIMATED ERROR:

Nothing specified.

REFERENCES:

COMPONENTS:

(1) Benzenesulfonamide, 4-amino-N-2-pyrimidinyl- (sulfadiazine); $C_{10}H_{10}N_4O_2S$; [68-35-9]

(2) Benzenesulfonamide, 4-amino-N-(4-methyl-2-pyrimidinyl)- (sulfamethylpyrimidine); $C_{11}H_{12}N_4O_2S$; [127-79-7]

(3) Phosphoric acid, disodium salt; Na_2HPO_4; [7558-94-4]

(4) Phosphoric acid, monopotassium salt; KH_2PO_4; [7778-77-0]

(5) Water; H_2O; [7732-18-5]

ORIGINAL MEASUREMENTS:

Frisk, A. R.; Hagerman, G.; Helander, S.; Sjögren, B.

Hygiea 1946, *108(12)*, 639-51

VARIABLES:

One temperature: 37°C; one pH: 6.1

PREPARED BY:

R. Piekos

EXPERIMENTAL VALUES:

Solubility of a mixture of sulfadiazine in sulfamethylpyrimidine in M/30 phosphate buffer of pH 6.1 at 37°C is 56 mg/100 ml solvent.

AUXILIARY INFORMATION

METHOD/APPARATUS/PROCEDURE:

An excess of sulfadiazine and sulfamethylpyrimidine in the phosphate buffer was shaken at 37°C for 24 h. The concn of the sulfonamides was detd by the Bratton and Marshall method (1) using a photoelec colorimeter.

SOURCE AND PURITY OF MATERIALS:

Neither source nor purity of the materials was specified.

ESTIMATED ERROR:

Soly: precision ±6 mg/100 ml (authors).

Temp and pH: not specified.

REFERENCES:

1. Bratton, A. C.; Marshall, E. K., Jr. *J. Biol. Chem.* 1939, *128*, 537.

COMPONENTS:

(1) Benzenesulfonamide, 4-amino-N-2-pyridinyl- (sulfapyridine); $C_{11}H_{11}N_3O_2S$; [144-83-2]

(2) Acetamide, N-[4[(2-pyridinylamino)sulfonyl]phenyl]- (acetyl sulfapyridine); $C_{13}H_{13}N_3O_3S$; [19077-98-6]

(3) Water; H_2O; [7732-18-5]

ORIGINAL MEASUREMENTS:

Langecker, H.

Arch. Exptl. Path. Pharmakol. 1948, *205*, 291-301.

VARIABLES:

One temperature: 37°C

PREPARED BY:

R. Piekos

EXPERIMENTAL VALUES:

Solubility of sulfapyridine and acetyl sulfapyridine in a saturated solution of both compounds in water at 37°C is 131 mg% (5.25×10^{-3} mol dm^{-3}, compiler) and 48 mg% (1.6×10^{-3} mol dm^{-3}, compiler), respectively.

AUXILIARY INFORMATION

METHOD/APPARATUS/PROCEDURE:

A mixt of sulfapyridine and acetyl sulfapyridine was boiled with water and the components were detd colorimetrically by the method of Bratton and Marshall (1) using a Havemann colorimeter (2).

SOURCE AND PURITY OF MATERIALS:

Nothing specified.

ESTIMATED ERROR:

Nothing specified.

REFERENCES:

1. Bratton, A. C.; Marshall, E. K., Jr. *J. Biol. Chem.* 1939, *128*, 537.
2. Havemann, R. *Klin. Wochenschr.* 1940, p. 503.

COMPONENTS:

(1) Acetamide, N-[4-[[(2-thiazolylamino)-sulfonyl]phenyl]- (acetyl sulfathiazole); $C_{11}H_{11}N_3O_3S_2$; [127-76-4]

(2) Acetamide , N-[4-[(2-pyrimidinylamino)-sulfonyl] phenyl]- (acetyl sulfapyrimidine) ; $C_{12}H_{12}N_4O_3S$; [127-74-2]

(3) Phosphoric acid, disodium salt; Na_2HPO_4; [7558-94-4]

(4) Phosphoric acid, monopotassium salt; KH_2PO_4; [7778-77-0]

(5) Water; H_2O; [7732-18-5]

ORIGINAL MEASUREMENTS:

Frisk, A. R.; Hagerman, G.; Helander, S.; Sjögren, B.

Hygiea, 1946, *108(12),* 639-51.

VARIABLES:

One temperature: 37°C; one pH: 6.1

PREPARED BY:

R. Piekos

EXPERIMENTAL VALUES:

Solubility of a mixture of acetyl sulfathiazole and acetyl sulfapyrimidine in M/30 phosphate buffer of pH 6.1 at 37°C is 44 mg/100 ml solvent.

AUXILIARY INFORMATION

METHOD/APPARATUS/PROCEDURE:

An excess of acetyl sulfathiazole and acetyl sulfapyrimidine in the phosphate buffer was shaken at 37°C for 24 h. The concn of the acetyl sulfonamides was detd by the Bratton and Marshall method (1) using a photoelec colorimeter.

SOURCE AND PURITY OF MATERIALS:

Neither source nor purity of the materials was specified.

ESTIMATED ERROR:

Soly: precision ±4 mg/100 ml (authors).

Temp and pH: not specified.

REFERENCES:

1. Bratton, A. C.; Marshall, E. K., Jr. *J. Biol. Chem.* 1939, *128,* 537.

COMPONENTS:

(1) Acetamide, N-[4-[(2-thiazolylamino)sulfonyl]phenyl]- (acetyl sulfathiazole); $C_{11}H_{11}N_3O_3S_2$; [127-76-4]

(2) Acetamide, N-[4-[[(4-methyl-2-pyrimidinyl)amino]sulfonyl]phenyl]- (acetyl sulfamethylpyrimidine); $C_{13}H_{14}N_4O_3S$; [127-73-1]

(3) Phosphoric acid, disodium salt; Na_2HPO_4; [7558-94-4]

(4) Phosphoric acid, monopotassium salt; KH_2PO_4; [7778-77-0]

(5) Water; H_2O; [7732-18-5]

ORIGINAL MEASUREMENTS:

Frisk, A. R.; Hagerman, G.; Helander, S; Sjögren, B.

Hygiea 1946, *108(12)*, 639-51.

VARIABLES:

One temperature: 37°C; one pH: 6.1

PREPARED BY:

R. Piekos

EXPERIMENTAL VALUES:

Solubility of a mixture of acetyl sulfathiazole and acetyl sulfamethylpyrimidine in M/30 phosphate buffer of pH 6.1 at 37°C is 62 mg/100 ml solvent.

AUXILIARY INFORMATION

METHOD/APPARATUS/PROCEDURE:

An excess of acetyl sulfathiazole and acetyl sulfamethylpyrimidine in the phosphate buffer was shaken at 37°C for 24 h. The concn of the acetylated sulfonamides was detd by the Bratton and Marshall method (1) using a photoelec colorimeter.

SOURCE AND PURITY OF MATERIALS:

Neither source nor purity of the materials was specified.

ESTIMATED ERROR:

Soly: precision ±6 mg/100 ml solvent (authors).

Temp and pH: not specified.

REFERENCES:

1. Bratton, A. C.; Marshall, E. K., Jr. *J. Biol. Chem.* 1939, *128*, 537.

COMPONENTS:

(1) Acetamide, N-[4-[(2-pyrimidinylamino)-sulfonyl]phenyl]- (acetyl sulfapyrimidine); $C_{12}H_{12}N_4O_3S$; [127-74-2]

(2) Acetamide, N-[4-[[(4-methyl-2-pyrimidinyl)amino]sulfonyl]phenyl]- (acetyl sulfamethylpyrimidine); $C_{13}H_{14}N_4O_3S$; [127-73-1]

(3) Phosphoric acid, disodium salt; Na_2HPO_4; [7558-94-4]

(4) Phosphoric acid, monopotassium salt; KH_2PO_4; [7778-77-0]

(5) Water; H_2O; [7732-18-5]

ORIGINAL MEASUREMENTS:

Frisk, A. R.; Hagerman, G.; Helander, S.; Sjögren, B.

Hygiea 1946, *108(12)*, 639-51.

VARIABLES:

One temperature: 37°C; one pH: 6.1

PREPARED BY:

R. Piekos

EXPERIMENTAL VALUES:

Solubility of a mixture of acetyl sulfapyrimidine and acetyl sulfamethylpyrimidine in M/30 phosphate buffer of pH 6.1 at 37°C is 80 mg/100 ml solvent.

AUXILIARY INFORMATION

METHOD/APPARATUS/PROCEDURE:

An excess of acetyl sulfapyrimidine and acetyl sulfamethylpyrimidine in the phosphate buffer was shaken at 37°C for 24 h. The concn of the acetylated sulfonamides was detd by the Bratton and Marshall method (1) using a photoelec colorimeter.

SOURCE AND PURITY OF MATERIALS:

Neither source nor purity of the materials was specified.

ESTIMATED ERROR:

Soly: precision ±8 mg/100 ml (authors).

Temp and pH: not specified.

REFERENCES:

1. Bratton, A. C.; Marshall, E. K., Jr. *J. Biol. Chem.* 1939, *128*, 537.

COMPONENTS:

(1) Benzamide, N-[(4-aminophenyl)sulfonyl]- (sulfabenzamide); $C_{13}H_{12}N_2O_3S$; [127-71-9]

(2) Benzenesulfonamide, 4-amino-N-2-pyrimidinyl- (sulfadiazine); $C_{10}H_{10}N_4O_2S$; [68-35-9]

(3) Benzenesulfonamide, 4-amino-N-(4-methyl-2-pyrimidinyl)- (sulfamerazine); $C_{11}H_{12}N_4O_2S$; [127-79-7]

(4) Water; H_2O; [7732-18-5]

ORIGINAL MEASUREMENTS:

Bhattacharyya, R.; Basu, U. P.

Indian Pharmacist 1950, *6(3)*, 77-8, 86.

VARIABLES:

Composition of the sulfonamide mixture

PREPARED BY:

R. Piekos

EXPERIMENTAL VALUES:

Composition of the sulfonamide mixture (1) : (2) : (3)	Solubility at 30°C mg/ml solution	Final pH
7 : 8 : 5	30.8	3.8
5 : 8 : 7	94.6	5.54

AUXILIARY INFORMATION

METHOD/APPARATUS/PROCEDURE:

A weighed sample of the sulfonamides was placed in a clean reagent bottle and a known vol of water was added. The mixt was shaken in a mech shaker at 80-100 strokes/min. After at least 24 h the mixt was filtered through a clean, dried and weighed sintered-glass crucible. At the end of the filtration the crucible was washed with about 1 ml of water, dried at 105°C for 2-3 h, cooled, and weighed to const wt. The pH was detd with a Cambridge bench type pH meter using a glass electrode.

SOURCE AND PURITY OF MATERIALS:

Neither source nor purity of the sulfonamides was specified.

Doubly distd water was used.

ESTIMATED ERROR:

Soly: not specified.

Temp: ±0.2°C (authors).

pH : ±0.01 unit (authors).

REFERENCES:

COMPONENTS:

(1) Benzamide, N-[(4-aminophenyl)sulfonyl]- (sulfabenzamide); $C_{13}H_{12}N_2O_3S$; [127-71-9]

(2) Benzenesulfonamide, 4-amino-N-2-pyrimidinyl- (sulfadiazine); $C_{10}H_{10}N_4O_2S$; [68-35-9]

(3) Benzenesulfonamide, 4-amino-N-(4-methyl-2-pyrimidinyl)- (sulfamerazine); $C_{11}H_{12}N_4O_2S$; [127-79-7]

(4) Phosphoric acid, monopotassium salt; KH_2PO_4; [7778-77-0]

(5) Sodium hydroxide; NaOH; [1310-73-2]

(6) Water; H_2O; [7732-18-5]

ORIGINAL MEASUREMENTS:

Bhattacharyya, R.; Basu, U. P.

Indian Pharmacist 1950, *6(3)*, 77-8, 86.

VARIABLES:

pH

PREPARED BY:

R. Piekos

EXPERIMENTAL VALUES:

Composition of the sulfonamide mixture (parts) (1) : (2) : (3)	Initial pH	Solubility at 30°C in M/20 KH_2PO_4 solution of pH corrected with M/20 NaOH solution (mg/ml solution)	Final pH
7 : 8 : 5	6.18	134.5	5.50
7 : 8 : 5	7.05	482.8	6.45
7 : 8 : 5	7.45	486.6	6.7
5 : 8 : 7	6.18	111	5.75
5 : 8 : 7	7.05	353.6	6.71
5 : 8 : 7	7.45	382.8	6.9

AUXILIARY INFORMATION

METHOD/APPARATUS/PROCEDURE:

A weighed sample of the mixture of sulfonamides was placed in a clean reagent bottle and a known vol of the buffer soln was added. The mixt was shaken in a mech shaker at 80-100 strokes/min. After at least 24 h the mixt was filtered through a clean, dried and weighed sintered-glass crucible. At the end of the filtration the crucible was washed with about 1 ml of water, dried at 105°C for 2-3 h, cooled, and weighed to const wt. The pH was detd with a Cambridge bench type pH meter using a glass electrode.

SOURCE AND PURITY OF MATERIALS:

Neither source nor purity of the materials, with the exception of water, was specified. The water was doubly distilled.

ESTIMATED ERROR:

Soly: not specified.

Temp: ±0.2°C (authors).

pH : ±0.01 unit (authors).

REFERENCES:

COMPONENTS:

(1) Benzamide, N-[(4-aminophenyl)sulfonyl]- (sulfabenzamide); $C_{13}H_{12}N_2O_3S$; [127-71-9]

(2) Benzenesulfonamide, 4-amino-N-2-thiazolyl- (sulfathiazole); $C_9H_9N_3O_2S_2$; [72-14-0]

(3) Benzenesulfonamide, 4-amino-N-2-pyrimidinyl- (sulfadiazine); $C_{10}H_{10}N_4O_2S$; [68-35-9]

(4) Benzenesulfonamide, 4-amino-N-(4-methyl-2-pyrimidinyl)- (sulfamerazine); $C_{11}H_{12}N_4O_2S$; [127-79-7]

(5) Water; H_2O; [7732-18-5]

ORIGINAL MEASUREMENTS:

Bhattacharyya, R.; Basu, U. P.

Indian Pharmacist 1950, *6(3)*, 77-8, 86.

PREPARED BY:

R. Piekos

VARIABLES:

One temperature: 30°C

EXPERIMENTAL VALUES:

Solubility of a mixture containing 3 parts of sulfabenzamide, 4 parts of sulfathiazole, 8 parts of sulfadiazine, and 5 parts of sulfamerazine in water at 30°C is 93.4 mg per ml solution[a].

[a]The final pH was 3.5

AUXILIARY INFORMATION

METHOD/APPARATUS/PROCEDURE:

A weighed sample of the sulfonamides was placed in a clean reagent bottle and a known vol of water was added. The mixt was shaken in a mech shaker at 80-100 strokes/min. After at least 24 h the mixt was filtered through a clean, dried and weighed sintered-glass crucible. At the end of the filtration the crucible was washed with about 1 ml of water, dried at 105°C for 2-3 h, cooled, and weighed to const wt. The pH was detd with a Cambridge bench type pH meter using a glass electrode.

SOURCE AND PURITY OF MATERIALS:

Neither source nor purity of the sulfonamides was specified.

Doubly distd water was used.

ESTIMATED ERROR:

Soly: not specified.

Temp: ±0.2°C (authors).

pH : ±0.01 unit (authors).

REFERENCES:

COMPONENTS:

(1) Benzenesulfonamide, 4-amino-N-2-pyrimidinyl- (sulfadiazine, SD); $C_{10}H_{10}N_4O_2S$; [68-35-9]

(2) Benzenesulfonamide, 4-amino-N-(4-methyl-2-pyrimidinyl- (sulfamerazine, SM); $C_{11}H_{12}N_4O_2S$; [127-79-7]

(3) Benzenesulfonamide, 4-amino-N-2-thiazolyl- (sulfathiazole, ST); $C_9H_9N_3O_2S_2$; [72-14-0]

(4) Sodium hydroxide; NaOH; [1310-73-2]

(5) Water; H_2O; [7732-18-5]

ORIGINAL MEASUREMENTS:

Garcia Onandia, A.; Holz, E.; Holz, S. *Acta Cient. Venezolana* 1955, *6(4)*, 157-63.

VARIABLES:

SD/SM/ST ratio

PREPARED BY:

R. Piekos

EXPERIMENTAL VALUES:

Compostion of sulfonamide mixt SD/SM/ST	Volume of 1.5N NaOH soln required to dissolve the mixture at 26°C	Composition of sulfonamide mixt SD/SM/ST	Volume of 1.5N NaOH soln required to dis solve the mixt at 26°C
g/g/g	cm^3	g/g/g	cm^3
0.1/0.1/0.1	0.825	0.1/0.1/0.2	1.1
0.1/0.2/0.2	1.375	0.1/0.1/0.3	1.375
0.1/0.3/0.3	1.9	0.1/0.1/0.4	1.675
0.1/0.4/0.4	2.425	0.1/0.1/0.5	1.925
0.1/0.5/0.5	2.975	0.1/0.2/0.1	1.075
0.2/0.1/0.2	1.375	0.1/0.3/0.1	1.35
0.3/0.1/0.3	1.925	0.1/0.4/0.1	1.6
0.4/0.1/0.4	2.45	0.1/0.5/0.1	1.875
0.5/0.1/0.5	3.025	0.2/0.1/0.1	1.1
0.2/0.2/0.1	1.375	0.3/0.1/0.1	1.375
0.3/0.3/0.1	1.9	0.4/0.1/0.1	1.675
0.4/0.4/0.1	2.475	0.5/0.1/0.1	1.95
0.5/0.5/0.1	3.025		

AUXILIARY INFORMATION

METHOD/APPARATUS/PROCEDURE:

Nothing specified.

SOURCE AND PURITY OF MATERIALS:

Nothing specified.

ESTIMATED ERROR:

Nothing specified.

REFERENCES:

COMPONENTS:

(1) Benzenesulfonamide, 4-amino-N-2-pyrimidinyl- (sulfadiazine, DS); $C_{10}H_{10}N_4O_2S$; [68-35-9]

(2) Benzenesulfonamide, 4-amino-N-(4-methyl-2-pyrimidinyl)- (sulfamerazine, SM); $C_{11}H_{12}N_4O_2S$; [127-79-7]

(3) Benzenesulfonamide, 4-amino-N-2-thiazolyl- (sulfathiazole, ST); $C_9H_9N_3O_2S_2$; [[72-14-0]

(4) Sodium hydroxide; NaOH; [1310-73-2]

(5) Water; H_2O; [7732-18-5]

ORIGINAL MEASUREMENTS:

Holz, E.; Garcia Onandia, A.; Holz, S.

Acta Cient. Venezolana 1955, *6(2)*, 68-73.

VARIABLES:

SD/SM/ST ratio

PREPARED BY:

R. Piekos

EXPERIMENTAL VALUES:

Composition of sulfonamide mixt SD/SM/ST	Volume of 1N NaOH soln required to dissolve the mixt at 26°C	Composition of sulfonamide mixt SD/SM/ST	Volume of 1N NaOH soln required to dissolve the mixt
g/g/g	cm^3	g/g/g	cm^3
0.1/0.1/0.1	1.25	0.4/0.1/0.4	3.725
0.1/0.1/0.2	1.625	0.5/0.1/0.5	4.5
0.1/0.1/0.3	2.025	0.2/0.1/0.1	1.65
0.1/0.1/0.4	2.425	0.3/0.1/0.1	2.075
0.1/0.1/0.5	2.85	0.4/0.1/0.1	2.5
0.1/0.2/0.2	2.0	0.5/0.1/0.1	2.9
0.1/0.3/0.3	2.825	0.2/0.2/0.1	2.05
0.1/0.4/0.4	3.65	0.3/0.3/0.1	2.85
0.1/0.5/0.5	4.425	0.4/0.4/0.1	3.7
0.1/0.2/0.1	1.625	0.5/0.5/0.1	4.45
0.1/0.3/0.1	2.025		
0.1/0.4/0.1	2.425		
0.1/0.5/0.1	2.825		
0.2/0.1/0.2	2.05		
0.3/0.1/0.3	2.975		

AUXILIARY INFORMATION

METHOD/APPARATUS/PROCEDURE:

Nothing specified.

SOURCE AND PURITY OF MATERIALS:

Not specified. Distd water was used.

ESTIMATED ERROR:

Nothing specified.

REFERENCES:

COMPONENTS:

(1) Benzenesulfonamide, 4-amino-N-2-thiazolyl- (sulfathiazole); $C_9H_9N_3O_2S_2$; [72-14-0]

(2) Benzenesulfonamide, 4-amino-N-2-pyrimidinyl- (sulfadiazine); $C_{10}H_{10}N_4O_2S$; [68-35-9]

(3) Benzenesulfonamide, 4-amino-N-(4-methyl-2-pyrimidinyl)- (sulfamerazine); $C_{11}H_{12}N_4O_2S$; [127-79-7]

(4) Phosphoric acid, monopotassium salt; KH_2PO_4; [7778-77-0]

(5) Sodium hydroxide; NaOH; [1310-73-2]

(6) Water; H_2O; [7732-18-5]

ORIGINAL MEASUREMENTS:

Bhattacharyya, R.; Basu, U. P.
Indian Pharmacist 1950, *6(3)*, 77-8, 86.

VARIABLES:

pH

PREPARED BY:

R. Piekos

EXPERIMENTAL VALUES:

Initial pH	Solubility at 30°C of a mixture of 1 part each of sulfathiazole, sulfadiazine, and sulfamerazine in M/20 KH_2PO_4 solution of pH corrected with M/20 NaOH solution	Final pH
	mg/ml solution	
6.18	114.8	6.24
7.05	212	7.49

AUXILIARY INFORMATION

METHOD/APPARATUS/PROCEDURE:

A weighed sample of the sulfonamides was placed in a clean reagent bottle and a known vol of the M/20 KH_2PO_4 soln was added, and the pH was adjusted to the desired value with M/20 NaOH soln. The mixt was shaken in a mech shaker at 80-100 strokes/min. After at least 24 h the mixt was filtered through a clean, dried and weighed sintered-glass crucible. At the end of the filtration the crucible was washed with about 1 ml of water, dried at 105°C for 2-3 h, cooled, and weighed to const wt. The pH was detd with a Cambridge bench type pH meter using a glass electrode.

SOURCE AND PURITY OF MATERIALS:

Neither source nor purity of the materials, with the exception of water, was specified. The water was doubly distilled.

ESTIMATED ERROR:

Soly: not specified.
Temp: ±0.2°C
pH : ±0.01 unit (authors).

REFERENCES:

COMPONENTS:

(1) Benzenesulfonamide, 4-amino-N-2-thiazolyl- (sulfathiazole); $C_9H_9N_3O_2S_2$; [72-14-0]

(2) Benzenesulfonamide, 4-amino-N-2-pyrimidinyl- (sulfapyrimidine); $C_{10}H_{10}N_4O_2S$; [68-35-9]

(3) Benzenesulfonamide, 4-amino-N-(4-methyl-2-pyrimidinyl)- (sulfamethylpyrimidine); $C_{11}H_{12}N_4O_2S$; [127-79-7]

(4) Phosphoric acid, disodium salt; Na_2HPO_4; [7558-94-4]

(5) Phosphoric acid, monopotassium salt; KH_2PO_4; [7778-77-0]

(6) Water; H_2O; [7732-18-5]

ORIGINAL MEASUREMENTS:

Frisk, A. R.; Hagerman, G.; Helander, S.; Sjögren, B.

Hygiea 1946, *108(12)*, 639-51.

VARIABLES:

One temperature: 37°C; one pH: 6.1

PREPARED BY:

R. Piekos

EXPERIMENTAL VALUES:

Solubility of a mixtrue of sulfathiazole, sulfapyrimidine and sulfamethylpyrimidine in M/30 phosphate buffer of pH 6.1 at 37°C is 160 mg/100 ml solvent.

AUXILIARY INFORMATION

METHOD/APPARATUS/PROCEDURE:

An excess of sulfathiazole, sulfapyrimidine and sulfamethylpyrimidine in the phosphate buffer was shaken at 37°C for 24 h. The concn of the sulfonamides was detd by the Bratton and Marshall method (1) using a photoelec colorimeter.

SOURCE AND PURITY OF MATERIALS:

Neither source nor the purity of the materials was specified.

ESTIMATED ERROR:

Soly: precision ±0.7 mg/100 ml (authors).

Temp and pH: not specified.

REFERENCES:

1. Bratton, A. C.; Marshall, E. K., Jr. *J. Biol. Chem.* 1939, *128*, 537.

COMPONENTS:

(1) Acetamide, N-[4-[(2-thiazolylamino)sulfonyl]phenyl]- (acetyl sulfathiazole); $C_{11}H_{11}N_3O_3S_2$; [127-76-4]

(2) Acetamide, N-[4-[(2-pyrimidinylamino)-sulfonyl]phenyl]- (acetyl sulfapyrimidine); $C_{12}H_{12}N_4O_3S$; [127-74-2]

(3) Acetamide,N-[4-[[(4-methyl(2-pyrimidinyl) amino]sulfonyl]phenyl]- (acetyl sulfamethylpyrimidine); $C_{13}H_{14}N_4O_3S$; [127-73-1]

(4) Phosphoric acid, disodium salt; Na_2HPO_4; [7558-94-4]

(5) Phosphoric acid, monopotassium salt; KH_2PO_4; [7778-77-0]

(6) Water; H_2O; [7732-18-5]

ORIGINAL MEASUREMENTS:

Frisk, A. R.; Hagerman, G.; Helander, S.; Sjögren, B.

Hygiea 1946, *108(12)*, 639-51.

VARIABLES:

One temperature: 37°C; one pH 6.1

PREPARED BY:

R. Piekos

EXPERIMENTAL VALUES:

Solubility of a mixture of acetyl sulfathiazole, acetyl sulfapyrimidine, and acetyl sulfamethylpyrimidine in M/30 phosphate buffer of pH 6.1 at 37°C is 89 mg/100 ml solvent.

AUXILIARY INFORMATION

METHOD/APPARATUS/PROCEDURE:

An excess of acetyl sulfathiazole, acetyl sulfapyrimidine and acetyl sulfamethylpyrimidine in the phosphate buffer was shaken at 37°C for 24 h. The concn of the acetylated sulfonamides was detd by the Bratton and Marshall method (1) using a photoelec colorimeter.

SOURCE AND PURITY OF MATERIALS:

Neither source nor purity of the materials was specified.

ESTIMATED ERROR:

Soly: precision ±9 mg/100 ml (authors).

Temp and pH: not specified.

REFERENCES:

1. Bratton, A. C.; Marshall, E. K., Jr. *J. Biol. Chem.* 1939, *128*, 537.

COMPONENTS:

(1) Benzenesulfonamide, 4-amino-N-(4,6-dimethyl-2-pyrimidinyl)- (sulfamethazine); $C_{12}H_{14}N_4O_2S$; [57-68-1]

(2) Benzenesulfonamide, 4-amino-N-(4 - methyl-2-pyrimidinyl)- (sulfamerazine); $C_{11}H_{12}N_4O_2S$; [127-79-7]

(3) Benzenesulfonamide, 4-amino-N-2-pyrimidinyl- (sulfadiazine); $C_{10}H_{10}N_4O_2S$; [68-35-9]

(4) Phosphoric acid, disodium salt; Na_2HPO_4; [7558-94-4]

(5) Phosphoric acid, monopotassium salt; KH_2PO_4; [7778-77-0]

(6) Water; H_2O; [7732-18-5]

ORIGINAL MEASUREMENTS:

Bandelin, F. J.; Malesh, W.

J. Am. Pharm. Assoc., Sci. Ed.

1959, *48*, 177-81.

VARIABLES:

pH at 37°C

PREPARED BY:

R. Piekos

EXPERIMENTAL VALUES:

Solubility of a 1:1:1 (by wt.) mixture of the three sulfonamides (triple sulfonamide) in buffers of varying mixtures of the following solutions:

$Na_2HPO_4 \cdot 7H_2O$ 71.6 g/l of distilled water (0.27 mol dm^{-3}, compiler);

KH_2PO_2 36.3 g/l of distilled water (0.27 mol dm^{-3}, compiler) at 37°C

Initial pH	Solubility in mg/100 ml buffer
4.5	96
5.0	98
5.5	102
6.0	109
6.5	130; 139 [a]
7.0	192
7.5	209

[a] obtained by extrapolation of the solubility curve.

AUXILIARY INFORMATION

METHOD/APPARATUS/PROCEDURE:

Solns were prepd by adding an excess of the sulfonamides to 10 ml of buffer soln at each pH level in 18 x 150-mm test tubes, stoppering the tubes, placing in a water bath at 37° C with gentle agitation for 24 h. The mixt was then filtered and a 1-ml aliquot was accurately pipetted into a volumetric flask for diln and analysis. The balance was retained for pH detn to ascertain any change in pH value. The sulfonamides were assayed colorimetrically at 545 nm by the method of Bratton and Marshall as described in detail by Biamente and Schneller (1). Standard curves were prepd for individual sulfonamides using accurately prepd std solns.

SOURCE AND PURITY OF MATERIALS:

Neither source nor the purity of the reagents were specified. Distilled water was used.

ESTIMATED ERROR:

Soly: duplicate samples were used for analysis (authors).

Temp: nothing specified.

pH : nothing specified.

REFERENCES:

1. Biamonte, A. R.; Schneller, G. E. *J. Am. Pharm. Assoc., Sci. Ed.*, 1952, *41*, 341.

COMPONENTS:

(1) Benzenesulfonamide, 4-amino-N-2-pyrimidinyl- (sulfapyrimidine); $C_{10}H_{10}N_4O_2S$; [68-35-9]

(2) Acetamide, N-[4-[(2-pyrimidinylamino)-sulfonyl]phenyl]- (acetyl sulfapyrimidine); $C_{12}H_{12}N_4O_3S$; [127-74-2]

(3) Water; H_2O; [7732-18-5]

ORIGINAL MEASUREMENTS:

Langecker, H.

Arch. Exptl. Path. Pharmakol. 1948, *205*, 291-301.

VARIABLES:

One temperature: 37°C

PREPARED BY:

R. Piekos

EXPERIMENTAL VALUES:

Solubility of sulfapyrimidine and acetyl sulfapyrimidine in a saturated solution of both compounds in water at 37°C is 176 mg% (7.0 x 10^{-3} mol dm^{-3}, compiler) at 47 mg% (1.6 x 10^{-3} mol dm^{-3}, compiler), respectively.

AUXILIARY INFORMATION

METHOD/APPARATUS/PROCEDURE:

A mixt of sulfapyrimidine and acetyl sulfapyrimidine was boiled with water and the components were detd colorimetrically by the method of Bratton and Marshall (1) using a Havemann colorimeter (2).

SOURCE AND PURITY OF MATERIALS:

Nothing specified.

ESTIMATED ERROR:

Nothing specified.

REFERENCES:

1. Bratton, A. C.; Marshall, E. K., Jr. *J. Biol. Chem.* 1939, *128*, 537.
2. Havemann, R. *Klin. Wochenschr.* 1940, p. 503.

COMPONENTS:

(1) Benzenesulfonamide, 4-amino-N-2-pyrimidinyl- (sulfadiazine); $C_{10}H_{10}N_4O_2S$; [68-35-9]

(2) Benzenesulfonamide, 4-amino-N-(4-methyl-2-pyrimidinyl)- (sulfamerazine); $C_{11}H_{12}N_4O_2S$; [127-79-7]

(3) Benzenesulfonamide, 4-amino-N-(4,6-dimethyl-2-pyrimidinyl)- (sulfamethazine); $C_{12}H_{14}N_4O_2S$; [57-68-1]

(4) Phosphoric acid, disodium salt; Na_2HPO_4; [7558-94-4]

(5) 1,2,3-Propanetricarboxylic acid, 2-hydroxy- (citric acid); $C_6H_8O_7$; [77-92-9]

(6) Water; H_2O; [7732-18-5]

ORIGINAL MEASUREMENTS:

Biamonte, A. R.; Schneller, G. H.

J. Am. Pharm. Assoc., Sci. Ed. 1952, *41*, 341-5.

VARIABLES:

pH

PREPARED BY:

R. Piekos

EXPERIMENTAL VALUES:

Solubility of a mixture of equal portions of sulfadiazine, sulfamerazine, and sulfamethazine in McIlvaine's disodium phosphate - citric acid buffer solutions at 37°C

Initial pH of buffer	Solubility (mg/100 ml solution)	Final pH
4.5	95.9	4.5
5.0	98.4	5.0
6.0	108.9	5.8
7.0	186.0	6.9
8.0	476.0	7.5

AUXILIARY INFORMATION

METHOD/APPARATUS/PROCEDURE:

A sample large enough to supply an excess of each constituent was equilibrated in buffer solns at 37°C for 18 h with agitation. The suspension was then immediately filtered through a Whatman No. 1 paper. The filtration time was approx 2 min. The sulfonamides in the filtrate were assayed spectrophotometrically by the method of Bratton and Marshall (1) using a Beckmann DU spectrophotometer.

SOURCE AND PURITY OF MATERIALS:

The source and purity of the materials was not specified. The mp of sulfadiazine, sulfamerazine and sulfamethazine was 253.5-4.5°C, 235.5-6.5°C, and 197.7-8.6°C, resp.

ESTIMATED ERROR:

pH and temp: not specified. Accuracy of the anal method was illustrated by the following values: expected 2.003, 3.004, 4.006, 5.007 mg/100 ml; found: 2.08, 3.06, 4.12, 5.10.

REFERENCES:

1. Bratton, A. C.; Marshall, E. K., Jr. *J. Biol. Chem.* 1939, *128*, 537.

COMPONENTS:

(1) Acetamide, N-[4-[[(4,6-dimethyl-2-pyrimidinyl)amino]sulfonyl]phenyl]- (acetyl sulfamethazine); $C_{14}H_{16}N_4O_3S$; [100-90-3]

(2) Acetamide, N-4-[[(4-methyl-2-pyrimidinyl)amino]sulfonyl]phenyl]- (acetyl sulfamerazine); $C_{13}H_{14}N_4O_3S$; [127-73-1]

(3) Acetamide, N-[4-[(2-pyrimidinylamino)sulfonyl]phenyl]- (acetyl sulfadiazine); $C_{12}H_{12}N_4O_3S$; [127-74-2]

(4) Phosphoric acid, disodium salt; Na_2HPO_4; [7558-94-4]

(5) Phosphoric acid, monopotassium salt; KH_2PO_4; [7778-77-0]

(6) Water; H_2O; [7732-18-5]

ORIGINAL MEASUREMENTS:

Bandelin, F. J.; Malesh, W.

J. Am. Pharm. Assoc., Sci. Ed. 1959, *48*, 177-81.

VARIABLES:

pH at 37°C

PREPARED BY:

R. Piekos

EXPERIMENTAL VALUES:

Solubility of acetylated 1:1:1 (by wt.) mixture of sulfonamides in buffers of varying mixtures of the following solutions: $Na_2HPO_4 \cdot 7H_2O$ 71.6 g/l of distilled water (0.27 mol dm^{-3}, compiler), KH_2PO_4 36.3 g/l of distilled water (0.27 mol dm^{-3}, compiler) at 37°C

Equilibrium pH	mg/100 ml of buffer based on free sulfonamides
4.5	116
5.0	121
5.5	132
6.0	158
6.4	216
6.5	230[a]
7.0	420[a]
7.1	490

[a]obtained by extrapolation of the solubility curve.

AUXILIARY INFORMATION

METHOD/APPARATUS/PROCEDURE:

Solns were prepd by adding an excess of the acetyl derivs to 10 ml of buffer soln at each pH level in 18 x 150-mm test tubes, stoppering the tubes, placing in a water bath at 37°C with gentle agitation for 24 h. The mixt was then hydrolyzed with 5% H_2SO_4 for 1 h to liberate the free sulfonamides. One-ml aliquot was accurately pipetted into a volumetric flask for diln and analysis. The sulfonamides were assayed colorimetrically at 545 nm by the method of Bratton and Marshall as described in detail by Biamonte and Schneller (1). Standard curves were prepd for individual sulfonamides using accurately prepd std solns.

SOURCE AND PURITY OF MATERIALS:

Neither source nor the purity of the reagents were specified.

Distilled water was used.

ESTIMATED ERROR:

Soly: duplicate samples were used for analysis (authors).

Temp: Nothing specified.

pH : Nothing specified.

REFERENCES:

1. Biamonte, A. R.; Schneller, G. E. *J. Am. Pharm. Assoc., Sci. Ed.*, 1952, *41*, 341.

COMPONENTS:

(1) Benzenesulfonamide, 4-amino-N-(4,6-dimethyl-2-pyrimidinyl)-, (sulfamethazine); $C_{12}H_{14}N_4O_2S$; [57-68-1]

(2) Benzenesulfonamide, 4-amino-N-(4-methyl-2-pyrimidinyl)-, (sulfamerazine); $C_{11}H_{12}N_4O_2S$; [127-79-7]

(3) Benzenesulfonamide, 4-amino-N-2-pyrimidinyl-, (sulfadiazine); $C_{10}H_{10}N_4O_2S$; [68-35-9]

(4) Calcium chloride; $CaCl_2$; [10043-52-4]

(5) Magnesium chloride; $MgCl_2$; [7786-30-3]

(6) Phosphoric acid, monoammonium salt; $NH_4H_2PO_4$; [7722-76-1]

(7) Potassium chloride; KCl; [7447-40-7]

(8) Sodium chloride; NaCl; [7647-14-5]

(9) Urea; CH_4N_2O; [57-13-6]

(10) Water; H_2O; [7732-18-5]

ORIGINAL MEASUREMENTS:

Bandelin, F. J.; Malesh, W.,

J. Am. Pharm. Assoc., Sci. Ed. 1959, *48*, 178-81.

PREPARED BY:

R. Piekos

VARIABLES:

pH at 37°C

EXPERIMENTAL VALUES:

Continued on the next page.

AUXILIARY INFORMATION

METHOD/APPARATUS/PROCEDURE:

Excess sulfonamides was added to aliquots of synthetic urine solutions and 1% H_3PO_4 or 1% NaOH solns were used to adjust the pH to the required value. The solns were agitated for 24 h with addn of acid or base to keep them at the desired pH level until equilibrium in pH and concn was attained. The solns were filtered and in aliquots the sulfonamides were assayed spectrophotometrically by a method described by Biamonte and Schneller (1).

SOURCE AND PURITY OF MATERIALS:

Nothing specified.

ESTIMATED ERROR:

Soly: average values of 2 detns are given.

Temp: not specified.

pH : not specified.

REFERENCES:

1. Biamonte, A. R.; Schneller, G. E., *J. Am. Pharm. Assoc., Sci. Ed.* 1952, *41*, 341.

COMPONENTS:

(1) Sulfamethazine; $C_{12}H_{14}N_4O_2S$; [57-68-1]

(2) Sulfamerazine; $C_{11}H_{12}N_4O_2S$; [127-79-7]

(3) Sulfadiazine; $C_{10}H_{10}N_4O_2S$; [68-35-9]

(4) - (10) Synthetic urine see previous page for details

ORIGINAL MEASUREMENTS: (CONTINUED)

Bandelin, F.; Malesh, W., *J. Am. Pharm. Assoc., Sci. Ed.* 1959, *48*, 177-81.

PREPARED BY:

R. Piekos

EXPERIMENTAL VALUES:

Solubility of a 1:1:1 (by wt.) mixture of the three sulfonamides (triple sulfonamide) in a solution containing $CaCl_2$ 0.143, $MgCl_2$ 0.121, $NH_4H_2PO_4$ 0.300, KCl 1.660, NaCl 2.950 and urea 20 g/dm^3 (synthetic urine, Mosher Vehicle) at 37°C

Equilibrium pH	Solubility (mg/100 ml synthetic urine)
4.5	100
5.0	108
5.5	118
6.0	136
6.4	182
7.1	275

AUXILIARY INFORMATION

METHOD/APPARATUS/PROCEDURE:

SOURCE AND PURITY OF MATERIALS:

ESTIMATED ERROR:

REFERENCES:

COMPONENTS:

(1) Acetamide, N-[4-[(2-pyrimidinylamino)-sulfonyl]phenyl]- (acetyl sulfadiazine); $C_{12}H_{12}N_4O_3S$; [127-74-2]

(2) Acetamide, N-[4-[(4-methyl-2-pyrimidinylamino)sulfonyl]phenyl]- (acetyl sulfamerazine); $C_{13}H_{14}N_4O_3S$; [127-73-1]

(3) Acetamide, N-[4-[[(4,6-dimethyl-2-pyrimidinyl)amino]sulfonyl]phenyl]- (acetyl sulfamethazine); $C_{14}H_{16}N_4O_3S$; [100-90-3]

(4) Phosphoric acid, disodium salt; Na_2HPO_4; [7558-94-4]

(5) 1,2,3-Propanetricarboxylic acid, 2-hydroxy- (citric acid); $C_6H_8O_7$; [77-92-9]

(6) Water; H_2O; [7732-18-5]

ORIGINAL MEASUREMENTS:

Biamonte, A. R.; Schneller, G. H.

J. Am. Pharm. Assoc., Sci. Ed. 1952, *41*, 341-5.

VARIABLES:

pH

PREPARED BY:

R. Piekos

EXPERIMENTAL VALUES:

Solubility of a mixture of equal parts of acetyl sulfadiazine, acetyl sulfamerazine and acetyl sulfamethazine in McIlvaine's disodium phosphate - citric acid buffers at 37°C

Initial pH of buffer	Solubility (mg/100 ml solution)	Final pH
4.5	119.0	4.5
5.0	119.0	5.0
6.0	152.7	6.0
7.0	390.2	6.8

AUXILIARY INFORMATION

METHOD/APPARATUS/PROCEDURE:

A sample large enough to supply an excess of each constituent was equilibrated in buffer solns at 37°C for 18 h with agitation. The suspension was then immediately filtered through a Whatman No. 1 paper. The filtration time was approx. 2 min. The compds were assayed in the filtrate spectrophotometrically by the Bratton and Marshall method (1) after deacetylation with concd HCl_{aq}. The instrument used was a Beckmann DU spectrophotometer.

SOURCE AND PURITY OF MATERIALS:

Acetyl sulfadiazine, acetyl sulfamerazine, and acetyl sulfamethazine had mp of 261.2-2.4°C, 248.2-9.4°C, and 249.5-50.6°C, resp, and were supplied by the American Cyanamid Co, Calco Chem. Div, Bound Brook, NJ.

The source and purity of the remaining materials were not specified.

ESTIMATED ERROR: pH and temp: not specified. Accuracy of the anal method was illustrated by the following values: expected 2.003, 3.004, 4.006, 5.007 mg/100 ml; found 2.08, 3.06, 4.12, 5.10, resp.

REFERENCES:

1. Bratton, A. C.; Marshall, E. K., Jr. *J. Biol. Chem.* 1939, *128*, 537.

COMPONENTS:

(1) Acetamide, N-[4-[[(4,6-dimethyl-2-pyrimidinyl)amino]sulfonyl]phenyl]- (acetyl sulfamethazine); $C_{14}H_{16}N_4O_3S$; [100-90-3]

(2) Acetamide, N-[4-[[(4-methyl-2-pyrimidinyl)amino]sulfonyl]phenyl]- (acetyl sulfamerazine); $C_{13}H_{14}N_4O_3S$; [127-73-1]

(3) Acetamide, N-[4-[(2-pyrimidinylamino)sulfonyl]phenyl]- (acetyl sulfadiazine); $C_{12}H_{12}N_4O_3S$; [127-74-2]

(4) Calcium chloride; $CaCl_2$; [10043-52-4]

(5) Magnesium chloride; $MgCl_2$; [7786-30-3]

(6) Phosphoric acid, monoammonium salt; $NH_4H_2PO_4$; [7722-76-1]

(7) Potassium chloride; KCl; [7447-40-7]

(8) Sodium chloride; NaCl; [7647-14-5]

(9) Urea; CH_4N_2O; [57-13-6]

(10) Water; H_2O; [7732-18-5]

ORIGINAL MEASUREMENTS:

Bandelin, F. J.; Malesh, W.

J. Am. Pharm. Assoc., Sci. Ed. 1959, *48*, 177-81.

VARIABLES:

pH at 37°C

PREPARED BY:

R. Piekos

EXPERIMENTAL VALUES:

see next page

AUXILIARY INFORMATION

METHOD/APPARATUS/PROCEDURE:

Excess acetyl sulfonamides was added to aliquots of synthetic urine solns and 1% H_3PO_4 or 1% NaOH solns were used to adjust the pH to the required value. The solns were agitated for 24 h with addn of acid or base to keep them at the desired pH level until equilibrium in pH and concn was attained. Then the solns were filtered and in aliquots the sulfonamides were assayed spectrophotometrically by the method described by Biamonte and Schneller (1). Before detn the soln was refluxed with 5% H_2SO_4 for 1 h to liberate the free amino compounds.

SOURCE AND PURITY OF MATERIALS:

Nothing specified.

ESTIMATED ERROR:

Soly: average values of 2 detns are given.

Temp: not specified.

pH : not specified.

REFERENCES:

1. Biamonte, A. R.; Schneller, G. E., *J. Am. Pharm. Assoc., Sci. Ed.* 1952, *41*, 341.

COMPONENTS:

(1) Acetamide, N-[4-[[(4,6-dimethyl-2-pyrimidinyl)amino]sulfonyl]phenyl]- (acetyl sulfamethazine); $C_{14}H_{16}N_4O_3S$; [100-90-3]

(2) Acetamide, N-[[(4-methyl-2-pyrimidinyl)-amino]sulfonyl]phenyl]- (acetyl sulfamerazine); $C_{13}H_{14}N_4O_3S$; [127-73-1]

(3) Acetamide, N-[4-[(2-pyrimidinylamino)-sulfonyl]phenyl]- (acetyl sulfadiazine); $C_{12}H_{12}N_4O_3S$; [127-74-2]

(4) - (10) Synthetic urine

ORIGINAL MEASUREMENTS: (CONTINUED)

Bandelin, F. J.; Malesh, W.,

J. Am. Pharm. Assoc., Sci. Ed.

<u>1959,</u> *48*, 177-81.

PREPARED BY:

R. Piekos

EXPERIMENTAL VALUES:

Solubility of a 1:1:1 (by wt.) mixture of the three acetyl sulfonamides (acetyl triple sulfonamide) in a solution containing $CaCl_2$ 0.143, $MgCl_2$ 0.121, $NH_4H_2PO_4$ 0.300, KCl 1.660, NaCl 2.950 and urea 20 g/dm^3 (synthetic urine, Mosher Vehicle) at 37°C

Equilibrium pH	Solubility (mg/100 ml synthetic urine)
4.5	218
5.0	223
5.5	231
6.0	254
6.5	163
7.0	630

AUXILIARY INFORMATION

METHOD/APPARATUS/PROCEDURE:

SOURCE AND PURITY OF MATERIALS:

ESTIMATED ERROR:

REFERENCES:

COMPONENTS:

(1) Benzenesulfonamide, 4-amino-N-2-thiazolyl- (sulfathiazole); $C_9H_9N_3O_2S_2$; [72-14-0]

(2) Benzenesulfonamide, 4-amino-N-2-pyrimidinyl- (sulfadiazine); $C_{10}H_{10}N_4O_2S$; [68-35-9]

(3) Benzenesulfonamide, 4-amino-N-(4-methyl-2-pyrimidinyl)- (sulfamerazine); $C_{11}H_{12}N_4O_2S$; [127-79-7]

(4) Water; H_2O; [7732-18-5]

ORIGINAL MEASUREMENTS:

Bhattacharyya, R.; Basu, U. P.

Indian Pharmacist 1950, *6(3)*, 77-8, 86.

VARIABLES:

One temperature: 30°C

PREPARED BY:

R. Piekos

EXPERIMENTAL VALUES:

Solubility of a mixture of 1 part each of sulfathiazole, sulfadiazine, and sulfamerazine in water at 30°C is 98.3 mg per ml of solution[a].

[a] The final pH was 4.4

AUXILIARY INFORMATION

METHOD/APPARATUS/PROCEDURE:

A weighed sample of the sulfonamides was placed in a clean reagent bottle and a known vol of water was added. The mixt was shaken in a mech shaker at 80-100 strokes/min. After at least 24 h the mixt was filtered through a clean, dried, and weighed sintered-glass crucible. At the end of the filtration the crucible was washed with about 1 ml of water, dried at 105°C for 2-3 h, cooled and weighed to const wt. The pH was detd with a Cambridge bench type pH meter using a glass electrode.

SOURCE AND PURITY OF MATERIALS:

Neither source nor purity of the sulfonamides was specified.

Doubly distd water was used.

ESTIMATED ERROR:

Soly: not specified.

Temp: ±0.2°C (authors).

pH : ±0.01 unit (authors).

REFERENCES:

COMPONENTS:

(1) Benzamide, N-[(4-aminophenyl)sulfonyl]- (sulfabenzamide); $C_{13}H_{12}N_2O_3S$; [127-71-9]

(2) Benzenesulfonamide, 4-amino-N-2-thiazolyl- (sulfathiazole); $C_9H_9N_3O_2S_2$; [72-14-0]

(3) Benzenesulfonamide, 4-amino-N-2-pyrimidinyl- (sulfadiazine); $C_{10}H_{10}N_4O_2S$; [68-35-9]

(4) Benzenesulfonamide, 4-amino-N-(4-methyl-2-pyrimidinyl)- (sulfamerazine); $C_{11}H_{12}N_4O_2S$; [127-79-7]

(5) Phosphoric acid, monopotassium salt; KH_2PO_4; [7778-77-0]

(6) Sodium hydroxide; NaOH; [1310-73-2]

(7) Water; H_2O; [7732-18-5]

ORIGINAL MEASUREMENTS:

Bhattacharyya, R.; Basu, U. P.

Indian Pharmacist 1950, *6(3)*, 77-8, 86.

VARIABLES:

pH

PREPARED BY:

R. Piekos

EXPERIMENTAL VALUES:

Initial pH	Solubility at 30°C of a mixture of sulfabenzamide 3, sulfathiazole 4, sulfadiazine 8, and sulfamerazine 5 parts in M/20 KH_2PO_4 solution of pH corrected with M/20 NaOH solution (mg/ml solution)	Final pH
6.18	188	6.23
7.05	469.4	6.93

AUXILIARY INFORMATION

METHOD/APPARATUS/PROCEDURE:

A weighed sample of the mixture of sulfonamides was placed in a clean reagent bottle and a known vol of the buffer soln was added. The mixt was shaken in a mech shaker at 80-100 strokes/min. After at least 24 h the mixt was filtered through a clean, dried and weighed sintered-glass crucible. At the end of the filtration the crucible was washed with about 1 ml of water, dried at 105°C for 2-3 h, cooled, and weighed to const wt. The pH was detd with a Cambridge bench type pH meter using a glass electrode.

SOURCE AND PURITY OF MATERIALS:

Neither source nor purity of the materials, with the exception of water, was specified.

ESTIMATED ERROR:

Soly: not specified.

Temp: ±0.2°C (authors).

pH : ±0.01 unit (authors).

REFERENCES:

COMPONENTS:

(1) Benzenesulfonamide, 4-amino-N-2-thiazolyl- (sulfathiazole); $C_9H_9N_3O_2S$; [72-14-0]

(2) Benzenesulfonamide, 4-amino-N-2-pyrimidinyl- (sulfapyrimidine); $C_{10}H_{10}N_4O_2S$; [68-35-9]

(3) Benzenesulfonamide, 4-amino-N-(4-methyl-2-pyrimidinyl)- (sulfamethylpyrimidine); $C_{11}H_{12}N_4O_2S$; [127-79-7]

(4) Acetamide, N-[4-[(2-thiazolylamino)sulfonyl]phenyl]- (acetyl sulfathiazole); $C_{11}H_{11}N_3O_3S_2$; [127-76-4]

(5) Acetamide, N-[4-[(2-pyrimidinylamino)-sulfonyl]phenyl]- (acetyl sulfapyrimidine); $C_{12}H_{12}N_4O_3S$; [127-74-2]

(6) Acetamide, N-[4-[[(4-methyl-2-pyrimidinyl)amino]sulfonyl]phenyl]- (acetyl sulfamethylpyrimidine); $C_{13}H_{14}N_4O_3S$; [127-73-1]

(7) Phosphoric acid, disodium salt; Na_2HPO_4; [7558-94-4]

(8) Phosphoric acid, monopotassium salt; KH_2PO_4; [7778-77-0]

(9) Water; H_2O; [7732-18-5]

ORIGINAL MEASUREMENTS:

Frisk, A. R.; Hagerman, G.; Helander, S.; Sjögren, B.

Hygiea 1946, *108(12)*, 639-51.

PREPARED BY:

R. Piekos

VARIABLES:

One temperature: 37°C; one pH: 6.1

EXPERIMENTAL VALUES:

Solubility of a mixture of sulfathiazole, sulfapyrimidine, sulfamethylpyrimidine, acetyl sulfathiazole, acetyl sulfapyrimidine and acetyl sulfamethylpyrimidine in M/30 phosphate buffer of pH 6.1 at 37°C is 283 mg/100 ml solvent.

AUXILIARY INFORMATION

METHOD/APPARATUS/PROCEDURE:

An excess of the three sulfonamides and their acetyl derivatives in the phosphate buffer was shaken at 37°C for 24 h. The concn of the dissolved compds was detd by the Bratton and Marshall method (1) using a photoelec colorimeter.

SOURCE AND PURITY OF MATERIALS:

Neither source nor purity of the materials was specified.

ESTIMATED ERROR:

Soly: precision: ±28 mg/100 ml (authors).

Temp and pH: not specified.

REFERENCES:

1. Bratton, A. C.; Marshall, E. K., Jr. *J. Biol. Chem.* 1939, *128*, 537.

SYSTEM INDEX

Page numbers preceded by E refer to evaluation texts whereas those not preceded by E refer to compiled tables.

REGISTRY NUMBER INDEX

Page numbers preceded by E refer to evaluation texts whereas those not preceded by E refer to compiled tables.

AUTHOR INDEX

Page numbers preceded by E refer to evaluation texts whereas those not preceded by E refer to compiled tables.

SOLUBILITY DATA SERIES

Volume 1	H. L. Clever, *Helium and Neon*
Volume 2	H. L. Clever, *Krypton, Xenon and Radon*
Volume 3	M. Salomon, *Silver Azide, Cyanide, Cyanamides, Cyanate, Selenocyanate and Thiocyanate*
Volume 4	H. L. Clever, *Argon*
Volume 5/6	C. L. Young, *Hydrogen and Deuterium*
Volume 7	R. Battino, *Oxygen and Ozone*
Volume 8	C. L. Young, *Oxides of Nitrogen*
Volume 9	W. Hayduk, *Ethane*
Volume 10	R. Battino, *Nitrogen and Air*
Volume 11	B. Scrosati and C. A. Vincent, *Alkali Metal, Alkaline Earth Metal and Ammonium Halides. Amide Solvents*
Volume 12	C. L. Young, *Sulfur Dioxide, Chlorine, Fluorine and Chlorine Oxides*
Volume 13	S. Siekierski, T. Mioduski and M. Salomon, *Scandium, Yttrium, Lanthanum and Lanthanide Nitrates*
Volume 14	H. Miyamoto, M. Salomon and H. L. Clever, *Alkaline Earth Metal Halates*
Volume 15	A. F. M. Barton, *Alcohols with Water*
Volume 16/17	E. Tomlinson and A. Regosz, *Antibiotics: I. β-Lactam Antibiotics*
Volume 18	O. Popovych, *Tetraphenylborates*
Volume 19	C. L. Young, *Cumulative Index: Volumes 1–18*
Volume 20	A. L. Horvath and F. W. Getzen, *Halogenated Benzenes, Toluenes and Phenols with Water*
Volume 21	C. L. Young and P. G. T. Fogg, *Ammonia, Amines, Phosphine, Arsine, Stibine, Silane, Germane and Stannane in Organic Solvents*
Volume 22	T. Mioduski and M. Salomon, *Scandium, Yttrium, Lanthanum and Lanthanide Halides in Nonaqueous Solvents*
Volume 23	T. P. Dirkse, *Copper, Silver, Gold, and Zinc, Cadmium, Mercury Oxides and Hydroxides*
Volume 24	W. Hayduk, *Propane, Butane and 2-Methylpropane*
Volume 25	C. Hirayama, Z. Galus and C. Guminski, *Metals in Mercury*
Volume 26	M. R. Masson, H. D. Lutz and B. Engelen, *Sulfites, Selenites and Tellurites*
Volume 27/28	H. L. Clever and C. L. Young, *Methane*
Volume 29	H. L. Clever, *Mercury in Liquids, Compressed Gases, Molten Salts and Other Elements*
Volume 30	H. Miyamoto and M. Salomon, *Alkali Metal Halates, Ammonium Iodate and Iodic Acid*
Volume 31	J. Eysseltová and T. P. Dirkse, *Alkali Metal Orthophosphates*
Volume 32	P. G. T. Fogg and C. L. Young, *Hydrogen Sulfide, Deuterium Sulfide and Hydrogen Selenide*
Volume 33	P. Franzosini, *Molten Alkali Metal Alkanoates*
Volume 34	A. N. Paruta and R. Piekos, *4-Aminobenzenesulfonamides. Part I: Non-cyclic Substituents*
Volume 35	A. N. Puruta and R. Piekos, *4-Aminobenzenesulfonamides. Part II: 5-Membered Heterocyclic Substituents*
Volume 36	A. N. Puruta and R. Piekos, *4-Aminobenzenesulfonamides. Part III: 6-Membered Heterocyclic Substituents and Miscellaneous Systems*
Volume 37	D. G. Shaw, *Hydrocarbons with Water and Seawater. Part I: Hydrocarbons C_5 to C_7*
Volume 38	D. G. Shaw, *Hydrocarbons with Water and Seawater. Part II: Hydrocarbons C_8 to C_{36}*